Global Satellite Meteorological Observation (GSMO) Theory

Stojče Dimov Ilčev

Global Satellite Meteorological Observation (GSMO) Theory

Volume 1

By

Stojče Dimov Ilčev
(Стойчо Димов Илчев)
Durban University of Technology (DUT)
Durban, South Africa

Stojče Dimov Ilčev
Durban University of Technology (DUT)
Durban, South Africa

ISBN 978-3-319-67118-5   ISBN 978-3-319-67119-2   (eBook)
DOI 10.1007/978-3-319-67119-2

Library of Congress Control Number: 2017953101

Printed on acid-free paper

This Springer imprint is published by Springer Nature
The registered company is Springer International Publishing AG
The registered company address is: Gewerbestrasse 11, 6330 Cham, Switzerland

Dedicated to the memory of my late father

Prof. Dimo Stoev Ilčev
(Димэ Стоев Илчев)

Preface

The new Springer book *Global Satellite Meteorological Observation (GSMO) Theory: Volume 1* presents in its introduction the development of radio and satellite communications including the history of satellite meteorological observation. The book includes major aspects of the space segment with satellite platforms and orbital mechanics, baseband signals and types of transmission systems, atmospheric electromagnetic radiation and radiative transfer, satellite meteorological parameters, main satellite meteorological instruments, and antenna systems with propagation.

Moreover, in *Global Satellite Meteorological Observation (GSMO) Applications: Volume 2*, topics about satellite meteorological networks, satellite imagery interpretation, satellite weather remote sensing, satellite meteorological applications, ground segments, user segments, and integrations in space meteorology will be introduced.

Today satellite meteorological instruments present as the "eyes from space," which can "see" the Earth and its atmosphere with all weather phenomena and, from a global perspective, can send all necessary observations, data, and images via satellite onboard transponders to the ground Earth stations (GES) or readout stations for processing facilities. On the other hand, it is very important to realize the idea and possibility for the integration of a meteorological satellite service, such as geostationary Earth orbit (GEO), polar Earth orbit (PEO), and new high elliptical orbit (HEO) satellite systems, together with other space meteorological facilities, under one worldwide observation umbrella.

For basic and principal technical information, the author of this book has drawn heavily mostly on the following sources:

- Kidder S. Q. and Ham T. C., "Satellite Meteorology – An Introduction," Academic Press, San Diego, CA, USA, 1995.
- Tan S. Y., "Meteorological Satellite Systems," Springer, New York, 2014.
- Menzel P., "Remote Sensing with Meteorological Satellites," University of Wisconsin Academic Press, Madison, WI, USA, 2012.

- Kelkar R. R., "Satellite Meteorology," BS Publications, Hyderabad, India, 2007.
- Sportisse B., "Fundamentals in Air Pollution – From Processes to Modelling," Springer, Dordrecht, Netherlands 2008.
- Liu G., "Satellite Meteorology," Florida State University, Florida, 2005.
- Ilcev D. S., "Global Mobile Satellite Communications for Maritime, Land and Aeronautical Applications – Theory Volume 1," Springer, Boston, 2016.
- Ilcev D. S., "Satellite Meteorological Observation System," Manual, DUT, 2015.
- Ilcev D. S., "Global Aeronautical Communication, Navigation and Surveillance," Theory Volume 1 and Applications Volume 2, AIAA, Reston, 2013.
- Ilcev D. S., "Space Meteorology," Manual, SSC, DUT, 2017.
- NOAA/NESDIS, "User's Guide for Building and Operating Environmental Satellite Receiving Stations," Suitland, MD, USA, 2009.
- Eumetsat, "The Meteosat System – Satellite Ground Segment Mission and Global Coordination," Darmstadt, Germany, 2000.
- Roscosmos/Russian Space Agency, "Status of Current and Future Russian Satellite Systems," Roshydromet, Moscow, 2010.
- Group of Authors, "Radiowave Propagation Information for Predictions for Earth-to-Space Path Communications," edited by C. Wilson and D. Rogers, ITU, Geneva.

Readers will find that this book has been written using up-to-date systems, techniques, and technology in satellite meteorological observation. The material has been systematized in such a way to cover developments in satellite meteorology space and ground segments, transmission systems, electromagnetic radiation in the atmosphere, radiative transfer, satellite meteorological parameters and instruments, antenna systems with practical solutions, and radio wave propagation.

The two volumes of this book were written in order to form a bridge between potential readers and current global satellite meteorological observation (GSMO) trends, system concepts, and transmission network architecture by using a very simple style with easily comprehensible technical information, characteristics, graph icons, figures, illustrations, and mathematical equations. The special approach in the two volumes of the GSMO books is the introduction of a complete space and ground segment with all significant segments introducing modern meteorological imaging systems on board satellites, transmission systems, and ground readout and processing facilities.

Volume 1 of this new book, nominated as "Theory," consists of seven chapters on the following particular subjects:

Chapter 1, "Introduction," depicts a background to the development of space systems and concepts of satellite communications in the function of transfer meteorological observation data and images. The special retrospective presents the evolution of space meteorological observations, history of early radio, evolution of satellite mobile and fixed communication systems, definition of VSAT broadcasting, international coordination organizations and regulatory procedures, space systems and radio communication frequency assignment, and history of satellite meteorology.

Chapter 2, "Space Segment," discusses the fundamental principles and theories of space platforms and orbital mechanics, laws of satellite motion, satellite parameters, new types of launching systems and station-keeping techniques, satellite orbits and geometric relations, spacecraft configuration, payload structure, types of orbits for meteorological and other satellite systems, meteorological satellite payloads on board satellite antenna systems, and components of a satellite bus.

Chapter 3, "Baseband and Transmission Systems," gives an essential basic knowledge of baseband signals and processing techniques. analog and digital transmissions, modulation and demodulation, channel coding and decoding, error corrections, multiple access techniques, fixed and mobile DVB-RCS standards, MPEG multimedia standards, satellite audio and video broadcasting, direct-to-home and other satellite digital broadcast systems, transmission standards, and new DVB-S2/S3 architectures.

Chapter 4, "Atmospheric Electromagnetic Radiation," presents all the fundamentals of atmospheric radiative transfer, energy emissions, radiative properties of matter, Earth's atmosphere applications and radiative transfer equation (RTE) with prime of radiations for infrared and visible imaging, radiative budget for the Earth atmosphere system, solar constant and emission effective temperature, and energy budget for the Earth or atmosphere systems.

Chapter 5, "Satellite Meteorological Parameters," explains very important satellite activities; such as, satellite weather observation, satellite meteorological instruments for observation and monitoring, parameters for temperature and trace gases, wind flow, clouds and aerosols, precipitation measuring technique, Earth radiation budget, measurements and monitoring of other Earth observation parameters with special review of hydrological analysis, sea waves and ocean dynamics, sea surface temperatures, pollution and ecosystem, cryosphere detection, agricultural and forestry, global land cover mapping, and desertification monitoring.

Chapter 6, "Satellite Meteorological Instruments," introduces all weather instruments on board PEO satellite meteorological systems from space; such as, the US POES and European spacecraft. All types of meteorological instruments on board GEO satellites; such as, the US GOES, European Meteosat, Russian Electro, Chinese Fengyun, Indian INSAT, and Japanese GMS satellites are also discussed here.

Chapter 7, "Antenna Systems and Propagation," includes research and introduction of current and new proposed prototypes of antenna solutions for satellite and other radio meteorological communications and broadcasting fixed, semi-fixed, and mobile systems: such as, low-gain omnidirectional antennas. medium-gain directional antennas, and high-gain directional aperture antennas. Moreover, this chapter introduces ground antennas for particular satellite meteorological systems: such as, directional antennas for PEO direct readout stations (DRS), multidirectional antennas for PEO DRS, directional antennas for GEO DRS, meteocast DVB-RCS GES antennas for GEO DRS, user shipborne antennas, and user vehicleborne antennas. This chapter also comprises all particulars about propagation effects significant for satellite transmission requirements in meteorology and weather observation data: such as, propagation fundamentals, refraction, absorption and non-LOS radio propagation, sky wave propagation, atmospheric effects on propagation, sky noise temperature contributions, path depolarization causes, propagation effects important for space communications and broadcasting, and so on.

Acknowledgments

Above all, the author of this book would like to express his very special appreciation and gratitude to Prof. Ahmed Cassim Bawa, former vice chancellor (VC) and principal of the Durban University of Technology (DUT), who gave him huge support in space science research and postgraduate studies. The author also expresses his special gratitude to the new VC Prof. Thandwa Mthembu, DVC Prof. Sibusiso Moyo, and DUT staff for the support and encouragement to establish the Space Science Centre (SSC) for Research and Postgraduate Studies in Space Science and for the moral assistance in completing this book.

The author is the chair of SSC, a research professor, and a supervisor at DUT for research and postgraduate studies. The author has a very important multinational project, the African Satellite Augmentation System (ASAS), for the entire African continent and Middle East and also many other proposals in radio and satellite CNS, digital video broadcasting-return channel via satellite (DVB-RCS), global radio and satellite tracking of mobiles and living beings, satellite SCADA (M2 M), stratospheric platform systems (SPS), and space solar power (SSP); he also had one significant GADSS project developed in 2000. He also would like to express his special appreciation to DUT for the generous contribution as a sponsor of this book.

The Durban University of Technology prides itself on the commitment to academic excellence.

The over 24,000 students who pass through the doors everyday are testament to a growing ethos of learning, research, and community engagement. DUT is a multi-campus university of technology at the cutting edge of higher education, renowned for technological training and academic prowess. The university is characterized by being research-driven with a focus on strategic and applied research that can be translated into professional practice. Furthermore, research output may be commercialized, thus providing a source of income for the institution. In striving to create a new and dynamic ethos, the university builds upon current strengths and celebrates the expertise of its staff. DUT is providing Web pages for its SSC for Research and Postgraduate Studies in Space Science at www.dut.ac.za/space_science – where the full study program, projects for instant developments, and research and supervisor staff are all presented.

The author is also very grateful to the group of authors for the various manuals, brochures, and pamphlets issued by IMO, ICAO, ITU, WMO, ESA, ETSI, ETRI, NOAA, Roscosmos, Roshydromet, China Meteorological Administration (CMA), ISRO, Japan Meteorological Agency (JMA) Sea Launch, Advantech Wireless,

Kongsberg, Dartcom, SeaSpace, Orbit, SCISYS, and other regulatory bodies and operators.

This book is dedicated to all his friends working in the shipping industry, to his newest friend Prof. Felix Mora-Camino, and to his present postgraduate students at DUT. He also wishes specially to acknowledge the valuable support and understanding from the publisher of this book, Springer, especially to Ms. Mary E. James, senior editor in applied sciences, and her assistants, Ms. Zoe Kennedy, Ms. Rebecca R. Hytowitz, and Mr. Brian Halm, and all Springer staff in India.

Finally, he would like to express a very heartfelt appreciation and gratitude to his lovely wife Svetlana M. Ilčeva and his family for their help and understanding while the manuscript was being written, especially to his dear children and grandchildren living in Montenegro: his son Marijan with his wife Vanja and their children Daria and Martin; his daughter Tatjana with her husband Boško and their children Anja and Stefan; his stepdaughter Olga and her husband Boris; his stepgranddaughter Bažena and stepson Lev; his sister Prof. Tatjana Ilčeva and niece Ivana in Belgrade, Serbia; and his cousin Valentin Boyadžiev and his family in Sofia, Bulgaria.

Durban, South Africa   Stojče Dimov Ilčev

Contents

About the Author

Stojče Dimov Ilčev is the chair of the Space Science Centre (SSC) for Research and Postgraduate Studies in Space Science for maritime, land, and aeronautical applications, global satellite augmentation systems (GSAS), GNSS systems, satellite tracking systems, mobile broadcasting, and meteorological observation systems at the Durban University of Technology (DUT), Durban, South Africa. He studied both maritime radio engineering and nautical science at the University of Montenegro in Kotor, maritime electronics and communications at the University of Rijeka in Croatia, and postgraduate satellite engineering at the University of Skopje in Macedonia and the University of Belgrade in Serbia. Ilčev holds bachelor (BSc), master in electrical engineering (MSc), and doctor of science (PhD) degrees. He also holds certificates for first-class radio operator (Morse), for GMDSS first-class radio electronic operator and maintainer, and for master mariner without limitations. Since 1969, Ilčev worked on board merchant ships, in a satellite Earth station, at a coast radio station, in a shipping company, in a nautical school, and at a maritime faculty. Since 2000, he worked at IS marine radio and CNS system companies on research and projects relating to modern communication, navigation, and surveillance (CNS) for maritime, land, and aeronautical applications. He has written five books on CNS engineering and satellite systems for mobile applications, and he has many projects and inventions in this field including DVB-RCS standards and stratospheric platforms.

Chapter 1
Introduction

Meteorological satellites have become essential for both meteorological observation and weather forecasting, continuing the fundamental concepts of conventional data exchange and international cooperation, which have been traditional for more than 150 years. They provide vital data at frequent intervals over wide observation areas, in the context of the international cooperation needed to ensure adequate worldwide data coverage. This kind of cooperation can be at local level provided by one country, continental level that integrates group of countries providing and global availability of data over nearly one quarter of the Earth surface and third level of cooperation is providing integration of previous two levels providing meteorological data available globally.

This chapter expresses the importance and reasons why satellite meteorological observation and weather forecasting are so important to the global economy, transportation, industry, trade, security and every day life on our Planet. It presents the principal structure and developments of radio, evolution of satellite fixed and mobile communication systems and history of satellite meteorology. The development history of satellite metrological systems is described briefly and this is followed by a summary of the various regional and global satellite programs and an introduction to the meteorological services.

In addition, this chapter also introduces history of early wired and radio communications, evolution of satellite communications, experiments with active communications satellites, Global Mobile Satellite Communications (GMSC), Global Navigation Satellite Systems (GNSS), Stratospheric Platform Systems (SPS), Satellite Meteorological Observation Systems (SMOS), Fixed Satellite Communications (FSC), Mobile VSAT Service for broadcast and broadband, International Coordination Organizations and Regulatory Procedures, Space Systems and Radiocommunication Frequency Assignment, History of Satellite Meteorology and Mobile Satellite Meteorological Service (MSMS).

Today meteorological satellites simply represent "Eyes from the Space", which can "see" the Earth, its atmosphere, weather images and forecasting factors from global perspective and send all necessary observations, data and images via

© Springer International Publishing AG 2018   1
S.D. Ilčev, *Global Satellite Meteorological Observation (GSMO) Theory*,
DOI 10.1007/978-3-319-67119-2_1

onboard satellite transponders to the direct readout Ground Earth Stations (GES) and processing facilities. On the other hand, it is very important to realize project for integration of satellite constellations for all applications including meteorological, such as Geostationary Earth Orbit (GEO and Polar Earth Orbit (PEO), with new proposed Low Earth Orbits (LEO), High Elliptical Orbits (HEO) and SPS or High Altitude Platforms (HAP) and Unmanned Aerial Vehicles (UAV). In integration can be included Pico, Nano and Micro satellites and GNSS solutions.

However, the important link in this chain is radio communication system as predecessors to television, satellite and other wireless systems, which have been used initially as media for transmission of meteorological data, images and other observations for local, regional and global weather forecast services. Namely, the weather radio communication is a network of radio stations in any hypothetical country that broadcast continuous weather information directly from a nearby national weather forecast office of the service's operator and service within service area.

The exploitation of modern radio system began from the history of implementation of first radio equipment invented by the great Russian professor Aleksandr Stepanovich Popov till today modern radio systems. For the first time in the world Popov's radio system was applied at shore and onboard ships for radio communication purposes as a part of today VHF, UHF and HF radio systems. Afterwards were developed radio navigation and radar systems as onboard radio equipment working on special frequencies for determination position of mobiles on the Earth surface, such as radiolocation and radio determination.

Officially, the equipment for radiolocation known as radar was invented by the British Sir Robert Watson-Watt in 1935, so it then became the first radar used in battle. There was a race between Britain and Germany who could produce the best radar for their national defense forces. Nevertheless, the Germans were unable to unlock its full potential.

The British had won the race and utilized the full power of the radar technology. Radiolocation is the process of finding the location of some objects like ships or aircraft through the use of radio waves, such today is surveillance radar. Thus, radar is measuring instrument in which the echo of a pulse of microwave radiation is used to detect and locate distant objects. Ground and space radars are serving today for Earth observation, however here will be not included any type of ground or satellite radars for remote sensing and meteorological observations.

1.1   Evolution of Meteorological Observations

The meteorological observation and information has been an important part of life as long as history of mankind exists and weather measurements have been done for centuries and up to the present days. It is claimed that the first measurements of rainfall were reported as early as from ancient Greece 500 B.C. In such a way, these first measurements were aimed to estimate the expected growth of wheat and used as basis for taxation of the farmers and other productions. These primitive measurements were made by using bowls.

Sometime later, humans have been continuously looking for ways to forecast the weather data for centuries. The Greek natural philosopher Theophrastus wrote a Book of Signs, in about 300 B.C. listing more than 200 practical ways of knowing when to expect rain, wind, fair conditions and other kinds of weather.

The first actual rain gauges were developed much later in Korea. The first early forecasts services were established in England, France, Germany and USA during 1800s to provide information about possible storms for seafarers. Although there is a long history of weather observations, their importance to everyday use for meteorological forecasting has grown enormously during the last century and especially during the last few decades.

On the other hand, modern technological developments of the last few decades, such as improved telecommunication and satellite systems, Automatic Weather Stations (AWS) and various long range measuring systems, such as satellites meteorological observation and weather radars, have enabled a range of modern possibilities for utilizing weather information. For a common person, these became evident in real time weather services tailored for individual customers, as well as the improved quality of the weather forecasts.

In the more than 45 past years since the first meteorological satellites were launched and deployed, they have become indispensable for study of the Earth's atmosphere, climate and meteorological phenomenons for weather synopsis and prognoses. Indeed, together with their land and ocean-sensing forerunners, meteorological satellites view the Earth from a global perspective, which is unmatched and incomparable by any other observing system.

Reliable weather services have high impact on society, government, corporate and private organizations, although their benefits are not always seen directly. Recent studies suggest that investing to weather services will benefit communities up to 8–20 times for each new projects, but these benefits needs time to be effective. For example, weather forecasting is helping for more cost effective agriculture, farming, mining, construction, transportation, trade, tourism, general safety and other business infrastructures. According to the practical indicators, for instance, accurate and on daily basis weather forecasting is providing much more safety and security in navigation of oceangoing ships at sea, for road and rail vehicles, short and longhaulage aircraft flying, and as is very important to reduce mortality on the roads. Thus, it is enough evident that many persons lose their lives every year due severe weather conditions. In this report, the benefits are achieved by improving the early warning and prediction systems in the areas, which are important to economics of these countries, which are very weather sensitive.

In addition to the benefits in the medium and long range planning, weather information is crucial if people are faced with environmental hazards, hurricanes, such as floods or heavy thunderstorms. These kinds of disasters often cause tremendous problems, especially in poorer countries affected by the tropical storms. Having good weather services and warning systems is one of the steps to help

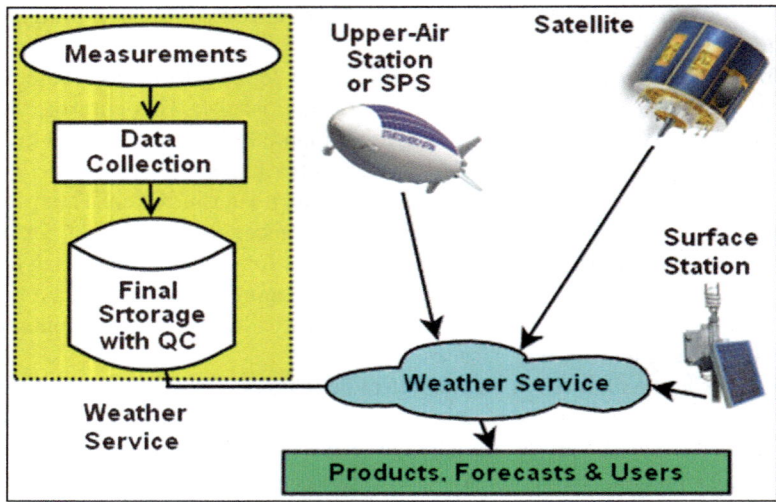

Fig. 1.1 Different meteorological integrated services (Courtesy of Manual: by Ilcev)

nations from poverty. However, in many countries, the public image of weather services is still quite low, and their importance is not understood.

Climate change is another important topic that needs more accurate observations, although those observations do not need to be done at high intervals or received in real time. At the moment, there are several projects going on concerning climate change observations. Some of the projects are digitalizing old observations from logbooks of the weather stations and, a bit more indirectly, weather information from logbooks of ships.

A typical weather service contains a number of different sources of forecasting data, which are analyzed and processed to ready-made products. Thus, these sources include traditional surface weather observations (Surface Stations), Upper-air stations (SPS, UAV or Radio Sounding), Meteorological satellites and many others in real time Quality Control (QC), shown in Fig. 1.1. These sources can be integrated in one modern meteorological and weather forecast model, and these products can be global surface weather observation and forecasts, maritime, aviation and other types of fixed and mobile weather predictions.

In the similar way new integration of different type of satellite transmissions in satellite meteorological observation infrastructures, such as traditional fixed and mobile satellite communications with Digital Video Broadcasting-Return channel via Satellite (DVB-RCS) can provide more cost effective and reliable transfer of meteorological data from Surface Stations, Upper-Air Stations and Weather Satellites for local, regional and global coverage including both poles.

The already developed GNSS networks, such as US GPS, Russian GLONASS, Chinese Compass (BeiDou) and other satellite navigation and determination systems in developing phase provide precise positioning data for ocean going vessels, land vehicles and aircraft. These systems are improving all mobile traffic control and management and enhance safety and security at sea, on the ground and in the air. Because of the need for new reliable and more effective service, these GNSS solutions are augmented with satellite communications and ground surveillance facilities. In addition, the GNSS signals can be used for Integrated Water Vapor, serve weather monitoring, humidity filed, participation and so on.

1.2 History of Early Radio Communication Systems

The word **communications** is derived from the Latin phrase "**communication**", which stands for the social process of information exchange and covers the human need for direct contact and mutual understanding. The word "**telecommunications**" means to convey and exchange information at a distance (**tele**) by the medium of electrical signals.

In general, telecommunications are the conveyance of intelligence in some form of signal, sign, sound or electronic means from one point to a distant second point. In ancient times, that intelligence was communicated with the aid of audible callings, fire and visible vapor or smokes and image signals. We have come a long way since the first human audio and visual communications, in case you had forgotten, used during many millennia. In the meantime primitive kinds of communications between individuals or groups of people were invented. Hence, as impressive as this achievement was, the development of more reliable communications and so, wire and radio, had to wait a couple of centuries more.

The invention of the telegraph in 1844 and the telephone in 1876 harnessed the forces of electricity to allow the voice to be heard beyond shouting distance for the first time. The British physicist M. Faraday and the Russian academic E. H. Lenz made experiments with electric and magnetic phenomenon and formulized a theory of electromagnetic (EM) induction at the same time. The British physicist James C. Maxwell published in 1873 his classical theory of electromagnetic radiation, proving mathematically that electromagnetic waves travel through space with a speed precisely equal to that of light.

The famous German physicist Heinrich Rudolf Hertz during 1886 experimentally proved Maxwell's theoretical equations. Thus during Hertz's studies in 1879 his mentor physicist Hermann von Helmholtz suggested that Hertz's doctoral dissertation has be on testing Ma well's theory of electromagnetism, published in 1865, which predicted the existence of electromagnetic waves moving at the speed of light, and predicted that light itself was just such a wave.

During his research and studies, Hertz demonstrated that High Frequency (HF) oscillations produce a resonant effect at a very small distance away from the source and that this phenomenon was the result of electromagnetic waves. Practically, Hertz used the damped oscillating currents in a dipole antenna, triggered by a high-voltage electrical capacitive spark discharge, as his Transmitter (Tx) source of Radiowaves (RW).

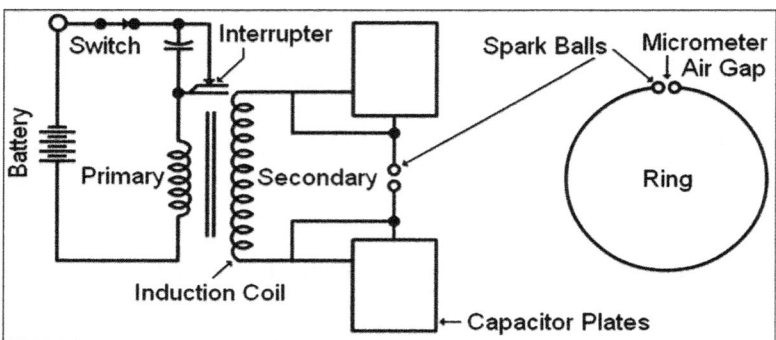

Fig. 1.2 Hertz's spark gaps Tx and Rx (Courtesy of Book: by Ilcev)

His detector in some experiments was another dipole antenna connected to a narrow spark gap as Receiver (Rx). A small spark in this gap signified detection of the radio waves. When he added cylindrical reflectors behind his dipole antennas, Hertz could detect radio waves about 20 m from the Tx in his laboratory. In Fig. 1.2 (Left) is illustrated the electronic diagram of spark generator as a main component of his radio Tx (Ruhmkorff coil with linear vibrator). In fact, the wave character of the spark generator discharge is creating short-lived electromagnetic waves. The waves thus produced were received by a resonator located a short distance away from the generator's aerial, which radio Rx is illustrated in Fig. 1.2 (Right). At the moment when the resonator picked up the waves, hardly perceptible sparks were produced in the resonator gap that could be detected using a magnifying glass. Therefore, it is after Hertz that the new discipline of Radio technology is sourced and after whom the frequency and its measuring unit Hertz (Hz) are named.

An English physician, Sir Oliver J. Lodge using the ideas of others, realized that the EM resonator was very insensitive and because of this phenomenon he invented a "coherer". A much better coherer was built and devised by a Parisian professor, Edouard Branly, in 1890. He put metal filings (shut in a glass tubule) between two electrodes and so a great number of fine contacts were created. This coherer suffered from one disadvantage, it needed to be "Shaken before use". Owing to imperceptible electric discharges, it always got "baked" and blocked.

In early 1889 the Russian professor of physics Aleksandar Stepanovich Popov conducted experiments along the lines of Hertz's research and successfully realized the first practical experiments with EM waves for the transmission of radio signals. Soon after professor Popov attended a meeting of the Russian Physical-Chemical Society at which the St. Petersburg Professor N. G. Yegorov reproduced Hertz's experiments, but in a manner that Popov felt to be insufficiently graphic. It took him only a few months to build a more compact and effective device to demonstrate Hertz's experiments, which he then gradually improved on, so that by 1894 he had constructed a working transmitter that generated electromagnetic waves based on Hertz's vibrator and using a Ruhmkorff coil.

This was the first transmission and reception system for remittent electronic waves suitable for the reliable communication of information in the history of telecommunications. The method of Popov invention became the prototype for the subsequent first generation of radio communication systems. At that time pioneer of radio equipment Popov built his first radio Receiver (Rx) with improved version of coherer design by both Sir Lodge and Branly, so using new inventions he practically realized the first experiments ever in the world with EM waves for the transmission of radio signals.

Like Hertz's early experiments, hardly had Lodge's paper been published in the journal Electrician in July 1894 than his experiments with the coherer were repeated by physicists around the world, but only Popov was able to obtain a practical result. In 1895 he created a system in which the device was mounted on a wooden base and placed in a metal cage to screen it from electrical interference. The aerial used was either a vertical length of wire or an asymmetrical vibrator with a parabolic mirror. The circuit proposed by Popov was also used for thunderstorm detectors and telegraph transmission.

In 1895, Popov upgraded the coherer's sensitivity and invented a special mechanism to automatically re-set the device. In fact, he improved Branly's radio Rx by the insertion of choke coils on each side of the relay to protect the coherer and also by replacing the spark gap with a vertical antenna insulated at its upper end and connected to the ground through the coherer. He then mounted a small bell in serial connection with the coherer's relay anchor, whose ringing effected its automatic destabilization and successive unblocked function of the receiver system. On 7 May 1895, he demonstrated his new apparatus to the members of the Russian Physic-Chemical Society: a lightning conductor as an antenna, a metal filings coherer and detector element with telegraph relay and a bell. The relay was used to activate the bell, which is announcing the occurrence of transmitting signals and in this way serving as a decoherer (tapper) to prepare the receiver to detect the next signals. This was the first telegraph station in the world, which could work without any wires.

Popov succeeded in making a reliable generator of EM waves, when the detecting systems in common use still were not at all satisfactory. So his system was extended to function as a wireless telegraph with a Morse telegraph key attached to the transmitter (Ruhmkorff coil and blocking capacitor), shown in Fig. 1.3 (Left). In May the same year, instead of a bell he contrived to use a clock mechanism to realize direct, fast destabilization of metal filings in the coherer upon receipt of the signals. In the same year, he succeeded in making a more reliable generator of EM waves, when the receiving (detecting) systems in common use still were not at all satisfactory. Thus, using the inventions of his predecessors and on the basis of his proper experiments, Popov elaborated the construction of the world's first radio receiver (coherer, amplifier, electric bell, aerial) with a wire-shaped antenna system in the air attached to a balloon, which sample is shown in Fig. 1.3 (Right).

In December 1895, he officially announced the success of a regular radio connection and on 21 March 1896 at the St. Petersburg University, he demonstrated

Fig. 1.3 Popov's spark gaps Tx and Rx (Courtesy of Book: by Ilcev)

it in public. Finally, on 24 May 1896 Popov installed a pencil instead of the bell and sent the first wireless message in the world at the distance of 250 m between two buildings, conveying the name "GENRICH GERZ" (the name of Hertz in Russian) by Morse code using his homemade transmitter and receiver.

In March 1897, Popov equipped a coastal radio station at Kronstadt and the Russian Navy cruiser Africa with his wireless apparatus and in summer 1897, Popov started to make experiments at sea, using radios on board ships before the entire world. In 1898, he succeeded in relaying information at a distance of 9 km and in 1899, a distance of 45 km between the island of Gogland and the city of Kotka in Finland. With all his inventions Popov made advances on the discoveries of Hertz and Branly and created the groundwork for the development of maritime radio.

In 1895, a few months later following the work of Popov, a young Italian experimenter, Guglielmo Marconi, started to use radio and was the first to put EM theory into business application. A detailed report on the Marconi's experiments was presented by chief of UK engineer telegraph agency V. Pris (1834–1913), who refused to assist him in the works in England (Forum – Yakimenko S). This report shows that the transmitter of G. Markoni was designed by his Italian teacher A. Rigi and that he used the receiver of professor Popov. Apparently, V. Pris in his report about the invention G. Markoni was forced to point out that it has been said previously: "G. Markoni did not do anything new". In conclusion can be stated that unlike Popov, Marconi was not inventor of radio, but good businessman able to compose something already done and was able to turn his research into a financial and manufacturing empire.

By the next year, he had sent Morse code messages at a distance of two miles. Moving to England to obtain patents on his equipment, he demonstrated radio reception over eight miles, see photo of Marconi with his first Radio Tx and Rx Equipment in Fig. 1.4. In 1897, he exhibited the use of radio between ship and shore and, according to Western literature practically started the use of maritime radio. By the same year, Marconi succeeded in getting his wireless telegraphy transmissions officially patented for the first time in the world. The owners of the Dublin Daily Express in Ireland invited Marconi to conduct wireless reports of the Kingston Regatta of July 1898 from the steamer Flying Huntress, the first ship equipped

Fig. 1.4 Marconi and his radio (Courtesy of Book: by Ilcev)

with a commercial wireless system. Using an antenna hung from a kite to increase the effective height of his masts and a LF of 313 kHz at 10 kW of power, on 12 December 1901 Marconi crackled out the first wireless message to span an ocean in the form of Morse code; three dots forming the letter "s". This telegraph signal was sent from Newfoundland in Canada and received in Cornwall on the west coast of England.

In the meantime, an American inventor Samuel Morse in 1836 demonstrated the ability of a telegraph system to transmit information over wires and radio as a series of electrical signals. Short signals are referred to as dots and long signals are referred to as dashes.

With the advent of radio communications, an international version of Morse code became widely used. In 1900, R. Fessenden made the first transmission of voice via radio in the USA, Fleming in 1904 discovered the diode valve, while their countryman and pioneer Lee de Forest developed and used a triode valve, which made it possible to use radio not only for radiotelegraphy but for voice communications. As early as 1907, he installed a triode valve mobile radio on a ferryboat operating on the Hudson River near New York City.

Philo Taylor Farnsworth was an American inventor of first television (TV) in 1927, when he demonstrated the first fully functional all-electronic image pickup device (video camera tube), the image dissector, as well as the first fully functional and complete all-electronic television system. That invention was persecutors of Television Receive Only (TVRO) and modern Direct Broadcast Satellite (DBS). All mentioned inventions were used in earlier transmission of meteorological data and weather forecast information locally and globally.

1.2.1 Evolution of Satellite Communications

The first known annotation about devices resembling rockets have been used by Archytus of Tarentumin, who invented in 426 B.C. a steam-driven reaction jet rocket engine that flew a wooden pigeon around his room. Apparently, the first rockets owe their origin to the invention of gunpowder in China around the tenth century. Devices similar to rockets were also used in China during the year 1232, during the siege of Beijing (according to another source: town of Kai-fung-fu) by the Mongols; the city's defenders fired missiles. It is believed that around the thirteenth century, knowledge of rocketry reached Italy and France.

All through the thirteenth to the fifteenth centuries there were reports of many rocket experiments. In England, a monk named Roger Bacon worked on improved forms of gunpowder that greatly increased the range of rockets. In France, Jean Froissart found that more accurate flights could be achieved by launching rockets through tubes. Its idea was the forerunner of the modern bazooka. Joanes de Fontana of Italy designed a surface-running rocket-powered torpedo for setting enemy ships on fire.

Human space travel had to wait almost a millennium, until Sir Isaac Newton's time, when we understood gravity and how a projectile launched at the right speed could go into Earth orbit. By the sixteenth century rockets fell into a time of disuse as weapons of war, though they were still used for fireworks displays, and a German fireworks maker, Johann Schmidlap, invented the "step rocket", a multi-staged vehicle for lifting fireworks to higher altitudes. The use of rockets near the Russian (today Ukrainian) city of Belgorod is recorded in 1516 and the first appearance of rockets in the Russian city of Usury dates from around 1675. Following the development of military missiles in Europe, the "Rocket Enterprise" was founded in Moscow around 1680 to provide the similar experiments. A signaling rocket developed in Russia in 1717 could reportedly reach an altitude of several hundreds meters.

In 1815, Russian artillery engineer Alexander Zasyadko started the development of 3 size types of battlefield missiles for the Russian army with range up to 2700 m. In 1849, another artillery engineer Konstantin Konstantinov developed rockets reaching a range of 4–5 km. Rocket experimenters in Germany and Russia began working with rockets with a mass of more than 45 kg. Some of these rockets were so powerful that their escaping exhausts flames bored deep holes in the ground even before lift-off.

Finally, the twentieth century came with its great progress and the historical age of space communications began to unfold. Russian scientist Konstantin Tsiolkovsky (1857–1935) published a scientific book on virtually every aspect of space rocketing. He propounded the theoretical basis of liquid propelled rockets, put forward ideas for multi-stage launchers and manned space vehicles, space walks by astronauts and a large platform system that could be assembled in space for normal human habitation. Although the Russian scientific elite was primarily concentrated in centers such as Moscow and St. Petersburg, far from two capitals, a modest physics teacher conceived some of the most remarkable ideas about the future of the human race. Almost unknown to its contemporaries, Tsiolkovskiy's work decades later was universally accepted as a theoretical foundation of modern astronautics.

In 1917, the Bolshevik revolution spurred profound changes in Russia. The new leaders of Soviet Russia conducted a ruthless policy of turning an agrarian peasant society into an industrial power. At the same time, the Soviet government spared no effort, in equipping the Red Army with new weapons. In line with this strategy, on 1 March 1921, the Soviet authorities created Gas-Dynamic Laboratory (GDL) for research in rocketry, to be led by Nikolai Tikhomirov. Tikhomirov had started studying problems of solid and liquid-fueled rockets as early as 1894 and in 1915 he patented "Self-propelled aerial and water-surface mines". Soviet researchers at GDL worked tirelessly on perfecting military missiles and developing new types of solid rocket fuel, which would allow the new weapons to compete with artillery.

In early 1924, two new Soviet (Russian) rocket programs were founded. They were the Central Bureau for the study of the problems of rockets (TsBIRP), and the All-Union Society for the Study of Interplanetary Flight (OIMS). The main leader of these early Russian rocket efforts was Fridrikh Tsander, who developed liquid propellant rockets in the 1920s and 1930s.

A little later, the American Robert H. Goddard launched in 1926 the first liquid propelled engine rocket. The rocket an 1.22 m high named "Nell," the world's first liquid-fueled rocket got an altitude of 12 m, a speed of 60 miles per hour and a distance of 46 m. The German society for space travel or VfR was founded in June of 1927, and it soon had around five hundred members and the first its journal "the Rocket".

At the beginning of the 1930s, the Soviet government sanctioned the creation of several research groups, which united rocket enthusiasts in organizations known as GIRD (Group for Investigation of Reactive Movement) were established in Moscow, Leningrad and other Soviet cities. In Moscow, thanks to the efforts of Sergei Korolev and Fridrikh Tsander, the government-sponsored Society for the Advancement of Defense, Aviation and Chemical Technology, Osaviakhim, agreed to fund GIRD. However, after Stalin took over the Soviet country, these rocket programs became Len-GRID and Mos-GRID, or "Group for the study of reactive motion" based in Leningrad and Moscow, respectively. These two centres, in turn, soon became the State Reaction Scientific Research Institute.

In January 1933 Tsander began development of the GIRD-X missile. It was originally to use a metallic propellant, but after various metals had been tested without success it was powered by the Project 10 engine, which was first bench tested at the beginning of March 1933. This design burned liquid oxygen and gasoline and was one of the first engines to be regeneratively cooled by the liquid oxygen, which flowed around the inner wall of the combustion chamber before entering it. However, problems with burn-through during testing prompted a switch from gasoline to less energetic alcohol.

Tsander died unexpectedly from an illness on 28 March 1933 and his engineer, Leonid Konstantinpvich Korneev, became the new leader of his research Brigade.

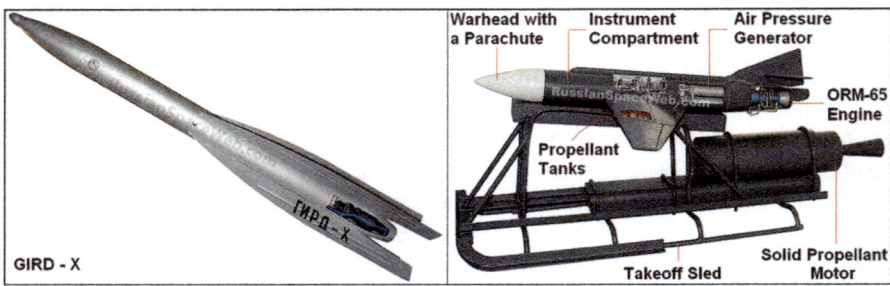

Fig. 1.5 Russian early rocket and winged rocket (Courtesy of Manual: by Ilcev)

According to the proposal of Stalin the Revolutionary Military Board established on 21 September 1933 a brand new institution RNII – Реактивный научно-исследовательский институт (РНИИ). The activities of the new institute began on October 31 by merging of GDL and GIRD. In the beginning the works on a rocket-glider were not a part of RNII activities and also the development of rocket engines using a liquid propellant was also not in the focus of activities, in fact the main activities were focused on military rockets, using solid fuel.

The first official Soviet rocket launch was the GIRD-9, on 17 August 1933, which reached the modest altitude of 400 m. Then, on 25 November 1933, GIRD-X (ГИРД-Х) became the first Soviet liquid-propellant rocket launched, shown in Fig. 1.5 (Left). The final missile was 2.2 m long by 140 mm in diameter, had a mass of 30 kg and it was anticipated that it could carry a 2 kg payload to an altitude of 5.5 km. On the other hand Soviet rocket scientists brought closest to the development of an unmanned cruise missile, thus On 16 June 1936 Korolev presented project of the "Object 218" powered by liquid fuel rocket engine and intended for stratospheric flights. On 29 January 1939, a small winged rocket took to the sky at a test range near Moscow as Code named Vehicle 212, shown in Fig. 1.5 (Right). Vehicle 212 sported many elements of its future successors, such as a three-axis gyroscopic autopilot.

The development of this rocket was also a backdrop for crucial debates among Soviet rocket pioneers and the missile's arrival on the launch pad coincided with the most horrific events in the Soviet history and in the life of Korolev, the founder of the Soviet space program. He was the main designer responsible for the rocket and who would later go on to mastermind all the early Soviet space triumphs, including wing rockets RP-318, RP-318-1, SK-9, Project 05 used the ORM-50 engine developed and rocket R-7, later Soyuz rocket, Sputnik, Yuri Gagarin's Vostok, and many more. His team attempted to modify Vehicle 212 into an aircraft-launched missile, capable of hitting "primarily" aerial targets as well as those on the ground.

As Tsiolkovsky was the first to discover that the reaction principle could reach space, it would make sense that the Russian pioneers and rocketry experts would provide very significant developments in rocket technology and technique before all other nations. He published a scientific book on virtually every aspect of space rocketing. He propounded the theoretical basis of liquid propelled rockets, put forward ideas for multi-stage launchers and manned space vehicles, space walks by astronauts and a large platform system that could be assembled in space for normal human habitation.

Between the two World wars, Russian and former USSR scientists and constructors used the great experience of Tsiolkovsky to design many models of rockets and in 1939 to build the first reactive weapon rockets ever, so called "Katyusha", which Soviet Red Army used against German troops at the beginning of the Great Patriotic War (WW2). Towards the end of the WW2, military constructors in Germany started with experiments to use their series V1, similar to Soviet wing rockets, and V2, similar to already developed Russian rockets, to attack targets in England. Among the people working at these projects was main constructor Wernher von Braun, who after WW2 leaded the US Space Program.

In 1947, Sergey Korolev created one of the most innovative management mechanisms in the early Soviet missile program known as the Council of Chief Designers shown in Fig. 1.6. This photo, a still from a rare film from the postwar years, shows the members of the Council at a meeting. From the left, Boris Chertok, Vladimir Barmin, Mikhail Ryazanskiy, Korolev, Viktor Kuznetsov, Valentin Glushko and Nikolay Pilyugin (standing), all Russian brains who put for the first time in history of human being a ballistic rocket in the Space.

After that, in October 1945, the British radar expert and writer of science fiction books Arthur C. Clarke proposed that only three communications satellites in Geostationary Earth Orbit (GEO) could provide global coverage for TV broadcasting.

The work on rocket techniques in Russia and the former USSR was much extended after the Great Patriotic War, thanks to enthusiastic and productive work of the Russian great rocket constructor and spacecraft designer Sergei Korolev. In 1951, Soviet Union realized the first launch of the "geophysical" rocket carrying live animals onboard.

The satellite era began when the Soviet Union shocked the globe with the launch of the first in the world artificial satellite, Sputnik 1, on 4 October 1957, shown in Fig. 1.7 (Left). Sputnik contained two radio transmitters, which sent back the "beep-beep-beep" signals heard round the world, and in such a way it provided the first Satellite Communication link with the Control Centre on the ground. Then a month later only, on 3 November 1957 an even larger and heavier satellite, Sputnik 2, carried the dog Laika into orbit. This launch marked the beginning of the use of artificial satellites to extend and enhance the horizon for communications, navigation, weather monitoring, meteorological observation and remote sensing.

This event signified the announcement of the space race and the development of satellite communications and navigation. That was followed on 31 January 1958 by the launch of US satellite Explorer 1, presented in Fig. 1.7 (Right). Explorer contained a cosmic ray detector, radio transmitter and temperature and micrometeoroid sensors. Therefore, the development of launchers and satellite systems for future cosmic explorations started, and the space race began

On 1–2 January 1959, is launched Luna 1 as the first artificial spacecraft to escape Earth orbit. On 12 September 1959 is launched Luna 2 the first man-made

Fig. 1.6 Council of Soviet Chief Designers (Courtesy of Manual: by NATO)

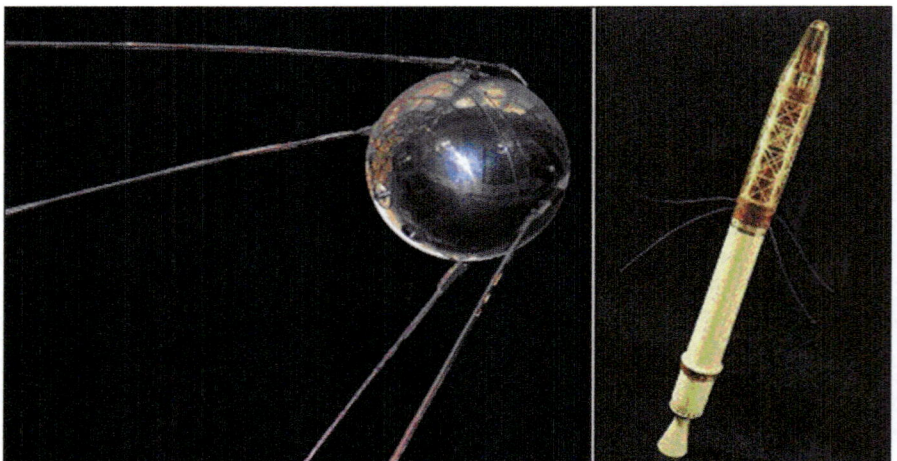

Fig. 1.7 Sputnik-1 and Explorer-1 satellites (Courtesy of Book: by Ilcev)

object to impact the Moon, and on 3 October the same year Luna 3 photographs far side of the Moon. On 19 August 1960, two dogs, Belka and Strelka, landed onboard the prototype of the spacecraft Sputnik 5, becoming first animals returning from orbit.

The most significant progress in space technology and engineering was on 12 April 1961, when Yuri Gagarin, officer of the former Soviet Union Air Forces lifted off aboard the Vostok-1 spaceship from Baikonur Cosmodrome and made the first historical manned orbital flight in space.

Fig. 1.8   Gagarin and Korolev (Courtesy of Book: by Ilcev)

In Fig. 1.8 (Left) is illustrated fist man flying in the Cosmos Soviet pilot Gagarin and in the same figure (Right) is picture of the Chief Designer of Soviet missile and space systems and the founder of modern cosmonautics Sergey Korolev. The twentieth century came with its great progress and the historical age of space rocketry and satellite communications began to unfold.

1.2.2   Experiments with Active Communications Satellites

After the launch of Sputnik-1, a sustained effort by the USA to catch up with the USSR was started. This was reflected in the first active communications satellite named SCORE launched on 18 December 1958 by the US Air Force. The second satellite, Courier, was launched on 4 October 1960 in High-inclined Elliptical Orbit (HEO) with its perigee at about 900 km and its apogee at about 1350 km using solar cells and a frequency of 2 GHz.

The maximum emission length was between 10 and 15 min for every successive passage. The third such satellite was Telstar-1 designed by Bell Telephone Laboratories experts and launched by NASA on the 10 July 1962 in HEO configuration with its perigee at about 100 km and apogee at about 6000 km, see Fig. 1.9 (Left). The plane of the orbit was inclined at about 45° to the equator and the duration of the orbit was about 2.5 h. Because of the rotation of the Earth, the track of the satellite as seen from the Earth stations appeared to be different on every successive orbit. Thus, over the next 2 years, Telstar-1 was joined by Relay-1, Telstar-2 and Relay-2. All of these satellites had the same problem, they were visible to widely

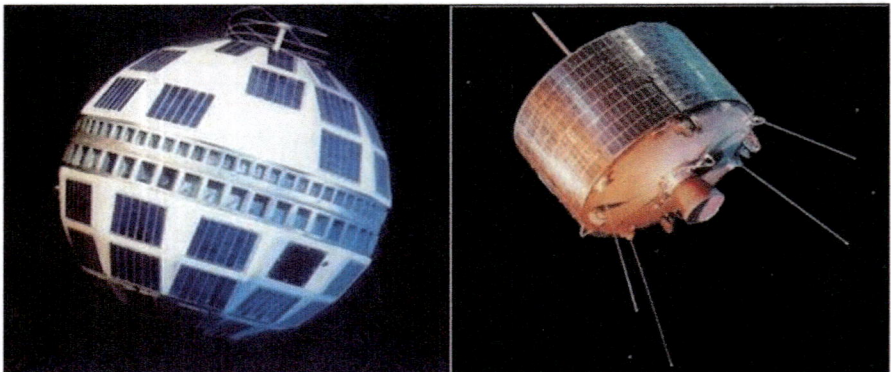

Fig. 1.9   Telstar-1 and Intelsat-1 (Courtesy of Book: by Ilcev)

separated Ground Earth Station (GES) for only a few short daily periods, so numbers of GES were needed to provide full-time service.

On the other hand, GEO satellites can be seen 24 h a day from approximately 40% of the Earth's surface, providing direct and continuous links between large numbers of widely separated locations. The World's first GEO satellite Syncom-1 was launched by NASA on 14 February 1963, which presented a prerequisite for the development of MSC systems. This satellite failed during launch but Syncom-2 and 3 were successfully placed in orbit on 26 July 1963 and 19 July 1964, respectively.

Both satellites used the military band of 7.360 GHz for the uplink and 1.815 GHz for the downlink. Using Frequency Modulation (FM) or Phase Shift Keying (PSK) mode, the transponder could support two carriers at a time for full duplex operation. Syncom-2 was used for direct TV transmission from the Tokyo Olympic Games in August 1964. These spacecraft continued successfully in service until some time after 1965 and they marked the end of the experimental period.

Technically, all these satellites were being used primarily for Fixed Satellite Service (FSS) experimental communications, which were used only to relay signals from Fixed Earth Stations (FES) at several locations around the world. Hence, one FES was actually located aboard large transport vessel the USNS Kingsport, home-ported in Honolulu, Hawaii. The ship had been modified by the US Navy to carry a 9.1 m parabolic antenna for tracking the Syncom satellites. The antenna dish was protected, like present mobile antennas, from the marine environment by an inflatable Dacron radome, requiring access to the 3-axis antenna through an air lock within the ship. Otherwise, the Kingsport ship terminal was the world's first true MES and could be considered the first Ship Earth Station (SES).

The ITU authorized special frequencies for Syncom communication experiments at around 1.8 GHz for the downlink (space to Earth) and around 7.3 GHz for the uplink (Earth to space). This project and trial was an unqualified success proving only the practically of the GEO system for satellite communications but because of the large size of the Kingsport SES antenna, some experts in the 1960s concluded that MSC at sea would never really be practical. However, it was clear that the potential to provide a high quality line-of-sight path from a ship to the land and vice versa via the satellite transponder existed at this time.

Intelsat was founded in August 1964 as a global FSS operator. The first commercial GEO satellite was Early Bird (renamed as Intelsat-1) developed by Comsat for Intelsat, see Fig. 1.9 (Right). It was launched on 6 April 1965 and remained active until 1969. Routing operations between the US and Europe began on 28 June 1965, a date that should be recognized as the birthday of commercial FSS. The satellite had 2×25 MHz width transponder bands, the first with 2 Rx uplinks (centered at 6.301 GHz for Europe and 6.390 GHz for the USA) and the second 2 Tx downlinks (centered at 4.081 GHz for Europe and 4.161 GHz for the USA), with maximum transmission power of 10 W for each Tx. This GEO system used several GES located within the USA and Europe and so, the modern era of satellite communications had begun.

In the meantime, considerable progress in satellite communications had been made by the former USSR, the first of which the Molniya-1 (Lightning) satellite was launched at the same time as Intelsat I on 25 April 1965. The Molniya satellites were put into an HEO, very different to those used by the early experiments and were used for voice, Fax and video transmission from central FES near Moscow to a large number of relatively small receive only stations.

In other words, that time became the era of development of the international and regional FSS with the launch of many communications spacecraft in the USSR, USA, UK, France, Italy, China, Japan, Canada and other countries. At first all satellites were put in GEO but later HEO and Polar Earth Orbits (PEO) were proposed, because such orbits would be particularly suitable for use with MES at high latitudes. The next step was the development of MSC for maritime and later for land and aeronautical applications. The last step has to be the development of the Non-GEO systems of Little and Big Low Earth Orbits (LEO), HEO and other GEO constellations for new MSS for personal and other applications.

1.2.3   Early Progress in Mobile Satellite Communications and Navigation

The first successful experiments were carried out in aeronautical MSC. The Pan Am airlines and NASA program in 1964 succeeded in achieving aeronautical satellite links using the Syncom-3 GEO spacecraft. The frequencies used for experiments were the VHF band (117.9–136 MHz), which had been allocated for Aeronautical MSC (AMSC).

The first satellite navigation system, called Transit was developed by the US Navy and become operational in 1964. The great majority of the satellite navigation receivers has worked with this system since 1967 and has already attracted about 100,000 mobile and fixed users worldwide. The former USSR equivalent of the

Transit was the Cicada system developed almost at the same time. Bots systems were successfully many years for their military and civilian applications.

Following the first American AMSC avionics experiments, the Radiocommunications Subcommittee of the Intergovernmental Maritime Consultative Organization (IMCO), as early as 1966 discussed the applicability of a Maritime MSC (MMSC) to improve ships communications. This led to further discussions at the 1967 ITU WARC for the where it was recommended that detailed plan and study be undertaken of the operational requirements and technical aspects of Maritime Mobile Satellite System (MMSS) by the IMCO and CCIR administrations.

A little bit later, the International Civil Aviation Organization (ICAO) performed a similar role to that of IMCO (described earlier) by the fostering of interest in Aeronautical Mobile Satellite System (AMSS) for Air Traffic Control (ATC) purposes. The majority of early work was carried out by the Applications of Space Technology to the Requirement of Aviation (Astra) technical panel. In the proper manner, this panel considered the operational requirements for and the design of, suitable systems and much time was spent considering the choice of frequency band. At the 1971 WARC, 2×14 MHz of spectrum, contiguous with the MMSC spectrum, was allocated at L-band for safety use.

Hence, the work of the Astra panel led to the definition of the Aerosat project, which aimed to provide an independent and near global AMSC, navigation and surveillance system for ATC and Airline Operational Control (AOC) purposes. The Aerosat project unfortunately failed because, whereas both the ICAO authority and world airlines of the International Air Transport Association (IATA) agreed on the operational benefits to be provided by such a system, there was total disagreement concerning the scale, the form and potential cost to the airlines. Finally, around 1969, the project failed for economic reasons.

The first experiment with Land MSC (LMSC) started in 1970 with the MUSAT regional land mobile satellite program in Canada for the North American continent. However, in the meantime, it appeared that the costs would be too high for individual countries and that in the future some sort of international cooperation was necessary to make this MSS project globally available.

In 1971, the ICAO recommended an international program of research, development and system evaluation. Before all, the L-band was allocated for Distress and Safety satellite communications and 2×4 MHz of frequency spectrum for MMSS and AMSS needs, by the WARC held in 1971. According to the recommendations, Canada, FAA of USA and ESA signed a memorandum of understanding in 1974 to develop the Aerosat system, which would be operated in the VHF and L-bands. Although, Aerosat was scheduled to be launched in 1979, the program was cancelled in 1982 because of financial problems.

The first GEP global MSC system was begun with the launch of the three Marisat satellites in 1976 by Comsat General. Marisat spacecraft had a hybrid payload: one transponder for US Navy ship's operating on an UHF-band and another one for commercial merchant fleets utilizing newly-allocated MMSC-bands. The first official mobile satellite telephone call in the world was established between vessel-oil platform "Deep Sea Explorer" operating close to the coast of Madagascar and the Phillips Petroleum Company in Bartlesville, Oklahoma, USA on 9 July 1976, using AOR CES and GEO of the Marisat system.

The IMCO convened an international conference in 1973 to consider the establishment of an international organization to operate the MMSC system. The International Conference met in London 2 years later to set up the structure of the International Maritime Satellite (Inmarsat) organization. The Inmarsat Convention and operating agreements were finalized in 1976 and opened for signature by states wishing to participate. On 16 July 1979 these agreements entered into force and were signed by 29 countries. The Inmarsat officially went into operation on 1 February 1982 with worldwide maritime services in the Pacific, Atlantic and Indian Ocean regions, using only Inmarsat-A SES at first. Moreover, the Marecs-1 B2A satellite was developed by nine European states in 1984 and launched for the experimental MCS system Prodat, serving all the mobile applications.

In 1985, the Cospas-Sarsat satellite Search and Rescue (SAR) mobile system was declared operational. Three years later the international Cospas-Sarsat Program Agreement was signed by Canada, France, USA and the former USSR. In 1992, the Global Maritime Distress and Safety System (GMDSS), developed by both Inmarsat and International Maritime Organization (IMO), before IMCO, began its operational phase. Hence, in February 1999, the GMDSS became fully operational worldwide as an integration of Radio MF/HF/VHF (DSC), Inmarsat and Cospas-Sarsat LEOSAR, GEOSAR and new developed MEOSAR systems.

The Transit system was switched off in 1996–2000 after more than 30 years of reliable service. By then, the US Department of Defense was fully converted to the new Global Positioning System (GPS). However, the GPS service could not have the market to itself, the ex-Soviet Union developed a similar system called Global Navigation Satellite System (GLONASS) in 1988 and ceased the previous Cicada system. While both, the Transit or Cicada system provides intermittent two-dimensional (latitude and longitude when altitude is known) position fixes every 90 min on average and was best suited to marine navigation, the GPS or GLONASS system provides continuous position and speed in all three dimensions, equally effective for navigation and tracking at sea, on land and in the air. In the meantime China started development of own satellite navigation system known as Compass (BeiDou), which is today operational for entire China only. It consists of two separate satellite constellations that have been operating since 2000 and a full-scale global system that is currently under construction and to be worldwide operational by 2020.

While both, the Transit or Cicada system provides intermittent two-dimensional (latitude and longitude when altitude is known) position fixes every 90 min on average and was best suited to marine and aeronautical navigation, the GPS or GLONASS system provides continuous position and speed in all three dimensions, equally effective for navigation and tracking at sea, on land and in the air.

The USA Federal Communications Commission (FCC) is reasonably encouraging toward private development of the Radio Determination Satellite System (RDSS), which would combine positioning fixing with short messaging. Thus, in

1985, Inmarsat developed the Standard-C system and later examined the feasibility of adding navigational capability by integrating GPS or GLONASS receiver in the mobile unit.

Although, the ESA satellite navigation concept, called Navsat, dates back to the 1980s. Moreover, the proposed new project has received relatively little attention and even less financial support from investors. Since 1988, the US-based Company Qualcomm for satellite communications and determination has established the OmniTRACS service for mobile messaging and tracking. Soon after, Eutelsat promoted a very similar system named EuroTRACS integrated with GPS and the Emsat communications system.

At the beginning of this millennium two Satellite Augmentation Systems were developed for Aeronautical Communications, Navigation and Surveillance (CNS): the American WAAS, European EGNOS, Japanese MTSAT, Russian SDCM, Chinese SNAS and Indian GAGAN. The next similar project is ASAS network for Africa and Middle East region, as first real venture into global satellite navigation systems for Southern Hemisphere.

Hence, it will augment the two military satellite navigation systems now operating, the US GPS and the Russian GLONASS and make them suitable for safety critical applications such as flying aircraft or navigating ships through narrow channels and port approaches. This system is a joint project of the ESA, the European Commission (EC) and Eurocontrol (the European Organization for the Safety of Air Navigation) and will become fully operational for commercial usage in 2004. The European Union contribution is the Global Navigation Satellite System (GNSS) as a precursor to a new system known as Galileo. This full GNSS, under development in Europe, is a joint initiative of the EC and the ESA in order to reduce dependency on the GPS service. The target of new the Galileo project is to start with operations by 2005 and still is not clear when will be fully operational.

In the meantime several very interesting projects are developing in Europe, Japan and the USA for new mobile and fixed multimedia Stratospheric Platform Communication Systems powered by fuel or the Sun's energy and manned or unmanned aircraft or airships equipped with transponders and antenna systems at an altitude of about 20–25 km.

1.2.4 Early Development in Meteorological Observation Systems

Meteorological observation and weather forecasting is the attempt by meteorologists to predict the state of the atmosphere at some future time and the weather conditions that may be expected. Weather forecasting is the single most important practical reason for the existence of meteorology as a science. It is obvious that knowing the future of the weather can be important for individuals and organizations. In fact, accurate weather forecasts can tell a farmer the best time to plant, when to carry out domestic animals to pasture, which route that sailors should choose, driving situation on the raods, an airport control tower what information to send to planes that are landing and taking off, how different companies to plan their works, residents of a coastal region when a hurricane might strike, and so on.

Many scientifically based weather forecasting organizations were not possible to provide any prediction until meteorologists were able to collect different data about current weather conditions from a relatively widespread system of observing stations and organize that data in a timely fashion. By the 1930s, these conditions had been met in US. Vilhelm and Jacob Bjerknes developed a weather station network in the 1920s that allowed for the collection of regional weather data. The weather data collected by the network could be transmitted nearly instantaneously by use of the telegraph, invented in the 1836 by S. Morse. Thus, the age of scientific forecasting, also referred to as synoptic forecasting, was under way.

In the US weather forecasting is the responsibility of the National Weather Service (NWS), a division of the National Oceanic and Atmospheric Administration (NOAA) of the state Department of Commerce. NWS maintains more than 400 field offices and observatories in all 50 states and overseas. The future modernized structure of the NWS will include 116 Weather Forecast Offices (WFO) and 13 river forecast centers. The WFO stations collect data from, forecast centres and other and from meteorological satellites circling Earth. Each year the US meteorological service collects nearly four million pieces of information about atmospheric conditions from these sources.

The data collected by WFO is used in the weather forecasting work of NWS. The data is processed by nine National Centers for Environmental Prediction (NCEP) and forwarded to seven of the centres focus on weather prediction, such as Aviation, Hydrometeorological, Climate, Marine, Space Environment, Storm and Tropical centres.

1.3 Development of Global Mobile Satellite Systems (GMSS)

Once the principles of radio systems were understood, mobile radiocommunications have been a matter of steadily developing and perfecting radio technologies and innovations, extending accessibility and enhancing range, improving efficiency, reducing the size, cost and power consumption of devices. With further MSC innovations an age long barrier was eliminated between vessels and shore, vehicles (road and rails) and dispatch centres, aircraft and airports and thus, facilities were created to provide mobile offices in ships, land vehicles and aircraft and to communicate with GES independently of space, place and time. The world is going to reduce communications barriers and move people across borders and so existing FSC and new MSC system including multipurpose satellite

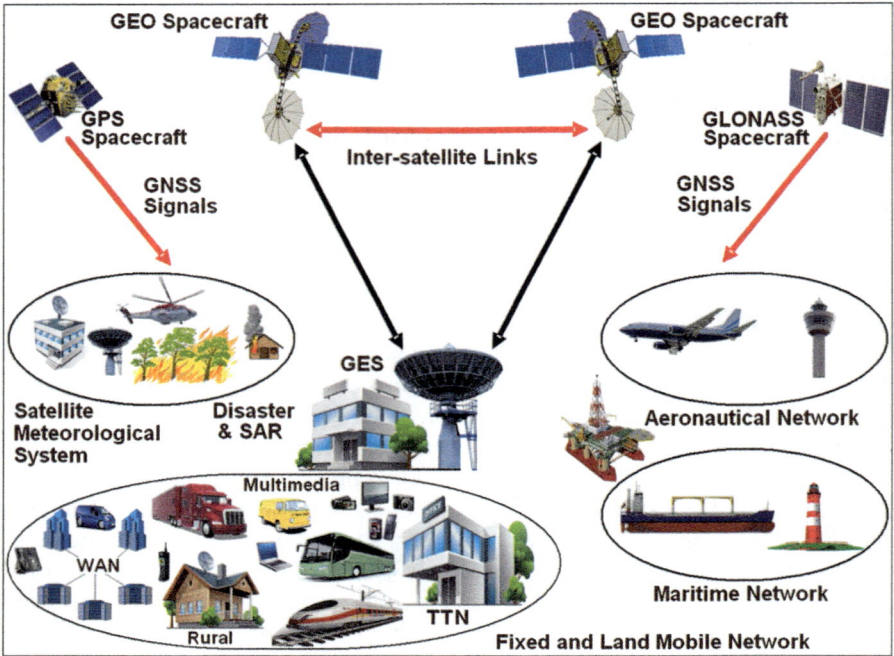

Fig. 1.10 Global space multipurpose systems (Courtesy of Manual: by Ilcev)

communications, navigation and meteorological services will be responsive to all these extraordinary changes and globalization trends, which scenario is shown in Fig. 1.10.

The MSC systems and technology also offer other benefits and perspectives. In many developing countries telephone density is still at a low level in urban and non-urban areas, because the cost of upgrading such facilities through wireless or TTN means is prohibitive for much of the world areas. Remote, rural and mobile service sectors in many regions are outside the reach of communications facilities, so the new MSS technology, with its instant ubiquitous coverage, may provide cost-effective solutions for developing countries.

1.3.1   Global Mobile Satellite Communications (GMSC)

The GMSC are GEO or Non-GEO satellite systems, which refers to all communications solutions that provide global MSC service directly to end users from a satellite segment, ground satellite network and TTN landline and/or radio infrastructures. The term GMSC means not only global coverage but also includes local or regional MSS solutions as an integral part of the worldwide telecommunications

Table 1.1 GMSC position within the telecommunication structure

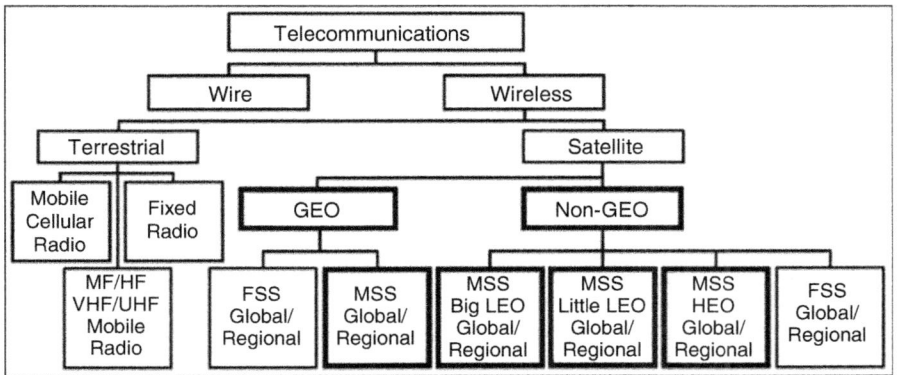

village. Namely, some of the regional or local MSS can be afterwards integrated to establish an GMSC network. In Table 1.1 is introduced an overview of telecommunication systems showing the respective satellite fit useful for any fixed and mobile satellite systems.

The GMSC systems are communication structures, which initiated to provide mobile links to vessels initially in the 1970s and later to aircraft and all kinds of land vehicles. It must be noted that GMSC providing global and regional coverage represents a new technology era in which wire terrestrial and wireless cellular voice, image, video and data systems are combined with MSC applications to provide services available anytime and anywhere. New satellite systems, such as Global Mobile Personal Satellite Communication (GMPSC) system and Very Small Aperture Terminals (VSAT), have also allowed global personal and commercial mobility for both satellite communication and determination.

New GEO or Non-GEO GMPSC systems have entered the field of MSC solutions, which for the past 30 years has been occupied predominately by Intergovernmental Satellite Organizations. In recent years, a growing number of private entities have been prepared to develop satellite technology, such as Iridium, Globalstar, Orbcomm, and son on. Recently, the modern DVB-RCS solutions are implemented more than decade and are serving both fixed and mobile applications.

At the same time, satellite technology and applications continues to advance; so satellite mobile terminals have become smaller, better and cheaper. Some GMSC systems now being developed are the initiative of the private sector or consortiums. This implies that there should be changes in policy, particularly in countries that do not foresee sufficient private participation in the telecommunication sectors to allow these systems to thrive and to realize their potential. As mentioned, GMSC systems can provide global or regional coverage. This capability has raised questions about national sovereignty, integrity and security of a country covered by a particular GMSC network.

Generally speaking, communication networks in the concerned country must always comply with national regulations that govern integrity and assistance to law enforcement and security agencies. These typically have requirements for national routing, location determination, call monitoring and legal interception. Therefore, seven categories of players in the GMSC community can be identified:

1. **National Regulatory Authorities** – The international community has to recognize the sovereign right of each country to adopt its telecommunication regulations and that the authority acts in the name of and on behalf of a certain state. It is the responsibility of the National Regulatory Authorities, according to their national laws, regulations and policies, to grant the appropriate authorization to allow GMSC services in a country.
2. **GMSC System Operators** – These are the owners or operators of the space segment, who have assumed all the financial, technical and commercial risks of developing GMSC systems and seek the harmonization of procedures governing the provision of services to avoid a proliferation of administrative impediments liable to constrain the development of the market. In this category can be included Satellite Meteorological Operators.
3. **GMSC Gateway/GES Operators** – Gateways are Ground Earth Stations (GES) satellite links between the space segment and TTN, from which, as main sources, GMSC terminal traffic is drawn. The GES terminals in some cases, depending on the business structure of the GMSC system, can be considered as a part of the space segment and can be managed by the GMSC satellite network operator.
4. **PSTN Operators** – The traditional PSTN operators provide most telecommunication services and networks, both wired and wireless, in a certain country. They are business partners and responsible for interconnection with terrestrial landline networks.
5. **Local and/or Regional Service Providers** – Local service providers are responsible for the local or regional provision of GMSC services, distributing GMSC terminals and billing GMSC customers. Otherwise, GMSC system or Gateway operators could also be local or regional service providers.
6. **GMSC Terminal Manufacturers** – These are companies that manufacture MSC and semi-fixed terminals for mobiles, including GEO and Non-GEO satellite networks.
7. **GMSC Terminal Users** – These are the customers whom all the other players are called upon to serve. They should receive good quality service at the best possible price, within the strict confines of the laws and regulations of the host countries.

1.3.2   Global Navigation Satellite Systems (GNSS)

The first GNSS generation GNSS offers augmentation systems in addition to the basic GPS and GLONASS constellations in order to achieve the level of performance suitable for enhanced maritime routing applications worldwide, especially in narrow passages, coastal navigation and approaching ports, for augmented civil aviation applications in oceanic and approaching flight, for land vehicles and also for ground positioning and surveying.

The GPS as a first generation of GNSS network is a satellite all-weather, full jam resistant, continuous operation radio navigation system, which utilizes precise range measurements from the GPS satellites to determine exact position and time anywhere in the world. This global system provides military, civilian and commercial maritime, land, aeronautical and ground users with highly accurate three-dimensional, common-grid, position and location data, as well as velocity and precision timing to accuracies that have not previously been easily attainable.

The GPS service is based on the concept of triangulation from known points similar to the technique of "resection" used with a map and compass, except that it is done with radio signals transmitted by satellites. The GPS receiver must determine when a signal is sent from selected GPS satellites and the time it is received. Nothing except a GPS receiver is needed to use the system, which does not transmit any signals; therefore they are not electronically detectable.

Because they only receive RF satellite signals, there is no limit to the number of simultaneous GPS users. The GPS satellite constellation is also providing facilities for new augmentation system known as Global Satellite Augmentation System (GSAS). Today are developed three Regional Satellite Augmentation Systems (RSAS) such as the US Wide Area Augmentation System (WAAS), the Japanese MTSAT Satellite-based Augmentation System (MSAS) and the European Geostationary Navigation Overlay Service (EGNOS). Recently, were projected or are in development stage other four RSAS infrastructures such as the Russian System of Differential Correction and Monitoring (SDCM), the Chinese Satellite Navigation Augmentation System (SNAS), Indian GPS/GLONASS and GEOS Augmented Navigation (GAGAN), and African Satellite Augmentation System (ASAS) designed by author.

The Russian Federation (former USSR) provides the GLONASS service from space for accurate determination of position, velocity and time for mobile or fixed users worldwide and in all weather conditions anywhere. Therefore, three-dimensional position and velocity determinations are based upon the measurement of transit time and Doppler Shift of RF signals transmitted by GLONASS satellites. The GLONASS satellite constellation as a first generation of GNSS will also provide signals for all RSAS infrastructures.

At this point, the GPS and GLONASS are parts of the first generation of GNSS known as GNSS-1, and there are in development phase other two GNSS of second generation known as GNSS-2 systems, such as the European Galileo and Chinese Compass.

The GNSS consists in many players with similar GMSC systems and three major segments:

1. The Space Segment has 24 satellites (21 functioning satellites and 3 on-orbit spares) and is controlled by a proprietary satellite operator or service provider.
2. The Control Segment is operated by Master Control and Monitor Stations.

3. The User Segment is represented by the military and civilian authorities for maritime, land and aeronautical users located worldwide. This segment offers Standard Positioning Service (SPS) and Precise Positioning Service (PPS), available to all users around the world. Access to the SPS does not require approval by a certain service provider but PPS is only available to authorized users via the service provider administration.

1.3.3   Stratospheric Platform Systems (SPS)

Aerial telecommunications have been investigated for three decades through the design and evaluation of stratospheric platforms able to offer multiple types of wireless services. The SPS stations may be airplanes or airships and manned or unmanned with autonomous operation coupled with remote control from the ground. There is increasing interest in the development of airspace platforms in recent years, so multiple type of SPS may carry multipurpose equipment at different altitudes for telecommunications, meteorological observation, remote sensing or digital broadcasting. Regarding the altitude of SCP aerial communication, there are three categories of balloons, High Altitude Platforms (HAP), Medium Altitude Platform (MAP) and Low Altitude Platform (LAP).

Above all, the SPS systems provide an excellent option for broadcast and broadband fixed and mobile communications including derived for DVB-RCS standards and technique. In addition, SPS stations can be deployed for onboard meteorological observation and Earth monitoring, they are able to provide GNSS solutions and can be very reliable and cost effective for rural communications as backbone to cellular and TTN. The SPS stations are officially not used for weather observation, but can be implemented for local scenario.

In the similar way as satellites, SPS stations can receive transmissions of meteorological data and images from Data Collection Platform (DCP) and provide their retransmissions to the Direct Readout Service (DRS) ground receiving terminals. They may also outline alternative network architecture scenarios for provision of cellular (wireless) access to the broadband communication services, and can coexist with WiMAX and WiFi service in same coverage area. Finally, SPS stations are suitable as well as for disaster, emergency and security communications, their survivability during a disaster and mobile ability to be continuously on station offer an ideal solution for an emergency communications capability.

The SPS systems are the newest space technique using top technologies for fixed and mobile applications including military solutions. This system use solar or fuel energy for propulsion and can be manned or unmanned. Similar to spacecraft, SPS station is carrying solar cells, payloads with transponders and antennas. The SCP network can be classified as the third layer of space communication infrastructure after satellite and terrestrial systems.

The SPS systems are using advanced digital transmission technologies offering reliable solutions such as fixed and mobile Voice, Data and Video over IP

(VDVoIP) services with speed over 150 Mb/s and up to 10 GB/s. The SPS structures are at about 15–30 km above the Earth with the coverage between 600 and 1000 km, each SPS acts as a very low orbiting satellite, providing high density, capacity and speed of service with low power requirements and no latency to an entire coverage area.

The SPS aircraft or airships do not require a launch vehicle, they can move under their own power throughout the space or can remain stationary and finally, while satellites cannot, they can be brought down to the Earth, refurbished and re-deployed without any service interruption. Same as satellite, each platform is independent and dedicated for certain area of coverage. The altitude enables the SPS system to provide a higher frequency reuse and a higher capacity then other wireless system.

1.3.4   Satellite Meteorological Observation Systems (SMOS)

Meteorological satellites have become essential for both meteorological and climatological observations, and continuing the two fundamental concepts of data and image exchange and international cooperation, which have been traditional for more than 150 years. They provide vital data at frequent intervals over wide areas, in the context of the international cooperation needed to ensure adequate worldwide data coverage. Cooperation and integration exist at two the following levels:

First at the worldwide level through those countries, which have come together to establish global Satellite Meteorology or Satellite Meteorological Observation Systems (SMOS) by integration of the USA, European, Russian, Chinese, Indian and Japanese meteorological satellite programs. In fact, this ensures the continuity of the SMOS constellations and ground infrastructures in availability of data and images over the entire Earth. The SMOS network contribution to the global observing system required for both meteorology and climatology data and imaging system.

The second level of cooperation is on a regional scale, which ensures the availability of the individual meteorological satellite data and images of each mentioned country alone and their availability over nearly one quarter of the Planet. Some countries are providing PEO and GEO meteorological satellite coverage, while in addition the US is providing LEO and Russia is providing LEO and HEO, so in total four satellite meteorological constellations.

The improvement in satellite technologies and advancements in satellite communication technologies have improve Earth observation, processing and transmission of data. The new generation satellites are regenerative as they have onboard processing capability that enables the satellite to amplify, or reformat received uplink data and route the data to specified locations, or actually regenerate data onboard as opposed to simply acting as a relay station between two or more ground stations.

Both PEO and GEO meteorological satellites are the most important parts of the Global Observation System of the World Weather Watch program for WMO. Meteorological satellites help to visualize atmospheric condition, including Earth surfaces, sea surfaces and clouds in a real-time basis. The GEO satellites are looking as stationary platforms for observers, while PEO satellites for observers are looking as a moving in circle around the Earth. Observation instruments are installed on satellites for meteorological and remote sensing purposes. The meteorological satellite observation has been used not only for weather forecasting, but also for climate change detection and atmospheric research.

Both SPS stations and weather observation satellite operates through the space. Thus the difference between this two is the altitude where they operate, cost to install and operation. The SPS stations operate at the low altitude compare to satellite. Introducing SPS as part of the future weather observation network can improve the quality of observation from the fact that it is closer to the Earth, offers cost effective solution, better, faster and greater speed of transmission. The use of SPS can improve weather observation especially in areas where it's not possible to have ground infrastructure especially in rural areas. This will highly contribute to understanding climate change issues and improve weather forecast.

1.4   Definition of Fixed Satellite Communications (FSC)

The FSC system enables a radiocommunications link between two or more FES or GES terminals at given positions, when one or more satellites are used, which is illustrated in Fig. 1.11. In this sense, the given position may be a specified position or any fixed point within a particular area. In some cases this particular service includes satellite-to-satellite links, which may also be operated in certain inter-satellite services.

Moreover, the FSS solutions may also include feeder links for other space communication services, including MSS or MSS GES can provide service for FES as well. The FSS signals are relayed between many FES, which are relatively large, complex and expensive systems. The FES terminals are connected to the conventional TTN and the service is intended for long distance VDV communications. According to the WARC-85/88 principle plan the FSS shares frequency bands with terrestrial networks in the 4/6, 12/14 and 20/30 GHz, which guarantees every country equal access to the GEO space constellations.

A typical example of FSS is the Intelsat, one of the pioneers in satellite communications, which first generation operates in the C band (4/6 GHz). At present, many global systems, such as Intersputnik, Telesat, Eutelsat, PanAmSat and others VDV and VSAT networks. Regional systems such as Optus in Australia and JCSAT in Japan operate in the Ku band (30/14 GHz), which provide coverage throughout most of Europe and Japan, respectively.

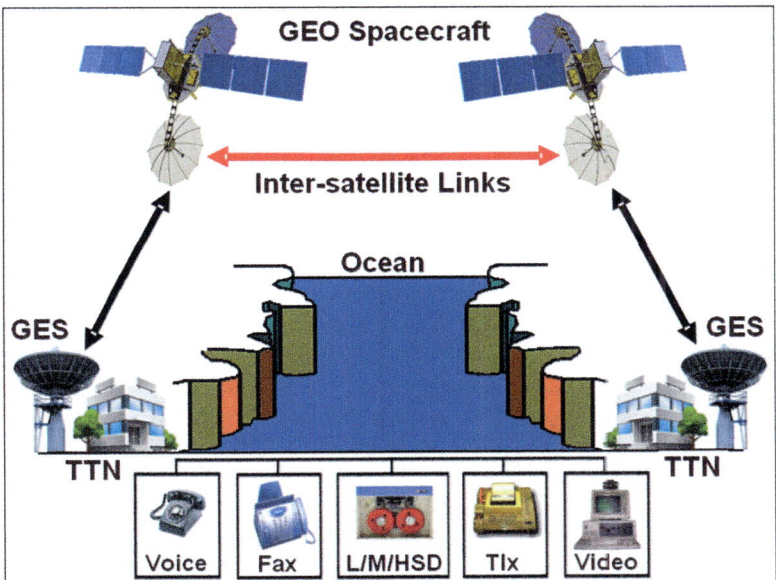

Fig. 1.11 Fixed satellite service (Courtesy of Book: by Ilcev)

1.4.1 Satellite Voice Network

Voice service is inherently interactive communication in nature providing global telephone infrastructures. In fact, telephony system represents first form of two way wire or wireless communications on distance.

Voice band channels are useful for relatively low data rate applications such as Fax, E-mail and low speed Internet access on less than 64 Kb/s. The digital standard for PSTN access at the subscriber level is ISDN service which provides 144 Kb/s of active data subdivided into two 64 Kb/s circuit switched bearer channels plus one data channel (2B + D).

The FSS voice network uses GEO satellite configuration and a bandwidth per channel from 8 to 64 Kb/s with FDMA and TDMA satellite transmissions. The system enables the international telephone trunks to extend the global coverage, thin route and to improve rural services in developing regions.

1.4.2 Satellite VSAT Network

The VSAT devices are quite similar to Inmarsat Mobile Earth Station (MES), which can be in fixed satellite service or installed onboard mobiles, such as small or large oceangoing ships, land vehicles (road and rail) and aircraft (airplanes and helicopters). This satellite equipment is small Earth stations capable of receiving (receivers only) from satellite and transmitting and receiving (transceivers) to or from spacecraft, providing two-way or interactive satellite communications. In such a way, fixed and mobile VSAT devices are classified as a communications media with either one-way or two-way facilities.

The term one-way VSAT includes the data terminals designed for the reception only of DBS transmission and conventional TVRO, using PAL and similar TV system. This device can also receive data using modulated sub carriers. Two-way VSAT can transmit and receive signals at rates to approximately 64 Kb/s. The low directivity of VSAT antennas limits the power and hence the boresight EIRP, which may be transmitted. The transceivers are usually completely solid state and can be highly integrated.

This equipment represents an important addition to the telecommunications world, because they can provide a service directly to their customers at virtually any geographic location covered by suitable satellite beam. They do not require any support from a local TTN and can even be run from portable or alternative power supply. This system is useful for private data network within countries and regions to promote business needs. The VSAT stations can be deployed for transmission meteorological data from Data Collection Platform (DCP) and provide their retransmissions to the Direct Readout Service (DRS) ground stations via certain meteorological satellite. In addition, the DRS ground station that deploys fixed VSAT equipment can be used for receiving observation data from meteorological satellites. Examples of such networks are Equatorial, Intelnet, Intelsat and other VSAT systems.

The VSAT system main applications are serving for data and documents distributions, rural communities, business utilizations and for disaster area communications. Data distribution can be in two-way between Central Hub stations for archive and data processing and all VSAT users. Documents distribution by VSAT can be only one-way satellite transmission from Hub library stations to all users. The users can be in touch with Hub via TTN.

Rural application of VSAT devices is very important for improving capability to transmit and receive much information from rural areas to central locations and vice versa. Rural communities would also like to provide speech service via VSAT systems, because of very limited telephone lines facilities, if they are available at all. Voice VSAT system operating at 4.8 Kb/s is already in use.

High quality speech can be carried on satellite channels with a data rate of 9.6 Kb/s using modern voice encodes and modems capable of operating at low C/N values, something like Inmarsat aeronautical voice system. A voice circuits set up between two VSAT devices include a double hop and accordingly a delay of about 0.5 s. A Hub located in a city can provide VDV and even VDVoIP to many fixed and mobile users.

Disaster areas and in generally for emergency needs mobile or transportable VSAT devices for alert and secure communications, almost regarding the terrain, locations and safety of life. The reason is similar that many ships, vehicles and aircraft now carry satellite distress beacons for alerting any distress situation.

Business VSAT applications are of an essential interest because of their large potential to provide persons and companies with the competitive edge. In generally, business wants to establish private networks to link their all locations and move their information in a safe manner and for the lowest possible cost. The service can be numerous like airlines and bus reservations, car rental conference facilities, insurance and newsgathering.

Although, the term VSAT equipment is generally used in connection with very small and fixed location terminals for business use, there are comparable developments in related fields should be considered. For example, very low cost TVRO system, which can receive MAC signals from high power satellites, could offer fixed and mobile VSAT type service in higher data rates and less cost.

The VSAT technology brings all of the features and benefits of unit or bi-directional FSS down to an extremely economical and usable for business data transmission. The system can also provide meteorological data transfer and bypass with TTN for VDV services using sophisticated digital technology and advanced communication network protocols.

The VSAT system enables use one or more 56 Kb/s data channels, each of which can be subdivided or applied directly. Voice communications is also possible using 16 or 32 Kb/s, depending on the compression algorithm. The VSAT network uses GEO configuration, a channel bandwidth from 64 to 512 Kb/s with FDMA and TDMA transmissions.

In Fig. 1.12 (Left) is shown GT&T Faraway SES100 VSAT cost-effective state-of-the-art satellite solution for companies having high communication flow to their clients, agents or branch offices located abroad or for remote and rural areas where communication services are unavailable, unreliable or too expensive.

This equipment is using V-sat communication network with African C-Band and European, African and Middle East Ku-Band coverage. The system is providing on demand High-quality voice up to 8 telephone lines, Group III Fax, Data transfer compatible via external Hayes compatible modem with rate from 4.8 up to 64 or 128 Kb/s, IP/X25/X400, Internet (E-Mail and Websites), Videoconference and Broadcasting (TV and Radio) services.

The Main unit with high capacity chassis allows several telephones (up to 8 interfaces), Fax and PCs to be connected simultaneously via parabolic antenna. The antenna is available with diameter of 65, 98, 120, 180 and 240 cm dishes depending on position in coverage area and distance from subsatellite point. This VSAT configuration employs a full-mesh, DAMA and PAMA network architecture that maximize the use of available space and ground-based resources.

Mobile satellite users can also be included in this category of satellite communications. Most probably in the near future VSAT system will offer full MSS similar to Inmarsat-C or mini-M systems. With an additional mobile antenna VSAT can offer MSS for maritime, land and aeronautical applications. Thus, another very last VSAT model is GT&T IPsky2 of V-Sat two-way transceiver system together with Internet modem and router lunched in January 2002 can be easily adapted in mobile unit with slight transformation of antenna system only, which is shown in Fig. 1.12 (Right).

Fig. 1.12 VSAT equipment for multipurpose applications (Courtesy of Manual: by GT&T)

This equipment is a low cost and highly compact Internet dedicated DVB-RCS-MPEG2 solution that also offers prepaid VoIP and Fax by satellite. In reality, customer will have an ultra-fast asymmetric Internet interface that is directly connected to the Internet backbone of 155 Mb and 34 Mb fiber using an E-Mai, FTP, TCP and Web Access service available on 24-h basis.

The maximum available download (outbound) speed of service via GEO is 2 Mb/s including very high level of transmission and reception coding for high speed and reliability. Return path or inbound speed is 33, 76.8 or 153.6 Kb/s depending on antenna size of 75, 96 or 1.2 m, respectively. Several PCs (from 5 to 13 or more if necessary) and Ethernet LAN can be connected to the IPsky2 modem/router through a Proxy server, public IP network, private FTP, VPN, etc.

1.5   Definition of Mobile VSAT Service

The mobile VSAT service is providing one-way transmission of TV program or transfer data, video and images from DBS TV Program, Direct and Cable TV and IPTV GES to the all kinds of MES terminals. The VSAT GES terminals are also connected via GEO spacecraft to the ground FSS VSAT via ground TTN or GEO satellite. While, another type of DVB-RCS VSAT service is providing two-way or interactive VDV, Fax, Tlx, IPTV and image transmission from and to mobiles, via GEO satellite and/or ISL, VSAT HUB GES and TTN. Scenarios for both VSAT mobile solutions are shown in Fig. 1.13, which can be also used for retransmission or reception of meteorological data and weather forecasts.

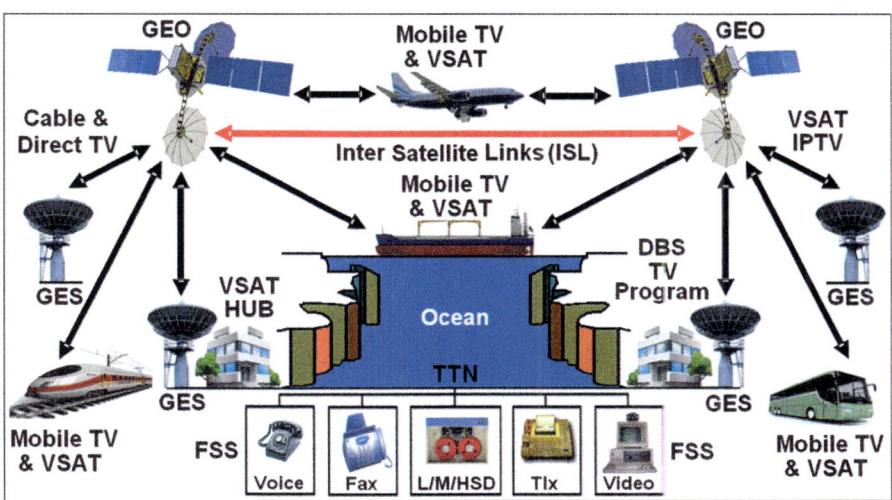

Fig. 1.13 Fixed and mobile satellite broadcasting (Courtesy of Manual by Ilcev)

1.5.1   Mobile Broadcast Satellite Service (MBSS)

The MBSS is one-way radiocommunication solution in which signals transmitted by Earth stations (GES) or retransmitted by space stations are intended for direct reception by the general public via satellite receivers and suitable antennas. The DBS TV or radio signals are transmitted from fixed positions on the ground to the satellite and then back to all MES via GES, community distribution by cable network or direct individual receivers. This service requires the generation of high RF power from the satellite to permit reception by small VSAT antennas onboard mobiles or on the ground, shown in Fig. 1.13.

Developments in digital audio/video broadcasting are, of course, not solely limited to the terrestrial domain of the home market. Digital services on offer by satellite are now making significant in-roads into the mobile entertainment market, which especially are using for maritime, land (road and rail vehicles) and aeronautical mobile applications. New digital broadcasting services (DVB-RCS) of CD-quality music programmes and DVD-quality video transmissions via GEO satellites is providing direct reception to many customers in mobiles have recently been attracting the attention of many people in transportation industries. The term "direct reception" in the MBSS shall encompass both individual fixed or mobile and community reception.

Individual direct reception is any reception of the emissions from space stations in the MBSS by simple domestic installations at home or on board mobiles and in particular those processing small antenna intended for private or common use onboard all type of mobiles. Community reception is any reception of the emissions from a space station in the MBSS by professional receiving equipment, which in some cases may be complex and can have antenna larger than those used for individual reception. This service is intended to be used by a group of the general public at one fixed or mobile location or through a distribution system covering a limited area.

The present MBSS operates at 12 GHz band and is designed for community reception, equipped with fixed satellite terminals and large antennas. Besides, if a satellite has enough power to transmit signals to be received by small antennas suitable for individual reception equipped with satellite VSAT, the system is called a DBS system.

Although the present system is designed for fixed service and terminals, new systems are also serving all three mobile applications equipped with auto tracking antennas. Some transportation means such as large ocean-going ships, cruisers, airplanes flying on intercontinental airlines, trains and buses traveling on international or interregional routes can install equipment to receive satellite TV programs from DVB-RCS and Direct Audio Broadcasting (DAB) stations. These mobile stations can be equipped with special tracking receiving only antenna systems as a separate unit or in combination with two-way communications satellite antennas. Therefore, this service may include audio (sound), video (television) and data transmission.

1. **Audio Broadcasting** – Audio broadcasting is a special radio emission via satellite for MSBS, in which signals transmitted by Earth stations or retransmitted by space stations are intended for direct reception of audio signals by the general public or mobile units via satellite receivers and corresponding antennas. In Europe, USA and Japan the satellite DAB systems in the frequency L and S-bands have been investigated to allow the direct transmission of high quality programs, comparable to CD, to be developed. Usually, road vehicles are equipped with such equipment.

2. **Video Broadcasting** – Video broadcasting is a special TV emission via satellite for MSBS, in which signals transmitted by Earth stations or retransmitted by space stations are intended for direct reception of DVB video and audio signals by the general public or mobile units via satellite TV receivers and corresponding antennas. All mobile units at sea, on land and in the air can be equipped with suitable video and tracking antenna equipment. The Inmarsat system also provides audio and videoconferencing via HSD MPDS or mobile ISDN interface using Fleet F77 service for ships and Swift64 for aircraft.

3. **Data Broadcasting** – Data broadcasting, using a different speed rate, is a special data transmission via satellite for MBSS, in which signals transmitted by Earth stations or retransmitted by space stations are intended for direct reception of data signals by the general public or MES terminal via special HSD satellite data receivers and corresponding omnidirectional or tracking antennas.

1.5.2 Mobile Satellite Broadband Service (MSBS)

The MSBS is a multimedia one or two-way radiocommunication service in which signals transmitted by mobile or fixed Earth stations or retransmitted by space stations are intended for direct reception by the general public or mobile units through satellite receivers and corresponding antennas.

This new broadcasting system, similar to the optical-fiber terrestrial network, will provide various advanced interactive multimedia services operating at higher transmission bit rates, such as voice (audio), video, different speeds of data, teletext, videoconferencing, distance learning, high resolution graphics, HiFi audio, high definition TV, color Fax and imaging, mobile Internet and PC communications. Such advanced networks will soon cover urban, fixed and mobile environments and most populations living in rural/remote areas, which cannot be covered by landline and optical fiber or wireless TTN.

These signals are transmitted from GES to the satellite and then back to all mobile or fixed applications via optical-fiber cable TTN. This service requires the generation of high RF power from the satellite to permit reception by receivers and small antennas on the ground. The MSBS has been developed by broadband MSC operators and will start to offer a multimedia service for personal and all three mobile applications, which allows any person or mobile to communicate anytime and anyplace. In fact, it will provide two categories of high-speed wireless access communications. The first will be serviced both outdoors and indoors, which can enable a high-speed rate up to 30 Mb/s and the second will provide ultra-high-speed indoors only, which can transmit high-speed signals up to 600 Mb/s. The second system cannot provide wide coverage areas or services in mobile environments, so the main application is limited to a "hot spot" of indoor premises.

The next example of MSBS is DVB-RCS for fixed and mobile solutions. This system ties: Terrestrial Networks (Broadband, Broadcasting, Internet, UMTS/GPRS Cellular, Private and Public) via satellite HUB (GES) with antenna and C, Ku or Ka-band satellites, with satellite Router Terminals for Remote Internet, VoIP, Videoconference, all E-services, Interactive TV/radio, Broadband LAN/WAN, Multicasting, Intranet/VPN, etc.

The newly designed Intelligent Satellite Transport System (ISTS) will be also part of the MBSS infrastructure, which comprises an advanced information and telecommunications network for users, roads and vehicles. The ISTS is expected to greatly contribute to solving problems such as traffic accidents and congestion. Not only solving such problems, ISTS will provide multimedia services for vehicle drivers and passengers. Otherwise, the ISTS consists in several development areas, including technical advances in MSC and navigation systems. control and tracking system, electronic toll collection system, assistance for safe driving and so forth, which is appreciated as one of the most promising mobile satellite multimedia businesses.

Digital networks used to operate 64 Kb/s channels with the switching (transmission) based on the so-called Synchronous Transfer Mode (STM). This mode was developed mainly for transmission; thus, switching functions are difficult to handle at different bit rates. At this point, in the interest of more flexible broadband switching, therefore, a new Asynchronous Transfer Mode (ATM) has been developed that can handle traffic relating to services that require widely differing bit rates. In ATM basically the information is put in fixed-length cells that are switched and transported to the broadband network and at the point of destination reconstituted in its original synchronous form. Typical new services requiring broadband switching equipment operating at higher speeds (bit rates) include: Desktop publishing; Multimedia service; Videoconferencing; Color Fax; HiFi music/HDTV, etc.

1.6   International Coordination Organizations and Regulatory Procedures

International coordination in space systems has been carried out by the world International Coordination Organizations, which include ITU, WMO, IHO and other.

1.6.1   *International Telecommunications Union (ITU) and Radio Regulations*

The ITU organization of the UN with all member governments has carried out the entire international coordination and regulation of mobile radio and satellite communications by the ITU RR. The ITU was inaugurated in 1932 and reorganized in 1992, with head office, all committees and departments located in Geneva, Switzerland. The ITU consists of three sectors: First, Radiocommunication (ITU-R), which ensures optimal, fair and rational use of the Radio Frequency (RF) spectrum, Second, Telecommunication Standardization (ITU-T), which formulates recommendations for standardizing telecommunication operations worldwide, and Telecommunication Development (ITU-D), which assists member countries in developing and maintaining internal communication operations, In fact, the World Administrative Radio Conference (WARC-92) covers the RF spectrum needs of numerous telecommunications services, from HF/VHF Broadcasting to satellite communications networks. The task of its International Radio Consultative Committee (CCIR) is to form study groups to consider and report on the operational and technical issues relating to the use of radio, space and all types of satellite communications.

1.6.2 World Meteorological Organization (WMO)

The WMO group coordinates all global scientific activity to allow prompt and accurate weather (WX) information, tropical storm forecasting and other services for public, private and commercial use, including the shipping and airline industries. World Meteorological Convention, by which body the WMO was created and adopted at the 12th Conference of Directors of the International Meteorological Organization, met in Washington in 1947. Although the Convention itself came into force in 1950, the WMO commenced operations as its successor in 1951 and, later that year, was established as an UN's specialized agency by agreement between the UN and the WMO.

The main purposes of the WMO World Weather Watch (WWW) structure are to facilitate international cooperation in the establishment of networks of stations for making meteorological, hydrological and other observations and to promote the rapid exchange of meteorological information by means of radio or satellite communications, the standardization of meteorological observations and the uniform publication of observations and statistics. The WMO also furthers specific weather and meteorology applications to maritime, aviation, agriculture, farming, industry and other human activities, climatology, atmospheric sciences, hydrology and instruments and methods of observation.

The main WMO task is to organize, control, collect and offer up-to-the-minute worldwide meteorological observations and weather information through many member-operated observation systems and telecommunication links with four PEO and five GEO satellites, about 10,000 land observation stations, 7000 ship stations and 300 moored and drifting buoys carrying automatic weather stations. Namely, each day, high-speed links transmit over 15 million data characters and 2000 weather charts through 3 World, 35 Regional and 183 National Meteorological Centres cooperating with each other in preparing weather analyses and forecasts in an elaborately engineered fashion. Thus, transoceanic ships and airplanes, research scientists on sea/air pollution or global climate change, the media and the general public are given a constant supply of timely data, which are very important for safe navigation and flight. Moreover, it is through the WMO that the complex agreements on standards, codes, measurements and communications are internationally established.

Data from all over the world are needed to provide weather (WX) forecasts and warnings. An aircraft does not take off, nor does a ship sails, without a WX received via agent or by means of radio or satellite communications for safety utilization. Combining facilities and services provided by all members, the program's primary purpose of the system is to make available meteorological, geophysical and environmental information enabling countries to maintain efficient meteorological services. Facilities in the regions outside any national territory are maintained by members on a voluntary basis.

Accordingly, WWW system comprises Global Observing System, Global Data Processing System, Global Telecommunication System, Data Management and

System Support Activities. The Hydrology and Water Resources Program concentrates on promoting worldwide cooperation in the evaluation of water resources and the development of hydrological networks and services, including data collection and processing, hydrological forecasting and warnings very important for the safe navigation of ships and the flight of aircraft and the supply of meteorological and hydrological data for design purposes.

1.6.3 International Hydrographic Organization (IHO)

The IHO is an international intergovernmental consultative and technical organization that was established in 1921 to support safety in navigation and the protection of the marine environment. The main IHO activity is standardization of nautical charts and documents, bathymetry and ocean mapping and related publications, technical assistance and training. The special activity of IHO related to MSS is radio navigational warnings implemented by the GMDSS as an integrated Radio and Satellite communication system of Inmarsat and Cospas-Sarsat systems. The GMDSS system has improved the dissemination of Maritime Safety Information (MSI), taking advantage of modern communications technology.

The far offshore navigation warnings are broadcast via Inmarsat-C Enhanced Group Call (EGC) SafetyNET Service, whilst polar areas warnings are covered by DCS HF radio services and coastal area warnings are transmitted via DCS MF/VHF radio and NAVTEX on a single frequency, timeshared and automatic broadcast. The NAVTEX and Digital Selective Call (DSC) VHF and HF radio services are already in operation in most parts of the world in framework of the GMDSS mission.

The advent of digital data transmission, computers and video display systems is having a considerable impact on hydrographic and navigation technology. This has made possible the development of ECDIS (Electronic Chart Display and Information Systems), which has become a major focus of activity. Several working groups have been established by the IHO to coordinate these new developments and ensure the standardization of systems and specifications. In 1992, the IHO adopted standards for the formatting of ECDIS data so that it can be readily exchanged between hydrographic offices. At this point, the Committee on ECDIS (COE) examines both the database requirements and the standardization and overall parameters of such systems.

1.7 Space Systems and Radiocommunication Frequency Assignment

Space systems are containing space and ground segment, which can provide global fixed or mobile satellite networks for communication facilities to and from GEO and Non-GEO for commercial and meteorological applications. Radio Frequency (RF) is EW range around 3 kHz to 300 GHz used for satellite communication and meteorological data transmission.

1.7.1 Meteorological Space and Ground Segments

The space segment provides the connection between the subscribers on shore and mobile or fixed users via GES or Gateways. It consists in one or more operational or spare spacecraft in a corresponding constellation. The satellite constellation can be formed by a particular type of orbit, such as GEO, Non-GEO (LEO, PEO and HEO) or their combinations.

The satellites can be independent or connected with each other through Inter-Satellite Link (ISL) or Inter-Orbit Link (IOL). The space segment can be shared among different radio networks in different areas, which include satellite and SCP networks. There are also constellations of multipurpose satellites, which platform can serve more than two payloads, such as a combination of Communication, Navigation and Surveillance (CNS) or GNSS, MSC and meteorological payloads.

The ground segment consists in three major network elements: GES terminals, ground support networks and user subsegments. Users subsystem can consists fixed and mobile customers including personal terminals. The user network comprises many main categories of user terminals connected to TTN interfaces, whose characteristics are highly related to its applications and operational environments. The main part of meteorological ground segment is DRS subsystem, which directly receives data and images from satellites or SPS and is known as GES with antenna system. The part of ground infrastructure is DCP as well, which is transmitting its collected meteorological data via satellites or SPS stations to the DRS ground receiving terminals. Then DRC terminals contain Weather Ground Processing of Satellite Meteorological Data and its distribution to the different customers.

1.7.2 Meteorological Frequency Designations and Classification of Services

The assignment of a radiocommunications frequency, band or channels is performed by an authorized administration for radiocommunications via platform, satellite or space stations to use an RF spectrum or frequency channels under

Table 1.2 Frequency bands designation

Band No.	Abbreviation	Band name	Name	Symbol	Frequency
4	VLF	Very Low Frequency	Myria m		3–30 kHz
5	LF	Low Frequency	Km		30–300 kHz
6	MF	Medium Frequency	Hm		300–3000 kHz
7	HF	High Frequency	Dam		3–30 MHz
8	VHF	Very High Frequency	m		30–300 MHz
9	UHF	Ultra High Frequency	dm		300–3000 MHz
				L-band	1–2 GHz
				S-band	2–4 GHz
10	SHF	Super High Frequency	cm		3–30 GHz
				C-band	4–8 GHz
				X-band	8–12 GHz
				Ku-band	12–18 GHz
				K-band	18–27 GHz
				Ka-band	27–40 GHz
11	EHF	Extremely High Frequency	mm		30–300 GHz
12	VEHF	Very Extremely High Frequency	deci mm		300–3000 GHz

specified conditions. At this point, in satellite communication fields the frequency bands are often denoted with alphabetical symbols such as L to Ka-bands. The radio spectrum shall be subdivided into nine RF bands and designated by progressive whole numbers, in accordance with Table 1.2.

Frequency band numbers and names are defined by the ITU RR, ITU Tables of Frequency Allocations in general or for a particular band and the mentioned alphabetic symbols by the IEEE Standard Radar Definitions. In a more general sense, frequency designations for all type of Meteorological Satellites (MetSat) constellations are used by a number of different administrations for their national or international MetSat WX (Weather) networks and can be systematized into three main categories:

1. **Allocated Frequency Bands for MetSat in the ITU Radio Regulations (RR)** – This is the first frequency allocation for MetSat transmissions presented in the Table 1.3. Thus, it provides RF allocations for Space Operations, GEO and Non-GEO MetSat systems, etc.
2. **Typical Applications in the Frequency Bands Allocated to the MetSat Service (1)** – It is the third frequency allocation for MetSat transmissions presented in the Table 1.4.

Table 1.3   Allocated frequency bands for MetSat in the ITU radio regulations (RR)

•	137 – 138 MHz (s-E)	
•	400.15 – 401 MHz (s-E)	
•	401 – 403 MHz (E-s)	
•	460 – 470 MHz (s-E)	⇨ Secondary allocation
•	1670 – 1710 MHz (s-E)	⇨ 1670 – 1675 MHz no new MetSat use (Res. 670)
•	2025 – 2110 MHz (E-s)	⇨ Space Operation (SO), Earth Exploration Satellite (EESS)
•	2200 – 2290 MHz (s-E)	⇨ SO and EESS
•	7450 – 7550 MHz (s-E)	⇨ RR Footnote 5.461A limits use to GEO Satellites
•	7750 – 7850 MHz (s-E)	⇨ RR FN 5.461B limits use to Non-GEO Satellites
•	8025 – 8400 MHz (s-E)	⇨ EESS with stringent PFD limits in Region 1 and 3
•	8175 – 8215 MHz (E-s)	
•	18.1 – 18.3 GHz (s-E)	⇨ Subject to extension to 300 MHz at WRC-2007
•	25.5 – 27 GHz (s-E)	⇨ EESS

Table 1.4   Typical applications in the frequency bands allocated to the MetSat service (1)

•	137 – 138 MHz (s-E)	
•	400.15 – 401 MHz (s-E)	
•	401 – 403 MHz (E-s)	
•	460 – 470 MHz (s-E)	⇨ Secondary allocation
•	1670 – 1710 MHz (s-E)	⇨ 1670 – 1675 MHz no new MetSat use (Res. 670)
•	2025 – 2110 MHz (E-s)	⇨ Space Operation (SO), Earth Exploration Satellite (EESS)
•	2200 – 2290 MHz (s-E)	⇨ SO and EESS
•	7450 – 7550 MHz (s-E)	⇨ RR Footnote 5.461A limits use to GEO Satellites
•	7750 – 7850 MHz (s-E)	⇨ RR FN 5.461B limits use to Non-GEO Satellites
•	8025 – 8400 MHz (s-E)	⇨ EESS with stringent PFD limits in Region 1 and 3
•	8175 – 8215 MHz (E-s)	
•	18.1 – 18.3 GHz (s-E)	⇨ Subject to extension to 300 MHz at WRC-2007
•	25.5 – 27 GHz (s-E)	⇨ EESS

3. **Typical Applications in the Frequency Bands Allocated to the MetSat Service (2)** – It is the third frequency allocation for MetSat transmissions presented in the Table 1.5.

To complement the RR satellite notification procedure of the ITU-R, an inter-agency RF coordination procedure was established in the framework of SFCG (RES A12-1R2).

Table 1.5 Typical applications in the frequency bands allocated to the MetSat service (2)

•	7450 – 7550 MHz (s-E) ⇨	Downlink of medium rate raw data from GSO MetSat to main Earth station (not used on current generation MetSat systems)
•	7750 – 7850 MHz (s-E) ⇨	Downlink of raw data from NGSO MetSat to main Earth station, but also dissemination
•	8025 – 8400 MHz (s-E) ⇨	Downlink of sensor data from GSO and NGSO MetSat to main Earth station
•	18.1 – 18.3 GHz (s-E) ⇨	Downlink of high rate raw data to main Earth station
•	25.5 – 27 GHz (s-E) ⇨	Downlink of high rate raw data to main Earth station (currently no plans for next generation MetSat)

1.8   History of Satellite Meteorology

In the more than four decades since the first meteorological satellites were launched, such as Vanguard 2 on 17 February 1959, they have become indispensable for study and observing of the Earth's atmosphere. Indeed, together with their land- and ocean-sensing cousins, meteorological satellites view the Earth from a global perspective that is unmatched and unmatchable by any other observing system.

It is human nature to want to find out about our surroundings, to explore our neighborhood, our planet Earth and beyond. Until the twentieth century, viewing Earth from a space-based perspective could only be accomplished by imagination. From ancient times, astronomers have looked up at the sky, recorded their observations and made up stories about how the universe was created and what it was like.

Ancient Greeks were more aware of the truth of their surroundings than other cultures in that time period. They discovered that Earth was a sphere and developed observational and mathematical techniques to measure the circumference of the planet. During recent times, with increasingly powerful ground-based telescopes came the discovery of the Milky Way and other galaxies and our understanding that the universe is expanding.

The practice of Earth observation involves the gathering of information about the planet's physical, chemical and biological systems, usually by remote sensing systems, which have grown technologically more and more sophisticated over time. Thus, the famous "Big Blue Marble" photograph of Earth, taken in 1972 by astronauts onboard Apollo 17 spacecraft, demonstrated the dramatic impact of viewing Earth from space, illustrated in Fig. 1.14. This emphasized the importance of minimizing the negative impact of modern human civilization to improve social and economic well-being.

Fig. 1.14 Blue Marble photography (Courtesy of Brochure: by NATO)

The art of weather observation and forecasting, which began with early civilizations and was based on recurring astronomical and meteorological events, was used to monitor and predict seasonal changes in the weather situation. In 650 BC, the Babylonians attempted to predict short-term weather based on cloud patterns and astrological observations, while Chinese weather prediction dates back to 300 BC, when annual calendars were developed according to repeated patterns of weather events. This experience accumulated over generations to produce weather lore.

In about 340 BC, Greek philosopher and scientist Aristotle described weather patterns in a treatise entitled "Meteorologica". This writing contained Earth science theories, such as on cloud formations, wind, rain, and other weather phenomena. This later led to his pupil, Theophrastus, compiling "The Book of Signs", which documented weather lore and forecast signs. These texts served as definitive weather forecasting references for more than 2000 years and helped to establish meteorology as a distinct discipline of study and developments. Weather forecasting advanced little from these ancient times until the Renaissance, despite many of Aristotle's claims being erroneous.

1.8.1 Early Meteorological Instrumentation

Over the centuries, it became apparent that forecasts based on weather lore, philosophical peculations and personal observations alone were not always reliable. In order to advance knowledge and understanding of the atmosphere, instruments were needed to measure properties, such as air, moisture, temperature and pressure. The first device to measure the humidity of air, called the hygrometer, was invented by Leonardo da Vinci in the fifteenth century. About 1593, Galileo Galilei, often deemed the father of modern observational astronomy, invented an early thermometer for temperature measurement using the expansion and contraction of air in a bulb to move water in an attached tube. His student Evangelista Torricelli invented the barometer for measuring atmospheric pressure in 1643.

In subsequent centuries, these meteorological instruments were refined and improved, and were being applied in association with observational platforms launched in the air, such as balloons and aircrafts, for taking from a height atmospheric meteorological measurements. In the meantime, a significant historical development was the invention of the first air balloon in 1783 by the French brothers Étienne and Joseph Montgolfier, which drawing of the first hot-air balloon is illustrated in Fig. 1.15. This discovery represents a precursor to the development of today facilities such as: dropsondes, radiosondes, air baloons, upper air stations, aircraft, stratospheric platforms and artificial satellites, which are representing the future integration components for space meteorological observations as well.

With advancements in meteorological measurement instrumentation in the eighteenth century came experimentation of many different airborne platforms to measure physical properties of the atmospheric column, including pressure, temperature, wind speed, wind direction and other properties.

They experimented with hydrogen-filled paper bags, which led to the correct notion that a buoyant force should cause ascent of the bags, if the inside gas was lighter than air. Since gas diffused out quickly and hydrogen was produced in small quantities, they subsequently tried "gas" produced by the combustion of a mixture of moistened straw and wool. This designed the first hot-air balloon in the world, which attained a height of about 1950 m.

In the same year, J.A.C. Charles and the Robert brothers designed and constructed a new hydrogen-filled balloon, but inflation was achieved only with great difficulty over a period of 4 days. Balloon flights began carrying animals and then subsequently men. Furthermore, balloons were improved to descend and ascend at will, and in time were improved with safe landing devices and better direction control. As techniques of the hydrogen manned balloon evolved rapidly it offered the possibility to investigate Earth's atmosphere. The first manned hydrogen balloon flight conducted by Charles and the Robert brothers carried a barometer and thermometer to measure the pressure and the temperature of the air, making it the first atmospheric meteorological measurements above Earth's surface. Also in 1784, Charles rose again and measured air temperature variation along with altitude in the atmosphere. After 1850, the application of balloons for measuring meteorological parameters was widely practiced. To monitor favorable weather conditions for balloon ascent, Charles also inaugurated the practice of launching a small pilot balloon prior to flight in order to determine the wind vector at different altitudes.

Subsequent technical advances in the use of unmanned air balloons made it possible to sound the atmosphere. Thus, colonel William Blaire in the US Signal Corps performed primitive experiments with weather measurements from a

Fig. 1.15   Drawing of the first hot-air balloon (Courtesy of Book: by Tan)

balloon, while the first really useful radiosonde was invented in France by Robert Bureau in 1929. This device sent precise encoded telemetry from weather sensors to the ground. Subsequent developments enabled radiosonde instruments to become smaller, lighter and more accurate. In Fig. 1.16 is shown early launch of radiosondes developed by the US Bureau of Standards in 1936 (Left) and that the US Army Air Force meteorologists preparing a hydrogen-filled balloon equipped with a radiosonde in 1944 (Right).

Radiosondes have also been used for exploring atmospheres of other planets, such as in the Soviet Union's Vega program, where probes dropped radiosondes to study the atmosphere of Venus. Up to present, radiosondes are still launched worldwide year-round. The US National Weather Service releases about 75,000 radiosondes each year, not including military soundings and for other specialized scientific purposes. Collective agreements have formed a global radiosonde station network worldwide (about 900 stations) that make an average of 1209 soundings each day to support weather forecast activities.

The modern age of weather forecasting began with the invention of the electric telegraph in 1835, which allowed for routine and instantaneous transmission of weather observations. It was possible to develop crude weather maps and to study surface wind patterns and storm systems. Synoptic weather forecasting was made possible by the compilation and analysis of data collected simultaneously from weather observing stations and conveyed across the globe via telegraph in the 1860s. Data collected by land locations are now conveyed worldwide via phone or wireless technology, enabling information to be communicated quickly for weather forecasts and studies of the atmosphere and climate.

Fig. 1.16 Early launch of radiosondes and preparing hydrogen-filled balloon in the hangar (Courtesy of Book: by Tan)

Thereafter, pilot balloons were superseded by free-flight sounding balloons, which carried sensors and radio telemetry transmitters. Sensors launched by weather balloons to measure atmospheric profiles of pressure, temperature, and relative humidity are carried in a unit commonly referred to as a radiosonde. Usually contained in a small, expendable instrument package suspended below a large balloon, the radiosonde provided an efficient way to systematically and regularly measure various atmospheric parameters to heights of over 30,000 m, without the necessity of considering weather conditions.

Nowadays, received meteorological information is transmitted to a ground-based station for data users via a radio transmitter and radiosondes are still used by national weather services for capturing high vertical resolution flight data. In addition, kites were frequently used for capturing meteorological observation data in the second half of the nineteenth century. A meteographic device, which is a chart recorder for measuring humidity and temperature, was usually attached. However, kites were linked to the ground and highly unstable due in windy conditions. During World War I, meteorological observations from kite flights were largely substituted for by new developed aircraft. Flying weather forecasts for aircraft were not required prior to World War I, since pilots mostly flew at low altitudes.

During wartime, however, aircraft were often required to fly in clouds, in bad weather conditions, and at high altitudes. Meteographs were often mounted on the wings of military aircraft to obtain meteorological information for monitoring flying conditions. Observations were recorded on a cylindrical chart that was retrieved after the landing of the aircraft, and meteorological parameters were read from the chart. Pilots were often required to reach a flying altitude of at least 4000 m, where they could black out from lack of oxygen, making this a very dangerous enterprise. It was often impossible to fly in bad weather, which unfortunately was when observations were needed the most.

1.8.2   *Evolution of PEO Meteorological Satellites*

The vast arrays of radiosonde stations, weather reconnaissance aircraft and newly developed observing systems have provided a significant amount of information about meteorological parameters and weather conditions. Sensors are measuring atmospheric constituents directly, such as thermometers, barometers, and humidity sensors, have been sent aloft on balloons, rockets or dropsondes for many years. Although precise in their measurements, these instruments have limited capabilities to provide regional or even global coverage, which is necessary for making accurate weather forecasts. Thus, the global network of radiosonde observing stations tend to have a highly concentrated dispersion in the northern hemisphere temperate zone for large land masses, whereas the density of observations for the southern hemisphere, tropical regions, the Arctic and most of the northern Pacific is relatively sparse.

Consequently, there is a high degree of uncertainty with tracking main tropical storms over the North Pacific Ocean, areas around Madagascar in Indian Ocean and Florida in US. Since the Earth's atmosphere is a single and closely interacting mass of air, disturbances can propagate throughout at a speed much faster than winds. The real-time synoptic monitoring of large areas of the Earth is necessary for improved meteorological data collection via satellites. Extended and long-range weather forecasts require data to be collected and distributed globally.

Earth-observing satellites are able to collect meteorological data at synoptic scales and in remote locations, tracking cloud cover, relative motion of storm systems and the jet stream, and maximum heights of clouds and vertical temperature profiles. Satellite imagery can identify cloud patterns associated with different types of weather conditions and patterns (e.g., spiral cloud patterns and convective cells), which are difficult to capture and monitor using conventional weather observations alone. Developments in satellite and other space technologies have resulted in enormous improvements in the accuracy of meteorological observation and weather forecasting. Satellites have particularly provided routine access to observations and data from remote areas of the globe.

As stated before, these efforts were intensified after the launch by the Soviet Union on 4 October 1957 of the first successful Earth satellite ever, Sputnik-1. The first successful US satellite, Explorer-1, was launched 123 days later on 31 January 1958. The cosmic era and race over the globe began.

However, many people were quite skeptical that meteorological satellites could become a significant aid to provide meteorological data and weather forecasting. In fact, the launch of Sputnik spurred new military, political, scientific and technological developments. As weather observing tools, satellites offered new capabilities for meteorologists, such as the ability to view Earth as a whole in a short period of time. Satellites could also provide very useful day-to-day monitoring capabilities, and information about remote regions where traditional data sources were not previously available. Collectively, weather satellites from the United States,

Fig. 1.17 Early meteorological satellites Vanguard-2 and Explorer-7 (Courtesy of Manual: by Ilcev)

Europe, India, China, Russia, and Japan can provide nearly continuous observations for a global weather watch.

The first weather satellite with meteorological instruments, Vanguard-2, was launched on 17 February 1959 by the US NASA, which is shown in Fig. 1.17 (Left). In fact, it was designed to measure cloud cover and resistance, but a poor axis of rotation kept it from collecting a notable amount of useful meteorological data. However, on 13 October 1959 NASA launched satellite Explorer-7 and made the first meteorological measurements from the satellite Suomi's net flux radiometer, shown in Fig. 1.17 (Right). The first real weather satellite to be considered a success was the US TIROS-1 (Television and Infrared Observational Satellite), launched by NASA on 1 April 1960, shown in Fig. 1.18 (Left). In 1964 an important series of meteorological satellites Nimbus was initiated. Nimbus-1 satellite was launched on 28 August 1964, shown in Fig. 1.18 (Right).

The meteorological satellite TIROS was operated for 78 days and proved to be much more successful than Vanguard-2. It released fuzzy images of thick bands and clusters of clouds over the United States, which one of the first images captured by the TIROS-1 satellite is shown in Fig. 1.19. This satellite would forever change weather forecasting and climate research in the field of Earth system science, which data was directly transmitted in real-time to ground stations within signal range of the satellite linked to forecasting centers. By 1965, TIROS imagery had been combined to generate the first global view of worldwide weather. The success of these meteorological observations proved to be effective for meteorological and environmental surveillance, paving the way for the Nimbus program, whose technology and findings are the heritage of most Earth-observing satellites.

To maintain their orientation in space many satellites spin. In the absence of external torques, angular momentum is conserved, and the spin axis points in a constant direction in space as the satellite orbits the Earth. Thus, TIROS spun at about 12 revolutions per minute (rpm). This caused a viewing problem with the first

Fig. 1.18 Meteorological satellites TIROS-1 and Nimbus-1 (Courtesy of Manual: by Ilcev)

Fig. 1.19 First TIROS-1 image (Courtesy of Brochure: by NASA)

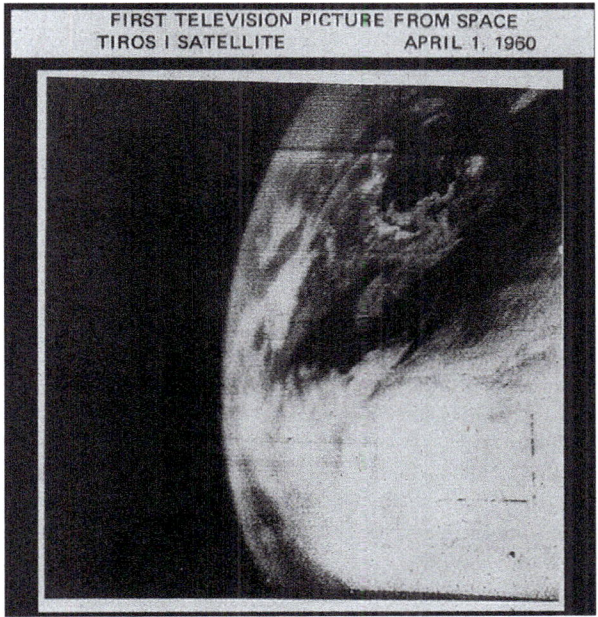

FIRST TELEVISION PICTURE FROM SPACE
TIROS I SATELLITE   APRIL 1, 1960

eight TIROS, however. The vidicon cameras on these satellites were on the "bottom" of the craft; that is, they pointed parallel to the spin axis and, therefore, in a constant direction in space. Thus, the Earth was in their field of view only about 25% of the time of each orbit. During the remaining 75% of each orbit, they viewed space and could produce images.

In 1964 an extremely important series of experimental meteorological satellites was initiated, the Nimbus series. The first in a series of second-generation meteorological research and development satellites Nimbus-1 had two notable firsts. It was the first three-axis stabilized meteorological satellite that use of momentum special wheels, namely flywheels inside the spacecraft controlled by horizon sensors, it rotated once per orbit (by placing torque on the appropriate momentum wheels), so that its instruments constantly pointed toward Earth. It was designed to serve as a stabilized, earth-oriented platform for the testing of advanced meteorological sensor systems and for collecting meteorological data. The polar-orbiting spacecraft consisted of three major elements, such as: a sensory ring, solar paddles, and the control system housing.

The Nimbus-1 spacecraft also was the first artificial sunsynchronous satellite, which means that it passed over any point on Earth at approximately the same time each day. This regularity increased its utility in operational forecasting, so the sunsynchronous orbit has been used ever since for US operational meteorological satellites in near-polar orbits. The High Resolution Infrared Radiometer (HRIR) of Nimbus-1 has a scanning radiometer quite similar to those in operational use today, provided night and day coverage. In Fig. 1.20 is shown an HRIR image of Hurricane Gladys. Its Automatic Picture Transmission (APT) camera was therefore much more useful than that of TIROS-8. This satellite only viewed the Earth 25% of the time, which a Nimbus-1 APT photo is shown in Fig. 1.21.

Hence, an important accomplishment of satellite meteorology is that since sometime in the mid-1960s when meteorological satellite coverage became continuous, from then there have been no undetected tropical cyclones anywhere on Earth. These ocean-born storms, which for centuries menaced seafarers and coastal and island dwellers, can no longer surprise potential victims. Lives are still lost to tropical cyclones, but many are now saved because of the warnings that meteorological satellites make possible.

The TIROS-9 is ninth in the group of 10 TIROS satellites, launched 22 January 1965, which introduced a new construction known as "cartwheel" configuration. The satellite's spin axis was tilted to be perpendicular to the orbital plane, and the cameras were reoriented to point out the side of the spacecraft. Thus, the satellite "rolled" like the wheel of a cart in its orbit about the Earth. With each rotation of the satellite, the cameras would point toward Earth, and a picture could be taken. This situation allowed global composite images to be made, which example is shown in Fig. 1.22.

In Fig. 1.23 are illustrated ground antenna (Left) and ground receiving equipment (Right) from inside building 9162 at the Camp Evans Diana site while receiving the first weather satellite photos. In fact, this is a sample one of the first NOAA Ground Earth Station (GES) serving meteorological satellites TIROS in 1962. These first pictures received from TIROS spacecraft and processed by ground infrastructures, were immediately flown to Washington where the head of NASA presented them to President Eisenhower for public release. Later, even more impressive images were obtained from many parts of the globe, among them

Fig. 1.20 Nimbus-1 HRIR image of Hurricane Gladys (Courtesy of Brochure: by NASA)

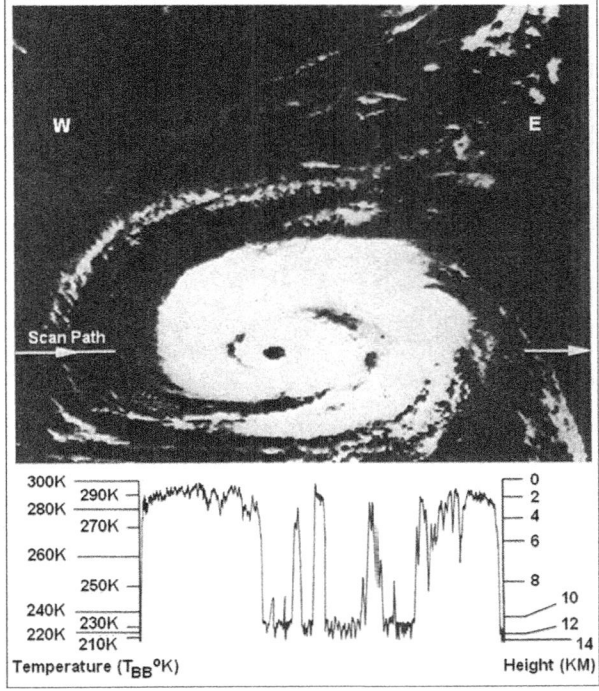

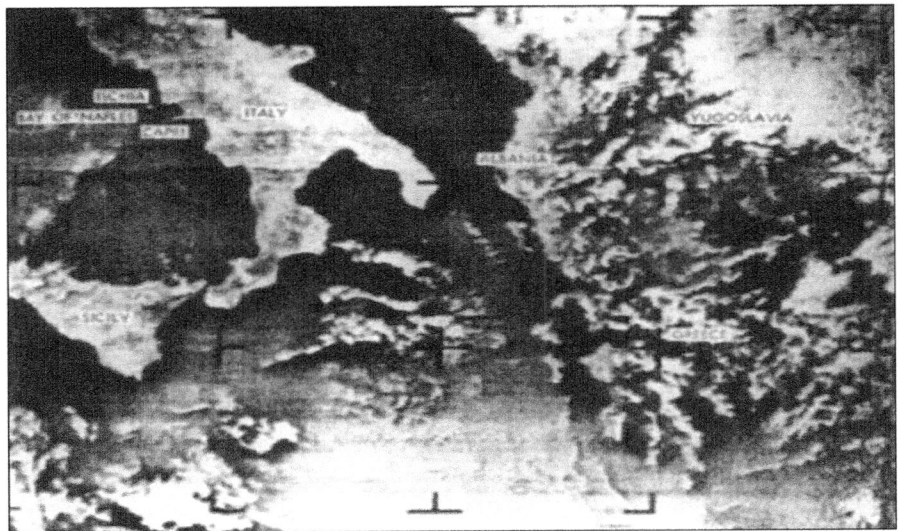

Fig. 1.21 Nimbus-1 APT photo (Courtesy of Brochure: by NASA)

FIRST COMPLETE VIEW OF THE WORLD'S WEATHER

Fig. 1.22 TIROS-9 Earth mosaic (Courtesy of Brochure: by NASA/NOAA)

Fig. 1.23 TIROS ground antenna and station (Courtesy of Brochure: by Camp Evans)

pictures of the Baja California Peninsula and the Suez-Canal-Red Sea area, which
are still vividly in the memory from this time.

In total, seven Nimbus satellites were launched this time. Some experiments on
the last one, Nimbus-7 launched 24 October 1978, were operational until late 1995.
In Fig. 1.24 is illustrated Nimbus-7 TOMS Total Ozone Distribution on 6 May
1993. The Nimbus series of PEO meteorological satellites tested many new concepts that have lead to the operational instruments in use today. This PEO satellite
served as a stabilized, Earth-oriented platform for the testing of advanced systems
for sensing and collecting data in the pollution, oceanographic and meteorological
disciplines. It had three major structures: (1) a hollow torus-shaped sensor mount,
(2) solar paddles, and (3) a control-housing unit that was connected to the sensor
mount by a tripod truss structure.

Similar in appearance to the Nimbus satellites, the Landsat series were designed
for land remote sensing. Landsat-1, also called the Earth Resources Technology

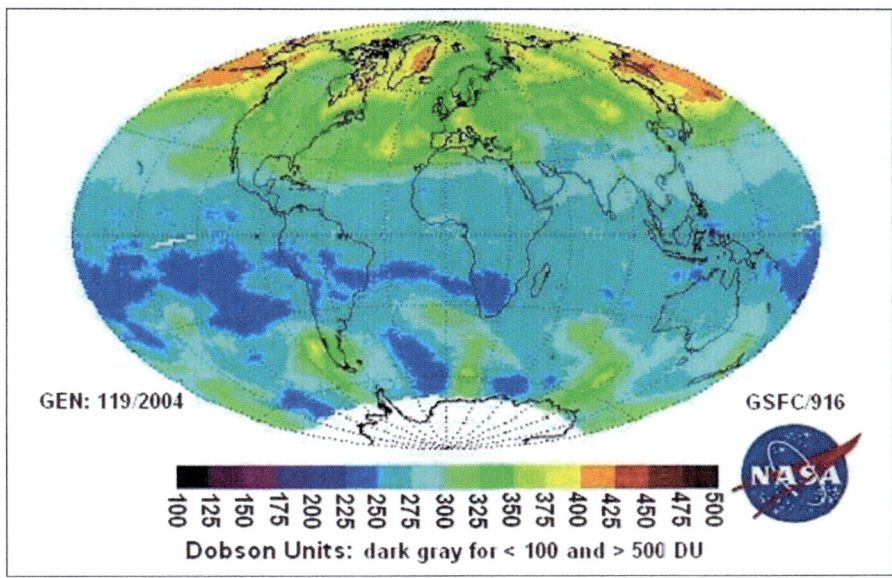

Fig. 1.24   Nimbus-7 total ozone distribution (Courtesy of Book: by Kramer)

Satellite (ERTS), was launched on 23 July 1972. Its sensors have extremely high resolution, 80 m in the first satellite and up to 30 m in the Landsat-5 satellite launched on 1 March 1984, which is depicted in Fig. 1.25 (Left). More modern Landsat-8 satellite was launched on 11 February 2013 with the Extended Payload Fairing (EPF) and its payload consists of two science instruments: Operational Land Imager (OLI) and Thermal Infrared Sensor (TIRS), shown in Fig. 1.25 (Right). The images of Landsat data are also used in meteorology primarily to study small clouds and surface features that may influence weather.

In the meantime, the Soviet Union (subsequently Russia) Space Agency developed series of PEO satellites for science, technology, radiation and meteorology, which first spacecraft to be given a Kosmos-1 (Sputnik-11) designation was launched on 16 March 1962. On 26 March 1969 the Soviet Union Meteor 1–1 was launched as first fully operational weather satellite, shown in Fig. 1.26 (Left). It weighed between 1200 and 1400 kg, and was originally placed in orbit at an altitude of 650 km. Two solar panels were automatically oriented toward the Sun. This spacecraft ceased operations in July 1970. It was the first of 25 similar spacecraft series from 1969 to 1977. The second series of operational Soviet meteorological satellite began on 11 July 1975 with the launch of Meteor 2–1.

India has been quite active in satellite meteorology, thus two polar orbiters have been launched, Bhaskara-1 on 7 June 1979 and Bhaskara-2 on 20 November 1981. Bhaskara-1, weighing 444 kg and was launched from Kapustin Yar aboard the Intercosmos launch vehicle. It was placed in an orbital perigee and apogee of 394 km and 399 km, respectively, at an inclination of 50.7°. It consisted of 2 TV

Fig. 1.25 Landsat-5 and Landsat-8 spacecraft (Courtesy of Brochure: by NASA)

Fig. 1.26 Soviet Meteor 1–1 and Chinese FY 1–1 polar orbiting meteorological satellites (Courtesy of Brochure: by Ilcev)

cameras operating in visible (600 nm) and near infrared (800 nm) and collected data related to hydrology, forestry and geology. SAtellite MIcrowave Radiometer (SAMIR) was operating at 19 and 22 GHz for study of ocean-state, water vapour, liquid water content in the atmosphere and so on.

In 1988 on 6 September and again in 1990 on 3 September the China launched FY 1–1 (Feng Yun – Wind and Cloud) meteorological satellites into approximately 900 km, 99° inclination orbits by CZ-4 boosters from Taiyuan, depicted in Fig. 1.26 (Right). The spacecraft was designed to be comparable to existing international PEO meteorological and remote sensing systems, including APT technique in the 137 MHz band. The satellite structure and support systems were designed and created by the Shanghai Satellite Engineering and Research Center of the China Space Technology Institute, whereas the payload was developed by the Shanghai Technical Physics Institute of the Chinese Academy of Sciences.

Fig. 1.27   S-1 and ATS-6 spacecraft (Courtesy of Brochure: by NASA)

1.8.3   Evolution of GEO Meteorological Satellites

On 7 December 1966, the US Applications Technology Satellite (ATS) was launched ATS-1 the first of six spacecrafts used to test the feasibility of placing a satellite into GEO Geosynchronous orbit. It was originally intended to be a communications satellite, but also provided a platform for meteorological and navigation equipment. It was designed to test GEO orbit techniques and applications at this special orbit that circles Earth once a day. This satellite orbit is exactly 22,236 miles (or 35,786 km) above Earth's surface.

In satellite GEO orbit within Earth's equatorial plane, the ATS-1 satellite transponder was able to transmit information to surface ground stations that are constantly pointed toward the satellite without tracking, which is shown in Fig. 1.27 (Left). This makes the orbit excellent for telecommunications, video broadcasting, and Earth observation. In Fig. 1.27 (Right) is illustrated the sixth ATS satellite, named ATS-6.

Temporally continuous GEO satellite images offer the ability to track clouds and tropical depressions. With this information wind speeds at cloud altitude can be inferred. Research into tracking clouds using image sequences began almost immediately, especially with the successful operation of the ATS-1 spin-scan camera. This imaging device provided the first high quality cloud-cover pictures and afforded a continuous watch of global weather patterns. The success of meteorological experiments carried aboard the ATS series of satellites led to NASA's development of two weather satellites designed specifically to make images of atmospheric observations, called the Synchronous Meteorological Satellites (SMS-1 and SMS-2). These two Geosynchronous meteorological satellites were launched in 1974 and 1975, respectively, and carried the first Data Collection Platform (DCP) repeater, which later evolved into the GOES program.

Data from meteorological or other platforms on the surface could be relayed by the satellite to a central receiving site, so data from meteorological remote ground sites could be easily obtained for the first time, which DCP sites are shown in

Fig. 1.28 DCP ground sites (Courtesy of Manual: by Ilcev)

Fig. 1.28. The cloud cameras on the ATS satellites made images in the visible portion of the spectrum only, and to provide images during the night SMS and the succeeding GOES have an infrared radiometer as well.

Since 27 June 1974, when SMS-1 became operational, the world meteorological consumers have had continuous, uninterrupted, 24-h-per-day observation and monitoring of most of the Western Hemisphere from space. The US Geostationary Operational Environmental Satellites (GOES) program was formally initiated with the first operational spacecraft GOES-1 launched on 16 October 1975, which is shown in Fig. 1.29 (Left). Its ability is to orbit synchronously with Earth's rotation, along with the Polar Operational Environmental Satellites (POES) and Defense Meteorological Satellite Program (DMSP) satellites enhanced NOAA forecasting capabilities. The next deployed GOES-2 and -3 were similar to GOES 1. Since the launch of SMS-2, the US has generally maintained two GEO satellites in orbit, one at 75° West longitude and one at 135° west longitude.

In the following year, 1977, Japan and Europe launched their first GEO weather satellites. These were respectively, the Geostationary Meteorological Satellite (GMS-1), shown in Fig. 1.29 **(Middle)** and the European Meteorological Satellite (Meteosat-1), shown in Fig. 1.29 (Right). Meteosat-1 provided visible imagery with a spatial resolution of 2.5 km and infrared window band imagery and water vapor band imagery, both at 5 km spatial resolution. Its water vapor imagery provided a very different view of Earth. It primarily observed upper tropospheric humidity and high cloud features, which indicated synoptic scale circulations. In 1979, three GOES and one Meteosat satellites were used as part of a Global Atmospheric Research Program (GARP) to define global atmospheric circulations.

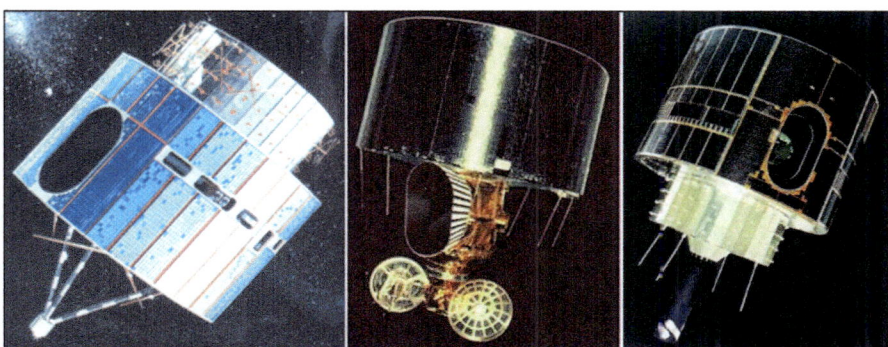

Fig. 1.29 US GOES-1, Japanese GMS-1 (Himawari-1) and European Meteosat-1 GEO spacecraft (Courtesy of Manual: by Ilcev)

Meteosat-1 spacecraft of European Space Agency (ESA) was the first GEO meteorological GEO satellite to make in 1977 images of mid to upper-troposphere water vapor at 6.7 m in addition to visible and 10–12 m infrared, shown in Fig. 1.30.

1.8.4   Evolution of Non-GEO Meteorological Satellites

The former Soviet Union used a Highly Elliptical Orbit (HEO) for Molniya communication satellites. The Molniya orbit has an inclination angle of 63.4° and period of 12 h. Its perigee is motionless at 7378 km (1000 km above the equator) and apogee, from which measurements are made, is at 45,730 km (39,352 km above the equator). The semimajor axis of 26,554 km with an eccentricity of 0.72 is chosen such that the satellite makes two orbits while the Earth turns once with respect to the plane of the orbit. The eccentricity is made as large as possible so that the satellite will stay near apogee longer, but it must not be so large that the satellite encounters significant atmospheric drag at perigee. Thus, the attractiveness of Molniya orbit is that it functions as a high-latitude, part-time, nearly GEO satellite. For about 8 h centered on apogee, the satellite is synchronized with the Earth, so that it is nearly stationary in the sky. For a meteorological satellite in a Molniya orbit, the rapid imaging capability, which is so useful from GEO, would be available in the high latitudes. Therefore, it has been suggested that Molniya satellite orbit would be useful for meteorological observations of the high latitudes, including Alaska, northern Europe and both Poles. Time and space sampling is an increasingly critical aspect of earth observation satellites. The highly eccentric orbit used by Soviet Molniya satellites functions much like a high-latitude GEO orbit, which Molniya 1–3 spacecraft is shown in Fig. 1.31 (Left).

Weather cameras are known to have been flown on several Molniya 1–3 communications satellites launched in the mid-1960s. Meteorological instruments placed on a new satellite in a Molniya orbit would improve the temporal frequency

Fig. 1.30 Meteosat-1 water vapor image (Courtesy of Manual: by SSEC)

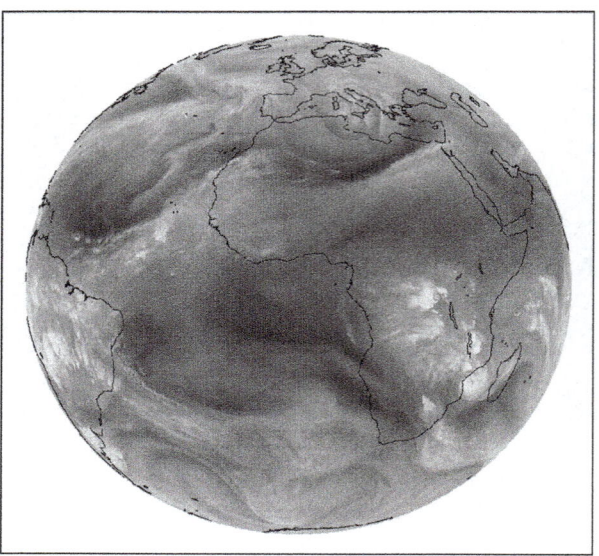

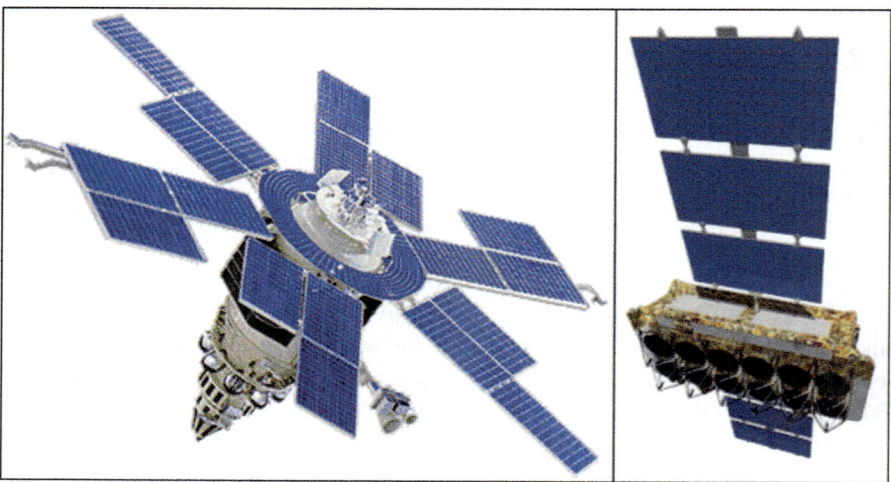

Fig. 1.31 Molniya and O3b satellites (Courtesy of Manual: by Ilcev)

of observation of high-latitude phenomena such as polar lows. Consideration of this new sampling strategy is suggested for future systems such as the Earth Probe satellites in the Mission to Planet Earth program as well as for operational meteorological satellite solutions. Russia already implemented new program of two HEO meteorological satellites named Arctic 1 and 2.

The next proposals for additional satellite meteorological orbits will be MEO and LEO satellite orbits. The MEO satellite orbit is already chosen for GNSS solutions, such as the US GPS and Russian GLONASS, while few years ago MEO is implemented for new O3b satellite communication, which MEO satellite is shown in Fig. 1.31 (Right). The initial constellation is made of 12 satellites at orbital height of 8062 km and fully scalable to meet market demand. This MEO satellite constellation with optimal coverage between 45° North and South latitudes will be ideal for satellite meteorological observation in this areas. The next LEO constellation is already built for meteorological observation by the US NOAA SMAP and OCO-2 Series, and Russian LEO Satellite Kanopus-V 1–2 and Resurs-P 1–2.

1.9 Mobile Satellite Meteorological Service (MSMS)

The MSMS is a special service for obtaining weather data via satellite and transmitted by Earth stations or retransmitted by space stations, which is intended for direct reception by the meteorological centres. This service is used for one-way transmission of meteorological data only from meteorological centres through the MSC system, including hydrological, observations and exploration particular to ships, oilrigs and aircraft mobile stations. In such a way, this meteorological information, forecast bulletins and weather warnings are very useful for the safe navigation and flight of ships and aircraft, respectively.

The current need for quick and easy access to marine and aeronautical WX data and forecast information has become increasingly important. The growing demand of new sources of energy has led the offshore oil exploration into more remote and hostile seas. Environmental constraints have narrowed the maneuvering margins of many at-sea operations, so there is a need for an operational WX/forecast system, which can quickly disseminate maritime and aeronautical WX to the users.

Much of this information is transmitted from meteorological observation satellites via direct readout to receiving ground stations where it can be displayed, analyzed and prepared for customers. These Direct Readout Service (DRS) were developed and operated by US NOAA, Russian (Former USSR countries – CIS), the ESA, Japan, China, India, and other countries worldwide. The most popular of these WX services are Wefax transmitted by the US Geostationary Operational Environmental Satellites (GOES), and the Automatic Picture Transmission (APT) from PEO NOAA satellites.

The GOES-8, 9 and 10 meteorological spacecraft series of US MOAA are currying as well as payloads with GEOSAR transponders of GMSC international Cospas-Sarsat system. Therefore, remotely sensed meteorological data are transmitted directly from GEO or PEO satellites in "real time" to many Ground Forecasting Centre (GFC) terminals and GES within signal range of the meteorological satellite. The meteorological images transmitted by meteorological observation satellites are designed with a format so that they could be received, processed and reproduced by relatively expensive ground DRS equipment, and retransmitted free of charge to anyone with the appropriate satellite receiving, display and printing equipment.

The former USSR (today Russia) has been attempting for decades to develop an effective GEO satellite observation program for meteorological applications. Namely, due in part to the very high latitude of their launch sites, it has been difficult for Russia to design, build and launch a dependable GEO satellite system for meteorological observation. After all, Russia had launched in 1994 a Geostationary Operational Meteorological Satellite (GOMS) called Elektro-1 that is placed in GEO at 76° E. During 1996 and 1997 the GOMS Wefax service has been operating erratically, and there been problems with the imaging sensors. An Elektro-2 satellite was scheduled for launch in 2000 and to improve GOMAS Wefax service. This satellite is also projected to be part of Cospas-Sarsat system, because is currying as well as GEOSAR transponder.

The European Space Agency (ESA) organization also operates a series of GEO and PEO satellites for meteorological observation. The constellation of GEO observation satellites is called Meteorological Satellite (Meteosat), which provide low resolution DRS similar to GOES Wefax, and called Secondary Users Data Station (SUDS). Thus, the Meteosat-6 was replaced by Meteosat-7 in 1988, and new series of MSG satellites were scheduled to be launched in 2001.

Japan has launched a series of Geostationary Meteorological Satellites (GMS). The last one GMS-5 was launched in 1994 at position of 140° E, which transmits both: a primary data stream and Wefax transmission on 1691 MHz. A new, advanced series of GEO MTSAT satellites for WX and Aeronautical Augmentation services will start with operation in 2003 and replace GMS-5 spacecraft, which is described in the same book of Volume 2.

The Indian INSAT-2A, 2B and 3 multipurpose satellites have several payloads; among the rest they carry both WX and GEOSAR (Cospas-Sarsat) transponder. The Chinese GEO Wefax program began with the launch of the FY-2 (Feng Yun) satellite in 1997.

The basic footprint for the current operational GEO satellites transmitting WEFAX data is presented in Fig. 1.32. There are many other MLMSC obtaining WX for mariners like AWT and others, and also for avionics such as METAR.

The meteorological service in many countries worldwide has launched a mobile weather website to make it easier for users of mobile devices to obtain weather information on the go. In the similar way, there is a special meteorological service that is providing synoptical weather charts, meteorological bulletin, weather forecasts and warnings for mariners and aviation. This service can be transmitted via special ground radio stations on VHF and HF frequencies or via GMSC service using Inmarsat or Iridium satellite systems.

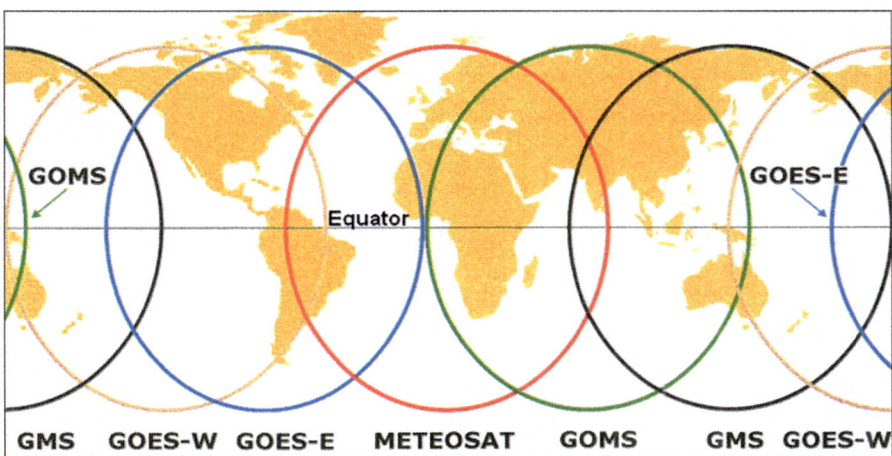

Fig. 1.32 Global GEO meteorological satellite coverage (Courtesy of Book: by Ilcev)

1.9.1 WEFAX System

The WEFAX (WX Fax) satellite images are designed with the format for the DRS provided by the US GOES system. Otherwise, similar services are transmitted from the European Meteosat, Russian GOMS, Japanese GMS and satellites from other countries. The Wefax system retransmits data consists of processed images produced by the primary imager on the GOES above mentioned GEO as well as other meteorological data and images relayed from PEO satellites. This System was first incorporated into the GOES satellites in 1975.

The format of the WEFAX signal was designed to be received and reproduced by low cost GES. This satellite delivered WEFAX signal should not be confused with the HF WX radio Fax transmission from Coast Radio Station (CRS), which some of service is still using. Namely, the WEFAX system is a line-of-sight satellite transmission, with different contents and uses specific receiving equipment. The GOES WEFAX data is transmitted as an analog signal at 1691 MHz and 240 lines per minute. A large amount and variety of WEFAX data can than be obtained.

In the current WEFAX schedule over 100 images can be received in a 24 h period. These consist of scheduled data transmissions of quadrants of the full Earth disc and equatorial regions in visible and infrared spectra, composite images from PEO, WX and ice charts, and operational messages. The WEFAX image is initially formatted by the special Ground Control Station (GCS), using high resolution GOES image data polar orbiter mosaics and weather charts, and rebroadcast through the GOES satellite back to DRS of GES. In fact, it is a delayed a DRS product or near real time. The basic structure of the GOES network of Tx (transmitted) from Weather Data Collection Platforms (WDCP) and Rx (received) weather information by mobiles is shown in Fig. 1.33.

In order to receive the Wefax transmissions most stations use a parabolic or dish antenna with stabilized and tracking platform for large ship MES. Thus, continuous

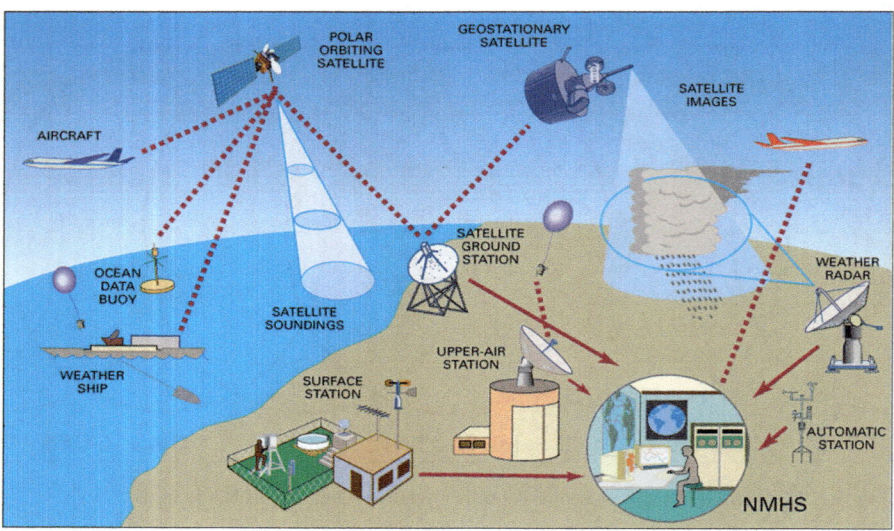

Fig. 1.33 WDCP meteorological satellite system (Courtesy of Paper: by Branski)

tracking of the antenna is not required for a fixed GES. Namely, once properly aligned to the satellite downlink signal, the antenna is locked into permanent position and rarely requires any further adjustment. Except the antenna, a basic WEFAX RDS station is typically comprised of the following electronic components: preamplifier, receiver, demodulator card to decode the satellite signals, PC for display the satellite imagery, a storage system (hard disc, CD writer or ZIP drive) to memorize and archive the satellite imagery, and computer software applications to manipulate the imagery (image processing and enhancement). The basic WEFAX receiving configuration system can be purchased for about 800/1500 US$ without PC configuration. The commercial DRS now available have been designed for a variety of computers like IBM compatible, Apple Macintosh and even some UNIX systems. While most of the details in the preceding article have emphasized the US NOAA GOES satellites and their WEFAX system, it is important to keep in mind that the WEFAX service is global in nature. The Wefax transmission of the Russian, Japanese and European GEO are nearly identical in their technical characteristics. The small differences between these systems are normally easily resolved in the computer software that comes with a commercial purchased receiving system, which will be usable even if moved to different locations.

1.9.2   Automatic Picture Transmission (APT)

The APT is using DRS from PEO satellites to provide WX imagery for world customers. The first APT was pioneered on TIROS-VIII satellite launched in

December 1963. Today, US TIROS PEO satellites continue to transmit images of the Earth by APT. These have been joined by Russian Meteor and the Chinese Feng Yun spacecraft, providing similar transmissions. Because of that, an international transmission standard has been agreed upon so a GES capable of receiving data from the US PEO can receive images from satellites of other countries as well. On the latest TIROS-N (ATN) series of PEO birds the APT images are produced by a special instrument called the Advanced Very High Resolution Radiometer (AVHRR). This instrument is designed to detect five channels of energy reflection from the Earth ranging from the visible spectrum, the near-infrared and infrared spectra. The analog APT signal is derived from the original five channel digital data and multiplexed so that only two of the original channels appear in the APT format.

These two images are selected from GCS, and during daylight passes they usually consist of the visual channel and one of the infrared channels. At night, two infrared images are usually found in the APT. The APT imagery consists of two pictures, side by side, representing the same view of the Earth in two different special bands. The Russian Meteor is transmitting daylight visible pictures only. The APT signal is transmitted continuously from the PEO satellites to the GES terminals. The result in image strip as long as the data transmission is received at the GES and as wide as the scanning instrument is designed to operate at a particular altitude. Radio reception of the APT signals is limited to line-of-sight from GES and can only be received when the PEO satellite is above the horizon, namely when is in view with GES. This is determined by both the altitude of the satellite and its particular path during the orbit across the GES reception range. At present, the US, Russian, and Chinese PEO operate at altitude between 810 and 1200 km. In such a manner, at these altitudes the maximum time of signal reception during an overhead pass is about 16 min. During this time a GES terminal can receive a picture strip equivalent to about 5800 km along the satellite path under the best reception conditions.

In order to obtain APT video data using direct reception, accurate information concerning locations, movements and times that the satellite can be received must be available. This is necessary because signal reception is possible only while the satellites are above that GES horizon. Although all PEO satellites have basic orbital characteristics in common, so each spacecraft is unique in its orbital parameters and needs to be tracked individually. The data necessary to locate and track the WX satellites is generally not difficult to obtain. At any rate, the generation of future orbits of a given satellite can be easily calculated and, if a directional antenna is used, determining the azimuth and elevation of the satellite as it passes over the GES is not difficult after the basic orbital pattern are understood.

The last generation of US ATN satellites represents the current PEO spacecraft available for receiving DRS data. The basic operational concept of this series is only to maintain two satellites in a polar orbit at all time. One will maintain an orbit so that it will pass over the GES, traveling from North to South during the morning, having a southbound Equator crossing at about 0830 local solar time. While, the next satellite will pass from South to North during the afternoon, having a north-bound Equator crossing at about 1430 local solar time. In fact, each of these satellites will also pass over the GES circa 12 h later traveling in the opposite direction. During Winter 1997/98 of North Hemisphere, the US is operating NOAA-12 as a morning spacecraft and NOAA-14 as the afternoon spacecraft.

With a two-satellite constellation such as the NOOA-12 and NOAA-14, one of which will pass over the observing/receiving station about every 6 h. Most stations can receive two consecutive passes (about 100 min apart) from each satellite, day and night. Being able to receive imagery about every 6 h is more than adequate to track the development and movement of large weather system within an 800–1000 km radius of the station.

Technological advances in microelectronics and computer software applications over the past two decades have made it rather simple to assemble and use a basic DRS GES suitable for APT. However, a basic DRS station consists the same components except is additional a method to predict when the PEO satellite has to be in view of the GES. This is because PEO satellites are not stationary like GEO over the adequate positions on the Earth.

The APT direct transmission service from PEO satellites is more suitable for smaller ships. The antenna is small and omni-directional therefore does not need to be aimed at the satellite direction. Several different types of antennas may be used for APT reception of slow speed data. One can be directional and requires tracking of the moving satellite, and the second type is omnidirectional. It has the advantage of being less expensive and not requiring tracking system, but will give a slightly reduced reception range. A turnstile reflector type of omnidirectional antenna is one of the simplest and least expensive antennas to use for APT.

The next quadrifilar helix (QFH) antenna is a special type of omnidirectional antenna that provides a much better radiation pattern compared to previous, and does not suffer from the loss of signal strength exhibited in simple turnstile antenna.

The APT signals from PEO satellites are transmitted on radio frequency between 137 and 138 MHz FM. Two US TIROS PEO satellites transmit APT on 137.5 and 137.62 MHz, and Russian Meteor has been using 137.85 MHz. Beginning in early 2003, the European Organization for the Exploration of Meteorological Satellites (EUMETSAT) will launch their first PEO meteorological satellite Metop-1. It will curry a suite of advanced sensors and DRS transmission system. Otherwise, of special note to the DRS user community is that the current APT service will be replaced by Low Rate Picture Transmission (LRPT) by Metop-1 spacecraft. The LRTP service will be digital rather than analog; requiring same adequate modification to present installed receiving stations.

1.9.3 Applied Weather Technology (AWT)

The AWT focus is on providing high quality maritime global weather routing data service through Inmarsat-A, B and M SES equipment. This includes providing the shipmaster with initial routing recommendation, continuous enroute weather forecast advisories (WX) and post-voyage analyses. The AWT is PC aided weather routing system and is committed to research and development in worldwide weather information system. Thus, a number of fully operational software packages are under continuous development in order to provide up-to-date technology solution to maritime and other clients:

1. **Voyage Simulation Engine** – Is PC-based computer software for dead reckoning vessel position commensurate with prospective WX and adjusted by true position reported by vessel. An integrated database engine enables instant access to ship's actual and potential performance enroute using MSS. On the other side, is possible to establish onboard ships a special voyage optimization for use by the bridge officers. In fact, this service from ashore provides the officers with the most optimal route, speed profile and engine configuration for any given voyage.
2. **Bonvoyage System (BVS)** – Is actually network compatible shore-based company fleet management and monitoring system or can be onboard ship graphical marine WX data briefing system. Color Enhanced WX maps on PC screen overlaying with vessels route information not only provides up-to-the minute fleet status at sea or in port, but also allows visual recognition of enroute WX conditions.

The master of vessel is routinely facing real time WX report, not only bearing the paramount burden of safety of his ship, cargo and crew, but also struggling for the most cost effective management and economical operations and voyages. For example, the forecast of an imminent storm system ahead of vessel track is in reality a very difficult task and huge responsibility of ship captain.

However, it is impossible for the traditional text-based WX routing to explain fully the detailed shape of the storm system, dangerous wave generating area, detail grid information, and to find out the best solution for safe and economical ship's routes. To explore the BVS is necessary to supply a modern PC configuration running Windows, minimum 8 Mb RAM and 10 Mb free disk space, Super VGA video card with at least 1 Mb RAM, and popular asynchronous 9600 baud modem or faster would be recommended. The BVS can be easily tailored for pleasure boat and yacht operators.

The AWT system is actively conducting research and development in area of ocean wave modeling, oceanographic data compression, optimum ship routing, high speed shipboard data transmission, and so on. These future developments, when available, will become an integral part of the overall weather information service.

1.9.4 Global Meteorological Technologies (GMT)

The GMT system has been in operation for approximately 8 years as a manufacturer and service provider of software supplying WX data. Such data is provided to ships or land based operations through the Internet, Inmarsat, HF/VHF radio or PSTRN/PSDN networks. This system utilizes the backup network used by NOAA and

Environment Canada for their WX service communication. The GMT WeatherWise software is a more economical and superior alternative to Weatherfax as a means of receiving WX charts.

Thus, charts are as near to real time WX conditions as can be technically be achieved, and are available within the hour of their measurement. Charts are presented in full colour using NOAA and WMO standard formats and codes, and are available 24 h per day. Users have to supply corresponding PC configuration and peripherals.

At all events, the faster is PC; the sooner analyses can be calculated. Otherwise, all action of the mentioned software can be activated by a mouse, trackball, joystick or touch screen equipped PC. A keyboard is not required to use GMT WeatherWise, other than to reconfigure active map and WX databases.

Data WeatherWise services are broadcast via Inmarsat-C or VHF radio and are available on GMT bulletin board via HF radio and Internet network. However, any GMDSS type approved Inmarsat-C SES can be used to receive Inmarsat FleetNET broadcast. A Hayes compatible 2400 baud, or higher, modem is required to log onto GMT service via Internet. Thus, the GMT WeatherWise is available in six modular regions: The North Pacific, North Atlantic, Eurasia, Far East, Good Hope and Horn.

The new Inmarsat GMSC systems, such as Fleet 33/55/77, Fleet Broadband, Fleet 1 and new Global Express can be used for receiving meteorological bulletin, weather forecasts, weather warnings and even navigation warnings via voice, data and video information.

1.9.5   Maritime Noble Denton Weather Services (NDWS)

The NDWS system is private initiative to provide special marine WX forecasting services to the international offshore and marine industry. Global meteorological data is gathered and processed from NDWS centre in London within 24 h to enable the production of site or route specific forecasts for any location worldwide. The WX forecasts are updated once or twice daily by Fax, Tlx or E-mail. Weather routing for ocean towages can also be provided by Noble Denton in-house Master Mariners.

1.9.6   Global Sea State Information via Internet (GSSII)

The ERS satellites operated by the ESA measure significant sea wave heights and period, 10 m wind speeds and direction benefits the satellites 24 h. These data are continuously received by UK Company Satellite Observing Systems where they are routinely processed and corrected before necessary calibration factors are applied. Results in the form of image maps and text summaries are generally available within 2–3 h of data acquisition for direct delivery over the Internet or retrieval on a daily subscription service via the World Wide Web (WWW). However, a Sea State Alarm system also operates to give immediate WX warning by E-mail messages of regional conditions in any part of the world in excess of 10 m significant wave height or 30 m/s wind speed or 6 m minimum swell.

1.9.7   Aeronautical Weather Applications

The most popular WX service for aircraft is provided by SITA and ARINC meteorological systems for aircraft in flight or in the airport as follows:

1. **SITA Aircom Weather System** – This WX system offer several alternatives via Radio, Inmarsat Aero terminals, Internet, On-site or Hosted solutions such as: Graphical WX charts created with country boundaries, airports and various other aeronautical features, thus providing maximum information within one viewable image; Graphical representation of WX data, mapped over airline routes, facilitates alternative route selection based on wind patterns and other critical WX information; Each instance of wind within user's environment can be programmed to receive a unique set of charts; Aircom surface WX with all meteorological parameters and NOTAM information important for safety flight; and Different WX charts adapted to pilot needs like Surface WX, Visibility, Satellite Imaginary, Radar Imaginary and Lighting. The Aircom WX charts service includes, but is not limited, the following types of weather information: significant weather, upper air weather (wind speed and direction), temperatures, icing, turbulence and precipitation.

2. **ARINC Value Added Service (VAS)** – Except other corporate and safety contribution the VAS system provides similar media as SITA to transfer all meteorological parameters and information to civil aircraft including Meteorological Aviation Reports (METAR) and Terminal Area Forecast (TAF). The ARINC network organize as well as Meteorological Data Collection and Reporting System (MDCRS), which collects information, organizes and disseminates real-time automated position and WX reports from participating airlines and forwards them to the National Weather Service (NWS) for input to their forecast models. Weather products include radar precipitation images, lighting, temperatures, icing, turbulence and accurate forecast of wind aloft are used to define areas of severe WX and contribute to flight planning efficiencies and aviation safety.

   This service is similar to GOES service for WX data collection from aircraft, ships or other mobile or stationary platforms and transfer them via satellite to NOAA/NESDIS stations, as is shown in Fig. 1.33. On the other hand, Terminal Weather Information for Pilots (TWIP) provides valuable situational awareness of weather conditions within 15 Nm of the airport. On the other hand, the TWIP service collects information from airport ground sensors and transmits severe

WX warnings such as wind shear microbursts, gust fronts, heavy-to-moderate precipitation to aircraft and ground operations computers.

3. **Aeronautical Weather Reports** – There are two types of weather reports known as a METAR and the aviation Selected Special Weather Report (SPECI). The METAR is observed hourly between 45 min after the hour till the hour and transmitted between 50 min after the hour till the hour. It will be encoded as a METAR even if it meets SPECI criteria. The SPECI criteria are a non-routine aviation weather report taken when any of the SPECI criteria have been observed. The SPECI criteria shall contain all data elements found in a METAR plus additional plain language information, which elaborates on data in the body of the report. All SPECI elements shall be made as soon as possible after the relevant criteria are observed.

1.9.7.1   Aviation Routine Weather Report (METAR)

The METAR format was introduced 1 January 1968 internationally and has been modified a number of times since. However, North American countries continued to use a Surface Aviation Observation (SAO) for current weather conditions until 1 June 1996, when this report was replaced with an approved variant of the METAR agreed upon in a 1989 Geneva agreement. The WMO publication No. 782 "Aerodrome Reports and Forecasts" contains the base METAR code as adopted by the WMO member countries.

The name METAR is commonly believed to have its origins in the French phrase: "message d'observation météorologique pour l'aviation regulars" or in English: "Aviation Routine Weather Observation Message" or simply "report", and would therefore be a contraction of MÉTéorologique Aviation Régulière (METAR).

In general, METAR reports typically come from airports or permanent weather observation stations. Reports are typically generated once an hour; if conditions change significantly, however, they can be updated in special reports called SPECI. Some reports are encoded by automated airport weather stations located at airports, military bases, and other sites. Some locations still use augmented observations, which are recorded by digital sensors, encoded via software, and then reviewed by certified weather observers or forecasters prior to being transmitted. However, observations may also be taken by trained observers or forecasters who manually observe and encode their observations prior to transmission.

A typical METAR report contains data for the temperature, dew point, wind speed and direction, precipitation, cloud cover and heights, visibility, and barometric pressure. A METAR report may also contain information on precipitation amounts, lightning, and other information that would be of interest to pilots or meteorologists such as a pilot report or PIREP, colour states and Runway Visual Range (RVR).

Although the METAR code is adopted worldwide, each country is allowed to make modifications or exceptions to the code for use in their particular country, e.g., the U.S. will continue to use statute miles for visibility, feet for RVR values, knots for wind speed, and inches of mercury for altimetry. However, temperature and dew point will be reported in degrees Celsius.

The US will continue reporting prevailing visibility rather than lowest sector visibility. Most of the current U.S. observing procedures and policies will continue after the METAR conversion date, with the information disseminated in the METAR code and format. The elements in the body of a METAR report are separated with a space. The only exceptions are RVR, temperature and dew point, which are separated with a solidus (/). When an element does not occur, or cannot be observed, the preceding space and that element are omitted from that particular report.

A METAR report contains the following elements: (1) Type of report, (2) ICAO station identifier, (3) Date and time of report, (4) Modifier (as required), (5) Wind, (6) Visibility, (7) Runway Visual Range (RVR) (as required) (8) Weather phenomena, (9) Sky condition, (10) Temperature and dew point group, (11) Altimeter, and (12) Remarks (RMK) (as required).

1.9.7.2 Aeronautical Weather Forecast

1. **Pilot Weather Reports (PIREP)** – No observation is more timely than the one made from the flight deck. In fact, aircraft in flight are the only means of observing icing and turbulence. Other pilots welcome PIREP as well as do the briefers and forecasters. A PIREP always helps someone and becomes part of aviation weather. Pilots should report any observation that may be of concern to other pilots. Also, if conditions were forecasted but were not encountered, a pilot should also provide a PIREP. This will help the WX to verify forecast products and create accurate products for the aviation community. Pilots should help themselves, the aviation public, and the aviation weather forecasters by providing PIREP. Required elements for all PIREP are type of report, location, time, flight level, aircraft type, and at least one weather element encountered. All altitude references are mean sea level (MSL) unless otherwise noted, distance in nautical miles and time is in Universal Coordinated Time (UTC). A PIREP is transmitted in a prescribed format as follows: Type of Report (UUA/UA), Location (OV), Time (TM), Altitude/Flight Level (FL), Aircraft Type (TP), Sky cover (SK), Flight Visibility and Weather (WX), Temperature (TA), Wind (WV). Turbulence (TB), Icing (IC) and Remarks (RM).

2. **Radar Weather Report** – General areas of precipitation including rain, snow, fogs, clouds and thunderstorms, can be observed by radar. Otherwise, on the radar screen cannot be visible anything during very bad weather conditions and in situation when deep clouds are spreading to the Earth surface.

3. **Satellite Weather Pictures** – Prior to weather satellites, WX observations were made only at distinct points within the atmosphere and supplemented by PIREP observation.

4. **Radiosonde Additional Data (RADAT)** – This data is obtained from the radiosonde observations that are conducted twice a day at 00 and 12Z, such as freezing level and the relative humidity associated with the freezing level.
5. **Aviation Weather Forecasts** – Good flight planning involves considering the following available weather information and weather forecasts: (1) Terminal Aviation Forecast (TAF), (2) Aviation Area Forecast (FA), (3) Inflight Aviation Weather Advisories, (4) Alaska, Gulf of Mexico, and International Area Forecasts (FA), (5) Transcribed Weather Broadcasts (TWEB) Text Products, (6) Winds and Temperatures Aloft Forecast (FD) and (7) Centres Weather Service Unit (CWSU), and including: (a) Hurricane Advisory (WH), (b) Convective Outlook (AC) and (c) Severe Weather Watch Bulletins (WW) and Alert Messages (AWW).

Chapter 2
Space Segment

The space segment of an artificial satellite system is one of its three main operational components, the other two being the ground and user segments. It comprises the satellite or satellite constellation and the uplink and downlink satellite links. The overall design of the operational satellite with payload, transponders and satellite bus, ground segment with transmission and antenna systems and end-to-end system of users is a complex task. Satellite communications payload design must be properly coupled with the capabilities and interaction with the spacecraft bus that provides power, stability and environmental support to the payload.

During the last six decades, commercial communication, meteorological, navigation and other satellite networks have utilized Geostationary Satellite Orbits (GEO) extensively to the point where portions of this orbit have become crowded and coordination between satellites is becoming constrained. Recently, Non-GEO satellite systems have grown in importance, both because of their orbit characteristics and Earth coverage capabilities in high latitudes. Some of their special features are described in this section.

This chapter describes orbital mechanics and their significance with regard to satellite use for weather observations and basic physical principles that reveal the shape of a satellite orbit and how to orient the orbital plane in space. Namely, to fully understand and use satellite meteorological data it is necessary to understand the orbits in which satellites are constrained to move and the space geometry with which they view the Earth. This knowledge allows us to calculate the position of a satellite at any time, thus orbit perturbations and their effects on meteorological satellite orbits are then discussed. Next the geometry of satellite tracking and Earth locations of the measurements made from the satellites are explored. This leads to a discussion of space-time sampling.

This theoretical section also describes satellite orbits and their significance with regard to the meteorological observation, the fundamental laws governing satellite orbits and the principal parameters that describe the motion of artificial satellites of the Earth. The types of orbits are also classified and compared from a

© Springer International Publishing AG 2018
S.D. Ilčev, *Global Satellite Meteorological Observation (GSMO) Theory*,
DOI 10.1007/978-3-319-67119-2_2

communication and meteorological system viewpoint in terms of Earth coverage performance and environmental and link constraints.

More than five decades ago were developed Polar Earth Orbit (PEO) satellites as a first solution for meteorological and navigation satellite applications. Soon later, many other type of satellite networks, such as communication and meteorological, have utilized GEO satellites extensively to the point where portions of these orbits have become crowded and coordination between satellites is becoming constrained. However, PEO as Leo Earth Orbit (LEO) satellite systems have grown in quite importance, both because of their orbit characteristics and coverage capabilities in high latitudes. In addition, can be used High Elliptic Orbits (HEO) and Medium Earth Orbits (MEO) for modern satellite meteorological observation. The chapter concludes with a brief overview of satellite launch vehicles and orbit insertion. Types of satellite orbits and perturbations are also classified and compared from the weather observation viewpoint in terms of coverage and link performances.

2.1   Platforms and Orbital Mechanics

The platform is an artificial object located in orbit around the Earth at a minimum altitude of about 20 km in the stratosphere and a maximum distance of about 36,000 km in the Space. The artificial platforms can have a different shape and designation but usually they have the form of aircraft, airship or spacecraft. In addition, there are special space stations and space ships, which are serving on more distant locations from the Earth's surface for scientific exploration and research and for cosmic expeditions.

Orbital mechanics is a specific discipline describing planetary and satellite motion in the Solar system, which can solve the problems of calculating and determining the position, speed, path, perturbation and other orbital parameters of planets and satellites. In fact, a space platform is defined as an unattended object revolving about a larger one. Although it was used to denote a planet's Moon, since 1957 it also means a man-made object put into orbit around a large body (planet), when the former-USSR launched its first spacecraft Sputnik-1. Accordingly, man-made satellites are sometimes called artificial satellites.

Orbital mechanics support a meteorological satellites project in the phases of orbital design and operations. The orbital design is based on a generic survey of orbits and at an early stage to identify the most suitable orbit for the objective metrological service. The orbital operation is based on rather short-term knowledge of the orbital motion of the satellite and starts with TT&C maintenances after the satellite is located in orbit.

2.1.1 Space Environment

The satellite service begins when a spacecraft is located as a platform in the desired orbital position in a space environment around the Earth. This space environment is a very specific part of the Universe, where many factors and determined elements affect the planet and satellite motions. The Earth is surrounded by a thick layer of many different gasses known as the atmosphere, whose density decreases as the altitude increases. There is no air and the atmosphere disappears at about 180 km above the Earth, where the Cosmos begins. The endless environment in space is not friendly and is extremely destructive, mainly because there is no atmosphere, the cosmic radiation is powerful, the vacuum creates high pressure on spacecraft or other bodies and there is the negative influence of low temperatures.

The Earth's gravity keeps everything on its surface. All the heavenly bodies such as the Sun, Moon, planets and stars have gravity and reciprocal reactions. Any object flying in the atmosphere continues to travel until it meets forces due to the Earth's gravity or until it has enough speed to surpass gravity and to hover in the stratosphere. However, to send an object into space, it first has to overcome gravity and then travel at least at a particular minimum speed to stay in space. In this case, an object traveling at about 5 miles/s can circle around the Earth and become an artificial spacecraft.

An enormous amount of energy is necessary to put a satellite into orbit and this is realized by using a powerful rockets or launchers, which are defined as an apparatus consisting of a case containing a propellant (fuel) and reagents by the combustion of which it is projected into the space. As the payload is carried on the top, the rocket is usually separated and drops each stage after burnout and brings a payload up to the required velocity and leaves it in orbit. A rocket is also known as a booster, as a rocket starts with a low velocity and attains some required height, where air drag decreases and it attains a higher velocity.

2.1.2 History of Motions in Space

The modern orbit types have been developed based on theories dating back centuries. The early Greeks initiated the orbital theories, postulating that the Earth was fixed, with the planets and other celestial bodies moving around it forming a geocentric universe. About 300 BC, however, Aristarchus of Samos suggested that the Sun was fixed and the planets, including Earth, were in circular orbits around it forming a heliocentric universe. Although Aristarchus was more correct (at least about a heliocentric solar system), his ideas were too revolutionary for that time. Other prominent astronomers and philosophers were held in higher esteem, and since they favored the geocentric theory, so Aristarchus's heliocentric theory was rejected, and the geocentric theory continued to be predominately accepted for many centuries.

In the year 1543, some eighteen centuries after Aristarchus proposed a helio-centric system, a Polish monk named Nicolas Koppernias (better known by his Latin name, Copernicus) revived the heliocentric theory when he published "De Revolutionibus Orbium Coelestium (On the Revolutions of the Celestial Spheres)". This work represented an advance, but there were still some inaccuracies. For example, Copernicus thought that the orbital paths of all planets were circles around the center of the Sun.

Tycho Brahe established an astronomical observatory on the island of Hven in 1576. For 20 years, he and his assistants carried out the most complete and accurate astronomical observations of the period. However, Brahe did not accept Copernicus's heliocentric theory and instead believed in a geo-heliocentric model that had the Moon and Sun revolving around the Earth, while the rest of the celestial bodies revolved around the Sun.

German astronomer Johannes Kepler, born in 1571, wondered why there were only six planets and what determined their separation. His theories required data from observations of the planets, and he realized that the best way to acquire such data was to become Brahe's assistant.

In 1600, Brahe set Kepler to work on the motion of Mars. This task was particularly difficult because Mars's orbit was the second most eccentric (of the then-known planets) and defied the circular explanation. After Brahe's death in 1601, Kepler finally discovered that Mars's orbit (and that of all planets) was represented by an ellipse with the Sun at one of its foci.

2.1.3   Laws of Satellite Motion

A satellite is an artificial object located by rocket in space orbit following the same laws in its motion as the planets rotating around the Sun or any star. Johannes Kepler, a German mathematician, has contributed a great deal to the field of astronomy and astrology. The Laws of Planetary Motion formulated by Kepler proves that the orbits of the planets are ellipses and not circles, as believed by many. The ellipse is a geometrical shape that has two foci, such that, the sum of the distance from the focus to any point on the surface of the ellipse is constant. The orbits of planets have small eccentricities (flattening of ellipse), and so, they appear as circles. Based on the properties of ellipses, Johannes Kepler devised three laws that explain the motion of planets around the Sun.

A satellite is an artificial object launched and located by rocket in orbit follows the same laws in its motion as the planets rotating around the Sun. Thus, three important laws for planetary motion were derived by Johannes Kepler, as follows:

1. **First Law** – The first Kepler law is also known as The Law of Orbits. As discussed earlier, an ellipse has two foci in which one can be a real with Sun or any star in the focus, and other is unreal. While studying the motion of planets around the Sun, Kepler explained that the path followed was elliptical, with the

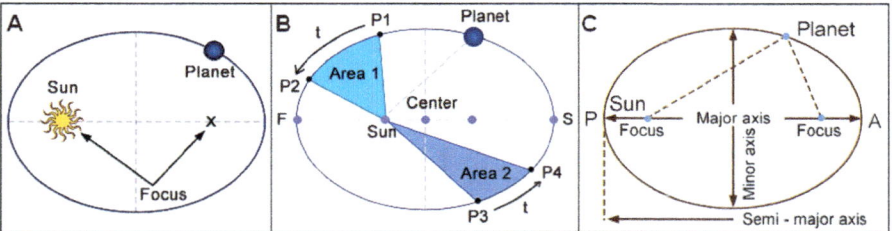

Fig. 2.1 Kepler's laws of satellite motion (Courtesy of Manual: by Ilcev)

Sun as one of the two foci. In simple terms, the law is stated as: "The orbit of each planet follows an elliptical path or all planets move in elliptical orbits, with the Sun at one focus", which is illustrated in Fig. 2.1 (a). This indicates that the Sun is in one real focus, while the other focus is known as the vacant or empty focus. As seen in the diagram, the Sun and the empty focus lie on the major axis of the ellipse, and the planet lies on the surface of the ellipse. As the planet is continuously moving around the Sun, and as the Sun is not at the center of the ellipse, the Planet-Sun distance will always keep on changing. The Law of Orbits proofs that planet motion lies in the plane around the Sun (1602).

2. **Second Law** – The second Kepler law is also known as The Law of Equal Areas, shown in Fig. 2.1 (b). As the Sun is one of the foci, it is clear that the Planet-Sun distance will be changing. But, the planet covers up for the increase in the distance by moving faster when it is closer to the Sun. This indicates that planets do not move at an uniform speed. This law states that: The line from the Sun to orbital planet or radius vector (r) sweeps out equal areas in equal intervals of time (t) as the planet travels around the ellipse. The point at which the speed of the planet is fastest is known as Perihelion or Perigee indicated with F (Fastest motion), while the distance with slowest speed is known as Aphelion (Apogee) indicated by S (Slowest motion). The distance measured from the Perihelion to the position of the Sun is known as Perihelion distance, while the distance from the Sun to the Aphelion is known as the Aphelion distance. The law says that, while moving in an elliptical path, the planet moves faster when it is closer to the Sun. This way, the radius sweeps equal areas in equal amount of time. If the planet is observed at successive times (P1, P2, P3, P4), it draw the radius vector during the first second observations, showing that the two radius vectors having the same area. So, the area swept during the time (t) by the planet to move from P1 to P2 is the same as the area swept while moving from P3 to P4. This is the Law of Equal Areas (1605).

3. **Third Law** – The third Kepler law of planetary motion in ellipse with Perigee (P) and Apogee (A) is alternatively known as The Law of Periods and Harmonic Law, depicted in Fig. 2.1 (c). This law relates the time required by a planet to make a complete trip around the Sun to its mean distance from the Sun. This law can be simply stated as: The square of the planet orbital period is directly proportional to the cube of the semi-major axis of its orbit. The square of the

planet's orbital period around the Sun (T) is proportional to the cube of the semi-major axis (a = distance from the Sun) of the ellipse for all planets in the Solar system (1618).

Kepler's laws only describe the planetary motion if the mass of central body insofar as it is considered to be concentrated in its centre and when its orbits are not affected by other systems. However, these conditions are not completely fulfilled in the case of Earth motion and its artificial satellites. Namely, the Earth does not have an ideal spherical shape and the different layers of mass are not equally concentrated inside of the Earth's body.

Because of this, the satellite motions are not ideally synchronized and stable, the motions are namely slower or faster at particular orbital sectors, which present certain exceptions to the rule of Kepler's Laws. Furthermore, in distinction from natural satellites, whose orbits are almost elliptical, the artificial satellites can also have circular orbits, for which the basic relation can be obtained by the equalizing the centrifugal and centripetal Earth forces.

Kepler's laws only describe the planetary motion if the mass of central body insofar as it is considered to be concentrated in its centre and when its orbits are not affected by other systems. These conditions are not completely fulfilled in the case of Earth motion and its artificial satellites. The Earth does not have an ideal spherical shape and the different layers of mass are not equally concentrated inside of the Earth's body. Because of this, the satellite motions are not ideally synchronized and stable, thus the motions are namely slower or faster at particular orbital sectors, which presents certain exceptions to the rule of Kepler's Laws.

Kepler's Laws were based on observational records and only described the planetary motion without attempting an additional theoretical or mathematical explanation of why the motion takes place in that manner.

In 1687, Sir Isaac Newton published his breakthrough work "Principia Mathematica" with own syntheses, known as the Three Laws of Motion:

1. **Law I** – Every body continues in its state of rest or uniform motion in a straight line, unless it is compelled to change that state by forces impressed on it.
2. **Law II** – The change of momentum per unit time of a body is proportional to the force impressed on it and is in the same direction as that force.
3. **Law III** – To every action there is always an equal and opposite reaction.

On the basis of Law II, Newton also formulated the Law of Universal Gravitation, which states that any two bodies attract one another with a force proportional to the products of their masses and inversely proportional to the square of the distance between them. This law may be expressed mathematically for a circular orbit with the relations:

In 1687, the English physicist British Sir Isaac Newton published his breakthrough work "Principia Mathematica" with own syntheses, known as the Three Laws of Motion, such as follows:

1. **Law I** – Every body continues in its state of rest or uniform motion in a straight line, unless it is compelled to change that state by forces impressed on it.

2. **Law II** – The change of momentum per unit time of a body is proportional to the force impressed on it and is in the same direction as that force.
3. **Law III** – To every action there is always an equal and opposite reaction.

On the basis of Law II, Newton also formulated the Law of Universal Gravitation, which states that any two bodies attract one another with a force proportional to the products of their masses and inversely proportional to the square of the distance between them. This law may be expressed mathematically for a circular orbit with the relations:

$$F = m(2\pi/t)^2(R + h) = G\left[M \cdot m/(R + h)^2\right] \qquad (2.1)$$

where parameter m = mass of the satellite body; t = time of satellite orbit; R = equatorial radius of the Earth (6.37816×10^6 m); h = altitude of satellite above the Earth's surface; G = Universal gravitational constant (6.67×10^{-11} N m^2 /kg^{-2}); M = Mass of the Earth body (5.976032×10^{24} kg) and finally, F = force of mass (m) due to mass (M). Force of mass can be also presented by the following relation:

$$F = ma = dv/dt \qquad (2.2)$$

where a = acceleration and v = velocity of satellite orbit.

The force of attraction between two distant point masses m_1 and m_2 separated by a distance r is giving the following relation:

$$F = Gm_1m_2/r^2 \qquad (2.3)$$

where G = Newtonian (or universal) gravitation constant.

In such a way, considering that the simple circular orbit and assuming that the Earth is a sphere, it is reasonable that it can be simply treated as a point mass. The centripetal force F_c required to keep the satellite in a circular orbit $= mv^2/r$, where v = orbital velocity of the satellite.

Therefore, the force of gravity that supplies this centripetal force is GMm/r^2, where M = mass of the Earth and m is the mass of the satellite. Equating these two forces will give the following relation:

$$F_c = mv^2/r = GMm/r^2 \qquad (2.4)$$

Division by m eliminates the mass of the satellite from the equation, which means that the orbit of a satellite is independent of its mass. Thus, the period of the satellite is the orbit circumference divided by the velocity: $T = 2\pi r/v$. Substituting in Eq. (2.3) gives the following relation:

$$T^2 = \left(4\pi^2/GM\right)r^3 \qquad (2.5)$$

The first generation NOAA meteorological satellites orbit at approximately 850 km above the Earth's surface. Since the equatorial radius of the Earth is about 6378 km, the orbit radius is about 7228 km. So, substituting in Eq. (2.4) shows that the NOAA satellites have a period of about 102 min.

However, radius required for a satellite in GEO has the same angular velocity as the Earth, so the angular velocity mean motion constant of a satellite shows:

$$\xi = 2\pi/T \tag{2.6}$$

Substituting Eq. (2.6) in Eq. (2.5) is giving the following formula:

$$r^3 = GM/\xi^2 \tag{2.7}$$

Inserting the angular velocity of the Earth, the required radius for an GEO is 42,164 km or about 35,786 km above the Earth's surface.

2.1.3.1 Geometry of Elliptical Orbit

The satellite in circular orbit undergoes its revolution at a fixed altitude and velocity, while a satellite in an elliptical orbit can drastically vary its altitude and velocity during one revolution. The elliptical orbit is also subject to Kepler's Three Laws of satellite motion.

Therefore, the characteristics of elliptical orbit can be determined from elements of the ellipse of the satellite plane with the perigee (Π) and apogee (A) and its position in relation to the Earth, see Fig. 2.2 (Left). The parameters of elliptical orbit are presented as follows:

$$e = c/a = \sqrt{\left[1 - (b/a)^2\right]} \text{ or } e = \left(\sqrt{a^2 - b^2}/a\right) \quad p = a\,(1 - e^2) \text{ or } p = b^2/a$$
$$c = \sqrt{(a^2 - b^2)} \qquad\qquad a = p/1 - e^2 \qquad\qquad b = a\sqrt{(1 - e^2)}$$

$$\tag{2.8}$$

where e = eccentricity (distance from the centre of the ellipse to one focus/ semimajor axis); a = large semi-major axis of ellipse; b = small semi-major axis of ellipse; c = axis between centre of the Earth and centre of ellipse and p = focal parameter. The equation of ellipse derived from polar coordinates can be presented with the resulting trajectory equation as:

$$r = p/1 + e\,\cos\,\Theta \;\; [m] \tag{2.9}$$

where r = distance of the satellites from the centre of the Earth (r = R + h) or radius of path; Θ = true anomaly or $\ni$ = eccentric anomaly. In this case, the position of the satellite will be determined by the angle called "the true anomaly", which can be counted positively in the direction of movement of the satellite from 0° to 360°, between the direction of the perigee and the direction of the satellite (S).

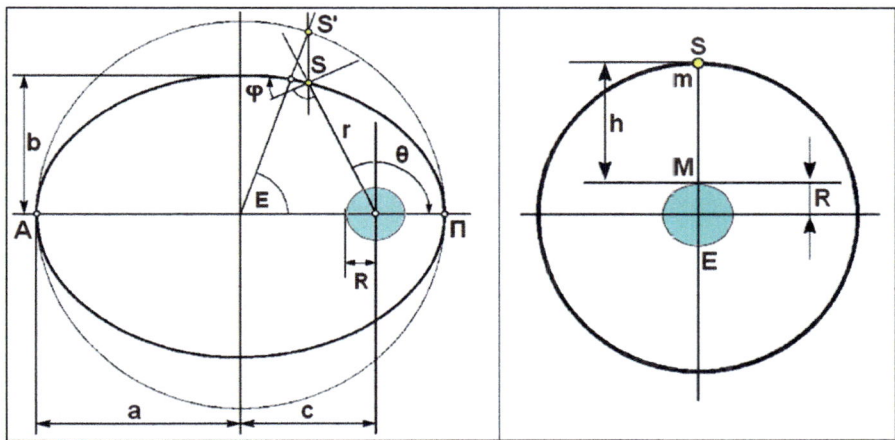

Fig. 2.2 Elliptical and circular satellite orbits (Courtesy of Book: by Galic)

The position of the satellite can also be defined by eccentric anomaly (Э), which is the argument of the image in the mapping, which transforms the elliptical trajectory into its principal circle, an angle counted positively in the direction of movement of the satellite from 0 to 360°, between the direction of the perigee and the direction of the satellite. The relations for both mentioned anomalies are given by the following equations:

$$\cos \Theta = \cos E - e/1 - e \cos E \qquad \cos E = \cos \Theta + e/1 + e \cos \Theta \quad (2.10)$$

The total mechanical energy of a satellite in elliptical orbit is constant; although there is an interchange between the potential and the kinetic energies. As a result, a satellite slows down when it moves up and gains speed as it loses height. Thus, considering the termed gravitation parameter μ = GM (Kepler's Constant μ = 3.99 × 10^5 km³/s²), the velocity of a satellite in an elliptical orbit may be obtained from the following relation:

$$v = \sqrt{[GM \ (2/r) - (1/a)} = \sqrt{\mu \ (2/r) - (1/a)]} \qquad (2.11)$$

Applying Kepler's Third Law the sidereal time of one revolution of the satellite in elliptical orbit is as follows:

$$t = 2\pi\sqrt{(a^3/GM)} = 2\pi\sqrt{(a^3/\mu)}$$
$$t = 3.147099647\sqrt{(26,628.16 \cdot 10^3)^3} \cdot 10^{-7} = 43,243.64 \ [s] \qquad (2.12)$$

Therefore, the last equation is the calculated period of sidereal day for the elliptical orbit of Russian-based satellite Molniya with apogee = 40,000 km, perigee = 500 km, revolution time = 719 min and a = 0.5 (40,000 + 500 + 2 × 6378.16) = 26,628.15 km.

2.1.3.2 Geometry of Circular Orbit

The circular orbit is a special case of elliptical orbit, which is formed from the relations $a = b = r$ and $e = 0$, see Fig. 2.2 (Right). According to Kepler's Third Law, the solar time (τ) in relation with the right ascension of an ascending node angle (Ω); the sidereal time (t) with the consideration that $\mu = GM$ and satellite altitude (h), for a satellite in circular orbit will have the following relations:

$$\tau = t/(1 - \Omega t/2\pi)$$
$$t = 2\pi\sqrt{(r^3/\mu)} = 3.147099647\sqrt{(r^3 \cdot 10^{-7})} \ [\text{s}] \tag{2.13}$$
$$h = \left[\sqrt[3]{(\mu t^2/4\pi^2)}\right] - R = 2.1613562 \cdot 10^4 \left(\sqrt[3]{t^2}\right) - 6.37816 \cdot 10^6 \ [\text{m}]$$

The time is measured with reference to the Sun by solar and sidereal day. Thus, a solar day is defined as the time between the successive passages of the Sun over a local meridian. In fact, a solar day is a little bit longer than a sidereal day, because the Earth revolves by more than 360° for successive passages of the Sun over a point 0.986° further. On the other hand, a sidereal day is the time required for the Earth to rotate one circle of 360° around its axis: $t_E = 23$ h 56 min 4.09 s. Therefore, a geostationary satellite must have an orbital period of one sidereal day in order to appear stationary to an observer on Earth. During rotation the duration of sidereal day $t = 85{,}164{,}091$ (s) and is considered in such a way for synchronous orbit that $h = 35{,}786.04 \times 10^3$ (m). The speed is conversely proportional to the radius of the path $(R + h)$ and for the satellite in circular orbit it can be calculated from the following relation:

$$v = \sqrt{(MG/R + h)} = \sqrt{(\mu/r)} = 1.996502 \cdot 10^{-7}/\sqrt{r} = 631.65\sqrt{r} \ [\text{m/s}] \tag{2.14}$$

From Eq. (2.8) and using the duration of sidereal day (t_E) gives the relation for the radius of synchronous or geostationary orbits:

$$r = \sqrt[3]{[(\mu t)/2\pi)^2]} \tag{2.15}$$

The satellite trajectory can have any angle of orbital planes in relation to the equatorial plane: in the range from PEO up to GEO plane. Namely, if the satellite is rotating in the same direction of Earth's motion, where (t_E) is the period of the Earth's orbit, the apparent orbiting time (t_a) is calculated by the following relation:

$$t_a = t_E \cdot t/t_E - t \tag{2.16}$$

This means, inasmuch as $t = t_E$ the satellite is geostationary ($t_a = \infty$ or $\tau = 0$). In Table 2.1 several values for times different than synchronous orbital time are presented.

According to Table 2.1 and Eq. (2.9) it is evident that a satellite does not depend so much on its mass but decreases with higher altitude. In addition, satellites in circular orbits with altitudes of a 1700, 10,400 and 36,000 km, will have t/ τ values

Table 2.1 The values of times different than the synchronous time of orbit

Parameter	Values of time					Unit
t	86,164.00	43,082.05	21,541.23	10,770.61	6052.00	s
h	35,786.00	20,183.62	10,354.71	4162.89	800.00	km
(R + h)	42,164.00	26,561.78	16,732.87	10,541.05	7178.00	km
v	3075.00	3873.83	4880.72	5584.12	7450.00	km/s^{-1}

2/2,18, 6/8 and 24/zero, respectively. In this case, it is evident that only a satellite constellation at altitudes of about 36,000 km can be synchronous or geostationary.

2.1.4 Horizon and Geographic Satellite Coordinates

The geographical and horizon coordinates are very important to find out many satellite parameters and equations for better understanding the problems of orbital plane, satellite distance, visibility of the satellite, coverage areas, etc. The coverage areas of a satellite are illustrated in Fig. 2.3 (a) with the following geometrical parameters: actual altitude (h), radius of Earth (R), angle of elevation (ε), angle of azimuth (A), distance between satellite and the Earth's surface (d) and central angle (Ψ) and sub-satellite angle or declination (δ), which is similar to the angle of antenna radiation.

The geographical and horizon coordinates of a satellite are presented in Fig. 2.3 (b) with the following, not yet mentioned, main parameters: angular speed of the Earth's rotation (ν), argument of the perigee (ω), moment of satellite pass across any point on the orbit (t_o), which can be perigee (Π), projection of the perigee point on the Earth's surface (Π'), spherical triangle (B'ΓP), satellite (S), the Point of the Observer or Mobile (M), latitudes of observer and satellite (φ_M and φ_S), longitudes of observer and satellite (λ_M and λ_S), inclination angle (i) of the orbital plane measured between the equatorial and orbital plane and the right ascension of an ascending node angle in the moment of t_o (Ω_o).

Otherwise, the right ascension of an ascending node angle (Ω) is the angle in the equatorial plane measured counter clockwise from the direction of the vernal equinox to that of the ascending node, while the argument of the perigee (ω) is the angle between the direction of the ascending node and the direction of the perigee.

2.1.4.1 Satellite Distance and Coverage Area

The area coverage or angle of view for each type of satellite depends on orbital parameters, its position in relation to the Ground Earth Station (GES) or Fixed Earth Station (FES) and geographic coordinates. Thus, this relation is very simple in the

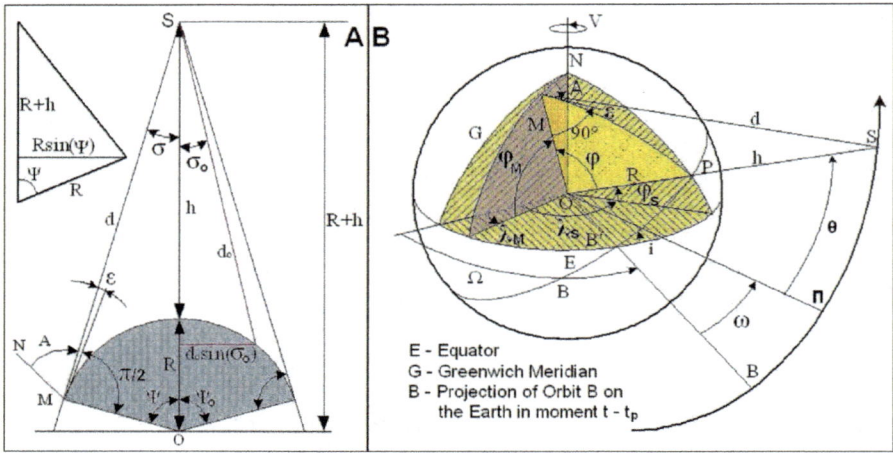

Fig. 2.3 Geometric projections of satellite orbits (Courtesy of Manual: by Solobev/Zhilin)

case where the sub-satellite point is in the centre of coverage, while all other samples are more complicated. Thus, the angle of an GEO satellite inside its range has the following regular reciprocal relation:

$$\delta + \varepsilon + \Psi = 90° \tag{2.17}$$

The circular sector radius can be determined by the following relation:

$$R_s = R \sin \Psi \tag{2.18}$$

When the altitude of orbit h is the distance between satellite and sub-satellite point (SP), the relation for the altitude of the circular sector can be written as:

$$h_s = R \left(1 - \cos \Psi\right) \tag{2.19}$$

From a satellite communications point of view, there are three key parameters associated with an orbiting satellite: **(1)** Coverage area or the portion of the Earth's surface that can receive the satellite transmission with an elevation angle larger than a prescribed minimum angle; **(2)** The slant range (actual LOS distance from a fixed point on the Earth to the satellite) and **(3)** The length of time a satellite is visible with a prescribed elevation angle.

Elevation angle is an important parameter, since communications can be significantly impaired if the satellite has to be viewed at a low elevation angle, that is, an angle too close to the horizon line. In this case, a satellite close to synchronous orbit covers about 40% of the Earth's surface. Thus, from the diagram in Fig. 2.3 (a) a covered area expressed with central angle (2δ or 2Ψ) or with arc (MP $\approx$ RΨ) as a part of Earth's surface can be derived with the following relation:

$$C = \pi \left(R_s^2 + h_s^2\right) = 2\pi R^2 (1 - \cos \Psi) \tag{2.20}$$

Since the Earth's total surface area is $4\pi R^2$, it is easy to rewrite C as a fraction of the Earth's total surface:

$$C/4\pi R^2 = 0,5 \ (1 - \cos \Psi) \tag{2.21}$$

The slant range between a point on Earth, such as Mobile Earth Station (MES), and a satellite at altitude (h) and elevation angle can be defined in this way:

$$z = \left[(R \sin \varepsilon)^2 + 2Rh + h^2\right]^{\frac{1}{2}} - R \sin \varepsilon \tag{2.22}$$

This determines the direct propagation length between GES, (h) and (ε) and will also find the total propagation power loss from GES to satellite. In addition, (z) establishes the propagation time (time delay) over the path, which will take an electromagnetic field as:

$$t_d = (3.33) \ z \ [\mu s] \tag{2.23}$$

To propagate over a path of length (z) km, it takes about 100 ms to transmit to GEO. If the location of the satellite is uncertain ± 40 km, a time delay of about ± 133 μs is always present in the Earth-to-satellite propagation path. When the satellite is in orbit at altitude (h), it will pass over a point on Earth with an elevation angle (ε) for a time period:

$$t_p = (2\Psi/360) \left(t/1 \pm (t/t_E)\right) \tag{2.24}$$

The quotations for right ascension of the ascending node angle (Ω) and argument of the perigee (ω) are as follows:

$$\Omega = 9,95 \ (R/r)^{3.5} \cos i \ \text{ or } \ \Omega = \Omega_o + \nu \ (t - t_o)$$
$$\omega = 4,97 \ (R/a)^{3.5} \left[5 \cos^2 i - 1/(1 - e^2)^2\right] \tag{2.25}$$

The limit of the coverage area is defined by the elevation angle from GES above the horizon with angle of view $\varepsilon = 0°$. In this case, the satellite is visible and its maximal central angle for GEO will be as follows:

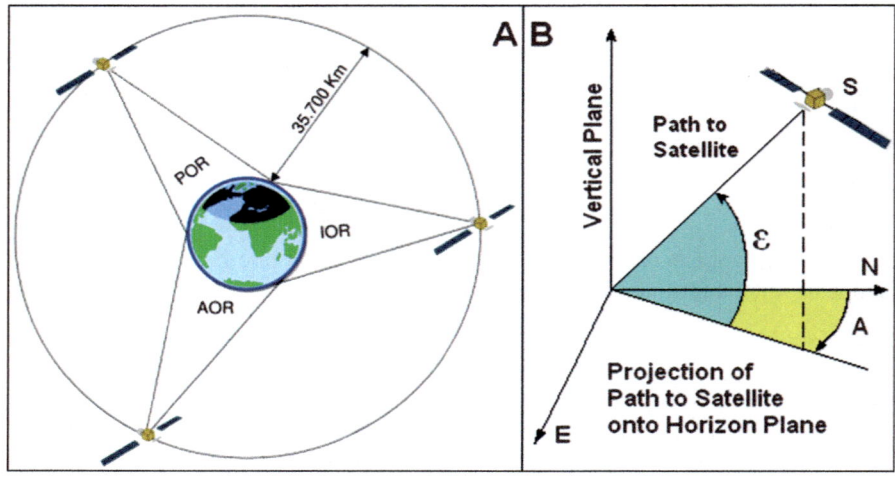

Fig. 2.4 GEO coverage and look angle parameters (Courtesy of Book: by Pratt)

$\Psi = \text{arc cos } (R \cos \varepsilon / r) - \varepsilon$ or
$\Psi = \pi/2 - \text{arc sin } (R/r) = \text{arc cos } (R/r) - \varepsilon = \text{arc cos } k - \varepsilon$
$\Psi = \text{arc cos } 6,376.16/42,164.20 = \text{arc cos } 0.15126956 = 81°17' \ 58.18"$
$C_{max} = 255.61 \cdot 10^6 (1 - 0.15126956) = 216.94 \cdot 10^6 \ (km^2)$

$$(2.26)$$

All GES with a position above $\Psi = 81°$ will be not covered by GEO satellites. Since the Earth's square area is 510,100,933.5 km^2 and the extent of the equator is 40,076.6 km, only with three GEO apart in the orbit by 120° is possible to cover a great area of its surface.

In Fig. 2.4 (a) are illustrated AOR (Atlantic), IOR (Indian) and POR (Pacific) satellite coverages. The zero angles of elevation have to be avoided, even to get maximum coverage, because this increases the noise temperature of the receiving antenna. Owing to this problem, an equation for the central angle with minimum angle of view between 5° and 30° will be calculated with:

$$\Psi_s = \text{arc cos } (k \cos \varepsilon) - \varepsilon \qquad (2.27)$$

The arch length or the maximum distant point in the area of coverage can be determined as:

$$l = 2\pi R \ (2\Psi/360 = 222.64\Psi \ [km] \qquad (2.28)$$

The real altitude of satellite over sub-satellite point is as follows:

$$h = r - R = 42,162 - 6,378 = 35,784 \ [km] \qquad (2.29)$$

The view angle under which an GEO satellite can see GES/MES is called the "sub-satellite angle". More distant points in the coverage area for GEO satellites are limited around $\varphi = 70°$ of North and South geographical latitudes and around $\lambda = 70°$ of East and West geographical longitudes, viewed from the sub-satellite's point. Theoretically, all Earth stations around these positions are able to see satellites by a minimum angle of elevation of $\varepsilon = 5°$. Such access is very easy to calculate, using simple trigonometry relations:

$$\delta_{\varepsilon=0} = arc \ sin \ k \approx 9° \qquad (2.30)$$

The angle (Ψ) is in correlation with angle (δ), which can determine the aperture radiation beam, which for satellite antenna in global coverage has a radiation beam of $2\delta = 17.3°$. According to Fig. 2.3 (a) it will be easy to find out relations for GEO satellites as:

$$tg \ \delta = k \ sin \ \Psi / 1 - k \ cos \ \Psi = 0.15126956 \ sin \ \Psi / 1 - cos \ \Psi / 1 - 0.15126956 \ cos \ \Psi)$$
$$\delta_s = 90° - \Psi_s = 8°42' \ 1.82''$$
$$(2.31)$$

Differently to say, the width of the beam aperture ($2\delta_s$) is providing the maximum possible coverage for synchronous circular orbit. The distance of GES with regard to the satellite can be calculated using Fig. 2.3 (a) and Eqs. (2.13) and (2.22) by:

$$d = R \ sin\Psi / sin \ \delta = r \ sin / cos \ \varepsilon \qquad (2.32)$$

The parameter (d) is quite important for transmitter power regulation of GES, which can be calculated by the following equation:

$$d = \sqrt{\left[(R+r)^2 - 2Rr \ cos \ \Psi\right]} = h\sqrt{\left[1 + 2\,(1/k)\,(R/h)^2(1 - cos \ \varphi \ cos \ \Delta\lambda)\right]} \ or$$
$$d = r\left[1 - (R \ cos \ \varepsilon/r)^2\right]^{1/2} - R \ sin \ \varepsilon$$
$$(2.33)$$

Accordingly, when the position of any observer is near the equator in sub-satellite point (P) or right under the GEO satellite, then its distance is equal to the satellite altitude and takes out value for $d = H$ of 35,786 km. Thus, every observer will have a further position from (P) when the central angle exceeds $\Psi = 81°$, when $d_{max} = 41,643$ km.

2.1.4.2 Satellite Look Angles (Elevation and Azimuth)

The horizon coordinates are considered to determine satellite position in correlation with an Earth observer, GES and user terminals. These specific and important horizon coordinates are angles of satellite elevation and azimuth, illustrated in Figs. 2.3 (a, b) and 2.4 (b), respectively.

The satellite elevation (ε) is the angle composed upward from the horizon to the vertical satellite direction on the vertical plane at the observer point. From point (M) shown in Fig. 2.3 (a) the look angle of ε value can be calculated by the following relation:

$$\text{tg } \varepsilon = \cos \Psi - k/\sin \Psi \tag{2.34}$$

The satellite azimuth (A) is the angle measured eastward from the geographical North line to the projection of the satellite path on the horizontal plane at the observer point. This angle varies between 0 and 360° as a function of the relative positions of the satellite and the point considered. The azimuth value of the satellite and sub-satellite point looking from the point (M) or the hypothetical position of observer can be calculated as follows:

$$\text{tg } A' = \text{tg } \Delta\lambda_M - k/\sin \Psi \tag{2.35}$$

Otherwise, the azimuth value, looking from sub-satellite point (P), can be calculated as:

$$\text{tg } A = \sin \Delta\lambda/\text{tg } \varphi \text{ or } \sin A = \cos \varphi \sin \Delta\lambda \text{ cosec } \Psi \tag{2.36}$$

However, parameter (A') is the angle between the meridian plane of point (M) and the plane of a big circle crossing this point and sub-satellite point (P), while the parameter (A) is the angle between a big circle and the meridian plane of point (P). Thus, the elevation and azimuth are respectively vertical or horizontal look angles, or angles of view, in which range the satellite can be seen.

In Fig. 2.5 (a) is presented a correlation of the look angle for three basic parameters (δ, Ψ, d) in relation to the altitude of the satellite. Inasmuch as the altitude of the satellite is increasing as the values of central angle (Ψ), distance between satellite and the Earth's surface (d) and duration of communication (t_c) or time length of signals are increasing, while the value of sub-satellite angle or declination (δ) indirectly proportional. An important increase of look angle and duration of communication can be realized by increasing the altitude to 30 or 35,000 km, while an increase in look angle is unimportant for altitudes of more than 50,000 km.

The duration of communication is affected by the direction's displacement from the centre of look angle, which will have maximum value in the case when the direction is passing across the zenith of the GES. The single angle of the satellite in circular orbit depends on the t/2 value, which in area of satellite look angle, can be found in the duration of the time and is determined as:

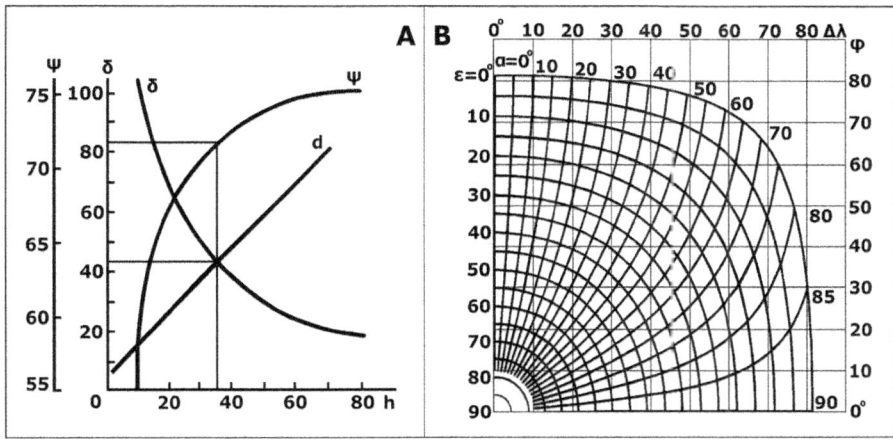

Fig. 2.5 Look angle parameters and graphic of geometric coordinates for GEO (Courtesy of Book: by Zhilin)

$$t_c = \Psi\, t/\pi \tag{2.37}$$

Practical determination of the geometric parameters of a satellite is possible by using many kinds of plans, graphs and tables. It is possible to use tables for positions of user (φ, λ), by the aid of which longitudinal differences can be determined between MES and satellite for four feasible ship's positions: N/W, S/W, N/E and S/E in relation to GEO.

One of the most important practical pieces of information about a communications satellite is whether it can be seen from a particular location on the Earth's surface. In Fig. 2.5 (b) a graphic design is shown which can approximately determine limited zones of satellite visibility from the Earth (user) by using elevation and azimuth angles under the condition that $\delta = 0$. This graphic contains two groups of crossing curves, which are used to compare (φ) and ($\Delta\lambda$,) coordinates of mobile positions.

In such a way, the first group of parallel concentric curves shows the geometric positions where elevation has the constant value ($\varepsilon = 0$), while the second group of fan-shaped curves starting from the centre shows the geometric positions where the difference in azimuth has the constant value (a $= 0$). This diagram can be used in accordance with Fig. 2.3 (b) in the following order:

1. First, it is necessary to note the longitude values of satellite (λ_S) and mobile (λ_M) and the latitude of the mobile (φ_M), then calculate the difference in longitude ($\Delta\lambda$) and plot the point into the graphic with both coordinates (φ_M & $\Delta\lambda$).
2. The value of elevation angle (ε) can then be determined by a plotted point from the group of parallel concentric curves.

Table 2.2 The form for calculation of Azimuth values

The GEO direction in relation to MES	Calculating of Azimuth angles
Course of MES towards S & W	A = a
Course of MES towards N & W	A = 180° − a
Course of MES towards N & E	A = 180° + a
Course of MES towards S & E	A = 360° − a

3. The difference value of azimuth (a) can be determined by a plotted point from the group of fan-shaped curves starting from the centre.
4. Finally, depending on the mobile position, the value of azimuth (A) can be determined on the basis of the relations presented in Table 2.2.

Inasmuch as the position of Ship Earth Station (SES) or any MES is of significant or greater height above sea level (if the bridge or ship's antenna is in a very high position) or according to the flight altitude of Aircraft Earth Station (AES), then the elevation angle will be compensated by the following parameter:

$$x = \arccos\left(1 - H/R\right) \tag{2.38}$$

where H = height above sea level of observer, FES or any MES. Let us say, if the position of GES is a height of H = 1000 m above sea level, the value of x ≈ 1°. This example can be used for the determination of AES compensation parameters, depending on actual aircraft altitude. In such a way, the estimated value of elevation angle has to be subtracted for the value of the compensation parameter (x).

2.1.4.3 Satellite Track and Geometry (Longitude and Latitude)

The satellite track on the Earth's surface and the presentation of a satellite's position in correlation to the MES results from a spherical coordinate system, whose centre is the middle of Earth, is illustrated in Fig. 2.3 (b). In this way, the satellite position in any time can be decided by the geographic coordinates, sub-satellite point and range of radius. Thus, the sub-satellite point is a determined position on the Earth's surface; above it is the satellite at its zenith.

More exactly, the longitude and latitude are geographic coordinates of the sub-satellite point, which can be calculated from the spherical triangle (B'ΓP), using the following relations:

$$\sin\varphi = \sin\left(\Theta + \omega\right)\sin i \quad \text{and} \quad \text{tg}\left(\lambda_S - \Omega\right) = \text{tg}\left(\Omega + \omega\right)\cos i \tag{2.39}$$

With the presented equation in previous relation it is possible to calculate the satellite path or trajectory of sub-satellite points on the Earth's surface. The GEO track breaks out at the point of coordinates $\varphi = 0$ and $\lambda = $ const.

Furthermore, considering geographic latitude (φ_M) and longitude (λ_M) of the point (M) on the Earth's surface presented in Fig. 2.3 (b), what can be the position of the user, taking into consideration the arc (MP) of the angle illustrated in Fig. 2.3 (a), the central angle can be calculated by the following relations:

$$\cos \Psi = \cos \varphi_S \cos \Delta\lambda \cos \varphi_M + \sin \varphi_S \sin \varphi_M \text{ or}$$
$$\cos \Psi = \cos \text{ arc } MP = \cos \varphi_M \cos \Delta\lambda \tag{2.40}$$

The transition calculations from geographic to spherical coordinates and vice versa can be computed with the following equations:

$$\cos \Psi = \cos \varphi \cos \Delta\lambda \text{ and } \text{tg } A = \sin \Delta\lambda/\text{tg } \varphi, \text{respectively}$$
$$\sin \varphi = \sin \Psi \cos A \text{ and } \text{tg } \Delta\lambda = \text{tg } \Psi \sin A \tag{2.41}$$

These relations are useful for any point or area of coverage on the Earth's surface, then for a centre of the area if it exists, as well as for spot-beam and global area coverage for MSC systems. The optimum number of GEO satellites for global coverage can be determined by:

$$n = 180°/\Psi \tag{2.42}$$

For instance, if $\delta = 0$ and $\Psi = 81°$ it will be necessary to put into orbit only 3 GEO satellites and to get a global coverage from 75° N to 75° S geographic latitude. Hence, in a similar way the number of satellites can be calculated for other types of satellite orbits.

The trajectory of radio waves on a link between an MES and satellite at distance (d) and the velocity of light ($c = 3 \times 10^8$ m/s) require a propagation time equal to:

$$T = d/c \text{ (s)} \tag{2.43}$$

The phenomenon of apparent change in frequency of signal waves at the receiver when the signal source moves with respect to the receivers (Earth) was explained and quantified by Johann Doppler (1803–53). The frequency of the satellite transmission received on the ground increases as the satellite is approaching the ground observer and reduces as the satellite is moving away. This change in frequency is called Doppler effect or shift, which occurs on both the uplink and the downlink. This effect is quite pronounced for LEO and compensating for it requires frequency tracking in a narrowband receiver, while its effect are negligible for GEO satellites. The Doppler shift at a transmitting frequency (f) and radial velocity (v_r) between the ground observer and the transmitter can be calculated by the following relation:

$$\Delta f_D = f v_r/c \text{ where } v_r = dR/dt \tag{2.44}$$

For an elliptical orbit, assuming that $R = r$, the radial velocity is given by:

$$v_r = dr/dt = (dr/\Theta)(d\Theta/dt) \tag{2.45}$$

The sign of the Doppler shift is positive when the satellite is approaching the observer and vice versa. Doppler Effect can also be used to estimate the position of an observer provided that the orbital parameters of the satellite are precisely known. This is very important for development of Doppler satellite tracking and determination systems.

2.1.5   Orientation in Space

As stated earlier, the position of the satellite is determining by the angle called "the true anomaly", which can be counted positively in the direction of movement of the satellite from $0°$ to $360°$, between the direction of the perigee and the direction of the satellite (S). In Eq. (2.9) is derived distance of the satellites (r) from the centre of the Earth (r = R + h) or radius of path; Θ = true anomaly or $\mathcal{I}$ = eccentric anomaly. The value of angle $90°$ for GEO satellite inside its range was also determined by Eq. (2.17) with main parameters such as: sub-satellite angle or declination (δ), angle of elevation (ε) and central angle (Ψ).

In such a way, by calculating r and Θ at any time t is possible to obtain position of orbital plane in the space. To provide this system requires the definition of a coordinate system.

The coordinate system of coordinates that is used in an inertial reference frame, normally in the special theory of relativity is called Earth-Centered Inertial (ECI) coordinate system. Here, the three space coordinates are usually Cartesian coordinates (X, Y, Z), and the time coordinate is the time as measured by an observer at rest in the coordinate system. On the other hand, in astrometry an inertial coordinate system is a reference frame formed by assigning coordinates to specific observable objects, such as the positions and proper motions of stars in a fundamental catalogue.

For orientation in the space, the coordinate system must be an inertial coordinate system, that is, a nonaccelerating system in which Newton's Laws of Motion are valid. Thus, a coordinate system fixed to the rotating Earth is not such a system. Here is adopted an astronomical coordinate system called the right ascension-declination coordinate system, shown in Fig. 2.6 (Left). In this system the axis X is aligned with the Earth's spin axis. The axis X is pointed from the center of the Earth to the Sun at the moment of the vernal equinox, when the Sun is crossing the equatorial plane from the Southern to the Northern Hemisphere. The axis Y is a right-handed coordinate system. The declination of a point in space is its angular displacement measured northward from the equatorial plane, and the right ascension is the angular displacement, measured counterclockwise from the axis X, of the projection of the point in the equatorial plane, depicted in Fig. 2.6 (Right).

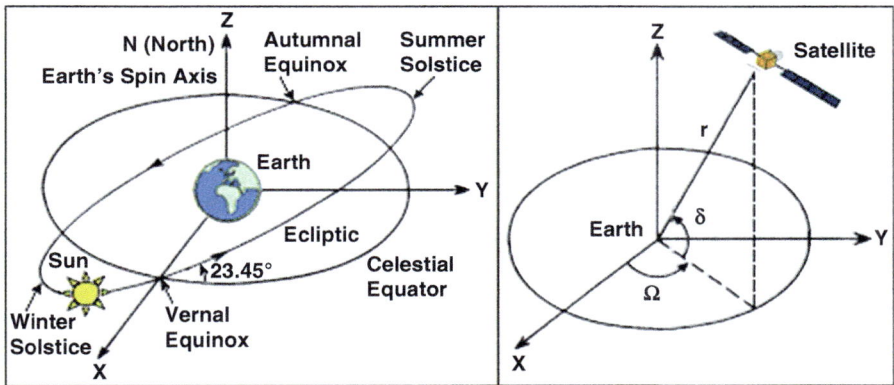

Fig. 2.6 Look angle parameters and coordinates (Courtesy of Book: by Kidder)

In Fig. 2.7 (Left) are shown three angles used to position an elliptical orbit in the right ascension declination (celestial) coordinate system between orbital and equatorial planes, inclination angle (i), right ascension of ascending node (Ω) and argument of perigee (ω).

1. **Inclination Angle** – It is the angle between the equatorial plane and the orbital plane. By convention, the inclination angle is 0 if the orbital plane coincides with the equatorial plane and if the satellite rotates in the same direction as the Earth. If the two planes coincide then the satellite rotates opposite to the direction of Earth and the inclination angle is 180°. In addition, prograde orbits are those with inclination angles less than 90°, while retrograde orbits are those with inclination angle greater than 90°.

2. **Ascending Node** – It is the point where the satellite crosses the equatorial plane going North (ascends). The right ascension of this point is the right ascension of ascending node, which is measured in the equatorial plane from the axis Z (vernal equinox) to the ascending node. In practice, the right ascension of ascending node has a more general meaning. It is the right ascension of the intersection of the orbital plane with the equatorial plane; thus it is always defined, not just when the satellite is at an ascending node.

3. **Argument of Perigee** – It is angle measured in the orbital plane, for Earth-centered orbits, between the ascending node (equatorial plane) and the perigee. For other specific types of orbits, words such as perihelion is for Sun-centered orbits, periastron is for orbits around stars and so on may replace the word periapsis. The argument of periapsis (also called perifocus or pericenter) is one of the orbital elements of an orbiting body. The angle from ascending node to its periapsis, measured in the direction of motion, is parametrical.

In Fig. 2.7 (Right) is depicted an example of the star position in the sky with observer in the coordination centre. The star is the point of interest, while the reference plane is the horizon or the surface of the sea and the reference vector

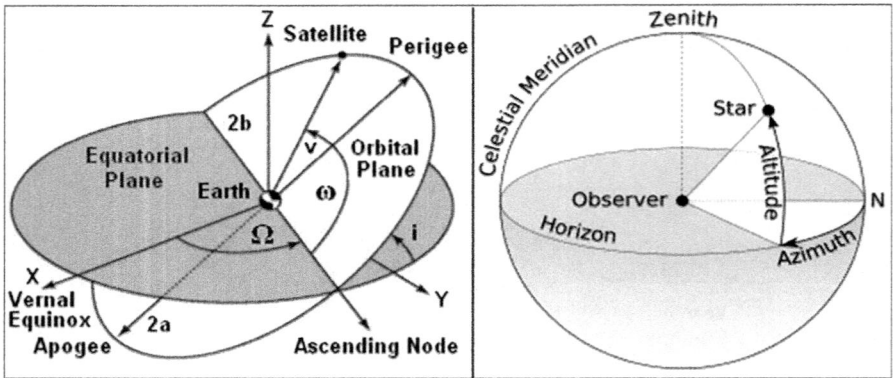

Fig. 2.7 Right ascension declination and star's position (Courtesy of Manual: by Ilcev)

points North (N). The azimuth is the angle between the North vector and the perpendicular projection of the star down onto the horizon. Thus, similar to the satellite platform, this is the direction of a celestial object, measured clockwise around the observer's horizon from North. In such a way, an object due North has an azimuth of 0°, one due East 90°, South 180° and West 270°. Altitude, sometimes referred to as elevation, is the angle between the star or other astronomical object and the observer's local horizon, while for visible objects it is an angle between 0° and 90°. Azimuth and altitude are usually used together to give the direction of an object in the topocentric coordinate system.

The system oriented by the spin axis of the Earth has special points at the North and South Poles, which uses lines of latitude and longitude as geocentric local coordinate system to pinpoint a location of Earth's surface or to demarcate the surface, which is illustrated in Fig. 2.8 (Left). In such a way, latitude is measured away from the equator, while the starting point for longitude is Greenwich meridian.

As stated earlier, here are used two orthogonal global coordinates to describe a position, such as declination, like a celestial latitude and right ascension, like a celestial longitude. As with latitude, declination is measured away from the celestial equator, while with longitude, right ascension has starting point known as "vernal equinox", which is shown in Fig. 2.7 (Left). This point at which the Sun appears to cross the celestial equator from South to North as it moves through the sky during the course of a year.

To understand better positioning of objects in the space will be necessary to describe the basic nomenclature of satellite orientation in the space, which is illustrated in Fig. 2.8 (Right) using Reference guide to the International Space Station (ISS). Alternatively, movement relative to these three axes could be described using shipping terms roll, pitch and yaw, to describe attitude of a satellite, but these don't really denote the sides, merely rotation of the body with respect to the X, Y and Z (depicted in this case as red, blue and green) axes in Cartesian coordinate system, respectively. A Cartesian coordinate system is a coordinate

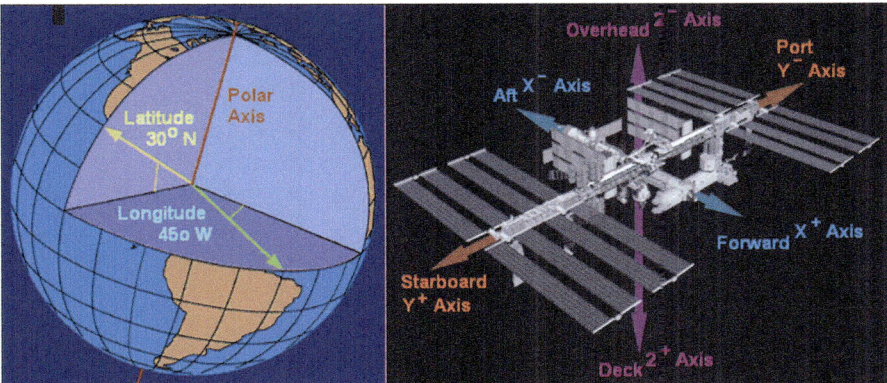

Fig. 2.8 Coordinates of the earth and ISS (Courtesy of Manual: by Pidwirny/NASA)

system that specifies each point uniquely in a plane by a pair of numerical coordinates, which are the signed distances to the point from two fixed perpendicular directed lines, measured in the same unit of length.

There might be other terms referring to the sides from the perspective of the spaceships and relative to its movement, such as nadir, which is direction directly below (opposite zenith), and zenith, which is directly above (opposite nadir). With regards to the maritime terms related to the vessels sides, there are additional sides, such as port, which is direction to the left side (opposite starboard) and starboard, which is direction to the right side (opposite port); And the remaining two sides mentioned in the figure, such as: ram (forward) or wake (aft) pointing, which also have maritime origin. NASA's Guide to the International Space Station Laboratory Racks Interactive however names direction towards nadir as deck and direction towards zenith overhead, and alternatively also the +/− axial values that follow the right-hand rule more commonly used by astronaut pilots during navigation or to describe station's attitude, such as during the docking.

2.1.6 Satellite Orbit Perturbations

After launching a satellite in circular orbits undergoes uniform angular velocity. By Kepler's Second Law, however, satellite in an elliptical orbit cannot have uniform angular velocity, so it must travel faster when is closer to Earth. The position of the satellite as a function of time can be found by applying Kepler's equation:

$$M_a = n\,(t - t_P) = e - \Im\,\sin\,e, \text{ where } n = 2\pi/T = \sqrt{\Im m_e/a^3} \qquad (2.46)$$

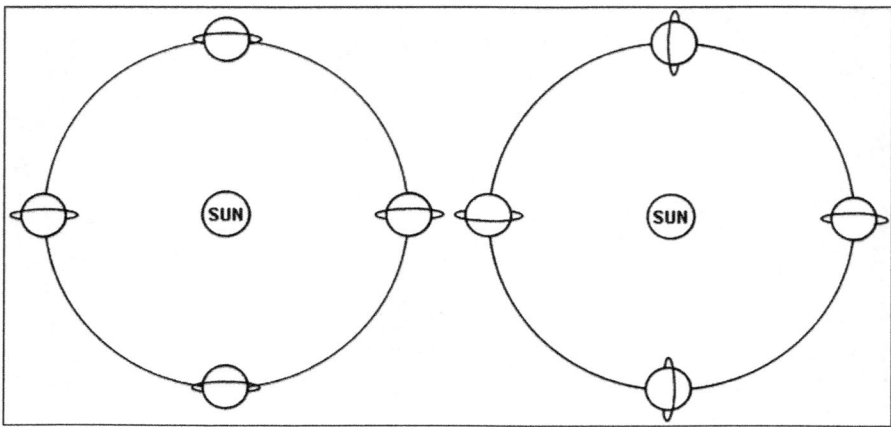

Fig. 2.9 Seasonal changes for Keplerian and sunsynchronous orbit (Courtesy of Book: by Kidder)

where values n = mean motion constant, t = time, t_P = time in perigeal passage, M_a = means anomaly, Э = eccentric anomaly, e = eccentricity and m_e = mass of the Earth.

In the other words, satellite orbits in which the classical orbital elements (except M_a) are constant are called Keplerian orbits. More exactly, viewed from space, Keplerian orbits are simple. The satellite moves in an elliptical path with the center of the Earth at one focus. The ellipse maintains a constant size, shape and orientation with respect to the stars, which seasonal changes are illustrated in Fig. 2.9, for Keplerian (Left) and Sunsynchronous Orbit (Right). Perhaps surprisingly, bit the only effect of the Sun's gravity on the satellite orbit is to move the focus of the ellipse (the Earth) in an elliptical path around the Sun (the Earth's orbit). Viewed from the Earth, Keplerian orbits appear complicated because the Earth rotates on its axis as the satellite orbits the Earth, which orbit of a representative satellite as viewed from a point rotating with the Earth is shown in Fig. 2.10.

The rotation of the Earth beneath a fixed orbit results in two daily passes of the satellite near a point on the Earth (assuming that the period is substantially less than a day and that the inclination angle is greater than the latitude of the point). Thus, one pass occurs during the ascending portion of the orbit; the other occurs during the descending portion of the orbit. This usually means that one pass occurs during daylight and one during darkness.

There is some variation in how the orbital elements are specified. For example, ESA system substitutes true anomaly for mean anomaly. Also, in less formal descriptions of satellite orbits, one frequently sees the height of the satellite above the Earth's surface substituted for the semimajor axis. Since the Earth is not round, the height of a satellite will vary according to its position in the orbit. Specifying the semimajor axis is a much better way to describe a satellite orbit.

The eccentric anomaly is useful to compute the position of a point moving in a Keplerian orbit. As for instance, if the body passes the periastron at coordinates

Fig. 2.10 Representative
satellite orbit (Courtesy of
Book: by Kidder)

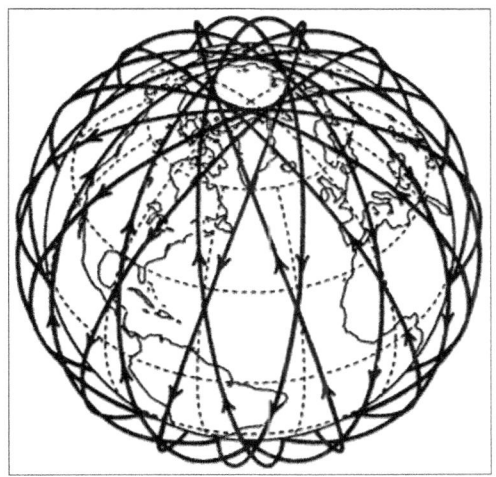

$x = a\,(1 - e)$, $y = 0$, at time $t = t_0$, then to find out the position of the body at any time, it will be at first to calculate the mean anomaly from the time and the mean motion n by the formula $M_a = n\,(t - t_0)$, then solve the Kepler equation above to get $Э$, then get the coordinates from:

$$x = a\,(\cos\,Э - e)\ \text{and}\ y = b\,\sin\,Э \qquad\qquad (2.47)$$

In reality, Kepler's equation is a transcendental equation because sine is a transcendental function, meaning it cannot be solved by $Э$ algebraically. More exactly, numerical analysis and series expansions are generally required to evaluate $Э$.

Although satellites travel in nearly Keplerian orbits, these orbits are perturbed by a variety of forces, see Table 2.3.

Forces arising from the last five listed processes are small and can be viewed as causing essentially random perturbations in the orbital elements. Operationally they are dealt with simply by periodically observing the orbital elements and adjusting the orbit with on-board thrusters. Forces due to the nonspherical Earth cause secular (linear with time) changes in the orbital elements.

These forces can be predicted theoretically and indeed are useful. The gravitational potential of Earth is a complicated function of the Earth's shape, the distribution of land and ocean, and even the density of crystal material. As a first-order correction to a spherical shape, we may treat the Earth as an oblate spheroid of revolution. In cross section the Earth is approximately elliptical.

The distance from the center of the Earth to the equator is, on average, 6378.140 km, whereas the distance to the poles is 6356.755 km. Thus, one can think of the Earth as a sphere with a 21 km thick "belt" around the equator. Therefore, the gravitational potential of the Earth is approximately given by the following equation:

Table 2.3 Forces and sources of orbital perturbations

Force	Source
Nonspherical gravitational field	Nonspherical, nonhomogeneous Earth
Gravitational attraction of auxiliary bodies	Moon, planets
Radiation pressure	Sun's radiation
Particle flux	Solar wind
Lift and drags	Residual atmosphere
Electromagnetic forces	Interaction of electrical current in the satellite with Earth's magnetic field

$$U = -GM/r \left[1 + \tfrac{1}{2} J_2 (r_e/r)^2 (1 - 3 \sin^2 Д) + \ldots \right] \tag{2.48}$$

where r_e = equatorial radius of the Earth, Д = declination angle and J_2 = coefficient of the quadrupole term. The higher-order terms are more than two orders of magnitude smaller than the quadrupole term and will not be considered here, although they are necessary for very accurate calculations.

How does this belt of extra mass affect a satellite's orbit? Therefore, one might expect it to cause the satellite to orbit at a different speed, and indeed it does. The time rate of change of the mean anomaly (dM_a/dt) is given by the mean motion constant n in the unperturbed orbit and by the anomalistic mean motion constant N, in a perturbed orbit. Considering only the quadrupole term anomaly can be formulated:

$$N = dM_a/dt = n \left[1 + 3/2 \, J_2 (r_e/a)^2 (1 - e^2)^{-3/2} (1 - 3/2 \sin^2 i) \right] \tag{2.49}$$

where a = semimajor axis and i = inclination.

When the inclination angle is less than 54.7°, N is greater than n, so the satellite orbits faster than it would in an unperturbed orbit. However, for larger inclinations, the satellite orbits more slowly than it otherwise would.

Because the belt exerts an equator ward force, one might also expect that it would have an effect on the inclination angle. In such a way, this force affects the right ascension of the ascending node rather than the inclination angle.

Just as the force of gravity causes a top to precess rather than to fall over, so the attraction of the belt causes the orbit to precess about the æ axis rather than to change its inclination angle. The rate of change of the right ascension of ascending node (Ω) can be determined by the following relation:

$$d\Omega/dt = -N \left[3/2 \, J_2 (r_e/a)^2 (1 - e^2)^{-2} \cos i \right] \tag{2.50}$$

The final effect of the belt is to cause the argument of perigee (ω) to rotate or precess, which relation is presented as:

$$d\omega/dt = -N \left[3/2\, J_2 (r_e/a)^2 (1 - e^2)^{-2} (2 - 5/2 \cos^2 i) \right] \qquad (2.51)$$

The other three orbital elements, a, e and i, undergo small, oscillatory changes that may be neglected. If SI Units are used, Eqs. (2.46), (2.49), (2.50) and (2.51) result respectively in values of n, N, $d\Omega/dt$ and $d\omega/dt$ whose units are radians per second. The anomalistic period of a perturbed orbit is simply presented as:

$$T_a = 2\pi/N \qquad (2.52)$$

However, because M_a is measured from perigee, the anomalistic period is the time for the satellite to travel from perigee to moving perigee. Of more use is the synodic or nodal period (T_s), which is the certain time for the satellite to travel from one ascending node to the next ascending node. An exact value of T_s must be calculated numerically; however, to very good approximation as follows:

$$T_s = 2\pi/[N + (d\omega/dt)] \qquad (2.53)$$

In summary, therefore, the first-order effects of the nonspherical gravitational potential of the Earth consist of a slow, linear change in two of the classical orbital elements, than the right ascension of ascending node and the argument of perigee, and a small change in the mean motion constant.

2.2 Spacecraft Launching and Station-Keeping Techniques

The launch of the satellite and controlling support services are a very critical point in the creation of space communications, meteorological or other systems and the most expensive phase of the total mission cost. The need to make a satellite body capable of surviving the stresses of the launch stages is a major element in their design phase. Satellites are also designed to be compatible with more then one model of launch vehicle and launching type. There are multi-stage expendable and manned or unmanned, reusable launchers. Owing to location and type of site there are land-based and sea-based launch systems. Additional rocket motors, such as perigee and apogee kick propulsion systems, may also be required.

The process of launching a satellite is based mostly on launching into an equatorial circular orbit and after in GEO, but similar processes or phases are used for all types of orbits. The processes involved in the launching technique depend on the type of satellite launcher, the geographical position of the launching site and constraints associated with the payload. In order to successfully put the satellite into the transfer and drift orbit, the launcher must operate with great precision with regard to the magnitude and orientation of the velocity vector. On the other hand, launching operations necessitate either TT&T facilities at the launching base or at the stations distributed along the trajectory.

2.2.1 Satellite Installation and Launching Operations

Satellites are usually designed to be compatible with more than one prototype of launchers. Launching, putting and controlling satellites into orbit is very expensive operation, so the expenses of launcher and support services can exceed the cost of the satellites themselves. The basic principle of any launch vehicle is that the rocket is propelled by reaction to the momentum of hot gas ejected through exhaust nozzles.

Thus, for a spacecraft to achieve synchronous orbit, it must be accelerated to a velocity of 3070 m/s in a zero-inclination orbit and raised a distance of 42,242 km from the centre of the Earth. Most rocket engines use the oxygen in the atmosphere to burn their fuel but solid or liquid propellant for a launcher in space must comprise both a fuel and an oxygen agent. There are two techniques for launching a satellite in the orbit, namely by direct ascent and by Hohmann transfer ellipse.

2.2.1.1 Direct Ascent Launching

A satellite may be launched into a circular orbit by using the direct ascent method, shown in Fig. 2.11 (a). The thrust of the launch vehicle is used to place the satellite in a trajectory, the turning point of which is marginally above the altitude of the desired orbit. The initial sequence of the ascent trajectory is the boost phase, which is powered by the various stages of the launch vehicle. This is followed by a coasting phase along the ballistic trajectory, the spacecraft at this point consisting of the last launcher stage and the satellite. As the velocity required to sustain an orbit will not have been attained at this point, the spacecraft falls back from the highest point of the ballistic trajectory.

When the satellite and final stage have fallen to the desired injection altitude, having in the meantime converted some of their potential energy into kinetic energy, the final stage of the launcher, called the Apogee Kick Motor (AKM) is activated to provide the necessary velocity increase for injection into the chosen circular orbit. In effect, the AKM is often incorporated into the satellite itself, where other thrusters are also installed for adjusting the orbit or the altitude of the satellite throughout its operating lifetime in space. The typical launch vehicles for direct ascent satellite launching are US-based Titan IV, Russian-based Proton and Ukrainian-based Zenit.

2.2.1.2 Indirect Ascent Launching

A satellite may be launched into an elliptical or synchronous orbit by using the successive or indirect ascent sequences, known as the Hohmann transfer ellipse method, illustrated in Fig. 2.11 (b). The Hohmann transfer ellipse method enables a satellite to be placed in an orbit at the desired altitude using the trajectory that

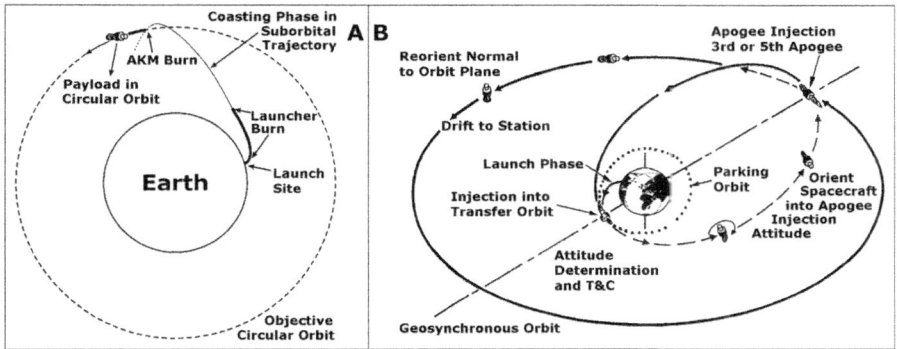

Fig. 2.11 Satellite installation in circular and synchronous orbit (Courtesy of Book: by Pascall)

requires the least energy. At the first sequence the launch vehicle propels the satellite into a low parking orbit by the direct ascent method.

The satellite is then injected into an elliptical transfer orbit, the apogee of which is the altitude of the desired circular synchronous orbit. At the apogee, additional thrust is applied by an AKM to provide the velocity increment necessary for the attainment of the required synchronous orbit. In practice it is usual for the direct ascent method to be used to inject a satellite into a LEO and for the Hohmann transfer ellipse to be used for higher types of satellite orbits.

2.2.2   Satellite Launchers and Launching Systems

Two major types of launch vehicles can be used to put a satellite into LEO, HEO and GEO constellation: Expendable and Reusable Vehicles. There are also two principal locations or site-based types of launching centres: Land-based and Sea-based launch systems.

2.2.2.1   Expendable Launching Vehicles

The great majority of communication satellites have been launched by expendable vehicles and this is likely to continue to be the case for many years to come. There are two types of these vehicles: expendable three-stage vehicles and expendable direct-injection vehicles.

1. **Expendable Three-Stage Vehicles** – Typical series of three-stage vehicles are Delta and Atlas (USA), Ariane (Europe), Long March (China), H-II (Japan), Soyuz (Russia) and so on. In addition, a new generation of launchers have already been developed with two-stages such as Delta III and Ariane 5. Both stages are propellant systems using cryogenic liquid fuel, while the first stage is

assisted by nine strap-on solid-fuel motors. The proven Soyuz launch vehicle is one of the world's most reliable and frequently used launch vehicle. After the US Space Shuttle program ended in 2011, Soyuz rockets became the only launch vehicle able to transport astronauts to the International Space Station (ISS).

The first and second stages of three-stage expandable launch vehicles are usually designed to lift it clear of the Earth's atmosphere, to accelerate horizontally to a velocity of about 8000 m/s and enters a parking orbit at a height of about 200 km. The plane of the parking orbit will be inclined to the equator at an angle not less than the latitude of the launch site. The most efficient way of getting from the parking orbit to a circular equatorial orbit is to convert the parking orbit into an elliptical orbit in the same plane, with the perigee at the height of the parking orbit and the apogee at about 36,000 km and then to convert the transfer ellipse to the GEO.

Thus, the third stage is fired as the satellite crosses the equator, which ensures that the apogee of the Geostationary Transfer Orbit (GTO) is in the equatorial plane. When the satellite is placed in the GTO, the third stage has completed its mission and is jettisoned. The final phase of the Hohmann transfer three-stage launch sequence is carried out by means of AKM built into the satellite. The propulsion of this motor is required to provide at the apogee of the GTO a velocity increment of such a magnitude and in such a direction as to reduce the orbit to zero and make the orbit circular. Once the satellite is in the GEO trajectory, the attitude is corrected, the antennas and solar panels are deployed and the satellite is drifted to the correct longitude (apogee position) for operation.

2. **Expendable Direct-Injection Vehicles** – Typical models of direct-injection launchers are the USA-based Titan IV and the Russian-based Proton, illustrated **in** Fig. 2.12 (**a** – **Left**) and (**a** – **Right**), respectively and also Russian Angara A-5 and Zenit (Ukraine). Otherwise, these types of vehicles do not need an AKM because direct-injection launchers have a fourth stage, which converts directly from GTO to GEO constellation. The Proton rocket is one of the most capable and reliable heavy lift launch vehicles in operation today. Proton D-1 and D-1-E launcher variants have three and four stages, respectively. At lift-off the total weight of Proton is about 688 tons and this vehicle has the capability of placing a maximum of 4500 and 2600 kg into GTO and GEO, respectively. The Proton SL12/D1e has four stages and is the largest currently available space launcher from the former-USSR since 1967. The Proton SL13/D1 variant has three stages.

2.2.2.2 Reusable Launch Vehicles

Reusable launch vehicles have already been developed in the USA (Space Shuttle) and former-USSR (Energia/Buran), which have as their aim the development of vehicles that could journey into space and return, all or much of their structure

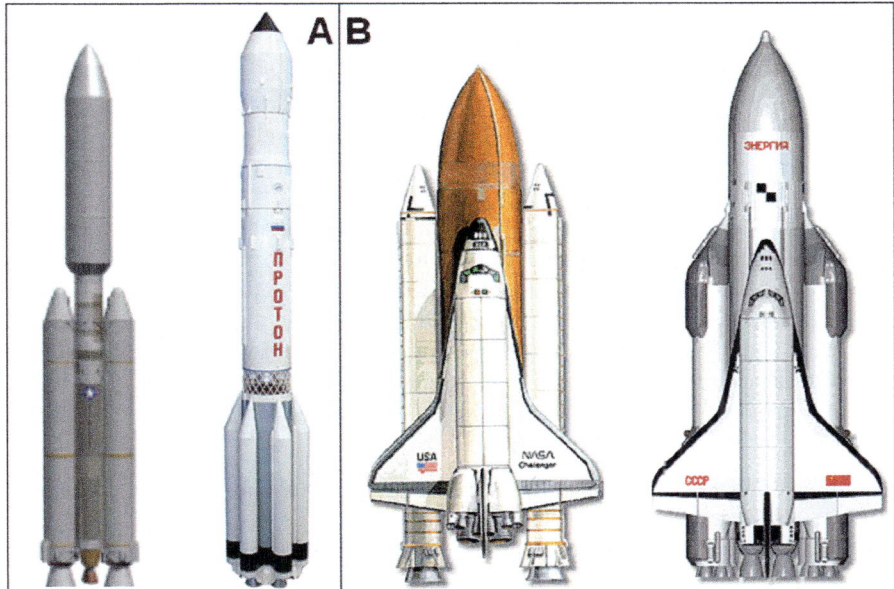

Fig. 2.12 Types of launch vehicles (Courtesy of Book: by Pascall)

being reusable and thus, the satellite launching will cost less. Moreover, in using these launchers there will be less burnt-out upper stages than with expendable vehicles. What remains in space, the small pieces in transfer orbits for many years and much small debris, remains in LEO for a long time, adding to the growing space junk hazard for operational satellites and future space operations. There are other projects for development of similar vehicles such as a small manned reusable space shuttle called Hermes (Europe) and Hope (Japan), unmanned space plane Hotol in UK is proposed, while in Germany and the USA two similar vehicles are projected: TAV (Trans Atmospheric Vehicle) and Sanger Space plane, respectively. Thus, in development of these small vehicles it is important to realize whether any of them could carry sufficient weight and be able to put communication satellites into the desired orbits.

1. **Space Shuttle** – The US-based NASA developed a fleet of manned reusable vehicles of Space Transportation System (STS) called Space Shuttle, which are capable of lifting a satellite of up to 29.5 tons into a parking orbit, inclined at 28.5°, with an altitude of up to 431 km, which is illustrated in Fig. 2.12 (**b – Left**). A Shuttle has three main elements: (**1**) the orbiter for carrying the satellite and crew; (**2**) a very large external tank containing propellant for the main engine of the orbiter and (**3**) two solid-propellant boosters. The reusable Space Shuttle plane is 37.2 m long, the fuselage is 4.5 m in diameter, the wingspan is 23.8 m and the mass is about 84.8 tons. This STS is designed to accommodate in total 7 crewmembers and passengers on board plane. The system came into

service in 1981 and made over twenty successful operational flights until January 1986, when the Shuttle Challenger was destroyed by a fault in the solid-propellant booster and all the crew were killed in a tragic accident. Following this disaster, NASA redesigned the booster but decided to use STS only for regular launches programme of government and scientific vehicles. The final shuttle mission was completed with the landing of Atlantis on 21 July 2011, closing the 30-year Space Shuttle program.

2. **Energia/Buran Spaceplane** – The launcher Energia is the most powerful operational reusable vehicle in the world, capable of carrying about 100 tons into space, whose four first-stage booster units are recoverable for reuse. In particular, it can launch the Buran space plane, enabling it to acquire a LEO and to land with the aid of its own rocket engine, shown in Fig. 2.12 (**b – Right**). The main purpose for which those very heavy lift vehicles were developed was to ferry personnel and supplies for the Russian space station Mir, and also to retrieve or repair satellites already in orbit.

The Energia vehicle can also carry into space a side-mounted canister containing an upper stage and a payload compartment suitable, for example, for a large heavy spacecraft or group of communication satellites to be placed in orbit. Energia flew for the first time on 15 May 1987, carrying a spacecraft mock-up and later on 18 November 1988 carrying an unmanned version of Buran space plane. The reusable Buran space plane is 36.3 m long, the fuselage is 5.6 m in diameter, the wingspan is 24 m and the mass is about 100 tons. It can be flown in automatic configuration or under the control of a pilot to place satellites in LEO or to retrieve them and come back to base for the next use. Up to ten people, crew and passengers, can be accommodated and it can carry in the cargo bay up to 30 tons into an orbit of 200 km altitude and 51.6° inclination. In fact, this plane enables large satellites to be put into orbit and construction of space stations to be considered for both for telecommunication purposes and for scientific missions.

The Energia Launch Vehicle was also the successor to the N-1 Moon Rockets, except that Buran was also used to launch Polyus from Baikonur Cosmodrome in Kazakhstan (former- Soviet Union). Energia was 60 m high and 18 m in diameter, consisting in a central core and four strap-on boosters, while the core was 58.1 m high and 7.7 m diameter. It used 4 RD-0120 rocket engines. The propellants were liquid hydrogen and oxygen. The strap-on boosters were then 38.3 m high and 3.9 m in diameter, with a single four-chamber RD-170 kerosene/liquid oxygen rocket engine.

In 1992, the Russian Space Agency (Roscosmos) decided to terminate the Energia/Buran Program due to Russia's economic difficulties after disintegration of former-Soviet Union. At that stage, the second Orbiter had been assembled and assembly of the third Orbiter with improved performance was nearing completion.

Although the Energia project has been abandoned, it may return to service if a market is found, or adequate partners. Consideration is being given in Russia to the development of a more compact winged space plane designed to ferry personnel and their luggage into space, such as new developed spaceplane Kliper by NPO

Fig. 2.13   Russian spaceplane Kliper (Courtesy of Brochure: by Zak)

Energia, see Fig. 2.13. This compact shuttlecraft could be placed atop of a Proton, Soyuz, Angara or any other launchers.

2.2.2.3   Land-Based Launching Systems

Most satellite launches have taken place from the following launch facilities:

1. **US-Based Launch Centres** – The USA launches satellites from two main locations, in Florida Cape Canaveral, suitable for direct equatorial orbit and the Vandenberg Air Force Base in California, suitable for polar orbit missions.
2. **Russian Launch Centres** – Russian satellites are launched from two main launch centres named Baikonur and Northern Cosmodrome. Baikonur lies north of Tyuratam in Kazakhstan, with the all launching support infrastructure for launching Proton and Energia heavy launchers. The Northern Cosmodrome is located near Plesetsk, south of the town Archangelsk, suitable for launching satellites for all purposes in high inclination orbits. This Cosmodrome is the world's busiest launch site. The newest cosmodrome in Russia is Vostochny Cosmodrome "Eastern Spaceport". This cosmodrome is under construction on the 51° North in the Amur Oblast, in the Russian Far East. When completed in 2018, it is intended to reduce Russia's dependency on the Baikonur Cosmodrome in Kazakhstan. The first launch took place on 28 April 2016 at 0201 UTC.
3. **European Launch Centres** – The main European launch Cosmodrome is the Guiana Space Centre in French Guiana, using Ariane vehicles. The position of

this Cosmodrome enables the best advantage to be taken of the Earth's rotation for direct equatorial orbit.

4. **Chinese Launch Centres** – The principal launch sites in China are Jiuquan and Xi Chang, for launching Long March vehicles. In the meantime, the Xi Chang launch centre has also become most used for launches into the GEO for the international market.

5. **Japanese Launch Centres** – The Japan's Tanegashima Space centre is situated in the prefecture of Kagashima. The facilities include the Takesaki Range for small rockets and the Osaki Range was used for the launch of H-I vehicles until the termination of program in 1992. After renovation the Osaki Range will be used as the launching for next generation of J-I Japanese vehicles. The new Yoshinobu launch complex has been constructed next to the Osaki centre to satisfy the requirements of the new H-II launcher.

2.2.2.4 Sea-Based Launch Systems

The Sea Launch Multinational Organization was developed in March 1996 to overcome the cost of land-based launch infrastructure duplication around the world. The newly formed Sea Launch system is owned by the Sea Launch Partnership Limited in collaboration with international partners such as US Boeing Commercial Space Company, Russian RSC Energia, Ukrainian KB Yuzhnoye/PO Yuzhmash, Shipping Anglo-Norwegian Kvaerner Group and Sea Launch Company, LLC.

The Sea Launch Company, partner locations and operating centres, has located at US-based headquarters in Long Beach, California and is manned by selected representatives of each of the partner companies.

The Sea Launch Partners have the following responsibilities and tasks:

1. Boeing responsibilities include designing and manufacturing the payload fairing and adapter, developing and operating the Home Port facility in Long Beach, integrating the spacecraft with the payload unit and the Sea Launch system, performing mission analysis and analytical integration, leading operations, securing launch licensing documents and providing range services.

2. RSC Energia is responsible for developing and qualifying the Block DM-SL design modifications, manufacturing the Block DM-SL upper stage, developing and operating the automated ground support infrastructure and equipment, integrating the Block DM-SL with Zenit-2S and launch support equipment, planning and designing the CIS portion of launch operations, developing flight design documentation for the flight of the upper stage and performing launch operations and range services.

3. KB Yuzhnoye/PO Yuzhmash are responsible for developing and qualifying Zenit-2S vehicle design modifications, integrating the launch vehicle flight hardware, developing flight design documentation for launch with respect to the first two stages, supporting Zenit processing and launch operations. Several

significant configuration modifications have been made to allow the basic Zenit design to meet Sea Launch's unique requirements.

4. The Anglo-Norwegian Kvaerner Group is responsible for designing and modifying the Assembly and Command Vessel (ACV), designing and modifying the Launch Platform (LP) and integrating the marine elements. Furthermore, Barber Moss Marine Management is responsible for marine operations and maintenance of both vessels.

The partner team of contractors has developed an innovative approach to establishing Sea Launch as a reliable, cost-effective and flexible commercial launch system. Each partner is also a supplier to the venture, capitalizing on the strengths of these industry leaders. The System consists in two main modules: Assembly (Command and Control Ship) and Launch Platform, both illustrated in Fig. 2.14 (a, b), respectively. However, transit for the ACV and the LP from Home Port in Long Beach to the launch site on the equator takes 10–12 days, based on a speed of 10.1 knots.

The Sea Launch Home Port complex is located in Long Beach, California. The Home Port site provides the facilities, equipment, supplies, personnel and other procedures necessary to receive, transport, process, test and integrate the spacecraft and its associated support equipment with the Sea Launch system. The Home Port also serves as the marine base of operations for both of the Sea Launch vessels.

The personnel providing the day-to-day support and service during pre-launch processing and launch conduct to Sea Launch and its customers are located at the Home Port. The ACV performs four important functions for Sea Launch operations: (1) It serves as the facility for assembly, processing and checkout of the launch vehicle; (2) It houses the Launch Control Centre (LCC), which monitors and controls all operations at the launch site; (3) It acts as the base for tracking the initial ascent of the launch vehicle and (4) It provides accommodation for the marine and launch crews during transit to and from the launch site. Therefore, the ACV is designed and constructed specifically to suit the unique requirements of Sea Launch. The ship's overall dimensions are nearly 200 m in length, 32 m in beam and a displacement of 34,000 tons.

Major features of the ACV infrastructure include: a rocket assembly compartment; the LCC with viewing room; helicopter capability; spacecraft contractor and customer work areas and spacecraft contractor and customer accommodation. Moreover, the rocket assembly compartment, which is located on the main deck of the ACV, hosts the final assembly and processing of the launch vehicle.

This activity is conducted while the vessels are at the Home Port and typically in parallel with spacecraft processing. The bow of the main deck is dedicated to processing and fuelling the Block DM-SL of the Zenit launch vehicle. After the completion of spacecraft processing and encapsulation the encapsulated payload is transferred into the rocket assembly compartment, where it is integrated with the Zenit-2S and Block DM. The launchers and the satellite are assembled horizontally in the ACV before sailing from the port of Long Beach to the designated launch site.

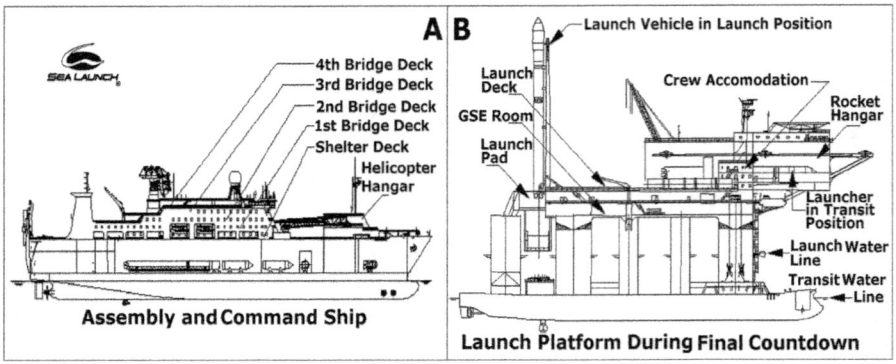

Fig. 2.14 Sea launch modules (Courtesy of Manual: by Sea Launch)

A launcher with a payload will then be transferred in the horizontal position to the launch pad on LP and raised to a vertical position for fueling and launching.

During the launch sequence, the crew of the LP will be transferred to the ACV, which will initiate and control the launch from a position about three miles away from the LP pad. The LP is an extremely stable sea platform from which to conduct the launch, control and other operations. The LP rides catamaran-style on a pair of large pontoons and is self-propelled by a four-screw propulsion system (two in each lower hull, aft), which is powered by four direct-current double armature-type motors, each of which are rated at 3000 hp. The LP in navigation has normal draft at sea water level but once at the launch location, the pontoons are submerged to a depth of 22.5 m to achieve a very stable launch position, level to within approximately 1°.

The ballast tanks are located in the pontoons and in the lower part of the columns. Six ballast pumps, three in each pontoon, serve them. The LP has an overall length of approximately 133 m at the pontoons and the launch deck is 78 by 66.8 m. The Zenit-3SL launcher is a two-stage liquid propellant launch vehicle solution capable of transporting a spacecraft to a variety of orbits.

The original two-stage Zenit rocket was designed by KB Yuzhnoye quickly to reconstitute former-Soviet military satellite constellations. The design emphasizes robustness, ease of operation and fast reaction times. The result is a highly auto-mated launch capability using a minimum complement of launch personnel. The launcher as an integrated part of the Sea Launch system is designed to place spacecraft into a variety of orbits and is capable of putting 5250 kg of payload into GEO.

The international Sea Launch mission provides a number of technical support systems that are available for the customer's use in support of the launch process, including most importantly the following:

1. **Communications** – Internal communications systems are distributed between the ACV and LP, which includes CCTV, telephones, intercom, videoconferencing, public address and vessel-to-vessel radiocommunications using Line-of-

Sight (LOS) direct system. It links external communication system and interconnects the various segments of the Sea Launch program. The external communication system includes Intelsat and two ground stations located in Brewster, Washington and Eik, Norway and provide the primary distribution Gateways to the other communication nodes. Customers can connect to the Sea Launch communication network through the convenient Brewster site. The Intelsat system ties the ACV and launch platform PABX systems to provide voice connectivity, while critical Voice, Fax, Tlx or data capability can be ensured by the Inmarsat satellite SES service.

2. **Tracking and Data Relay Satellite System (TDRSS)** – The Sea Launch system uses a unique dual telemetry stream with the TDRSS. Telemetry is simultaneously received from the Zenit stages, the Block DM upper stage and the payload unit during certain portions of the flight. The Block DM upper stage and payload unit data are combined but the Zenit data is sent to a separate TDRSS receiver. Zenit data is received shortly after lift-off at approximately 9 s and continues until Zenit Stage 2/Block DM separation, at around 9 min. These data are routed from the NASA White Sands GES to the Sea Launch Brewster GES and to the ACV. The data are also recorded at White Sands and Brewster for later playback to the KB Yuzhnoye design centre. When the payload fairing separates, the payload unit transmitter shifts from sending high-rate payload accommodation data by LOS to sending combined payload unit/Block DM by TDRSS. The combined data is again routed from White Sands to Brewster, where it is separated into Block DM and payload unit data and then sent on to the ACV. The data are received onboard ship through the Intelsat communications terminal and are routed to Room 15 for upper-stage data and Room 94 for PLU data. Simultaneously, Brewster routes Block DM data to the Energia Moscow control centre station. The TDRSS coverage continues until after playback of the recorded Block DM data.

3. **Telemetry System** – Sea Launch uses LOS telemetry systems for the initial flight phase, as well as the TDRSS for later phases. The LOS system, which includes the Proton antenna and the S-band system, is located on the ACV. Other telemetry assets include Russian ground tracking stations and the Energia Moscow control centre. The following subsections apply to launch vehicle and payload unit telemetry reception and routing.

4. **Weather (WX) Data System and Forecast** – The ACV unit has a self-contained WX station, which includes a motion-stabilized C-band Doppler radar equipment, surface wind instruments, wave radar, upper-atmospheric balloon release station, ambient condition sensors and access to satellite imagery and information from an on-site buoy.

2.3   Types of Orbits for Meteorological and Other Satellite Systems

An orbit is the circular or elliptical path that the satellite traverses through space. This path appears in the chosen orbital plane in the same or different angle to the equatorial plane. All communication satellites always remain near the Earth and keep going around the same orbit, directed by centrifugal and centripetal forces. Each orbit has certain advantages in terms of launching (getting satellite into position), station keeping (keeping the satellite in place), roaming (providing adequate coverage) and maintaining necessary quality of communication services, such as continuous availability, reliability, power requirements, time delay, propagation loss and network stability.

There is a large range of satellite orbits but not all of them are useful for metrological or communication systems. The one most commonly used orbit for satellite meteorology are GEO and PEO (LEO) constellations. In addition, can be also used Highly Elliptical Orbit (HEO) and latterly Geosynchronous Inclined Orbits (GIO) and Medium Earth Orbit (MEO), depicted in Fig. 2.15 (a). Thus, it is essential to consider that satellites can serve all communication, navigation, meteorological and observation systems for which they cannot have an attribute such as fixed or mobile satellites and the only common difference is which type of payload or transponder they carry onboard.

After many years of research and experiments spent on finding the global standardization for spatial communications, satellites remained the only means of providing near global coverage, even in those parts which other communications systems are not able to reach. There is always doubt about the best orbital constellation that can realize an appropriate global coverage and a reliable communications solution. Unfortunately, there is no perfect system today; all systems have some advantages or disadvantages. The best conclusion is to abridge the story and to say briefly that today the GEO system is the best solution and has only congestion as a more serious problem.

The track of the satellite varies from 0 to 360°, which is illustrated in Fig. 2.15 (b). The track of the GEO satellite is at a point in the centre of the coordinate system; two tracks are apparent movements of the GIO satellite with respect to the ascending node of both 30° and 60° inclination angles and the last is the track of the PEO satellite with an inclined orbit plane to the equator of 90°. The tracks of HEO Molniya (part of the track) and Tundra (complete track) orbits are depicted in Fig. 2.15 (c 1/2), respectively. These two tracks pass over the African Continent and almost all of Europe.

It is sufficient to see Table 2.4 and to understand that the major reasons for LEO problems are enormous satellite cost, complex network and short satellite visibility and lifetime. The LEO/PEO constellations are the same or similar and because of differences in inclination angle of orbital plane and type of coverage they will be considered separately.

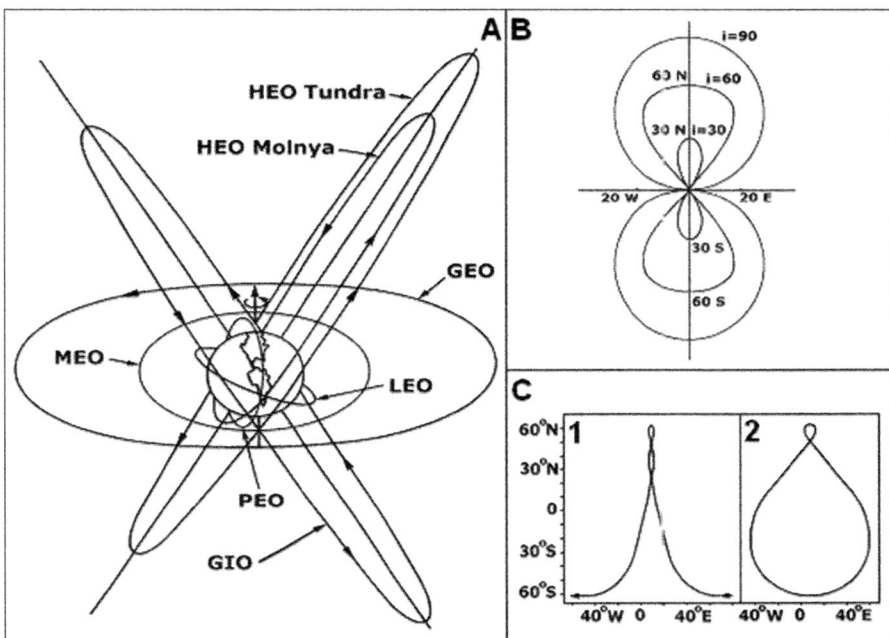

Fig. 2.15 Type of satellite orbits and tracks (Courtesy of Book: by Ilcev)

2.3.1   Low Earth Orbits (LEO)

The LEO systems are either elliptical or more usually circular satellite orbits between 500 and 2000 km above Earth surface and bellow the Inner Van Allen Belt. The orbit period at these altitudes varies between 90 min and 2 h, shown in Fig. 2.15 (a). The radius of the footprint of a communications satellite in LEO constellation varies from 3000 to 4000 km. The maximum time during which a satellite in LEO orbit is above the local horizon for an observer on the Earth is up to 20 min. The traffic to a LEO satellite has to be handed over much more frequently than all other types of orbit. When an LEO satellite, which is serving particular users, moves below the local horizon, it needs to be able to quickly handover the service to a succeeding one in the same or adjacent orbit.

Due to the relatively large movement of a satellite in LEO constellation with respect to an observer on the Earth, satellite systems using this type of orbit need to be able to cope with large Doppler shifts. Satellites in LEO are not affected at all by radiation damage, but are affected by atmospheric drag, which causes the orbit to gradually deteriorate. For LEO satellites the aerodynamic drag is likely to be significant and in general, some of the other perturbations, such as precession of the argument of the perigee, resolve to zero in the orbit is circular or polar. Moreover, a perturbation is unlikely to produce a serious effect on the operation of a multi-satellite constellation since it will usually affect all satellites of the

Table 2.4 The properties of four major orbits

Orbital properties	LEO/PEO	MEO	HEO	GEO
Development period	Long	Short	Medium	Long
Launch and satellite cost	Maximum	Maximum	Medium	Medium
Satellite life (years)	3–7	10–15	2–4	10–15
Congestion	Low	Low	Low	High
Radiation damage	Zero	Small	Big	Small
Orbital period	<100 min	8–12 h	½ Sidereal day	1 Sidereal day
Inclination	90°	45°	63.4°	Zero
Coverage	Global	Global	Near global	Near global
Altitude range (km^{-3})	0.5–1.5	8–20	40/A − 1/P	40 (i = 0)
Satellite visibility	Short	Medium	Medium	Continuous
Handover	Very much	Medium	No	No
Elevation variations	Rapid	Slow	Zero	Zero
Eccentricity	0 to high	High	High	Zero
Handheld terminal	Possible	Possible	Possible	Possible
Network complexity	Complex	Medium	Simple	Simple
Tx power/antenna	Low	Low	Low/high	Low/high
Gain	Short	Medium	Large	Large
Propagation delay	Low	Medium	High	High
Propagation loss	High	Medium	Low	Zero

configuration in equal measure. It is obvious that satellites in LEO and MEO constellation are subject to orbital perturbation.

Today, the US NOAA SMAP and OCO-2 Series and Russian Kanopus-V 1–2 and Resurs-P 1–2 are operational LEO meteorological satellites, which will be introduced in Chap. 6.

2.3.2 Circular Orbits

The GEO together with and geosynchronous satellite constellation has great advantages for meteorological observations, and in the future MEO satellites will have significant success in these fields.

2.3.2.1 Medium Earth Orbits (MEO)

The MEO satellite constellations, known also as Intermediate Circular Orbits (ICO), are circular orbits located at an altitude of around 10,000 to 20,0000 km between the Van Allen Belts. The MEO satellites are operated in a similar way to Big LEO systems providing global coverage, which is shown in Fig. 2.15 (a). Compared to a LEO system a MEO constellation can only be in circular orbit.

Doppler Effect and handover is less frequent and propagation delay is about 70 ms and free space loss is greater. These satellites are affected by radiation damages from the Inner Van Allen Belt only during the launching period and are subject to orbital perturbation.

Cosmic radiation in this orbit is lower, with subsequently longer life expectancy for the complete MEO configuration, fewer eclipse cycles means that battery lifetime will be more than 7 years, higher average elevation angle from users to satellite minimizes probability of LOS blockage and higher RF output power required for both indoor and handheld MSC terminals. There is in exploitation a special model of MEO constellation known in practice as Highly Inclined Orbit. This particular orbit is of interest because it has been chosen for existing and proposed GNSS systems such as existing US GPS and Russian GLONASS. If Galileo at all became operational, it will use MEO configuration, which would have 24 satellites in 3 orbital planes equidistant from each other, at an altitude of 20,000 km and at an inclination of 55°.

However, MEO satellite configuration is successfully used as MEOSAR of Cospas-Sarsat system and can be effective for metrological observations as well. At present is developed O3b new MEO satellite constellation for fixed and mobile communications, which can include meteorological instruments and transponders as well in satellite payload.

2.3.2.2 Geostationary Earth Orbits (GEO)

A GEO has circular orbit in the equatorial plane, with orbital period equal to the rotation of the Earth of 1 sidereal day, which is achieved with the orbital radius of 66,107 (Equatorial) Earth Radii, or orbital height of 35,786 km, which is depicted in Fig. 2.15 (a). This satellite will appear fixed above the surface of the Earth and remain in a stationary position relative to the Earth itself. For instance, this orbit is with zero inclination and track as a point but in practice, the orbit has small non-zero values for inclination and eccentricity, causing the satellite to trace out a small figure eight in the sky, which track is illustrated in Fig. 2.15 (b). The footprint or service area of a GEO satellite covers almost 1/3 of the Earth's surface or 120° in longitude direction and up to 75°–78° latitude North and South of the Equator but cannot cover the Polar Regions. In this way, near-global coverage can be achieved with a minimum of three satellites in orbit moved apart by 120°, although the best solution is to employ four GEO satellites for better overlapping. This type of orbit is essentially used for communication, metrological, GEOSAR Cospas-Sarsat system and other services with the following advantages:

(a) Satellite remains stationary with respect to one point on the Earth's surface and so the GES satellite antennas can be beamed exactly towards the focus of the GEO satellite without periodical tracking.

(b) Inmarsat GEO satellite constellation consisting in three or four satellites can cover all three-ocean regions with overlapping longitudes, except for the Polar Regions beyond latitudes of 75° North and South.
(c) Doppler shift, affecting synchronous digital systems caused by satellites to drift in orbit, affected by the gravitation of the Moon and to a lesser extent of the Sun is small for GES within satellite coverage.

The disadvantages of GEO compared with LEO and MEO operation are as follows:

(a) Long signal delay is due to the large distance of about 35,800 km if the satellite is in zenith for GES and about 41,000 km at the minimum elevation angle of about 5°. For the EM waves traveling at the speed of light this causes a round-trip signal delay of 240–270 ms and full duplex delay of 480–540 ms.
(b) Required higher RF output power and the use of directional satellite antennas.
(c) Launch procedure to put a satellite in GEO is expensive but the total cost of 4 satellites is less than the cost of a minimum of 12 or 40 for MEO and LEO, respectively.

An GEO satellite is at essentially fixed latitude and longitude, so even a narrow-beam Earth antenna can remain fixed. Satellites in GEO can use high and recently low-gain antennas, which helps to overcome the great distances in achieving the required Effective Radiated Isotropic Power (EIRP) at ground level. Using satellite spot beam antennas GEO coverage can be confined to smaller spot areas with bigger power and higher speed of transmission. The GEO satellites pass through both Van Allen Belts only on launch, so their effect is insignificant. After reaching the end of operational life a satellite has to be removed from its orbital slots into a graveyard orbit some 200 km above the GEO plane. The GEO satellite constellation seems likely to continue to dominate in the meteorological satellite observation world, including satellite communication and broadcasting systems, providing near global coverage with low and high-power transmission.

2.3.2.3 Geosynchronous Inclined Orbits (GIO)

The GIO constellations would consist in four satellites at 6-h intervals around the Earth an inclination of 45° to the equatorial plane, which is shown in Fig. 2.15 (a). This orbit provides polar coverage for 6 h either side of their most Northerly and Southerly movement and needs special GES with full tracking antennas. The GIO satellite has a period of orbit equal to or very little different from a sidereal day (23 h 56 min and 4.1 s), which is time for one complete revolution of the Earth. The satellite movement speed has only very little difference from the angular velocity of the Earth, so in such a way, this movement also has constant angular velocity.

Thus, the projection of this satellite movement on the equatorial plane is not at a constant velocity. There is an apparent movement of the satellite with respect to the reference meridian on the surface of the Earth and that of the satellite on passing through the nodes. The orbit may be inclined at any angle, which produces a repeating ground track. In Fig. 2.15 (b) is presented tracks of 30° and 60° inclined orbits. The coupled N–S and E–W motion of GIO satellites is shown as a figure eight pattern, while the patterns could also be distorted circles. Depending on the inclination angle, the GIO satellite shows points on the equator at various longitudes.

A satellite may operate in this orbit for several reasons. First, it is often desirable to save the inclination control fuel required for GEO circle. Second, usually there is no need to control inclination because tracking GES antennas are required for other reasons. Some GEO satellites may last beyond their planned lifetime if run low on fuel and cease inclination control. In effect, the GIO constellation with non-zero inclination can be chosen because of easy launching and placing of the satellite into orbit. This satellite must move with an angular velocity equal to the Earth and be in a prograde orbit, that is, revolving eastward in the same direction as the Earth rotates. Otherwise, the only requirements for a GIO constellation are the right period and direction of rotation.

2.3.3   Highly Elliptical Orbits (HEO)

Using inclined HEO configuration, both polar areas can be effectively covered with four satellites, two in each polar orbit. The elevation angle to the HEO satellites remains high for most of the 12-h period of visibility, which is especially required for continuous Euro-Asian regional coverage providing continuous service. At this point, blocking of the beam due to occlusion of the satellite by buildings, mountains, hills and trees is minimized. Besides, multiple trajectories caused by successive reflection of various obstacles are also reduced in comparison with systems operating with low elevation angles, like GEO.

The apogee altitude combines polar coverages with nearly synchronous advantages. Thus, minimum two special GES in both northern and southern Polar Regions are required to serve MES terminals. The GES tracking can be reached by a fairly directive fixed antenna while the satellite is in its slow apogee sector, the HEO space constellation is namely designed to cover the area under the apogee. Tracking of the satellite is facilitated on account of the small apparent movement and the long visibility duration. Otherwise, it is even possible to use antennas whose 3 dB bandwidth is a few tens of degrees, with fixed pointing towards the zenith, which permits the complexity and cost of the terminal to be reduced while retaining a high gain.

A satellite in HEO constellation near the apogee can also use a high gain antenna to overcome the great distances in achieving the required EIRP values. The noise captured by the GES antenna, from the ground or due to interference from other radio systems and atmosphere, is also minimized due to the high elevation angles. At any rate, these advantages have led the former-USSR to use these orbits for a long time in order to provide coverage of high latitude territories for mobile systems.

The HEO satellite two-way voice transmission has a similar delay as a GEO at the apogee of about 0.25 s. Therefore, free space loss and propagation delay for HEO is comparable to that of the GEO constellation. Compared with GEO, the launch and satellite cost of the HEO constellation is reasonably low; this constellation is free of congestion because of only a few current and projected new HEO systems and provides high elevation angles for GES, which reduces atmospheric losses. Due to the relatively large movement of a satellite in HEO with respect to an observer on Earth, satellite systems using this model of orbit need to be able to cope with large Doppler shifts, 14 kHz for Molniya and 6 kHz for Tundra orbits in L-band 1.6 GHz. However, as the former-USSR's experience has shown, satellites in this orbit tend to have rather a short lifetime due to the repetitive crossing of both Van Allen Belts. The rest of the disadvantages are the necessity of constant satellite tracking at the MES, compensation of signal loss variation, long eclipse periods and complex control system of MES and spacecraft.

The HEO satellite typically has a perigee at about 500 km above the Earth's surface and an apogee as high as 50,000 km. The orbits are inclined at 63.4° in order to provide services to locations at high northern latitudes. The particular inclination value is selected in order to avoid rotation of the apses, i.e., the intersection of a line from the Earth's centre to the apogee and the Earth's surface will always occur at latitude of 63.4°N. Orbit period varies from eight to 24 h. Owing to the high eccentricity of the orbit, a satellite will spend about two thirds of the orbital period near apogee and during that time it appears to be almost stationary for an observer on Earth (this is referred to as apogee dwell). After this period, a switchover needs to occur to another satellite in the same orbit in order to avoid loss of communications. There have to be at least three HEO satellites in orbit, with traffic being handed over from one to the next every 8 h at a minimum.

When there is an orbit in HEO plane of non-zero inclination, the satellite passes over the region situated on each side of the equator and will possibly cover the polar regions if the inclination of the orbit is close to 90°. By orienting the apsidal line, namely the line between perigee and apogee, in the vicinity of the perpendicular to the line of nodes (when ω is close to 90° or 270°), the HEO satellite at the apogee systematically returns above the regions of a given hemisphere. In this way, it is possible to establish satellite links with GES, FES or MES terminals located at high latitudes. Although the satellite remains for several hours in the vicinity of the apogee, it does move with respect to the Earth and after a time dependent on the position of the MES, the satellite disappears over the horizon as seen from the mobiles. However, to establish permanent links it is necessary to provide several suitably phased satellites in similar orbits, which are spaced around the Earth (with different right ascensions of the ascending node and regularly distributed between 0 and 2π) in such a way that the satellite moving away from the apogee is replaced (handover) by another satellite in the same area of the sky as seen from the MES. However, the problems of satellite acquisition and tracking by the MES are simplified. Finally, there only remains the problem of handover and switching the links from one satellite to other, so the RF link frequencies of the various satellites can be different in order to avoid interference.

Examples of HEO systems are Molniya, Tundra, Loopus Borealis of Ellipso system and Archimedes, which obits are shown in Fig. 2.15 (a). The ESA proposed Archimedes system employs a so-called "M-HEO" 8-h orbit. This produces three apogees spaced at 120°. Each apogee corresponds to a service area, which could cover a major population centre, for example the whole European continent, the Far East and North America.

2.3.3.1 Molniya Orbit

The first prototype HEO Molniya satellite was launched in 1964 and to date more than 150 have been deployed, primarily produced by the Applied Mechanics NPO in Krasnoyarsk, former-USSR. The HEO Molniya satellites weigh approximately 1.6 metric tons at launch and stand 4.4 m tall, with a base diameter of 1.4 m. Electrical energy is provided by 6 windmill-type solar panels, producing up to 1 kW of power. A liquid propellant attitude control and orbital correction configuration maintains satellite stability and performs orbital maneuvers, although the latter usage is rarely needed. Sun and Earth sensors are used to determine proper spacecraft attitude and antenna pointing. The first Molniya 3 spacecraft appeared in 1974, primarily to support civil communications (domestic and international), with a slightly enhanced electrical power system and a communications payload of three 6/4 GHz transponders with power outputs of 40/80 W.

The second stratum of the Russian space-based communications system consists of 16 HEO Molniya-class spacecraft in highly inclined 63° semi-synchronous orbit planes. With initial perigees between 450 and 600 km fixed deep in the Southern Hemisphere and apogees near 40,000 km in the Northern Hemisphere. In fact, Molniya satellites are synchronized with the Earth's rotation, making two complete revolutions each day with orbital period of 718 min. The laws of orbital mechanics dictate that the spacecraft orbital velocity is greatly reduced near apogee, allowing broad visibility of the Northern Hemisphere for periods up to 8 h at a time. Thus, by carefully spacing 3 or 4 Molniya spacecraft, continuous communications can be maintained. This type of orbit was pioneered by the USSR and is particularly suited to high latitude regions, which are difficult or impossible to service with GEO satellites.

The 16 operational Molniya satellites are divided into two types and four distinct groups. Namely, eight Molniya-1 satellites were divided into two constellations of four vehicles each. Both constellations consist of four orbital planes spaced 90° apart, but the ascending node of one constellation is shifted 90° from the other, i.e., the Eastern Hemisphere ascending nodes are approximately 65° and 155°E, respectively. Although the system was designed to support the Russian Orbita TV network, a principal function was to service Russian government and military communications traffic via a single 40 W 1.0/0.8 GHz satellite transponder.

Table 2.5 Molniya and Tundra orbit parameters

Characteristics	Molniya orbit	Tundra orbit
Orbital period (t)	12 h	24 h
Sidereal period	11 h 58 min 2 s (half day)	23 h 56 m 4 s (full day)
Semi-major axis (a)	26,556 km	42,164 km
Inclination (i)	63.4°	63.4°
Eccentricity (e)	0.6 to 0.75	0.25 to 0.4
Perigee altitude (h_p)	a(1 − e) − R	a(1 − e) − R
(e.g.: e = 0.71)	1250 km	25,231 km
Apogee altitude (h_a)	a(1 + e) − R	a(1 + e) − R
(e.g.: e = 0.71)	39,105 km	46,340 km

The hypothetical Russian Molniya network can employ minimum 3 HEO satellites in three 12-h orbits separated by 120° around the Earth, with apogee distance at 39,354 km and perigee at 1000 km. This orbit takes the name from the communication system installed by the former USSR, whose territories are situated in the Northern Hemisphere at high latitudes. The orbital period (t) is equal to (t_E /2), or about 12 h. The characteristics of an example Molniya orbit are given in Table 2.5.

The only one-track cycles of a total of two satellite tracks on the surface of the Earth is shown in Fig. 2.15 **(c-1)** for a perigee argument equal to 270°. The shape of this track is cycles of one orbit only near Greenwich Meridian, so the centre, of the next identical track is around 180° westward. Therefore, the satellite at apogee passes successively on each orbit above two points separated by 180° in longitude. The apogee is situated above regions of 63° latitude (the altitude of the vertex is equal to the value of the inclination and the apogee coincides with the vertex of the track when the argument of the perigee is equal to 270°). The large ellipticity of the orbit results in a transit time for the period of the orbit situated in the Northern Hemisphere greater than that in the Southern Hemisphere.

The value of inclination, which makes the drift of the argument of the perigee equal to zero, is 63.45°. A value different from this leads to a drift, which is non-zero but remains small for value of inclination, which does not deviate too greatly from the nominal value. By way of example, for an inclination angle i = 65°, that is variation of 1.55°, the drift of argument of the perigee has a value of around 6.5° per annum.

It is evident that the Molniya HEO satellite has the advantage of high-elevation-angle coverage of the Northern Hemisphere, because of a need to completely cover a great part of the Russian territory. Three satellites in this orbit and phasing are chosen so that at least one satellite is available at any time over the horizon. Thus, with three satellites, each satellite is used (or handover is) 8 h per day, while with four satellites handover is every 6 h. The GES must use tracking antenna systems, so a terminal with only one antenna will have an outage during handover (switching) from one satellite to another.

Using experience and efficiency of Molniya HEO satellites in 2009 Russia developed the first meteorological HEO satellite known as Arctica, which will be introduced in Chap. 6.

2.3.3.2 Tundra Orbit

The Russian Tundra HEO system employs 2 satellites in two 24-h orbits separated by 180° around the Earth, with apogee distance at 53,622 km and perigee at 17,951 km, which provides visibility duration of more than 12 h with high elevation angles. The Tundra orbit can be useful for regional coverage for both FSS and MSS applications. Similar to the Molniya orbit, this orbit is particularly useful for LMSS where the masking effects caused by surrounding obstacles and multiple path are pronounced at low elevation angles, ($> 30°$).

The period (t) of the orbit is equal to t_E, which is around 24 h. The characteristics of an example orbit of this type are given in Table 2.5. This orbit has only one track on the Earth's surface, as shown in Fig. 2.15 (c-2), for a perigee argument equal to 270°, inclination i = 63.4° and eccentricity e = 0.35. The latter parameter can have three values of eccentricity e = 15, e = 25 and e = 45.

According to the value of orbital eccentricity, the loop above the Northern Hemisphere is accentuated to a greater or lesser extent. For eccentricity equal to zero, the track has a form of a Fig. 8, with loops of the same size and symmetrical with respect to the equator. When the eccentricity increases, the upper loop decreases, while the lower loop increases and the crossover point of the track is displaced towards the North. This loop disappears for a value of eccentricity of the order of e = 0.37 and the lower loop becomes its maximum size. The transit time of the loop represents a substantial part of the orbital period and varies with the eccentricity. The position of the loop can be displaced towards the East or West, with respect to the point of maximum latitude, by changing the value of argument of the perigee (ω) and the eccentricity.

2.3.3.3 Loopus Orbit

The proposed Loopus system, which employs 3 satellites in three 8-h orbits separated by 120° around the Earth, has an apogee distance at 39,117 km and perigee at 1238 km. This orbit has similar advantages and disadvantages as for the Molniya orbit. One of the problems encountered by the GES is that of repointing the antenna during the handover (changeover) from one satellite to another. With orbits whose track contains a loop, it is possible to use only the loop as the useful part of the track in the trajectory.

2.3.4 Polar Earth Orbits (PEO)

The PEO constellation is today a synonym for providing coverage of both Polar Regions for different types of meteorological observation and satellite determination services. Namely, a satellite in this orbit travels its course over the geographical North and South Poles and will effectively follow a line of longitude. Certainly, this orbit may be virtually circular or elliptical depending upon requirements of the program and is inclined at about 90° to the equatorial plane, covering both poles. The orbit is fixed in space while the Earth rotates underneath and consequently, the satellite, over a number of orbits determined by its specific orbit line, will pass over any given point on the Earth's surface. Therefore, a single satellite in a PEO provides in principle coverage to the entire globe, although there are long periods during which the satellite is out of view of a particular ground station. Accessibility can of course be improved by deploying more than one satellite in different orbital planes. If two PEO satellite orbits are spaced at 90° to each other, the time between satellites passes over any given point will be halved, which orbit is shown in Fig. 2.15 (a).

The PEO system is rarely used for communication purposes because the satellite is in view of a specific point on the Earth's surface for only a short period of time. Any complex steerable antenna systems would also need to follow the satellite as it passes overhead. At any rate, this satellite orbit may well be acceptable for a processing store-and-forward type of communications system and for satellite determination and navigation.

There are four primary requirements for PEO systems as follows:

1. To provide total global satellite visibility for worldwide LEOSAR Cospas-Sarsat distress and safety satellite beacons EPIRB, PLB and ELT applications;
2. To provide global continuous coverage for current or newly developed and forthcoming satellite navigation systems;
3. To provide at L-band or any convenient spectrum the communication requirements of ships and aircraft in the Polar Regions not covered by the Inmarsat system; and
4. To provide global coverage for meteorological and synoptic observation stations.

The Inmarsat team has studied two broad ranges of orbit altitude of PEO for both distress and communication purposes, first, low altitudes up to 1400 km and second, high altitudes above 11,000 km. In reality, these two orbit ranges are separated by the Inner Van Allen radiation belt. In the regions of the radiation belt the radiation level increases roughly exponentially with height at around 1000 km, reaching a peak at about 5000 km altitude. Therefore, a critical requirement to reduce high-energy proton damage to the solar cell arrays of the satellite system constrains the PEO to low and high altitudes. As is evident, another Outer Van Allen Belt has no negative influence on these two PEO constellations because it lies far a way between MEO and GEO satellite planes.

These two specific systems studied by Inmarsat are Cospas-Sarsat Low PEO at 1000 km altitude and High PEO at 12,000 km altitude, similar to that studied by ERNO, named SERES (Search and Rescue Satellite) system. Thus, it is considered that these two systems demonstrate clearly the solutions tradeoff and constraints on a joint PEO distress, SAR and communication mission. Other possible orbits for polar coverage can be an inclined HEO Molniya constellation of four satellites; GIO 45° inclined orbit of four satellites and 55° inclined circular MEO at 20,000 km altitude for GPS and GLONASS satellite navigation systems. As stated, Cospas-Sarsat system has developed MEOSAR using MEO satellites, GEOSAR using three GEO satellites for global distress communications satellite beacons in combination with first developed LEOSAR systems using four PEO satellites.

For both the Low and High PEO systems the number of operational satellites required to provide adequate Earth coverage needs to be minimized in order to achieve minimum system costs. An IMO and ICAO requirement for the GMDSS/ Cospas-Sarsat mission is that there should be no time delay in distress alerting anywhere in the globe. This orbit was also suitable for the first satellite navigation systems Transit and Cicada, developed by the USA and the former-USSR, respectively.

Otherwise, with a limited number of low altitude PEO satellites it is impossible to provide continuous coverage to polar region, because the view of individual spacecraft is relatively small and their transit time is short. However, because the time for a single orbit is low, less then 2 h and a different section of the polar region is covered at each orbit due to Earth rotation, this drawback is somewhat offset. For a given number of satellites, preferably about eight, it is possible to optimize the constellation that maximizes total system coverage, to improve handover and minimize waiting time between transits. This problem can be solved with additional GES terminals over pole area.

In Fig. 2.16 is illustrated the Earth track of ten successive orbits of satellite in Low PEO with an altitude of 1000 km. The GES in shaded area A (4200 km in diameter) would see the satellite, in the absence of environmental screening, at an angle of elevation not less than 10°, while the satellite was passing through the equatorial plane. The coverage area has the same size and shape wherever the satellite is in the orbit but its apparent size and shape would change with latitude, being distorted by the map projection used in the figure. Thus, the South Pole coverage area at a single pass of the satellite is shown in figure by shaded area B.

The same figure shows that a single PEO satellite in a polar orbit will have a brief sighting of every part of the Earth's surface every day. There will be 2 or 3 of these glimpses per day near the equator, the number increasing as the poles approach.

The period of visibility as seen from the MES range from about 10 min, the satellite passing overhead, down to a few seconds when the satellite appears briefly above the horizon. If the orbital plane of the satellite is given an angle of inclination differing from 90° of the PEO, a similar Earth track is obtained but the geographical distribution of the satellite visibility changes. One LEO satellite with an orbital inclination of 50° would have better visibility between 60° N and 60° S latitude than a PEO satellite but it would have no visibility at all of the Polar Regions.

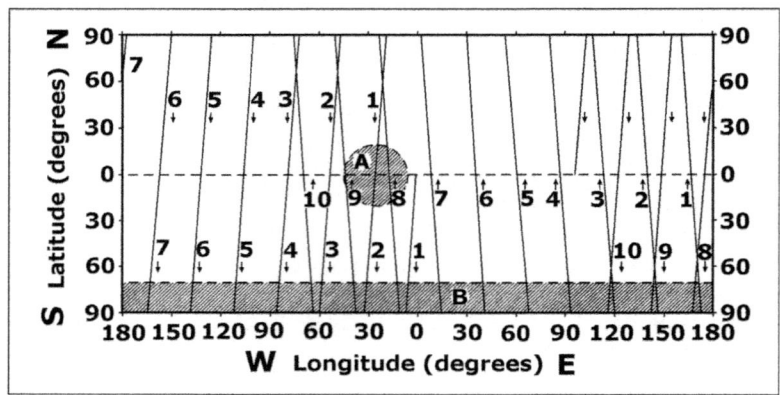

Fig. 2.16 Type of satellite orbits and tracks (Courtesy of Book: by Pascall)

2.4   Main Characteristics of Metrological Satellite Orbits

Nearly all-present meteorological satellites are using one of two satellite orbits, such as sunsynchronous (PEO) or geostationary (GEO), but other orbits are also useful, such as LEO and HEO.

2.4.1   Sunsynchronous Polar Orbits

The nonspherical gravitational perturbation of Earth, far from being a problem, has a very useful application. As shown in Fig. 2.9 (Left), the angle between the lines that join the Sun and the ascending node to the center of the Earth changes in a Keplerian orbit because the orbital plane is fixed while the Earth rotates around the Sun. This causes the satellite to pass over an area at different times of the day. For example, if the satellite passes over near noon and midnight in the spring, it will pass over near 6:00 am and 6:00 pm in the winter. In a such a manner, several additional problems result, among them are: (1) the data do not fit conveniently into operational schedules, (2) orientation of solar cell panels is difficult and (3) dawn or dusk visible images may not be as useful as images made at other times. Fortunately, the perturbations caused by the nonspherical Earth can be employed to keep the Sun-Earth-satellite angle constant.

The Earth makes one complete revolution about the sun (2π radians) in one tropical year (31,556,925.9747 s). Thus, the right ascension of the sun changes at the average rate of 1.991064×10^{-7} rad s^{-1} ($0.9856473°$ day^{-1}). If the inclination of the satellite is correctly chosen, the right ascension of its ascending node can be made to precess at this same rate. At this point, a satellite orbit that is so synchronized with the Sun is called a sunsynchronous orbit. For a satellite with a semimajor axis of 7228 km and zero eccentricity, Eq. (3.53) requires an inclination of 98.8° to be sunsynchronous. In Fig. 2.9 (Right) is shown the change with season of a sunsynchronous orbit.

Because the sun-Earth-ascending node angle is constant in a sunsynchronous satellite orbit, the satellite is often said to cross the equator at the same local time every day. Unfortunately, local time is an ambiguous term. This can be used to express equivalence as follows:

$$LT \equiv t_U + \lambda/15^0 \tag{2.54}$$

where t_U = coordinated universal time in hours and λ = longitude in degrees of a particular point. Equator Crossing Time (ECT) is the local time when a satellite crosses the equator:

$$ECT \equiv t_U + \lambda_N/15^0 \tag{2.55}$$

where λ_N = longitude of ascending or descending node. Therefore, if λ_N is constant, as it is for a sunsynchronous satellite, then ECT is constant.

Sunsynchronous satellites are classified by their ECT values. Thus, noon satellites (or noon-midnight satellites) ascend (or descend) near noon LT (local time). They must, therefore, descend (or ascend) near local midnight. Morning satellites ascend (or descend) between 06 and 12 h LT, and descend (or ascend) between 18 and 24 h LT. Afternoon satellites ascend (or descend) between 12 and 18 h LT, and descend (or ascend) between 00 and 06 h LT.

The highest latitude reached by the subsatellite point (in any orbit) is equal to the inclination angle (or the supplement of i, in the case of retrograde orbits). Since sunsynchronous orbits reach high latitudes, they are referred to as near polar orbits. This is frequently shortened to polar orbits, although they do not cross directly over the poles. These orbits are also called LEO to distinguish them from geostationary orbits (GEO). Note, however, that PEO is a general term for a satellite that passes near the poles, and LEO is a general term for a satellite that orbits not far above the Earth's surface.

While sunsynchronous satellites are of necessity polar orbiters and LEO, the converse is not necessarily true. The ground track of a satellite is the path of the point on the Earth's surface that is directly between the satellite and the center of the Earth (the subsatellite point). Thus, in Fig. 2.17 is illustrated the ground track for three orbits of the typical sunsynchronous NOAA 11 satellite, which main parameters are presenting the following values: a = 7229.606 km, i = 98.97446^0, e = 0.00110058, M_a = 192.28166^0, Ω = 29.31059^0, ω_0 = 167.747540, Epoch Time = 22 March 1990 1 h 15 min and 55.353 s UTC and Nodal Period = 102.0764 min.

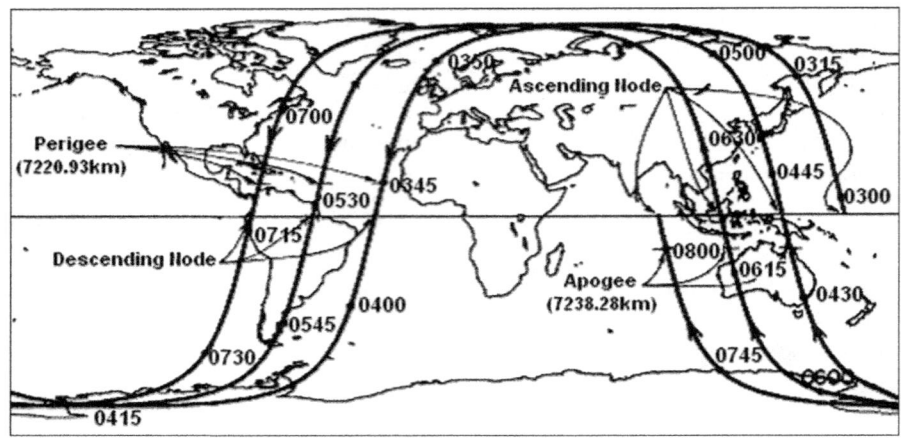

Fig. 2.17   Ground track of sunsynchronous satellite (Courtesy of Book: by Kidder)

2.4.2   Geostationary Circular Orbits

As stated before, using Eq. (2.7) is calculated the radius of a geosynchronous orbit to be 42,164 km. Perturbations due to the nonspherical Earth, however, require a slight adjustment.

The adjustment is small, because geosynchronous satellite orbit is about 6.6 of Earth radii, and the correction terms are inversely proportional to the square of this ratio. In fact, for an orbit with zero eccentricity and zero inclination, Eqs. (2.49), (2.51) and (2.53) require a semimajor axis of 42,168 km to be geosynchronous.

The terms geosynchronous and geostationary are often used interchangeably. In fact, they are not the same. Geosynchronous means that the satellite orbits with the same angular velocity as the Earth. However, geostationary orbit is geosynchronous, but it is also required to have zero inclination angle and zero eccentricity. In such a way, geostationary satellites, therefore, remain essentially motionless above a point on the equator. They are classified by the longitude of their subsatellite point.

Second-order perturbations cause an GEO satellite to drift from the desired orbit. Periodic maneuvers, performed as frequently as once a week, are required to correct the orbit. These maneuvers keep operational geostationary satellites very close to the desired orbit. For example, on 11 March 1990, the US GOES 7 satellite had an inclination angle of $0.05°$; therefore, it did not venture more than $0.05°$ latitude from the equator.

In Fig. 2.18 is illustrated the ground track for a geostationary satellite that is no longer used for imaging and therefore whose orbit is not so carefully maintained. Note that this satellite's orbit is not quite geostationary, because its orbit around the Earth is drifting westward slightly each day.

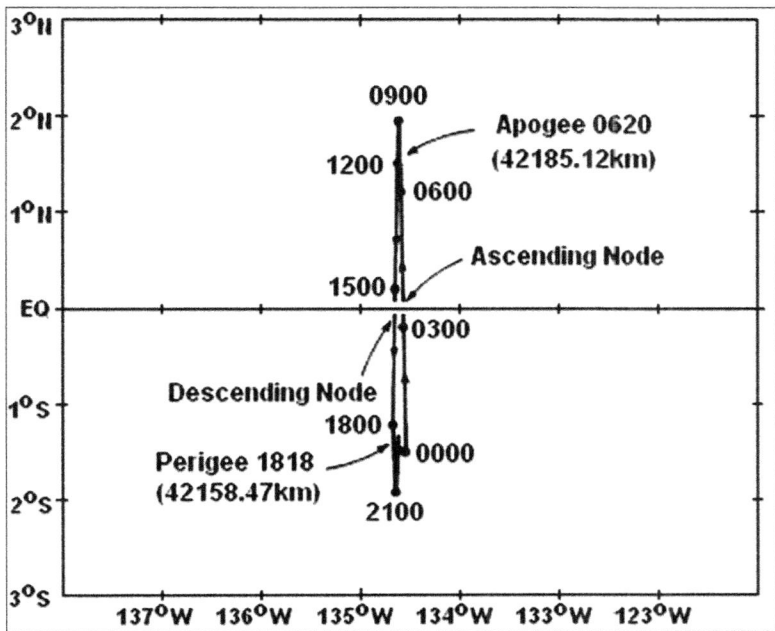

Fig. 2.18 Ground track of geostationary satellite (Courtesy of Book: by Kidder)

Thus, the principal GEO parameters are presenting by the following values: $a = 42171.798$ km, $i = 1.97310^{0}$, $e = 0.0003160$, $M_a = 147.020^{0}$, $\Omega = 80.259^{0}$, $\omega_0 = 223.89100$, Epoch Time = 6 March 1990 4 h 8 min and 20 s UTC and Nodal Period = 1431.297 min.

As stated above and in the other words, the satellite ground track is an imaginary line on the Earth's surface that traces the course of another imaginary line between Earth's centre (O) and an orbiting satellite (S).

In this sense, near-circular satellite orbits, which may be considered as circular in a first approximation, constitute a very important and frequently encountered case. Let now to study some notions developed specifically for these orbits, such as the equatorial shift or the apparent inclination. In this sense, when the orbit is circular, the motion of satellite is uniform with angular frequency (n), which is discussed earlier as mean motion.

The three Euler angles θ, Φ and ψ are serving to specify the orbit and its perigee in space. In this case is necessary to specify satellite (S). Obtaining the correspondence between the Euler angles and the orbital elements is giving as follows: $\theta = \Omega$, $\Phi = i$ and $\psi = \omega + \Im$.

Although they are fixed for the Keplerian orbit, the ascending node angle (Ω); argument of the perigee (ω) and $M - nt$ vary in time for a real orbit, however, the inclination i remain constant. The distance from S to the centre of attraction O or distance of the satellites from the centre of the Earth ($r = R + h$), known also as radius of path, is given by expressed in terms of semi-major axis (a), true anomaly (ϑ) and eccentricity (e) as:

$$r = a(1 - e^2)/1 + e\cos\vartheta \qquad (2.56)$$

Using the notation introduced above, stated value of $\psi = \omega + \vartheta$ can be replaced by:

$$\psi = n\,(t - t_{AN}) \qquad (2.57)$$

Therefore, the ground track of the satellite is determined by two quantities relating to the ascending node taken as origin, namely, its longitude λ_0 and the crossing time t_{AN}, which constitute the initial conditions of the uniform motion.

2.4.3 Other Satellite Orbits

Geostationary and sunsynchronous are only two of infinite possible orbits. Others have been and will become useful for meteorological satellites. The US LEO Earth Radiation Budget Satellite (ERBS) was launched in 1984 by Space Shuttle in orbit at an altitude of 600 km with an inclination angle of 57°. It was placed in this orbit so that it would precess with respect to the Sun and sample all local times over the course of a month.

The former Soviet Union placed its Meteor satellites in low Earth orbit with inclination angles of about 82°. The former Soviet Union also used a highly elliptical orbit for Molniya communications satellites. Thus, it has been suggested that this orbit would be useful for meteorological observations of the high latitudes (Kidder and Vonder Haar 1990). This satellite has an inclination angle of 63.4°, at which the argument of perigee is motionless thus the apogee, from which measurements are made, stays at a given latitude.

The semimajor axis is chosen such that the satellite makes two orbits while the Earth turns once with respect to the plane of the orbit. The eccentricity is made as large as possible so that the satellite will stay near apogee longer. The eccentricity must not be so large that the satellite encounters significant atmospheric drag at perigee. A semimajor axis of 26,554 km and an eccentricity of 0.72 result in a perigee of 7378 km (1000 km above the equator), an apogee of 45,730 km (39,352 km above the equator), and a period of 717.8 min. The attractiveness of this orbit is that it functions as a high-latitude, part-time, nearly GEO satellite. For about 8 h centered on apogee, the satellite is synchronized with the Earth and is nearly stationary in the sky. For a meteorological satellite in a Molniya orbit, the rapid imaging capability, useful from GEO, would be available in the high latitudes as well.

In such a way, as meteorological satellite and instruments become more specialized, more custom orbits are likely to be used in the future. which include GEO, PEO, LEO and HEO.

To provide global coverage of modern meteorological satellite observation system can be deployed special satellite constellation known as Hybrid Satellite Orbits (HSO) such as:

1. **Combination of GEO and HEO Constellations** – In development this kind of HSO can be included combination of minimum three GEO and four HEO satellites;
2. **Combination of GEO and PEO Constellations** – This HSO combination can include minimum 3 GEO and four PEO stateliest;
3. **Combination of GEO and LEO Constellations** – This HSO combination needs minimum three GEO and two LEO satellites in Sun synchronous orbit;
4. **Combination of MEO and HEO Constellations** – This HSO combination will need minimum eight MEO and four HEO satellites; and
5. **Combination of MEO and LEO Constellations** – This HSO combination needs minimum three GEO and minimum two LEO satellites in Sun synchronous orbit.

2.5 Meteorological Satellite Payloads and Antenna Systems

An onboard communication element of meteorological satellite consists many in two major functional units: payload and bus. The primary function of the payload is to provide communication between ground sensors known as Data Collection Platform (DCP) repeaters and ground segment of Direct Readout Ground (DRG) Earth Stations. However the main function of satellite payload is to transmit collected meteorological data from different onboard sensors to DRG. While the bus provides all the necessary electrical and mechanical support to the payload and all satellite missions illustrated in Fig. 2.19.

The payload is made up of the multipurpose repeaters and antenna systems. The repeater performs the required processing of the signal and the antenna receives signals from GES and to transmit signals to FES or MES in the coverage area and vice versa. Two main types of satellite repeaters are possible for onboard utilization: Transparent and Regenerative transponders however are developed many other types for different satellite applications.

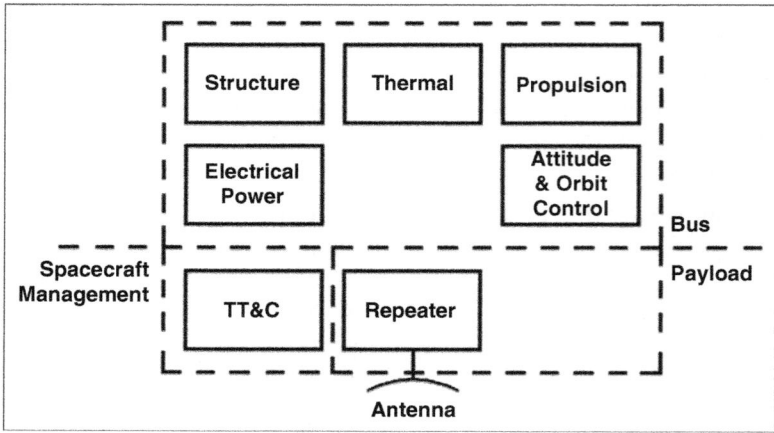

Fig. 2.19 Spacecraft sub-system (Courtesy of Book: by Richharia)

2.5.1   Transparent or Bent-pipe Communication Transponder

The basic function of the satellite transponder is to isolate individual carriers or groups of carriers of signals and to boost their power level before they are retransmitted to the ground stations. The carrier frequencies are also altered as the carriers pass through the satellite. Satellite repeaters that process the carrier in this way are typically referred to as transparent or bent-pipe transponders, which are illustrated in Fig. 2.20 (**Above**). Only the basic RF characteristics of the carrier (amplitude and frequency) are altered by the satellite.

The detailed signal carrier format, such as the modulation characteristics and the spectral shape, remains completely unchanged. Transmission via a transparent satellite transponder is often likened to a bent-pipe because the satellite simply channels the information back to the ground stations.

A bent-pipe is a commonly used satellite link when the satellite transponder simply converts the uplink RF into a downlink RF, with its power amplification. Initially, the received uplink signals from ground-satellite-ground by Receiver (Rx) antennas are filtered in an Input Bandpass Filter (IBF) prior to amplification in a Low Noise Amplifier (LNA). In addition, the output of the LNA is then fed into a Local Oscillator (LO), which performs the required frequency shift from uplink to downlink RF and the bandpass Channel Filter after the Mixer removes unwanted image frequencies resulting from the down conversion, prior to undergoing two amplification stages of signals in the Channel and High Power Amplifier (HPA).

Finally, the output signal of the HPA is then filtered in the Output Bandpass Filter (OBF) prior to transmission through Transmitter (Tx) antenna to the ground. The IFB is a bandpass filter which blocks out all other RF used in satellite communications. After that, the receiver converts the incoming signal to a lower

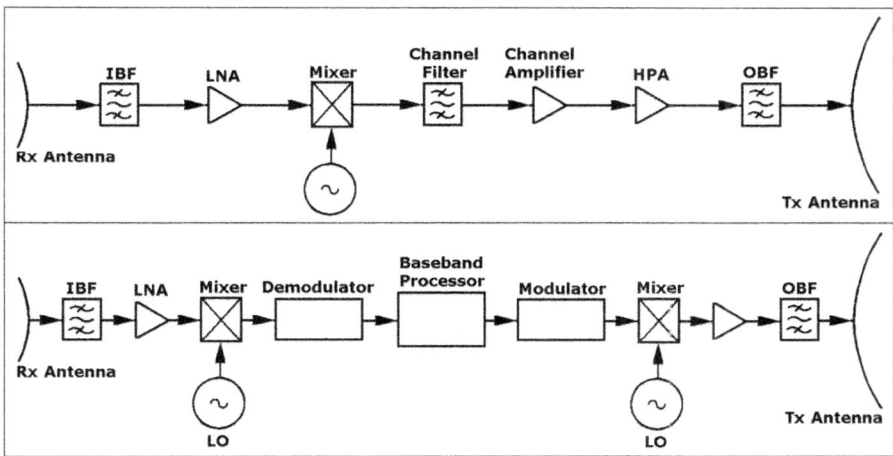

Fig. 2.20 Configuration of spacecraft transponders (Courtesy of Handbook: by ITU)

frequency, using an LO which is controlled to provide a very stable frequency source. This is needed to reduce all noises to facilitate processing of the incoming signal and to enable the downlink frequencies to be established.

The channel filter isolates the various communications channels contained in the waveband allowed through by the input filter. Filtering often leads to large power losses, creating a need for extra amplification, usually followed by a main amplifier. In order to attain the required gain of HPA this segment may employ either a Solid-State Power Amplifier (SSPA) or a Traveling Wave Tube Amplifier (TWTA). In a more complex transponder design, in order to achieve higher RF power, it may be possible to combine the output of several amplifiers.

2.5.2   Regenerative Communication Transponder

Other satellite system designs go through a more complex onboard process to manipulate the carrier's formats, by using on-board processing architecture. This payload architecture offers advantages over the transparent alternative, including improved transmission quality and the prospect of compact and inexpensive ground terminals.

A typical on-board processing system will implement some or all of the functions that are performed by the ground-based Tx and/or Rx in a transparent satellite system. Therefore, these functions may include recovery of the original information on board the satellite and the processing of this information into a different carrier format for transmission to the ground stations.

In fact, any satellite transponder that recreates the signals carrier in this way is usually referred to as a regenerative transponder, which is illustrated in Fig. 2.20

(**Below**). This type of satellite transponder provides demodulation and modulation capacity completely on board the satellite.

The received uplink signal goes along the down-converter segment prior to coming into the on-board demodulator, where it is demodulated and processed in the base band processor. This technology provides flexible functions, such as switching and routings. The downlink signal generated by an on-board modulator passes along the up-converter segment and is transmitted via the antenna. For this type of system link design can be separately conducted for the uplink and downlink because link degradation factors are decoupled between the uplink and downlink by the on-board demodulator and modulator, supported by the base band processor.

A regenerative transponder with base band processing permits reformatting of data without limitation to ground Rx, while the bent-pipe system requires a satellite link design for the entire link, involving both uplink and downlink, but the forward link burst rate is limited by the ground Rx G/T and demodulation performance.

2.5.3 Satellite Meteorological e Communication Transponder

The meteorological payload is made up of the multipurpose repeaters and antenna systems. The repeater performs the required processing of the signal and the antenna system is used to receive signals from ground stations and to transmit signals to the ground receiving and processing facilities in the coverage area and vice versa.

In Fig. 2.21 are illustrated main components of meteorological satellite payloads, which contain the following main components of instruments: Scan Motor Drive, Optical Sensors, Picture Element Buffer and additional radiometers.

On the other hand, meteorological satellite has communication payload, which contain two repeaters. First is Dissemination Repeater, which Receiver (Rx) and Transmitter (Tx) with both spacecraft antennas uses 2 GHz, while Rx and Tx with both antennas of onboard Data Collection Repeater uses 0.4 GHz.

The duties of the meteorological satellite payloads are to contain radio Rx, which acquire operational commands and meteorological satellite data from the ground and retransmit this data via Tx in Data Collection Repeater to the readout ground receiving and processing facilities. In such a way, second Tx in Dissemination Repeater can retransmit data from the ground and send data collected by the onboard instruments to Earth stations.

Then meteorological satellite may carry main components shown in Fig. 2.22, including computers that process the data from the onboard instruments and the tape recorders on which data are recorded for later transmission to Earth. Usually meteorological GEO satellites carry onboard several different payloads, not just one. The primary payloads are meteorological radiometer, which measures Earth

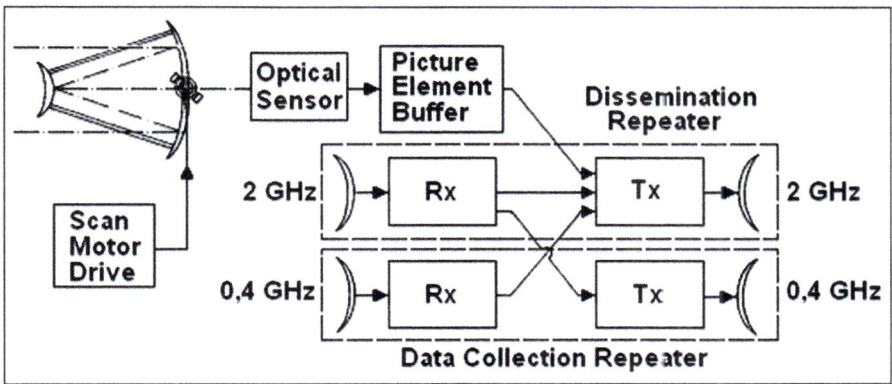

Fig. 2.21 Meteorological satellite payloads (Courtesy of Book: by Berlin)

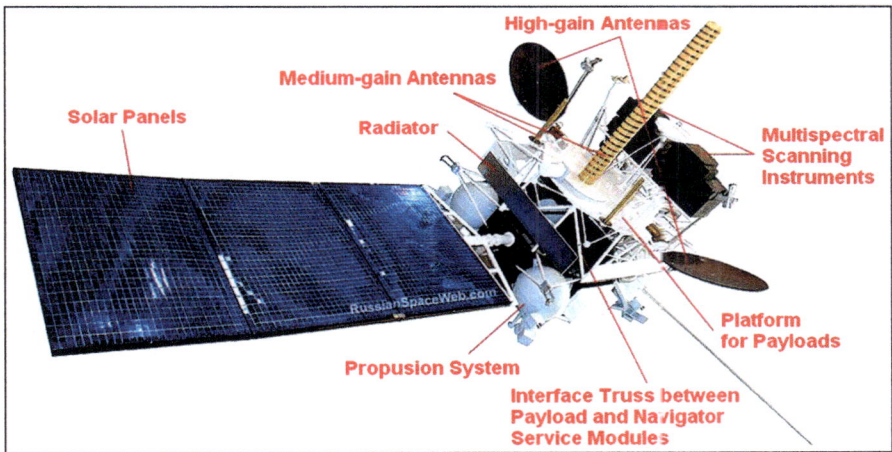

Fig. 2.22 Main components of Russian electro GEO satellite (Courtesy of Booklet: by Roscosmos)

radiances in the visible (0.4–1.0 μm) and thermal infrared (10–13 μm) spectral bands. Same satellites carry and additional payloads not related to meteorology.

For example, the Search and Rescue (SAR) payload of Cospas-Sarsat system detects signals from downed ships, land vehicles, persons and aircraft, giving the precise estimate of their location to aid in rescue operations. The Space Environment Monitor (SEM) measures energetic particles (protons, electrons and alpha particles) for solar and ionospheric studies. Finally, the Data Collection System (DCS) relays meteorological and other data transmitted from ground-based instruments.

2.5.4 Diagram of VSAT GEO Satellite Communication Repeater

As stated earlier, a transparent satellite payload makes no distinction between uplink carrier and uplink noise, and both signals are forwarded to the downlink facilities. On the other hand, however, at the Earth station receiver, one gets the downlink noise together with the uplink-retransmitted noise.

Moreover, a regenerative satellite payload entails onboard demodulation to process of the uplink carriers. Onboard satellite regeneration is most conveniently performed on digital carriers. In fact, the bit stream obtained from demodulation of a given satellite uplink carrier is then used to modulate a new carrier at downlink frequency. This signal carrier is noise-free; hence a regenerative satellite payload does not retransmit the uplink noise on the downlink. The overall link quality is therefore improved.

Moreover, intermodulation noise can be avoided, as the satellite Channel Amplifier (CA) is no longer requested to operate in a multicarrier mode. Indeed, several bit streams at the output of various demodulators can be combined into a Time Division Multiplex (TDM), which modulates a single high rate downlink carrier.

However, this carrier is amplified by the CA section, which can be operated at saturation without generating intermodulation noise, as the carrier it amplifies is unique. This concept of VSAT satellite transponder is illustrated in Fig. 2.23. It should be emphasized that today's commercial and other type of satellites can be used for VSAT services, which are not equipped with regenerative payloads but only with transparent ones. In reality, only a few experimental satellites such as NASA's Advanced Communications Technology Satellite (ACTS) and the Italian ITALSAT satellite have incorporated a regenerative payload, but they are no longer in operation. For instance, some satellites of the Eutelsat fleet are equipped with a regenerative payload (Skyplex) but can be used only by Earth stations operating according to the DVB-S standard.

In the more determined sense, it is important to underline that DVB-RCS technique and technology is very important and useful for the future more reliabble and succesful high speed file transfer of meteorological data and images using VSAT transponders onboard satellites meteorlooskim. The DVB-RCS standars will improve data and image transmissions with the following

1. **DVB-S Standard** – This standard provides DVB adopted technique nearly in all parts of the world (except US and Japan) for transfer of any kind of digital data using Multiplex of MPEG-II video/audio and IP packets. The cost of VSAT and other hardware is very low for about 50 $ available (PC card) and plugs into existing PC or laptop Dish, plus receiver front-end for another 50 $.
2. **DVB-S2 Standard** – This DVB standards benefits from recent developments and improvements in transmission Voice, Data and Video (VDV) VSAT technology, powerful error-correction coding, fade-mitigation to overcome impairments of the channel due to unfavourable propagation conditions (rainstorms)

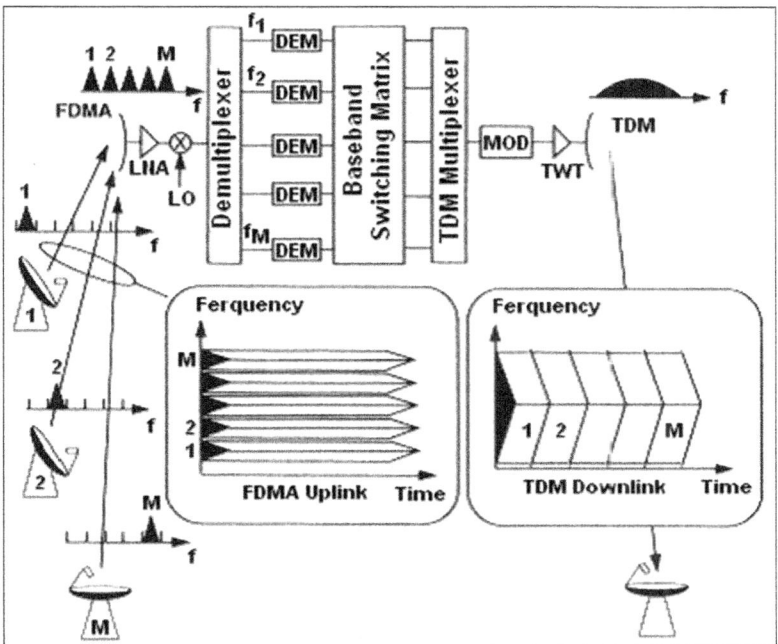

Fig. 2.23 Diagram of VSAT spacecraft transponders (Courtesy of Book: by Maral)

and provide typically 30–35% capacity increase over DVB-S under same transmission conditions.

2.5.5 Antenna System onboard Metrological Satellites

The antenna systems of onboard meteorological satellites can be many types and may have different shapes. The spacecraft antenna system mounted on the spacecraft structure similar is composed of two main integrated elements: the receiving antennas, which are detecting all signals from the ground stations, and transmitting antennas, which are sending data and video signals to the ground receiving and processing stations.

The transmit antenna systems are providing a global (wide) beam on the Earth's surface via GEO satellites, and local footprint via PEO satellites. However, narrow circular beams from GEO or Non-GEO can be used to provide spot beam coverage. For instance, from GEO the Earth subtends an angle of 17.4° Antenna beams 5.8° wide can reuse three frequency bands twice in providing Earth disc coverage.

For instance, from GEO the Earth subtends an angle of 17.4°. Thus, in this content will be explained four types only, such as follows:

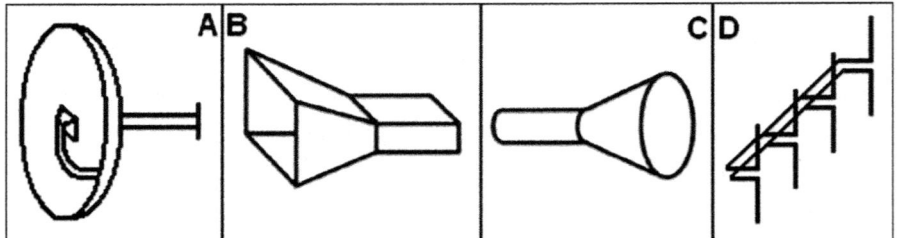

Fig. 2.24 Four types of spacecraft antennas (Courtesy of Book: by Ilcev)

2.5.5.1 Reflector Antennas

The parabolic reflector is a good example of reflectors at radio microwave frequencies, shown in Fig. 2.24 (a). In the past, these antenna were used mainly in space applications onboard spacecraft, but today they are very popular and used by almost everyone who wishes to receive the large number of television or broadcasting channels transmitted all over the globe providing gain in values of 20–30 dB. Reflector antennas are typically used when very high gain or narrow main beam is required. Gain is improved and the main beam narrowed with increase in the reflector size. Large reflectors are however difficult to simulate as they become very large in terms of wavelengths. Reflector antennas are usually illuminated by one or more horns and provide a larger aperture than can be achieved with a horn alone, which prototype made by the US company Lockheed Martin is shown in Fig. 2.25 (Left).

For maximum gain, it is necessary to generate a plane wave in the aperture of the reflector. This is achieved by choosing a reflector profile that has equal path lengths from the feed to the aperture, so that all the energy radiated by the feed and reflected by the reflector reaches the aperture with the same phase angle and creates a uniform phase front. One reflector shape that achieves this with a point source of radiation is the paraboloid, with a feed placed at its focus, shown in Fig. 2.25 (Right).

The paraboloid, however, is the basic shape for most reflector antennas, and is commonly used for earth station antennas. Satellite antennas often use modified paraboloidal reflector profiles to tailor the beam pattern to a particular coverage zone. At this point, the preferred way of making directional spacecraft antennas is to make a dish to reflect the waves rather then to refract them with a lens.

These dishes are easy to make and unlike a lens, which needs to be, solid and they need only a very thin reflective surface. When the reflector is given a specific shape called a parabola and the receiver is placed at the focus the waves received are parallel. This makes this antenna the preferred type for radio telescopes and other extreme range applications.

However, the reflector dish using focal point the waves are leaving the lens parallel to each other and therefore will go great distances with out scattering.

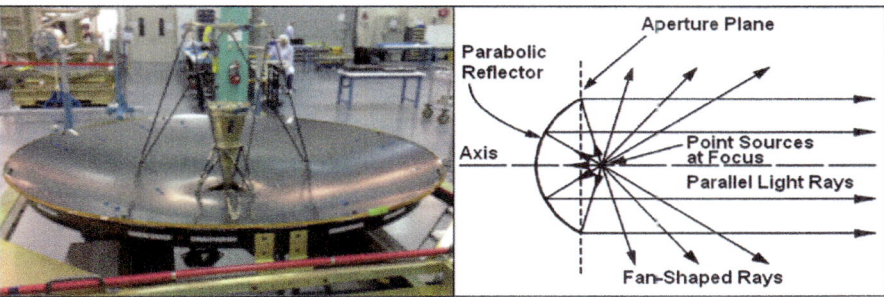

Fig. 2.25 Parabolic reflector antenna systems (Courtesy of Manual: by Ilcev)

While this works it has problems mostly that is transmission the waves moving to the left are unaffordable.

2.5.5.2 Aperture Antennas (Horn Antennas)

A horn is an example of aperture antennas, which are used in satellite spacecraft more commonly, which pyramidal horn is shown in Fig. 2.24 (b), while conical horn is shown in Fig. 2.24 (c). Rectangular or pyramidal horn antenna is one of the simplest and most widely used antennas. Horns have been used for more than a hundred years, and today they used in satellite communications, radio astronomy in communication dishes as feeders, in measurements, etc.

Horn antenna is used at MW when for global coverage relatively wide beams are required. A horn is a flared section of waveguide that provides an aperture several wavelengths wide and a good match between the waveguide impedance and free space. It is also used as feeds for reflectors, either singly or in clusters. Horns and reflectors are examples of aperture antennas that launch a wave into free space from a waveguide. In such a way, it is difficult to obtain gains much greater than 23 dB or beamwidths narrower than about 10° with horn antennas. In such a way, for higher gains or narrow beamwidths a reflector antenna or array must be used.

2.5.5.3 Array Antennas

A grouping of several similar or different antennas forms a single array antenna, shown in Fig. 2.24 (d). The control of phase shift from element to element is used to scan electronically the direction of radiation. Antenna arrays are able to produce radiation patterns that combined, have characteristics that a single antenna would not. The antenna elements can be arranged to form a 1 or 2 dimensional antenna array. A number of antenna array specific aspects will be outlined using 1dimensional arrays for simplicity reasons. Antennas exhibit a specific radiation pattern, which overall radiation pattern changes when several antenna elements are combined in an array.

The array factor quantifies the effect of combining radiating elements in an array without the element specific radiation pattern taken into account. The overall radiation pattern results in certain directivity and thus gain linked through the efficiency with the directivity. Directivity and gain are equal if the efficiency is 100%.

2.5.6   Characteristics of Spacecraft Antenna System

An antenna pattern is a plot of the field strength in the far field of the antenna when a transmitter drives the antenna. The gain of an antenna is a measure in dB of the antenna's capability to direct energy in one direction, rather than all around. A useful principle in antenna theory is reciprocity, which means that an antenna has the same gain and pattern at any given frequency whether it transmits or receives. An antenna pattern measured when receiving is identical to the pattern when transmitting.

As stated earlier, the antenna is providing global, spot and multiple beam coverages, but it can provide scanning and orthogonally polarized beams or coverage zones as well. The pattern is frequently specified by its 3-dB beamwidth, the angle between the directions in which the radiated (or received) field falls to half the power in the direction of maximum field strength.

However, a satellite antenna is used to provide coverage of a certain area or zone on the Earth's surface, and it is more useful to have contours of antenna gain with maximum strengths of the signal in the middle of the coverage area and with decreasing of signals to the peripheries.

When computing the signal power received by an GES from the satellite, it is important to know where the station lies relative to the satellite transmit antenna contour pattern, so that the exact EIRP can be calculated. If the pattern is not known, it may be possible to estimate the antenna gain in a given direction if the antenna boresight or beam axis direction and its beamwidth are known.

Furthermore, a greater power density per unit area for a given input power can be achieved very well, when compared with that produced by a global circular beam, leading to the use of much smaller receiving MES antennas. The equation that determines received power (P_R) is proportional to the power transmitted (P_T) separated by a distance (R), with gain of transmit antenna (G_T) and effective area of receiving antenna (A_R) and inverse proportional with 4π and square of distance. The relations for P_R and G_T are presented as follows:

$$P_R = P_T G_T A_R / 4\pi R^2 \text{ and } G_T = 4\pi A_T / \lambda^2 \tag{2.58}$$

where G_T = effective area of transmit antenna and λ = wavelength. The product of P_T and G_T is gain, generally as an increase in signal power, known as an EIRP.

Signal or carrier power received in a link is proportional to the gain of the transmit and receive antennas (G_R) presented as:

$$P_R = P_T G_T G_R \lambda^2 / (4\pi R)^2 \text{ or } P_R = P_T G_T G_R / L_P L_K \text{ [W]} \tag{2.59}$$

The last relation can be derived with the density of noise power giving:

$$P_R / N = P_T G_T (G_R / T_R)(1/K L_P L_K) \tag{2.60}$$

where L_P = coefficient of energy loss in free space, L_K = coefficient of EMW energy absorption in satellite channels, T_R = temperature noise of receiver, G_R/T_R is the figure of merit and K = Boltzmann's Constant (1.38×10^{-23} J/K or its alternatively value is -228.6 dBW/K/Hz).

At any rate, P_R has a minimum allowable value compared with system noise power (N), i.e., the Carrier and Noise (C/N) or Signal and Noise (S/N) ratio must exceed a certain value. This may be achieved by a trade-off between EIRP ($P_T G_T$) and received antenna gain (G_R). If the receive antenna on the satellite is very efficient, the demands on the GES/FES/MES are minimized. Similarly, on the satellite-to-Earth link, the higher the gain of the satellite transmit antenna, the greater the EIRP for a given transmitter power. Satellites often have onboard parabolic dish antennas, though there are also other types, such as phased arrays.

The principal property of a parabolic reflector is its ability to turn light from a point source placed at its focus into a parallel beam, mostly as illustrated in Fig. 2.25 (Right). In practice the beam can never be truly parallel, because rays can also be fan-shaped, i.e., a car headlamp is a typical example. In a microwave antenna the light source is replaced by the antenna feed, which directs waves towards the reflector. The length of all paths from feed to aperture plane via the reflector is constant, irrespective of their angle of parabolic axis. The phase of the wave in the aperture plane is constant, resulting in maximum efficiency and gain. The gain of an aperture (G_a) and parabolic (G_p) type of antennas are:

$$G_a = \eta \left(4\pi A / \lambda^2\right) = 4\pi A_E / \lambda^2 \text{ and } G_p = \eta (\pi D)^2 / \lambda^2 \tag{2.61}$$

where η = efficiency factor, A = projected aperture area of antenna, $A_E = \eta^A$ is the effective collecting area and D = parabolic antenna diameter. Thus, owing to correlation between frequency and wavelength, $f = c/\lambda$ is given the following relations:

$$G_p = \eta (\pi D f/c)^2 = 60,7 (D f)^2 \tag{2.62}$$

where the second relation comes from considering that $\eta \approx 0.55$ of numerical value. If this value is presented in decibels the gain of antenna will be calculated as follows:

$$G_T = 10 \log G_p \tag{2.63}$$

For example, a parabolic antenna of 2 m in diameter has a gain of 36 dB for a frequency at 4 GHz and a gain of 38 dB for a frequency at 6 GHz. Parabolic antennas can have aperture planes that are circular, elliptical or rectangular in shape.

Thus, antenna with circular shape and homogeneous illumination of aperture with a gain of -3 dB has about 47.5% of effective radiation, the rest of the power is lost. To find out the ideal characteristics it is necessary to determine the function diagram of radiation in the following way:

$$F\,(\delta_o) = s\,(\delta_o)/s\,(\delta_o = 0) \qquad (2.64)$$

where parameter s (δ_o) = flow density of radiation in the hypothetical satellite angle (δ_o) and s $(\delta_o = 0)$ = flow density in the middle of the coverage area. Looking the geometrical relations in Fig. 2.3 (a) follows the relation:

$$F\,(\delta_o) = d_o/h = \cos\,\delta\sqrt{(k^2 - \sin^2\delta_o)}/1 - k \qquad (2.65)$$

where, as mentioned, $k = R/(R + h) = \sin\,\delta$ and if $\delta_o = \delta$, the relation is defined by the following equation:

$$F\,(\delta) = k\,\cos\,\delta \qquad (2.66)$$

For GEO satellite the value of ΔL is given as a function of angle δ, which is the distance from the centre of the coverage area, where the function diagram of the radiation is as follows:

$$F\,(\delta) = \Delta L = 20\,\log = 20\,\log R/(R+h)\,\cos\,\delta$$
$$= 10\,\log R/(1 + 2R/h)\;[\text{dB}] \qquad (2.67)$$

Therefore, in the case of GEO satellites the losses of antenna propagation are greater around the periphery than in the centre of the coverage area for about 1.32 dB. The free-space propagation loss (L_P) and the input level of received signals (L_K) are given by the equations:

$$L_P = (4\pi\,d/\lambda)^2 \text{ and } P_R/S = P_T G_T/4\pi d^2 L_K \qquad (2.68)$$

The free-space propagation loss is caused by geometrical attenuation during propagation from the transmitter to the receiver.

2.6 Satellite Bus

The satellite bus is usually called a platform and consists in several sections, shown in Fig. 2.19. The function of the satellite platform is to support the payload operation reliably throughout the mission of primary construction section, such as

Structure Platform (SP), Electric Power (EP), Thermal Control (TC), Attitude and Orbit Control (AOC), Telemetry, Tracking and Command (TT&C) and Propulsion Engine.

2.6.1   Structure Platform (SP)

The structure has to house and keep together all components of bus and communications modules, enable protection from the environment and facilitate connection of the satellite to the launcher.

It comprises a skeleton on which the equipment modules are mounted and a panel, which covers and provides protection for sensitive parts during the operational phase from micrometers and helps to shield the equipment from extremes of heat, coldness, vacuum and weightlessness, including the relatively small dynamic forces produced by the station-keeping, attitude control engines and inertial momentum devices.

The spacecraft is protected during the launch phase with an enclosure, or nose cone. At the end, the nose cone is jettisoned, at which time the spacecraft must survive the inertial and thermal stress of an additional propulsion stage until it is inserted into orbit. In this sense, a spacecraft is virtually free of gravitational stress when in orbit, which allows the use of very large deployable arrays, which would collapse under their own weight on the Earth's surface without problems.

Thus, large stresses are developed during launch as a result of massive acceleration and intense vibration, so the SP body must be sufficiently strong to withstand all external forces. On the other hand, all large structures such as antenna and solar arrays have to be folded and protected during a launch sequence and must have a deployable mechanism. The deployment of structures requires a special technique in the vacuum of space because of the lack of a damping medium, such as air.

2.6.2   Electric Power (EP)

The primary source of power for a communications satellite is the Sun. Hence solar cells are used to convert energy received from the Sun into an electrical source. The principal components of the power supply system include: (1) Power electric generator, usually solar cell arrays located on the spinning body of a spin-stabilized satellite or on the paddles for a three-axes stabilized satellite; (2) Reliable electrical storage devices, such as batteries, for operating during periods of solar eclipses; (3) Electrical harness for conducting electricity to all of the devices demanding power; (4) The special converters and regulators delivering regulated voltage and currents to the devices on board the spacecraft and (5) The electrical control and protection section is associated with the remote TT&C satellite system.

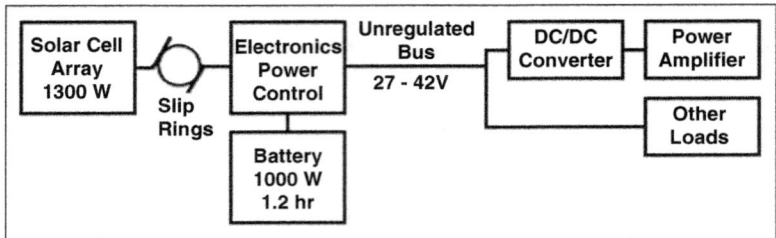

Fig. 2.26 Satellite electric power subsystem (Courtesy of Book: by Gordon)

The solar arrays are the motor during entire life of the satellite, providing sufficient power to all active components. Occasionally the Earth or Moon blocks the Sun, then researchable batteries provides secondary electrical power. A typical satellite power subsystem is shown in Fig. 2.26. Most components are duplicated for redundancy. On spinning satellites the rings are between the power subsystem, on the rotating part, and the satellite bus.

Each cell delivers about 150 mA at a few hundred millivolts and an array of cells must be connected in series or parallel to give the required voltage and current for operating the equipment until the end of its life and to recharge the batteries when the satellite moves out of an eclipse. Charge is applied via the main electric power bus or a small section of the solar cell. During exploitation, batteries are sometimes reconditioned by intentionally discharging them to a low charge level and recharging again, which prolongs their life.

The operational status of onboard batteries including recharge, in-service or reconditioning is remotely controlled by a special ground segment. Thus, the mass of a battery constitutes a significant portion of the total satellite mass. Therefore, a useful figure of merit to evaluate the performance of a battery is capacity in W/h per unit weight taken at the end of its life. Until recently virtually all satellites used Ni-Cd (Nickel-Cadmium) batteries because of their high reliability and long life-time. These batteries provide a low specific energy of about 30 to 40 Wh/kg. The latest type of Ni-H (Nickel-Hydrogen) batteries can store at least 50% more energy per kilogram.

When a satellite passes through the Earth's shadow, the solar arrays stop producing power and the satellite structures use the energy from batteries. The GEO satellite undergoes around 84 eclipses in a year, with a maximum duration of 70 min. Thus, the eclipse occurs twice a year for 42 consecutive days each time. The percentage of eclipses' duration for GEO and HEO is much less than for lower satellite orbits. The LEO satellites can undergo several thousand eclipses in a year.

For example, a LEO satellite in equatorial orbit at an altitude of 780 km can remain in the Earth's shadow for 35% of the orbital period. For a MEO under similar conditions, the maximum eclipse duration would be about 12.5% of the orbital period and a total duration of about 3 h a day, with about 4 eclipses per day.

Otherwise, the Sun can also be sometimes eclipsed by the Moon's shadow, which is less predictable.

2.6.3   Thermal Control (TC)

There is not air at satellite orbit surrounded by very harsh space environment. The average satellite temperature is determined by the absorbed solar energy, different thermal radiation into space and internal electric dissipation, what depends on the satellite shape and surface.

Thermal control of a communication satellite is very important factor during entire satellite lifetime, which is necessary to achieve normal temperature balance and proper performance of all subsystem. In spacecraft design, the TC system has the function to keep all parts of spacecraft within acceptable temperature ranges during all mission phases. It is essential to guarantee the optimum performance and success of the mission, because if a component encounters a temperature, which is too high or too low, so could be damaged or affected.

Thermal stress to the satellite constructions results from high temperature effects from the Sun and from low temperatures occurring during eclipse period. The obvious objective of the TC is to assure that the spacecraft structure and all equipment is maintained within temperatures that will provide successful operations. In such a way, a satellite undergoes different thermal and other conditions during the launch and operational phase. The vacuum in space limits all heat transfer mechanisms to and from a spacecraft and its external environment to that of radiation.

However, some main parts are usually in direct sunlight with a flux density of over 1 kW/m^2, while other parts are facing the shadow side at a temperature of about −270 °C. In addition, an eclipse causes temperature variation from around −180° to +60 °C, when the ambient temperature falls well below 0 °C and rises rapidly from the moment the satellite emerges from the eclipse. In fact, all these extremes in space have to be eliminated or moderated for normal satellite operations, especially because all electronic devices need optimum temperatures between −5° and +45 °C.

These problems can be solved by remote TC techniques, using both passive and active means of controlling and regulating the temperature inside spacecraft. The passive means are simple and reliable, using surface finishes, filters and insulation blankets. The active means are necessary to supplement the passive systems, which include louvers and blinds operated by bimetallic strips, heat pipes, thermal louvers and different electrical heaters.

Heat pipes are used to transfer heat from internal hot spots or devices to remote radiator surfaces or must be transported to the outside surface where it can be dissipated. On the other hand, special electric heaters are used to maintain minimum component or structure temperatures during cold conditions. Accordingly, the

TC subsystem ensures temperature regulation for optimum efficiency and satellite performance.

2.6.4 Attitude and Orbit Control (AOC)

The attitude and orbital control subsystem checks that a spacecraft is placed in its precise orbital position, and maintains, thereafter, the required attitude throughout its mission. Control is achieved by employing momentum wheels, which produce gyroscopic torques, combined with an auxiliary reaction control gas thruster system. Many various sensors are employed to detect attitude errors, including Sun's initial orientation purposes.

The satellite antennas require AOC system that will keep them pointed always at the Earth, frequently within 0.1° and 0.01°. In Fig. 2.27 is shown a block diagram of AOC system. The sensors direct any pointing errors and correct them by changing the speed or direction of a rotating wheel. The main performance specification of an AOC system is determined by the disturbance torques and the required pointing accuracy.

The AOC system performs satellite orientation and accurate orbital positioning throughout its lifetime, because loss of attitude renders a spacecraft useless. There are in use two common AOS, such as attitude control and orbit or station keeping control systems. The objective of attitude control is to keep the antenna RF beam pointing at the intended areas on the Earth, which procedure involves as follows:

(a) Measuring the attitude of the satellite by sensors;
(b) Comparing the results of measurements with the required values;
(c) Calculating the corrections to reduce eventual errors and
(d) Introducing these corrections by operating the appropriate torque units.

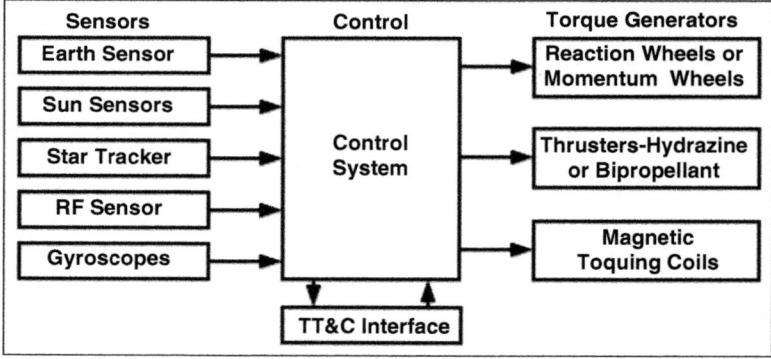

Fig. 2.27 Satellite AOC subsystem (Courtesy of Book: by Gordon)

1. **Attitude Control** – Currently, all types of attitude stabilization systems have relied on the conservation of angular momentum in a spinning element, which can be classified into the two categories already mentioned, such as spin-stabilized and three-axis stabilization. The satellite is rapidly spun around one of its principal axes of inertia. Thus, in the absence of any perturbing torque, the satellite attains an angular momentum in a fixed direction in an absolute frame of reference. For the GEO satellite, the spin (pitch direction) axis must be parallel to the axis of the Earth's rotation. The perturbation torques reduces the spin of the satellite and they affect the orientation of the spin axis. The second system of attitude control is a body-stabilized design in a three-axis stabilized satellite, whose body remains fixed in space. This solution is the simplest method of attitude control using a momentum wheel, which simultaneously acts as a gyroscope, in a combination of spin and drive stabilization. Certain perturbing torques can be resisted by changing its spin speed and the resulting angular momentum of the satellite.

2. **Orbit or Station-Keeping Control** – On-board propulsion requirements for both GEO and Non-GEO are important to keep a satellite in the correct orbital attitude and position. For this reason several types of propulsion systems are used, such as arc jet thrusters, ion and solar electrical propulsion, pulsed plasma thrusters, iridium-coated rhenium chambers for chemical propellants, etc. In order that the appropriate station-keeping corrections can be applied, it is essential that the orbit and position of a satellite are accurately determined. This may be done by making measurements of the angular direction and distance of the satellite from the Earth station, or a number of GES terminals. When the orbit and position of the satellite have been determined, it is possible to calculate the velocity increments required to keep the N–S and E–W excursion of the satellite within the tolerated limits. The frequency with which N–S correction must be made depends on the maximum allowable value of the orbital inclination but the total increment required each year to cancel out the attraction of the Sun and Moon is 40–50 m/s. Otherwise, E–W station-keeping is usually achieved by allowing the satellite to drift towards the nearest point of equilibrium until it reaches the maximum tolerable error in longitude, then the process is repeated on the other side of the nominal longitude and finally, the satellite drifts back once more towards the point of equilibrium and the process is repeated. The frequency and magnitude of the velocity increments required depend on the angular distance between the satellite and the points of equilibrium and on the tolerable error, which is a maximum of about 2 m/s.

2.6.5   Telemetry, Tracking and Command (TT&C)

The telemetry, tracking, command and communication equipment enables data to be send continuously to the Earth stations, received from these stations and allows

Fig. 2.28 Satellite TT&C
(Courtesy of Book: by
Maini)

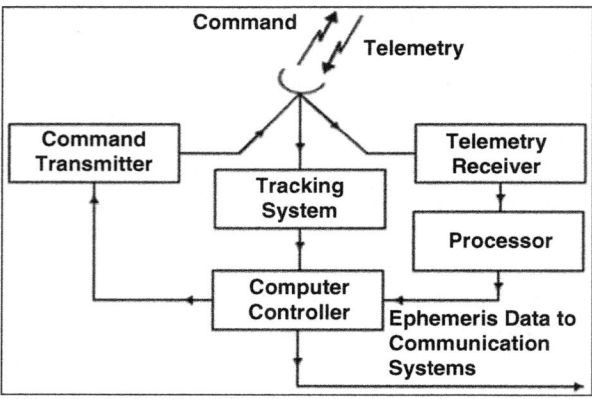

ground control stations to track the spacecraft and to monitor the health of the spacecraft and also to send commands to carry out various tasks like switching the transponders in and out of service, switching between redundant units, etc. The TT&C subsystem monitors and controls the satellite functions right from the lift-off stage and to the end of its operational life in space. So, it is very important, not only during orbital injection and the positioning phase, but also throughout the operational fife of the satellite. During the orbital injection and positioning phase, the telemetry link is primarily used by the tracking system to establish the satellite-to-Earth station communication channel. After the satellite is put into the desired slot in its intended orbit, its mission is to monitor the health of various subsystems onboard satellite. In Fig. 2.28 is shown the block schematic arrangement of the basic TT&C subsystem.

The TT&C system supports the function of spacecraft management for successful operation of payload and bus sub-systems. The main functions of a TT&C are as follows:

1. **Telemetry Sub-System** – The function of telemetry is to monitor various spacecraft parameters and performances such as voltage, current, temperature, output from attitude sensors, reaction wheel speed, pressure of propulsion tanks and equipment status and to transmit the monitored data to the Satellite Control Centre (SCC) on the Earth. At this point, the telemetered data are analyzed at the SCC and used for routine operational and failure diagnosis purposes, to provide data about the amount of fuel remaining, to support determination of orbital parameters, etc.

2. **Tracking Sub-System** – The function of tracking is to provide necessary sources to Earth stations for the tracking and determination of orbital parameters. In such a way, to maintain a satellite in its assigned orbital slot and provide look angle information to GES in the network, it is necessary to estimate the orbital parameters regularly. These parameters can be obtained by tracking the communications satellite from the ground and measuring its angular position

and range. Most SCC employ angular and range or range-rate tracking to control satellite orbits.

3. **Command Sub-System** – This sub-system receives commands transmitted from the ground SCC, verifies reception and executes commands to perform various functions of the satellite during its operational mission, such as follows: Satellite transponder and beacon switching, Antenna pointing control, Switch matrix reconfiguration, Controlling direction and speed of solar arrays drive, Battery reconditioning, Thruster firing and Switching heaters of the various systems.

2.6.6 *Propulsion Engine (PE)*

The functions of the propulsion motors are to generate the thrust required for the attitude and orbital control of errors caused by solar and lunar gravity and other influences, or possibly the adequate assistance of the satellite into its final orbit.

Hence, these errors are normally corrected at set intervals in response to commands from SCC. The necessary impulse is provided by thrusters, which operate by ejecting hot or cold gas under pressure. The thrust requirements for orbital control are provided by mono or bi-propellant fuels. The attitude control thrusters are positioned away from the centre of the mass to achieve the maximum thrust, the thrust being applied perpendicular to the direction of a spacecraft's centre of mass.

The orbit control thrusters are mounted so that the thrust vector passes through the centre of mass. The relocation of a satellite from transfer orbit into GEO may be performed by apogee boost motor. In some satellite configurations this is achieved by a solid or liquid fuel engine. Moreover, the choice between these two motors has a significant effect on the internal arrangements of the satellite.

Chapter 3
Baseband and Transmission Systems

This chapter is explaining the theoretical framework of baseband signals and transmission systems between meteorological satellites and ground Earth stations infrastructures as very important segment in reception and distribution of meteorological data. Different early satellite systems, primarily communication and later meteorological were used the analog transmission technique. Although modern satellite communication networks are still using the analog technique, the rapid development of high-speed digital satellite equipment is fostering a new trend toward completely digital satellite transmissions.

This chapter also includes technique and technology that enable data, facsimile and images signals to be transmitted from different satellite sensors to ground weather fixed and mobile infrastructures (Space-to-Earth), fixed service (point-to-point and point-to-multipoint) systems and data from meteorological cites or other platforms on the Earth surface to be relayed via meteorological satellite to a ground central receiving weather sites.

There are several methods of modulation, multiplexing and multiple access techniques used in satellite transmission and their reverse processes, with some overlapping. Modulation is the process by which the baseband signal in EM form can be impressed upon a carrier, so Phase (PM) and Frequency (FM) modulations are used heavily in satellite communications because of their positive ability to deal with nonlinear distortion, noise and interference. The amplitude of the PM and FM carrier is held constant, so there is no apparent change in the power level. Most nonlinear distortion is the result of amplitude variations on the carrier, so PM and FM is able to perform better in this environment than Amplitude Modulation (AM), which has a major limitation in directly using AM for satellite systems.

Multiplexing and multiple access require sharing the resources of the satellite. Facilities are shared based on spectrum assignment (frequencies), by time sharing (time domain) and by spatial separation (antenna beam and polarization). Theoretically, any method can be used for the transmission of analog and digital signals. In practice, frequency is easier used with analog signals, whereas time division is easier to use with digital signals. Multiplexing consists in combining the signals

© Springer International Publishing AG 2018   145
S.D. Ilčev, *Global Satellite Meteorological Observation (GSMO) Theory*,
DOI 10.1007/978-3-319-67119-2_3

from several users into a single signal, which then forms the signal used for modulation of the carrier. After demodulation, the individual signals are separated by an inverse operation called demultiplexing system. Then, all types of coding, decoding, error corrections and compression, including Digital Video Broadcasting-Return Channel via Satellite (DVB-RCS) system and protocols are discussed.

3.1   Baseband Signals

The information transmitted over an RF communications link consist in signals conveyed from one user to another one or two ways transmissions. Such signals are called baseband signals, which consist in basic electrical impulses carried by the radio path of a satellite or via other communication networks. If the baseband signal is analog, the voltage which represents it can take any value within a given range but if the signal is digital, the voltage takes discrete values within a given range.

The type of baseband signals is determined by the communication requirements of the final users and the nature of the network, which can be voice, data and video (image) signals. Note that voice (speech) transmission is not used in meteorological transmissions so far, but perhaps voice satellite channels can be used for transmissions of data, video (image) and facsimile transfer.

Each of these radio signals has to be arranged in a form suitable for transmission over some physical layer (air or wire) and such a technique is called baseband signal processing. The modified baseband signal is then superimposed onto a higher frequency carrier wave, when the signal modulates the carrier to a value suitable for propagation over the many different transmission links, such as radio, satellite, etc. The analog process at the transmit end of the link is called modulation and the circuit that performs modulation is the modulator, while at the receive end of the link, it is called demodulation and the circuit that recovers the baseband information is the demodulator. Digital systems use a modulator for transmission and demodulator for reception of the RF carriers within one unit, known as a modem.

The process of modulation/demodulation applies to radio, satellite and terrestrial links by applying analog or digital transmission methods. The encryption device can be attached to Tx providing secure and secret service from eavesdroppers almost anywhere in the world.

The bandwidth classifiers and a brief description of system characteristics are as follow:

1. **Narrowband** – It uses low data of 10 Kb/s by channel bandwidths of 12.5–25 KHz;
2. **Wideband** – It is developed to provide higher data rate services by channel bandwidths of 50, 100 or 150 KHz and provides data rates that are measured in 100 Kb/s; and

3. **Broadband** – It provides megabits per second of data. The current allocation is 50 MHz of spectrum at 4.9 GHz band.

Here are considered for usage most common baseband signals such as: Telephone or Voice Signal; Data and Multimedia Signals; Sound Signals; and Television or Video Signals.

3.1.1 Voice Signals

The first commercial voice service via satellites was established by Intelsat at a time when underwater cables could not keep pace with the rapid growth of telephone traffic. In spite of data traffic growth in recent years, voice traffic has become dominant and will remain in the future integrated with videophones. In general, signals from most telephone sets are analog but these signals can be transformed into digital mode and multiplexed at a local telephone exchange for transmission on trunk routes.

In addition, satellites are effective for broadcast distribution Voice, Data and Video (VDV) signals anywhere and anytime, including voice (Tel) modems are able to send digital data over satellite voice channels or via an 3.3 KHz analog voice channel interfaced to the landline voice equipment.

Based on extensive studies of voice signals, it is now well known that most of the speech energy during normal conversation lies between 0 and 3400 Hz and the range of voice frequency that can be registered by the human ear is up to about 20 kHz. Therefore, a bandwidth of 3–4 kHz is allocated to each voice channel. The CCITT recommends the range of telephony signals as 300–3400 Hz. The maximum energy of a signal representing speech is in the range of 800 Hz and about 99% of the energy is situated below 3000 Hz. The signal power of an average talker implied to a zero relative level point is given by:

$$P_m = P_a + 0.115\sigma^2 + 10\log\tau \quad [dBm_o] \tag{3.1}$$

where $P_a = -12{,}9$ dBm_o represents the average power of the speech signal; $\sigma = 5.8$ dB is the standard deviation of the normal distribution of the active speech power; $\tau = 0.25$ is the active factor of a talker. Thus, in total the value of $P_m = -15$ dBm_o (Article G.223 of CCITT Recommendation).

The quality of a received analog voice signal has been specified by the CCITT to give a worst-case baseband signal-to-noise ratio for voice (Tel) signals, for transmission over a long distance, as 50 dB and with the maximum allowable noise in the baseband of 10,000 picowatts. Speech is characterized by having a large, dynamic range of up to 50 dB to accommodate the volume difference between a whisper and a shout.

3.1.2 Data and Multimedia Signals

Data transmission is composed using digital signals consisting in a series of bits. Actually, bits are bi-state pulses with low state called "0" ($-V$) and high state "1" ($+V$). Information is superimposed on a digital stream by arranging groups of bits called words, which can be used for the transmission of analog signals from a telephone or alphanumeric characters from a PC keypad. Therefore, facsimile, telex, data and PC E-mail networking terminals are used for the transfer of data through the medium of different transmission techniques and applications that are expanding rapidly.

Data signals used by satellite systems can be broadly classified into three ranges:

1. Narrowband data of about 300 b/s can be transmitted via ground stations TDMA 24 Kb/s Tlx channel with possibly 16 data bursts directed to Ground Earth Stations (GES), Fixed Earth Station (FES) or User Earth Stations (UES) and PSTN via a modem.
2. Full duplex voice and 9.6 Kb/s data signals can be transmitted in the Single Channel Per Carrier (SCPC) data channel with a rate of 24 Kb/s in O-QPSK modulation scheme. In fact, the data channel enables packet data transfer using the CCITT X.25 recommendation for interface Data Terminal Equipment (DTE) and Data-Circuit Terminating Equipment (DCTE) operating in packet mode to the PSDN by dedicated circuits. This channel also supports CCITT Group-3 Fax, which is also available in the SCPC voice channel; using a 2.4 b/s data rate and APC voice codes.
3. Wideband one-way High Speed Data (HSD) transmission can be supported with a rate of 64/56 Kb/s via voice channels on a dedicated frequency with a special type of V.32 (9600 b/s) modem. This service can be used for the transfer of PC data, high quality digital audio and compressed video. The Duplex HSD will offer two-way 492 Kb/s via standard IP service for aircraft in both directions. The two-way HSD can be also used for digital multiplexing data channels and video conferencing between ships and shore.

There are several known techniques in transmitting digital data, facsimile, E-mail and PC networking data applications, whose utility is also expanding rapidly for fixed and mobile systems. However, the transmission systems using portfolio of the HSD service offers a cost-effective extension to corporate LAN and WAN even in the world's most rugged regions. Moreover, packet data transfer allows users to pay according to the amount of data they send rather than how long they spend online suited to Internet web-based applications such as Intranet access and E-commerce via satellite.

Meanwhile, the quality of data service provided by some satellite systems enables speed of transmission up to 7.2 Kb/s and the contributed Bit Error Rate (BER) is less than 1×10^{-6}. Hence, a higher satellite terminal rate of about 9.6 Kb/s can be processed if the equipment incorporates an elastic buffer to accommodate the required flow control.

Therefore, data is becoming the most common vehicle for information transfer related to a large variety of services, including voice telephony, video and PC-generated information exchange. One of the most appealing aspects of data is the ability to combine, onto a single transmission, support data generated by a number of individual sources, which resulting in a single data stream called aggregate traffic. This is paramount for transfer of multimedia traffic integrating VDV signals that often displays a reduced bit rate variance versus traffic generated by the individual sources due to the embedded statistical multiplexing.

A network operator uses this opportunity to dimension its links (and particularly satellite links) with a capacity that is less than the sum of the peak bit rate of the individual sources. This dimensioning considers both the burstiness of the data traffic and the multiplexing techniques in use. The traffic is typically transported by packets, so the data structure (transport stream) developed for conveying TV program components could be used to carry any type of data, taking advantage of the well-recognized standard and mass-market production of equipment.

For example, the Moving Picture Experts Group (MPEG-2) transport stream (MPEG-TS) packets are used with DVB-S data transmission. The MPEG solution is the name of a family of standards used for coding audio-visual information (e.g., movies, video, music) in a digital compressed format. As for video program, MPEG-2 transport stream (MPEG-TS) packet has a fixed size of 188 bytes of which 4 bytes are used for the packet header and 144 bytes for the payload. The header consists of a synchronization (sync) byte, a Packet Identifier (PID), transport error indication and a field of adaptation options.

The Asynchronous Transfer Mode (ATM) data format is used in high data rate, terrestrial networks. The ATM packet, called an "ATM cell", is a fixed size of 53 bytes, of which 5 bytes are used for the header and 48 bytes for the payload. The header consists of a Virtual Channel Identifier (VCI), a Virtual Path Identifier (VPI), a payload type, and priority and Header Error Check (HEC) fields.

3.1.3 Sound (Audio) Signals

A high-quality radio sound program occupies a band from 40 Hz to 15 kHz. The test signal is a pure sinusoid at a frequency of 1 kHz. Its power relative to the zero reference level for an impedance of 600 W is 1mW or 0 dB m0 s (the "s" suffix indicates that the value relates to the sound program test signal). The mean power of a sound program is 3.4 dBm0 and the peak power (exceeded for a fraction less than 10^{-5} of the time) is equal to 12 dBm0.

For the emission of digitally encoded audio signals, the analogue sound program must go through an analogue-to-digital converter. This implies sampling, quantization and source encoding. Two encoding techniques, Pulse Code Modulation (PCM) and adaptive delta modulation (ADM) are considered. Some formats have been defined using sampling rates of 32, 44.1 or 48 kHz of different audio signals.

For instance, where satellites are used for broadcasting of audio program to mobile receivers, such as Digital Audio Broadcasting (DAB) and Orthogonal Frequency Division Multiplexing (OFDM) can be used. This transmits data by dividing the stream into several parallel bit streams, which each at lower bit rate, then uses these sub streams to modulate several carriers. The OFDM time-domain waveforms are chosen such that mutual orthogonality is ensured, even though the spectra from several subcarriers may overlap. Thus, Coded OFDM (COFDM) is able to combat the time dispersion due to multipath, frequently encountered over mobile satellite channels.

3.1.4   Video and Television Signals

Prior to the advent of satellite communications it was very impracticable to relay video and TV programs across oceans because of the limited bandwidth of submarine coaxial cables. Today optical fiber cables have enough channels but cannot, like satellite systems, transmit Digital Broadcasting Satellite (DBS) TV video to many widely separated fixed or Very Small Aperture Terminal VSAT stations in rural or remote areas and especially to mobile terminals. Direct broadcast by satellite has been in use about 15 years, because the first such European satellite was launched in 1987. The demand for this type of satellite service continues to gain interest for the distribution of TV programs direct to homes (households) or other fixed stations and later, to provide a service for mobile stations. There are various inexpensive ways in which it is possible to add data transmissions to the DBS and to provide cheap information services and E-mail for both domestic and business purposes, to onshore or offshore TV subscribers.

More then a decade ago, some satellite systems started to provide still pictures mode from the oil fields to any office using a wide range of analog and digital portable devices for transmitting, receiving and processing high resolution color photographs in less then 5 min. In addition, these satellite systems also developed the HSD Store-and-Forward video system. Namely, with the aid of advanced video codecs it is possible to digitize and compress video material and then transmit it at 492 Kb/s to a receiver, which decompresses and buffers the data in almost full motion video. The store-and-forward technique also ensures that the received material is error-free, since the data transmission is achieved by using error detection and repeat transmission mode.

There is a new possibility recently to use Integrated Services for Digital Network (ISDN) interfaces as well for the transfer of intensive data interactive applications such as Video Conferencing (VC) and Digital Image Transfer (DIT). Therefore, utility of standard ISDN interfaces will enable one to easily connect mobile fleet offices with corporate applications. Video baseband allows satellite users to relay video from almost anywhere on the globe. This provides an effective flexible solution for media companies and the growing number of corporation who also use the VC system to facilitate "real time" discussion on a regular basis. Users benefit from ISDN speeds and easy-to-use portable equipment, which delivers immediate live action, training or medical and equipment diagnostic support, as required.

The Multiplexed Analogue Components (MAC) standard was proposed in the 1980s for satellite broadcasting (DBS) television and video, which was never developed into a successful commercial service. Fully digital techniques based on video compression were developed in the 1990s resulting in the widely recognized MPEG standards. The TV has a baseband signal of a few Mb/s, while transmission of digital TV and video is possible without requiring a huge amount of radio-frequency spectrum. At the same time, standards for DVB-RCS, and in particular the satellite versions of (DVB-S and DVBS2, have been adopted, making satellite broadcasting television a successful commercial service.

3.1.5 Basic Concept of Modulation

Modulation is when the signal message that is to be conveyed is transposed in baseband, signals, such as voice or raw PC data, into a suitable form so that it can be transmitted over the media involved, usually plain air with absorbing gases as described earlier.

The raw data or voice signal can be modulated onto a carrier frequency as shown in Fig. 3.1 (a), while both demodulated signals are illustrated in Fig. 3.1 (b). A frequency carrier is selected to suit propagation conditions, and then spectrum and frequency management constraints and ITU radio regulations and efficient and reliable transportation of the information across the media. There are mathematical equations that describe this modulation process. The reverse process demodulation is when the useful information is extracted from the unuseful carrier.

There are many different types of modulation schemes in transmission systems available to the radio designers and engineers. In principle, these can be broken down into analogue modulation such as Amplitude or Frequency Modulation (AM or FM) schemes or variants on these, as can be used to modulate voice signals. More recently, digital modulation such as Frequency Shift Keying (FSK), Phase Shift Keying (PSK) or advanced variants of these can be used to convey digital messages and/or digitized voice streams, or of course both. The selections of appropriate modulation schemes are usually dependent on. but not necessarily limited to, the following factors: (1) Required ranges of propagation; (2) Frequency of operation and propagation properties; (3) Spectral efficiency; (4) Equipment complexity, reliability, size weight and cost; and (5) Required data or voice throughput rate; regulatory constraints.

The Modulation Conundrum is illustrated in Fig. 3.2, how baseband unmodulated signal is transforming into modulated baseband signal in frequency domain. In such a way, there is a relationship between all of these aspects, mathematically it is presented by the known Claude Shannon's law, and it is usually necessary to trade off one requirement against the others to achieve the best overall

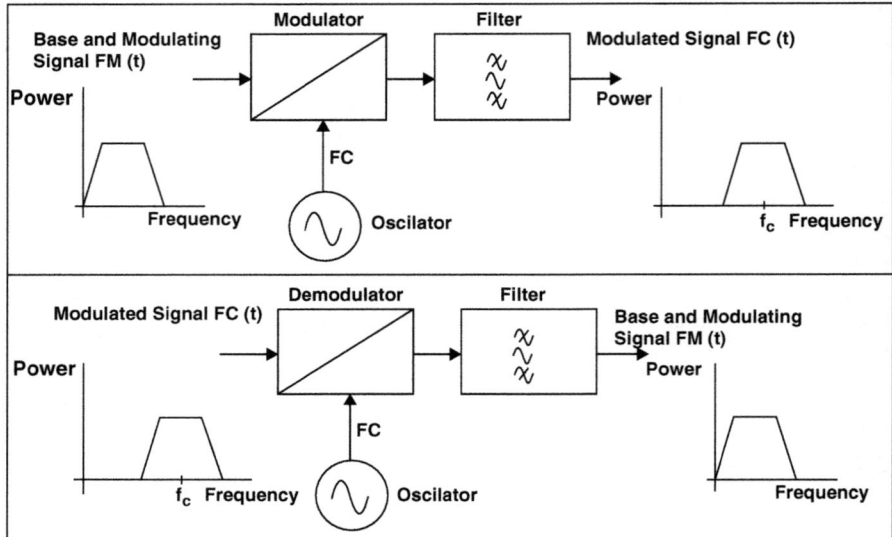

Fig. 3.1 Basic modulator and demodulator (Courtesy of Book: by Stacey)

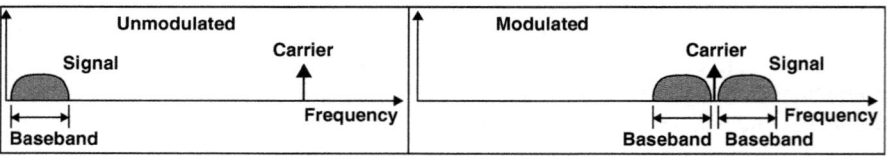

Fig. 3.2 Frequency domain (Courtesy of Book: by Stacey)

engineering solution. Thud, for low-specification systems where performance is not so critical, rugged and rudimentary low-order modulation schemes can work very efficiently, and this keeps cost and complexity down. For more sophisticated systems where data throughput in a limited bandwidth is premium, some of the more sophisticated and elaborate modulation schemes pushing the limits of Shannon's law become applicable if not essential.

3.1.6   Analog and Digital Domains

Before coming into the details of the modulation schemes it is worth momentarily looking at the Analog-to-Digital (A/D) aspects of signal modulation. Modern satellite systems generally digitize voice, data and images and transmit them via a digital medium, where quality can generally be kept, before putting them back in their original analogue domains. With data there is never any A/D and it stays

Table 3.1 Analog and digital domains

Analog-to-analog conversion (analog speech to analog radio)	Digital-to-analog conversion (data or digitized speech to radio)
Amplitude modulation	Amplitude shift keying
Frequency modulation	Frequency shift keying
	Phase shift keying

entirely in the digital domain. The modulation process can be considered as the inverse of this process; i.e. data is modulated onto a carrier conversion, which is akin to Digital-to-Analog (D/A) version, and then it is transmitted over the media and converted back to digital via A/D conversion at the end, shown in Table 3.1.

With voice over legacy radio systems, it is never even coming into the digital domain. It is analog when it starts, it goes through an analog-to-analog conversion as it is modulated and then at the end it is an analog-to-analog demodulation process.

Analog modulation is supporting emissions of Double Side Band-Amplitude Modulation (DSB-AM), Single Side Band-AM (SSB-AM), Suppressed Carrier-AM (SC-AM) and Frequency Modulation (FM) modulation scheme is supporting emissions of Phase Shift Keying (PSK), Frequency Shift Keying (FSK), Differential Phase Shift Keying (DPSK), Differential Frequency Shift Keying (DFSK), Quadrature Amplitude Modulation (QAM) and Trellis Coded Modulation (TCM).

3.2 Analog Transmission

Analog transmission in terrestrial or satellite telecommunication systems is characterized by processing performed on the baseband signal before and after modulation in order to improve the quality of the link. The carrier can support only one or few channels for the transmission of baseband signals. In the case of a carrier transmitted from the station representing only a single user channel, this is SCPC transmission. On the other hand, if the frequency carrier represents a number of multiplexed users, it is designated Multiple Channel Per Carrier (MCPC) and use several channels is Frequency Division Multiplexing (FDM) transfer.

A signal needs to be impressed on an RF carrier for transmission through the satellite and for this purpose uses a process known as modulation. The objective of any communications is to transmit the modulated carrier to Rx as reliably as possible, so that the demodulated signals can be satisfactorily recovered. In analog transmission systems, the information waveform in the form of voice, data or video signal are modulated directly from the source onto the carrier at the modulator of Tx, by using methods of Amplitude (AM), Frequency (FM) and Phase Modulation (PM). Frequency and phase modulation are used most widely for direct analog transmission in satellite communications, while amplitude modulation is a process used indirectly in the satellite link.

Modulation may also be used at very low frequencies, like the more common and various forms of PSK modulation scheme, which is often done on a carrier with an RF of about 70 MHz lower than the transmission RF. This RF is then up-converted to the transponder frequencies on 6/4 GHz for amplification and retransmission. Previous types of satellite do not change the received modulation before retransmission, while new satellites are being designed to allow only one modulation method to be employed in the uplink and another for the downlink, both optimized. The space link between two GES can be accomplished by the combination of modulation and multiplexing techniques. In the context of analog modulation will be included specific schemes as well as DSB-AM, SSB-AM and SC-AM.

As stated earlier, in satellite data communications, analog signals are used to transmit data over the telephone or voice systems. Analog signals can be converted to the digital using a codec (coder/decoder), which process is called digitizing. Phones that connect to all-digital communication links use codecs to convert analog voice signals to digital signals.

3.2.1 Baseband Processing

The purpose of baseband processing is to improve the quality of the space (satellite) link using different methods whose cost is less than that arising from modification of one of the parameters involved in the link budget. The principal methods of baseband processing for telephone transmission are speech activation, pre- and de-emphasis and companding and for TV transmission pre- and de-emphasis only.

1. **Speech Activation** – The principle of speech activation is to establish the space link only when the subscriber is actually speaking. As the activity factor is $\tau = 0.25$, its application to a multicarrier SCPC system should permit a reduction of the power required on the satellite by about 6 dB. In practice the reduction is only in the order of 4 dB, allowing guard times for activation and deactivation of the carrier and the sensitivity of S/N spikes. Therefore, the activation threshold parameter has to be from -30 to -40 dBm$_o$, the carrier activation time can be 6–10 ms and deactivation time can be 150–200 ms.
2. **Pre- and De-Emphasis** – The noise at the output of the demodulator of a FM transmission has a parabolic spectral density, when the high frequency components of the signal are more affected by noise than the low frequencies.
3. **Companding Process** – Companding is a process of compression and expansion for reducing the effect of noise on speech channels in accordance with the specification of CCITT Recommendations G.162 and article G.166. In effect, a compandor comprises a compressor and an expander and an improvement in the S/N ratio at the output of the demodulator is obtained by reducing the dynamic range of the signal before modulation (compression) and performing

the inverse operation after demodulation (expansion) in order to restore the original speech signal to its correct relative level.

3.2.2 Analog Modulation and Multiplexing

Source waveforms of analog signals are directly modulated onto RF or IF carriers at the transmitter, using any form of three types of modulation: AM, FM or PM.

Analog modulation is a process in which some characteristics of a HF carrier are varied in accordance with the baseband signal, which is the creation, transmission and reception of EM fields. In reality, EM fields and waves can be used to communicate all kinds of information from place to place. Therefore, an analog modulated signal can be represented as the following sinusoidal wave:

$$c(t) = A \cos[2\pi f \, t + \varphi] \tag{3.2}$$

where A = amplitude, $2\pi f$ = form of ω_c = angular frequency of the carrier, f = frequency and φ = carrier phase. Modulation can be achieved by altering the amplitude, frequency or phase of the wave in accordance with the information signals. Consequently, AM, FM and PM analog carriers are simply forms of modulated carriers in which either the amplitude, frequency or phase is modulated by the information waveform. In such a way, by changing the amplitude, frequency or phase results in AM, FM or PM, respectively, which are illustrated in Fig. 3.3. Thus, since EM fields are vector fields, it is also possible to modulate the polarization of a wave to communicate information. However, polarization modulation is only normally used for special purposes, as the polarization state of a wave can easily be "scrambled" during free space transmission by effects such as unwanted reflections from buildings.

The AM system is a type of linear modulation in which the baseband signal is linearly related to the modulated signal, while FM and PM are kinds of angle modulation in which the baseband signal is angularly related to the modulated signal. However, angle modulated FM and PM signals require more carrier bandwidth than AM but achieve a higher demodulated C/N ratio for the same carrier C/N value. The reverse process of recovering the information signal from the modulated carrier at the demodulator of the receiver is known as demodulation. Multichannel operation of FM for analog high-capacity satellite telephone transmission is normally accomplished by a system called FDM scheme. In such a system different channels are separated from one another by being assigned different subscribers, which are then combined to fill the total bandwidth of the transponder.

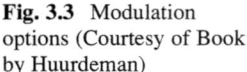
Fig. 3.3 Modulation
options (Courtesy of Book:
by Huurdeman)

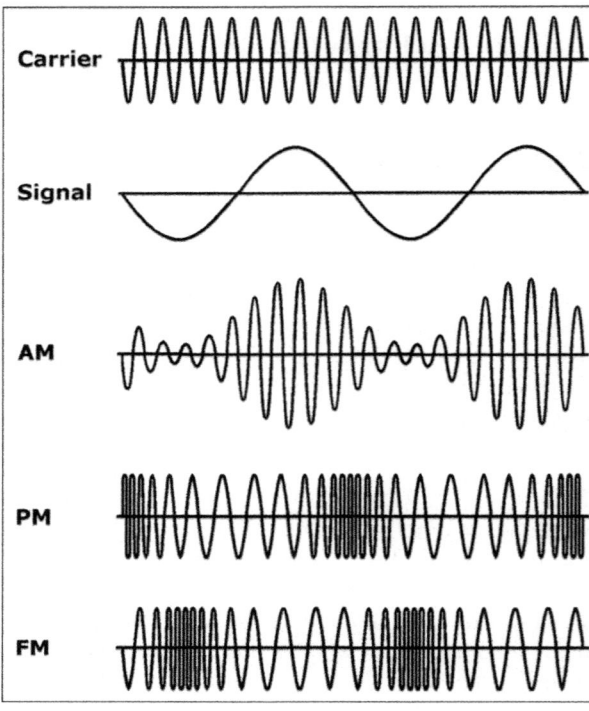

3.2.2.1   Amplitude Modulation (AM)

Amplitude modulation is a type of linear modulation and is not used as a transmission modulating process in the satellite link. However, it can be used to modulate individual voice channels before combining them using FDM technique.

In AM, the amplitude of the carrier frequency is modified by the amplitude of a modulating signal, namely a carrier has to be amplitude modulated when the amplitude of the carrier varies in accordance with the signals. The side bands are displaced at either side of the carrier frequency by the magnitude of the modulation frequency and the amplitude depends on the modulation amplitude. The fundamental equation of AM modulation signals is:

$$c(t) = A[1 + \Delta_a m_{AM}(t)] \, \cos{(\omega_c t + \varphi)} \tag{3.3}$$

where $m_{AM}(t)$ = AM signal represents the information waveform to be transmitted and Δa = AM index (coefficient giving the degree of modulation of angular frequency of the carrier).

The AM method is employed usually in AM radio broadcasting and radiocommunication transmissions. In these systems the intensity or amplitude of the carrier wave varies in accordance with the modulating signal. When the carrier is thus, modulated, a fraction of the power is converted to side band extending above and below the carrier frequency by an amount equal to the highest modulating frequency.

If the modulated carrier is rectified and the carrier frequency filtered out, the modulating signal can be recovered. This form of AM is not a very efficient way to send information because the power required is quite large and because the carrier, which contains no information, is sent along with the information. In AM, the information is carried only in the side bands and therefore power in the carrier remains unutilized. In a Double Side Band Suppressed Carrier (DSB-SC), modulation of the carrier is suppressed and only side bands are used. The amplitude of this wave does not follow the signal amplitude and consequently, the inherent simplicity of using envelope detection is lost. Therefore, this modulation is not used in satellite communications.

In a variant of amplitude modulation, called Single Side Band (SSB), the modulated signal contains only one side band and no carrier. Hence, the information can be demodulated only if the carrier is used as a reference. This is normally accomplished by generating a wave in the receiver (Rx) at the carrier frequency. The SSB is used for long-distance HF radiotelephony and telegraphy over land, such as in the maritime HF radio frequency bands and submarine cables. In the other words, the SSB scheme is called an SSB Suppressed Carrier (SSB-SC) when the carrier is suppressed.

In fact, the most common application of SSB modulation in satellite communications is to multiplex voice carriers into a composite baseband signal. Thus, the required C/N ratio and the occupied bandwidths are two aspects considered in assessing the suitability of SSB for satellite transmission. The occupied RF bandwidth of an SSB transmission is the same as the baseband bandwidth, which is 4–5 kHz for a single telephone channel transmission. Typical bandwidth of −30 kHz is necessary for FM and 20 kHz for the O-QPSK, so both schemes are widely used in satellite communications.

However, a typical satellite communication link can economically provide C/N ratios in the order of 10–12 dB, making SSB transmission inefficient from power considerations. That is to say, this disadvantage can be offset to a large extent by the use of compandors, which offer an S/N ratio advantage of about 15–20 dB and in such a manner, SSB transmission appears attractive. This scheme is called Amplitude Companded SSB (ACSSB). Bandwidth efficiency is essential for mobile satellite service and for this reason ACSSB has been considered favorably for such applications.

3.2.2.2 Frequency Modulation (FM)

In FM scheme, the amplitude of the carrier frequency is modified by the frequency deviation of a modulating signal. The side bands are displaced at either side of the carrier frequency with an infinite number of separate bands, most resolving near zero. The amplitude depends on the modulation frequency change and the frequency depends on the rate of frequency change of the modulation. Therefore, the fundamental equations of FM modulation are given below:

$$c(t) = A \, \cos \left[\omega_c t \, 2\pi\Delta_f \int m_{FM}(t)dt + \varphi \right] \qquad (3.4)$$

where $m_{FM}(t)$ = FM signal representing the information waveform to be transmitted and Δf = FM index (coefficient giving the degree of modulation angular frequency of the carrier).

Using FM the frequency of the carrier wave is varied in such a way that the change in frequency at any instant is proportional to another signal that varies with time. The FM band has become the choice of music listeners and for the audio portion of TV broadcasting because of its low-noise and wide-bandwidth qualities; however, the FM schemes are also extensively used in satellite communications. Examples of FM applications are multiplexed telephony, SCPC systems and TV broadcasting.

Moreover, FM satellite systems are well suited for those cases where the baseband signal is in analog form. An example is FM with companding used in Rx and Tx voice channels of the analog satellite system of the Earth station transceiver. For instance, this scheme also offers advantages for transmission of digital data in applications where simple receivers are essential, such as the satellite paging system. At any rate, an important requirement of a paging system is the need for simple, low cost and rugged receivers.

3.2.2.3 Phase Modulation (PM)

Phase modulation (PM) like frequency modulation is a form of angle modulation so-called because the angle of the sine wave carrier is changed by the modulating wave. The two methods are very similar in the sense that any attempt to shift the frequency or phase is accomplished by a change in the other. The PM relation can be expressed with:

$$c(t) = A \, \cos \left[\omega_c t + \Delta_p m_{PM}(t) + \varphi \right] \qquad (3.5)$$

where $m_{PM}(t)$ = PM signal representing the information waveform to be transmitted and Δp = PM index (coefficient giving the degree of modulation angular frequency of the carrier).

3.2.3 Double Side Band-Amplitude Modulation (DSB-AM)

As is known AM is when the amplitude of the carrier is directly proportional to the modulating signal. Probably the simplest form of AM is DSB-AM mode, which

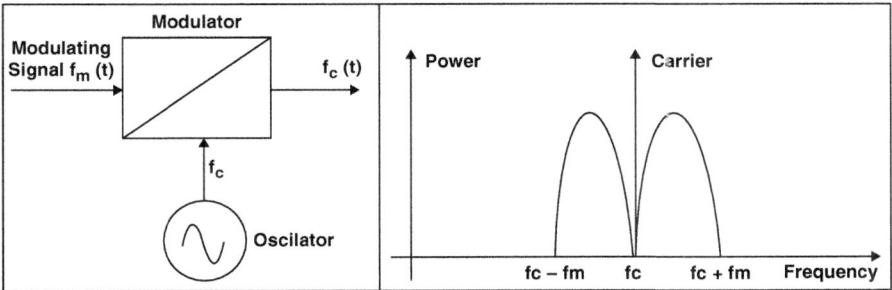

Fig. 3.4 Basic AM modulator and power spectrum for DSB-AM (Courtesy of Book: by Stacey)

diagram is illustrated in Fig. 3.4 (Left). The DSB-AM emission is described mathematically with the following equation:

$$f_c(t) = A \cos (2\pi f_c t + \varphi) \pi \tag{3.6}$$

where $f_c(t)$ is the modulated signal output. From this relation is following equation:

$$A = K + f_m(t) \tag{3.7}$$

where K is the unmodulated carrier amplitude and $f_m(t)$ is the Baseband signal, and than is getting the following relations:

$$f_m(t) = a \cos (2\pi f_m t) \tag{3.8}$$

$$f_c(t) = [K + a \cos (2\pi f_m t)] \cos (2\pi f_c t + \Phi) \tag{3.9}$$

When $\Phi = 0$, then multiplying out the cos function:

$$f_c(t) = K\{ \cos (2\pi f_c t) + 0.5 \ m \ \cos [2\pi \ (f_{c-}f_m) \ t] + 0.5 \ m \ \cos [2\pi \ (f_c + f_m)t] \} \tag{3.10}$$

At this point, dept. of modulation (m) = modulating signal amplitude (a)/ unmodulated carrier amplitude (K). In the frequency domain, these discrete components can be seen in Fig. 3.4 (Right), which are called upper and lower sidebands, respectively. Thus, it is interesting to note that the modulated bandwidth of DSB-AM is twice the size of the minimum baseband bandwidth or the frequency component, and that it is independent of depth of modulation.

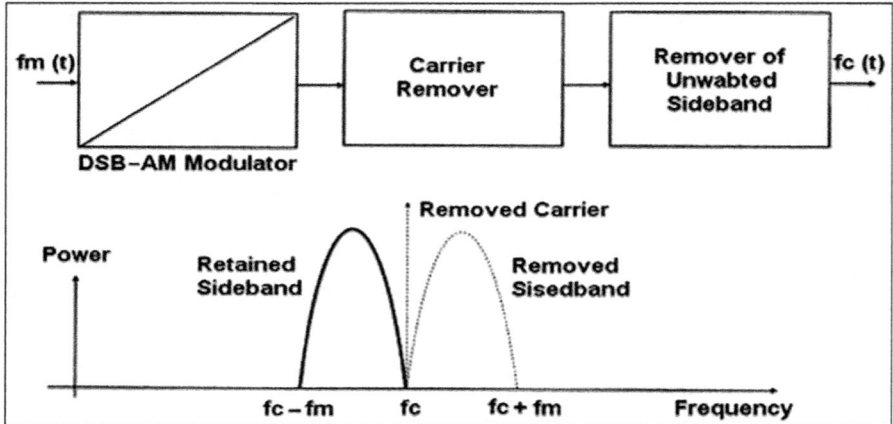

Fig. 3.5 SSB modulation block diagram and power spectrum for SSB-AM (Courtesy of Book: by Stacey)

3.2.4   Single Side Band-Amplitude Modulation (SSB-AM)

With DSB-AM practical work, it is found that the upper sideband is a perfect reflection of the lower sideband; i.e. the same information is carried by both. Thus, by removing one of the sidebands and by removing the carrier, therefore, no information is lost and the S/N (signal-to-noise radio) increases for a given transmit power as all the transmitted energy is concentrated into the information signal. This is the basis for SSB-AM block diagram and power spectrum illustrated in Fig. 3.5.

The construction of an SSB is fractionally more complicated than that of DSB-AM. It can be in synthesized two different ways.

Method 1 – As SSB indirect synthesis and filtering, a SSB signal can be synthesized either by using basic DSB-AM or DSB-SC-AM modulator and filtering out the unwanted carrier and sideband (usually done at a IF).

Method 2 – Arguably a more elegant way of modulation is to use Hilbert modulator as SSB direct synthesis, which essentially cancels out the unwanted sideband and carrier.

Mathematically, this is described with the following quotation:

$$V_{out} = 0.5k \, \cos \left[2\pi \left(2f \pm f_m \right)t + \varphi \right] \pm 0.5k \, \cos \left(2\pi \, f_m t - \varphi \right) \qquad (3.11)$$

The HF component is filtered out and expressing:

$$\begin{aligned} V_{out} &= 0.5k \, \cos \left(2\pi \, f_m t - \varphi \right) \\ &= 0.5k \, \cos \left(2\pi \, f_m \right)t \cos\Phi \pm 0.5k \, \sin \left(2\pi \, f_m \right)\sin\varphi \end{aligned} \qquad (3.12)$$

More sophisticated demodulation is required (coherent detector with PLL, costas loop).

It can be prone to phase distortions (due to Tx and Rx oscillators beating and Φ changing), so high oscillator specification is required at both ends. Not really heard by listeners (quality not high). The HF and other SSB systems are using SSB modulation for reasons of spectral efficiency and a limited resource. Also a number of earlier military systems are known to have used SSB as a basis of amplitude modulation, which uses transmitter power and bandwidth more efficiently. Suppressed Carrier Double Side Band AM (SCDSB-AM) is modulation technique also used in fixed and mobile transmissions.

Worth mentioning in passing, there is an intermediate version of modulation called suppressed carrier double side band AM, which lies between DSB-AM and SSB-AM. In its basic form it is the same as DSB-AM, with the carrier filtered out. It has the advantage of concentrating more power into the useful sidebands, so there is an S/N improvement; however, there is no bandwidth saving and it requires a more complex synthesis. It is rarely used, as most applications will do a full conversion to the SSB, where power and spectral efficiency become important and there are no mainstream applications of this in the some mobile bands.

3.2.5 Frequency Division Multiplexing (FDM)

The process of combining baseband signals and sharing the communication channels is known as multiplexing mode, while the reverse process of extracting individual baseband signals is called demultiplexing. In these circumstances, the CCITT (In French: Comité Consultatif International Téléphonique et Télégraphique) proposed all multiplexing standards including FDM, which is applicable only to telephony baseband signals. Whereas, the digital Time Division Multiplexing (TDM) standard is applicable to all types of baseband signals.

Each type of GES usually transmits and receives many kinds of signal transmissions to and from spacecraft. Multiplexing enables the division of channels or the combination of two or more input signals into a single output for transmission. The most common and known analog multiplexing method is FDM, used in satellite communication transmissions. The simplest approach to multiplexing is to assign a specific part of the available frequency or bandwidth spectrum to each signal.

If two signals initially have the same spectrum, the frequency of one or both is shifted, so in such a way they will not overlap. The FDM is a multiplexing solution where signals occupy the channel at the same time but on different frequencies, namely different message signals are separated from each other in frequency. The multiplexing scenario of FDM is shown in Fig. 3.6, which illustrates simultaneous transmission of three message signals over a common communication channel.

It is clear from the block schematic arrangement shown that each of the three messages signals modulates a different carrier.

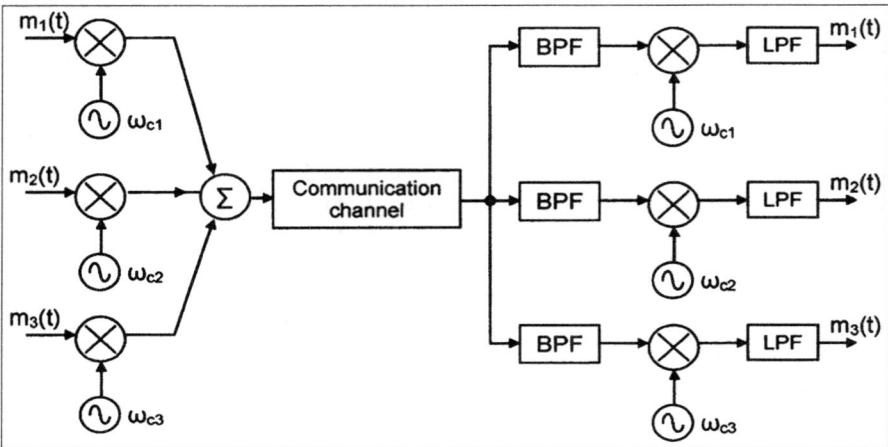

Fig. 3.6 Frequency division multiplexer (Courtesy of Book: by Maini)

As stated earlier, frequency modulation is when the carrier frequency is modulated by the baseband signal. The FM mode is used for some specialized military systems, in mobile telemetry and finally in the adjacent band to the VHF communications for broadcasting information. It is mainly used by broadcasting in the band between 88 and 108 MHz, just below the VHF navigation and communication bands 108.000–117.975 MHz and 117.975–137.000 MHz respectively. However, a clear understanding of FM is potentially useful for adjacent channel compatibility work. In addition, it is synthesized using a voltage-controlled oscillator. Also it is more susceptible to Doppler shifts, which even although these have been shown to be very minor in a previous section, make it impractical for mobile communications. Historically, FM gives a better quality of voice service. The FM modulation is used by entertainment systems on board aircraft where channels are FDM into time slots. Otherwise, the most common used modulation technique in fixed and mobile telephony is SSB.

3.3 Digital Transmission

Before providing any explanation about solutions of digital transmission will be necessary to introduce some relations:

1. **Shannon's Theory** – Shannon's theory provides a theoretical relationship between bits per second, baud and S/N for error-free transmission presented in Table 3.2.

Table 3.2 Shannon's law

M	2^M	Minimum S/N required	Maximum spectrum efficiency (maximum capacity)
1	2	3	1 bits/Hz
2	4	15	2 bits/Hz
3	8	63	4 bits/Hz

$$R \ (b/s) = M \ (\text{number of signal states}) \times r \ (\text{baud or signal rate}) \qquad (3.13)$$

$$C = B \ \log_2[(1 + S/N)] \qquad (3.14)$$

where C is channel capacity in b/s, B is bandwidth in Hz, S/N is signal to noise ratio, R is the data rate and is always $< C$ for errorless transmission. If $R > C$, errors will occur. Theoretical relationship between M and S/N is as follows:

$$M = [(1 + S/N)]^{0.5} \ \text{ or } \ S/N = M^2 - 1 \qquad (3.15)$$

2. **Non-Errorless Transmission** – Of course Shannon's law does not apply if it is decided to push past this theoretical limit of data throughput if errors can be tolerated. Actually it is sometimes a prudent approach to tolerate the errored environment and to make facilities for correcting these errors by Forward Error Correction (FEC) and Cyclic Redundancy Coding (CRC); of course FEC and CRC too are overhead and not useful data payload.

However, this can be a more efficient way to operate and maximize data throughput. Alternatively the errors can just be tolerated and the application layer of the data system can be designed to tolerate this or to correct it. Also in a number of digital systems, errors will be present from the propagation problems already present and these have to be tolerated.

Digital transmission relates to the link for which the GES is designed to produce the digital signals by PC or modem and to send them through Tx. It is possible to transmit analog signals (voice or broadcast) in digital form. Although this choice implies an increased baseband, it permits signals from diverse origins to be transmitted on the same satellite channels and the satellite link to be incorporated in ISDN, which implies the use of TDM. The digitization of analog signals implies the stages of sampling, quantization and coding. The simple digital transmission chain includes a transmitting and receiving segment. The first unit of the transmitting segment is TDM with input signals from digital (direct) and analog (via encoder) sources and after multiplexed signals pass devices such as: Data Encryption, Channel Encoding, Scrambling and Digital Modulation, where the digital signal is transmitted through up or down satellite links. In the receiving segment of reverse mode incoming signal goes through devices such as: Demodulator, Descrambling,

Channel Decoding, Data Decryption and TDM Demultiplexing in its direction to different users.

Digital signals use the same principle to modulate a carrier as analog signals. The AM, FM and PM schemes are all applicable to digital modulation; digital equivalents are ASK, FSK and PSK and special hybrid modulation solutions have also been developed to optimize digital modulation, such as QAM.

3.3.1 Delta Modulation (DM)

Analog signals, such as speech and video signals, generally have a considerable amount of redundancy, namely, there is a significant correlation between successive samples. Thus, when these correlated samples are coded as in the Pulse Code Modulation (PCM) system, the resulting digital stream contains redundant information. The redundancy in these analog signals makes it possible to predict a sample value from the preceding sample values and to transmit the difference between the actual sample value and the predicted sample value estimated from the past samples. This result in a technique called difference encoding. One of the simplest forms of this is DM, which provides a staircase approximation of the sampled version of the analog input signal.

This type of modulation in digital technique is a way of digitizing a voice waveform, transmitting the digits and reconstructing the original analog waveform that avoids the quantizer and the A/D and D/A converters employed in PCM. In Linear Delta Modulation (LDM) a circuit of DM determines the difference between an incoming waveform or message signal m(t) and estimated waveform or error signal e(t), where a difference of signal error voltage is as:

$$\Delta m(t) = m(t) - e(t) \qquad (3.16)$$

The quantizer output is a positive constant when $\Delta m(t)$ is positive and vice versa. Namely, the difference between the input and the approximation is quantized into two levels, $+\Delta$ and $-\Delta$, corresponding to a positive and a negative difference, respectively. Moreover, at any sampling instant the approximation can be increased or decreased by step size (Δ) value, depending on whether it is below or above the analog input signal.

Furthermore, a digital output of 1 or 0 can be generated according to whether the difference is $+\Delta$ or $-\Delta$. These pulses go to a conventional PSK digital modulator for transmission. Increasing the step size (Δ) will result in poor resolution and increasing the sampling rate will lead to a higher digital bit rate.

In LDM the step size is fixed at a value that provides performance near the peak, while better final performance may be achieved through a scheme of Adaptive Delta Modulation (ADM) in which the value of step size is varied during the modulation process. On the other hand, is difficult to make comparisons between the performance of PCM and DM because the latter is continually improving.

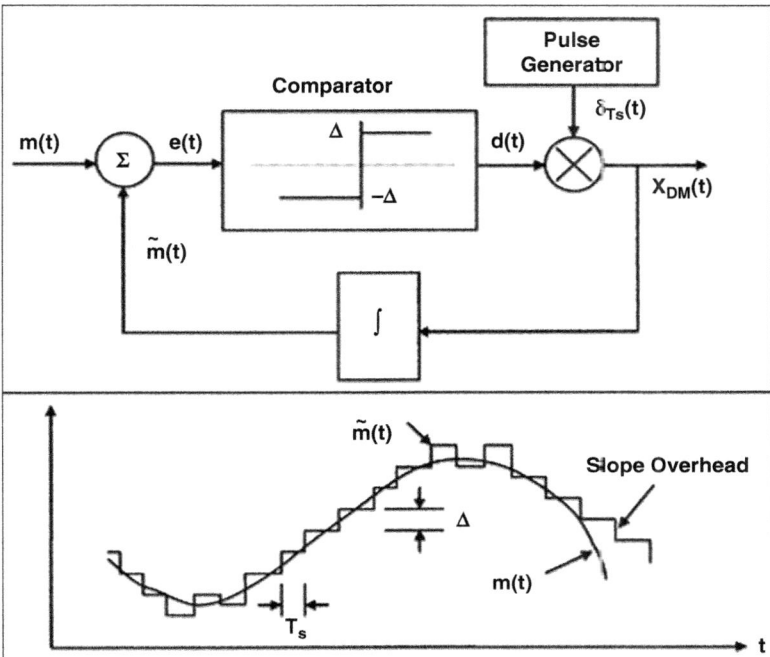

Fig. 3.7 DM and output waveform of DM system (Courtesy of Book: by Maini)

The DM mode has various forms, so in one of the simplest form, only one bit is transmitted per sample just to indicate whether the amplitude of the current sample is greater or smaller than the amplitude of the immediately preceding sample. It has extremely simple encoding and decoding processes but then it may result in tremendous quantizing noise in case of rapidly varying signal. In Fig. 3.7 (Above) is illustrated a simple delta modulator system, where m(t) is added to a reference signal with the polarity shown, while in Fig. 3.7 (Below) is presented its output waveform.

The reference signal is integral part of the delta modulated signal. The error signal is fed to comparator, and at output of comparator is $(+\Delta)$ for $e(t) > 0$ and $(-\Delta)$ for $e(t) < 0$. At this point, the output of DM techniques are a series of impulses with the polarity of each impulse depending upon the signal of $e(t)$ at the sampling instants of time. A DM signal can be demodulated by integrating the modulated signal to obtain the staircase approximation and then passing it through lower filter. The smaller the Δ value, the better is the reproduction of the message signal.

3.3.2 Coded Modulation (CM)

The CM is a combination of modulation and error correction codes without degrading the power of bandwidth efficiency. Thus, using FEC, such as block and convolutional codes, the bit error performance is improved by expanding the required bandwidth.

Obtaining the power efficiency requires twice the bandwidth of the original uncoded signal because of the increase in the symbol rate of modulation and complex implementation. This mode can be done by increasing the number of phases in PSK modulation without expanding signal bandwidth. In this case, the 8-PSK signals with 2/3 rate of convolutional code have the same bandwidth as uncoded 4-PSK. However, the bit error performance degrades by about 4 dB due to an increase in phase but will be referable if the coding gain becomes more than 4 dB. There are two practical CM in use for MSS: the Trellis Coded Modulation (TCM) and the Block Coded Modulation (BCM).

3.3.2.1 Trellis Coded Modulation (TCM)

The TCM scheme uses the combination of convolutional coding and expanded signal sets of 8-PSK to transmit two information bits per symbol. The modulation signals in TCM are assigned to each one- or two-satellite trellis branch, although binary code symbols are assigned in the convolutional codes. Here is a very important definition of measuring the distance between modulation signals assigned to each trellis branch. The modulation signal assignment in TCM can be designed either by the Euclidean or Hamming distances.

Initially the modulator states are based on a trellis shown in Fig. 3.8 (a). This scheme is intelligent and involves more signals processing theory than the simpler prior modulation schemes. This sample works on the principle of putting states that are likely to occur simultaneously as far apart on the constellation diagram shown in Fig. 3.8 (b), with maximum distance of separation employing Euclidean distance. This sample is showing variable phase and amplitudes determining Euclidean distance as the distance between two points in constellation. Similarly a binary sequence that is unlikely to occur can be put together on the constellation diagram with minimum Euclidean distance. This offers an effective coding gain over the simpler QAM and for a given BER and data throughput would consequently require a lower SNR than QAM.

The performance of TCM scheme can be improved by increasing the number of states or by modifying the signal constellation. Thus, one solution is provided by multidimensional signals and another is performed by multiple coded modulations, known as Multiple TCM (MTCM). A TCM 8-DPSK modem with a rate of 2/3 and 16 states for 4800 b/s has been implemented in the NASA MSAT-X experimental program for ground communications.

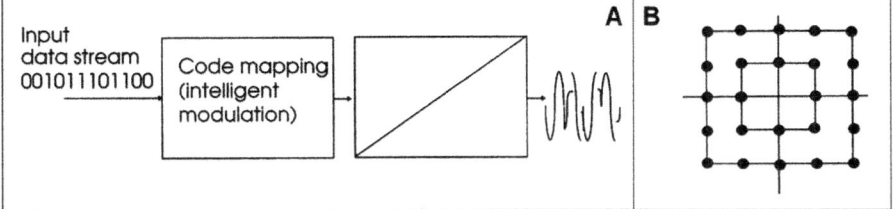

Fig. 3.8 TCM and constellation diagrams (Courtesy of Book: by Stacey)

3.3.2.2   Block Coded Modulation (BCM)

Instead of convolutional coding this scheme uses short binary block codes, which could be simpler and faster to decode. Similarly to TCM, this type of modulation can improve performance by using Multiple BCM (MBCM).

Thus, MBCM with two symbols per branch has a coding gain of 3 dB relative to conventional BCM. The two symbols per branch MBCM has a performance for coding gains of 1.1 dB and 2.2 dB at BER $= 10^{-3}$ relative to the BCM 8-PSK and the uncoded QPSK, respectively.

3.3.3   *Pulse Code Modulation (PCM)*

The digitization starts with the conversion of analog voice signals into a digital format. An analog can be converted into a digital signal of equal quality if the analog signal is sampled at a rate that corresponds to at least twice the signal's maximum frequency.

The peak-to-peak amplitude rate of the modulating signal in PCM is divided into a number of standard levels, which in the case of binary systems is an integral power of 2. The amplitude of the signal to be sent at any sampling instant is the nearest standard level.

So, if at a particular sampling instant, the signal amplitude is 3.2 V, it will not be sent as a 3.2 V pulse, as can be the case of Pulse Amplitude Modulation (PAM), instead it will be sent as the digit 3, if 3 V is the nearest standard amplitude. Thus, where the signal range has been divided into 28 levels, it will be transmitted as 0000011, which coded waveform is shown in Fig. 3.9 (a).

This scheme is known as quantizing process in PCM shown in Fig. 3.9 (b), where the number of bits for 2^n chosen standard levels per code group is n + 1. In fact, it is evident that the quantizing process distorts the signals. This distortion is referred to a quantizing noise, which is random in nature as the error in the signal's amplitude and that actually sent after quantizing is random. The maximum error can be as high as half of sampling interval, which number of level is 16 or 1/32 of total signal amplitude range.

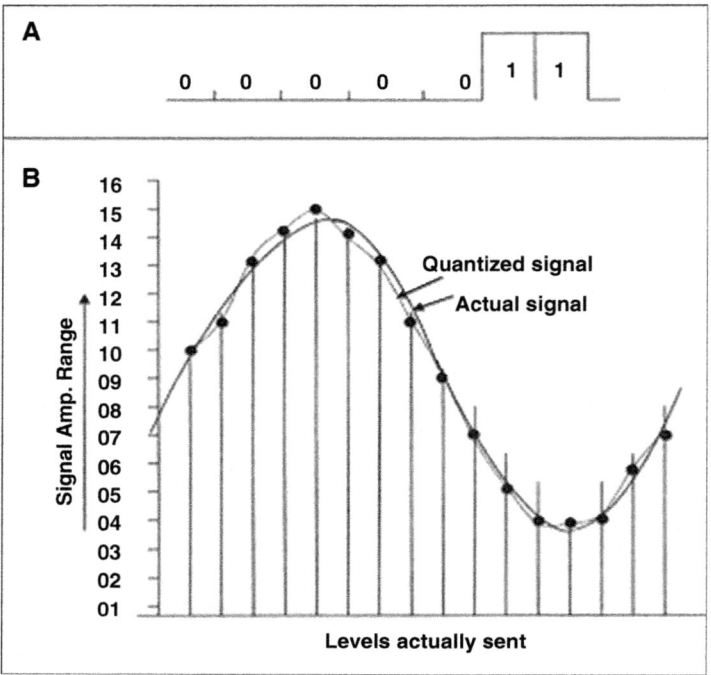

Fig. 3.9 Quantizing process in PCM (Courtesy of Book: by Maini)

A technique for converting an analog signal to a digital transmission form is PCM, which requires three operations:

1. **Sampling** – This operation converts the continuous analog signal into a set of periodic pulses, the amplitudes of which represent the instantaneous amplitudes of the analog signal at the sampling instant. Thus, the process of sampling involves reading of input signals at discrete points in time. Hence, the sampled signals consist in electrical pulses, which vary in accordance with the amplitude of the input signal. In accordance with the Nyquist sampling rate, an analog signal of bandwidth B Hz must be sampled at a rate of at least 1/2B to preserve its wave shape when reconstructed.

2. **Quantizing** – This technique is the process of representing the continuous amplitude of the samples by a finite set of levels. If V quantizer levels are employed to represent the amplitude range it take the $\log_2 V$ bit to code each sample. In voice transmission 256 quantized levels are employed, hence each sample is coded using $\log_2 256 = 8$ bits and thus, the digital bit rate is $8000 \times 8 = 64,000$ b/s. In such a way, the process of quantization introduces distortion into the signal, making the received voice signals raspy and hoarse. This type of distortion is known as quantization noise, which is only present during speech. However, when a large number of quantization steps, each of ΔS

volts, are used to quantize a signal having an rms signal level S_{rms}, the signal-to-quantization noise ratio is given by the following equation:

$$S_{qn} = (S_{rms})^2/(\Delta S)^2/12 \tag{3.17}$$

A large number of bits are necessary to provide an acceptable signal-to-quantization noise ratio throughout the dynamic amplitude range. Some analysis of speech signals shows that smaller amplitude levels have a much higher probability of occurrence than high levels.

3. **Coding** – This solution protects message signals from impairment by adding redundancy to the message signal. Another important approach in digital coding of analog signals is Differential PCM. This is basically a modification of DM where the difference between the analog input signals and their approximation at the sampling instant is quantized into V levels and the output of the encoder is coded into $\log_2$ V bits. In such a way, it combines the simplicity of DM and the multilevel quantizing feature of PCM mode and in many applications can provide good reproduction of analog signals comparable to PCM, with a considerable reduction in the digital bit rate.

3.3.4 Quadrature Amplitude Modulation (QAM)

Each higher PSK modulation requires a better S/N ratio performance, which is difficult to achieve without special schemes. Instead of 16 or higher PSK modulation QAM is used, which is a combination of amplitude and phase modulation. In effect, modulation can be achieved in a similar manner to that of QPSK, by which the in-phase and quadrature carrier components are independently amplitude modulated by the incoming data streams. The incoming signals are detected at the receiver using matched filters.

In terms of bandwidth, it is a highly efficient method for transmitting data flow over the satellite channels. However, the sensitivity of the QAM method to variation in amplitude, limits its applicability to satellite communication systems in practice, where non-linear payload characteristics may distort the waveform, which is resulting in the reception of erroneous messages.

The QAM scheme is a combination of the phase domain. The number of possible states = 360/number of phase changes used. The functional block diagram of modulator for QAM is shown in Fig. 3.10 (a). The demodulation and detection process for QAM signals shown in Fig. 3.10 (b) becomes quite complex, but is not unachievable with modern technology. It is used frequently in high-capacity data point-to-point radio links. 64QAM, 128QAM, 256QAM and even 518QAM are not uncommon.

Obviously, these have high spectral efficiency in the amount of data per unit RF bandwidth they are able to pass. But consequently a very good SNR is needed to

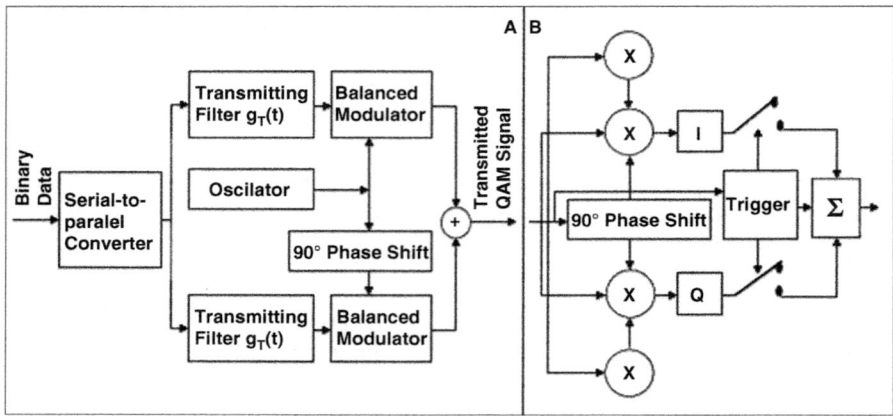

Fig. 3.10 Modulator and demodulator of QAM (Courtesy of Book: by Stacey)

retain a high BER. In Fig. 3.14 (d) is shown a 16-QAM signal space diagram with a 16-state grid that permits both carrier amplitude and phase change. This scheme is not yet considered favorable for satellite communication.

3.3.5 Time Division Multiplexing (TDM)

Satellite links normally relay many signals from different UES but to avoid interfering with each other it is necessary for some kind of separation or division. This separation is known as multiplexing and its common forms are FDM (already explained) and TDM. The TDM is easier to implement with digital modulation and to form hybrid solutions applicable to all type of baseband signals.

The TDM is a time multiplexing solution where a group of various transmission signals on the same frequency at different times take turns using a channel. In this way, a group of pulses from a number of channels may be interleaved to form a single high rate bit stream of multiplexed assembly directly modulated onto the RF carrier. Since digital signals are precisely timed and consist in short pulse groups with relatively long intervals between them, TDM is the only natural way to combine digital signals for transmission. This system has the advantage that less equipment is required than is needed to modulate each channel onto a separate carrier and that the transmission efficiency of a satellite is usually better when it is carrying a few transmissions on many channels and vice versa. Accurate timing is essential to the correct operation of digital systems.

Thus, for that reason the TDM system uses a synchronous clock, which controls the timing of all slave clocks and plesiochronous independent clocks with very good accuracy. If the transmission is without errors and breaks, in the synchronous

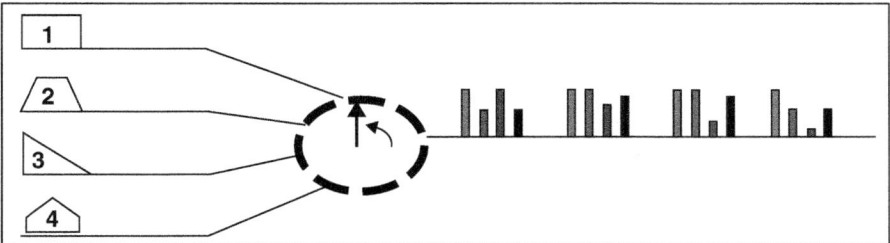

Fig. 3.11 Time division multiplexing (Courtesy of Book: by Stacey)

system it would be only necessary to provide single markers at the beginning of transmission where the decoder could identify all streams of bits.

The TDM is using when there is one RF channel shared in time between the different users. If a suitable TDM repetition frequency is chosen, the user does not need to be aware of the discontinuity shown in Fig. 3.11. This is the principle deployed in modern ground cable and fiber networks to which the radio systems described in this book need to interface.

However, in any practical communication system, regular markers must be provided in the bit stream so that the decoder can extract groups of digitally representative samples, identify the channels of a TDM assembly and resynchronize the system after errors, breaks of transmission or drift of the clock. Therefore, when plesiochronous time multiplexes are combined to form higher order of TDM, it is necessary, in order to preserve synchronism over the long term, to allow for the occasional addition of dummy or padding bits.

3.3.6 Types of Digital Shift Keying

Digital signals can be used to modulate the amplitude, frequency and phase and therefore the solutions of shift keying available for digital modulation are Amplitude Shift Keying (ASK), Frequency Shift Keying (FSK) and Minimum Shift Keying (MSK) as applications of FSK and Phase Shift Keying (PSK). In terms of performance, ASK and FSK, both illustrated in Fig. 3.12 (**a, b**), respectively, require twice as much power to attain the same BER performance as PSK, shown in Fig. 3.12 (c). At the top of the same figure a stream of a digital signal with 1 and 0 binary state is presented.

3.3.6.1 Amplitude Shift Keying (ASK)

This scheme can be accomplished simply by the on-off gating of a continuous carrier. The simplest ASK technique is to represent one binary level (binary 1) by a

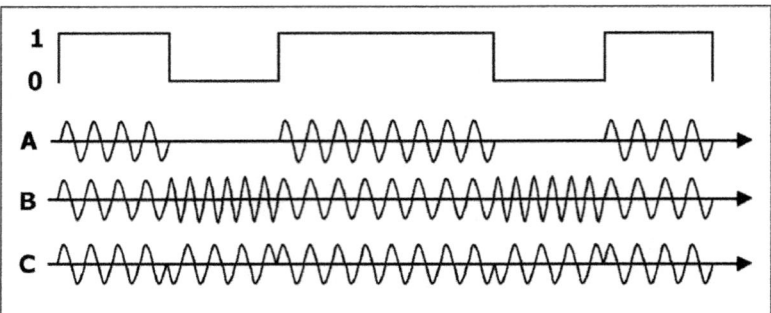

Fig. 3.12 Comparison of: (**a**) ASK; (**b**) FSK and (**c**) PSK (Courtesy of Book: Sheriff)

single signal of fixed amplitude and the other level (binary 0) by switching off the signal.

The absence of the signal for one of the binary levels has the disadvantage that if fault conditions exist it could be misinterpreted as received data. Waveform for ASK, using different amplitude signals for the logic levels, are an alternative method to prevent this disadvantage. As with speech telephony circuits, the upper side band and carrier may be suppressed to reduce the bandwidth requirement and concentrate the available power on the signal containing the information.

3.3.6.2   Frequency Shift Keying (FSK)

In FSK, the carrier frequency is changed between discrete values. The FSK is a frequency modulation scheme in which digital information is transmitted through discrete frequency changes of a carrier signal. The FSK technology is a method used of transmitting digital signals in different telecommunication systems.

This solution may be used whereby the carrier frequency has one value for a 1 bit and another for a 0 bit. The main difficulty in the use of this FM technique is that the gap between the frequencies used must be increased as the modulation rate increases. More exactly, for a restricted channel bandwidth, especially using in-band supervisory signaling, there is a limit to the maximum bit rate that is possible with this technique.

3.3.6.3   Minimum Shift Keying (MSK)

The MSK is a binary form of Continuous Phase Frequency Shift Keying (CPFSK), where the frequency deviation (Δf) from the carrier is set at half the reciprocal data rate of 1/2 T. The MSK scheme may also be viewed as a special form of Offset-QPSK, consisting in two sinusoidal envelope carriers, employing modulation at half the bit rate. It is for this reason the MSK demodulator is usually a coherent quadrature detector, similar to that for QPSK.

The error rate performance is the same as that of BPSK and QPSK. Similarly, differentially encoded data has the same error performance as D-PSK. This solution can also be received as an FSK signal using coherent or non-coherent methods, however, this will degrade the performance of the link. At any rate, the side lobes of MSK are usually suppressed, using Gaussian filters and the modulation method scheme adopted by GSM cellular systems.

3.3.6.4 Phase Shift Keying (PSK)

The PSK scheme is a technique using a multistate signaling stream in which the rate of data transmission can be increased without having to increase the bandwidth. Consequently, the vast majority of MSS employ a method of phase modulation known as PSK. In this shift keying system the phase of the carrier changes in accordance with the baseband digital stream or information content, which general form of a PSK modulation scheme is given by the following equation:

$$s_m(t) = A \cos (\omega t + \phi) \text{ and } \phi = (2m + 1) \pi/M \qquad (3.18)$$

where A = amplitude; ω = frequency angle; ϕ = phase angle varied in accordance with the information signal; m = integer in the range from 0 to (M − 1) and M = number of states. Depending on how many bits can be combined in a group of information as a symbol, there are a number of combination possibilities for PSK digital carriers.

3.3.7 Combinations of PSK Digital Carriers

There are several types of hybrid solutions used for the combination of PSK Digital Carriers. The PSK family is most popular for satellite communications and especially for mobiles. A real-valued band bass signal s(t), which has a common form for these types of PSK methods, is expressed as follows:

$$s(t) = A(t) \cos [2\pi f_c t + \phi(t)] = A(t) \cos [2\omega_c t + \phi(t)] \text{ [V]} \qquad (3.19)$$

where A (t) = amplitude; f_c = carrier frequency of signal s(t); ϕ (t) = phase angle and ω_c = frequency angle varied in accordance with the signal s(t). Including information waveform m(t) the previous relation can be determined as follows:

$$s(t) = \text{Re} [m(t)] e^{j2\pi f c t} \qquad (3.20)$$

In linear modulation methods such as pulse, PAM and PSK can be expressed by:

$$m(t) = A(t) \exp [j\phi(t)] = \sum_{n=0}^{\infty} a_n p(t) \ (t - nT) \qquad (3.21)$$

where Re = real part of the complex in the next bracket; a_n = information carrying symbols; $p(t)$ = signal pulse and T = time interval of symbol.

3.3.7.1 Binary PSK (BPSK)

The simplest form of PSK is Binary PSK (BPSK), where the digital information modulates a sinusoidal carrier. For a general case of M-ary PSK (number of states) in BPSK M = 2, so the baseband bit rate and the symbol rates are the same. Binary data is expressed by a_n = exp. $(j\phi_n)$ for $\phi_n = 0$ or π and the phase changes every data bit of duration period T_b. When $p(t)$ is a rectangular pulse over symbol duration $T = T_b$, the BPSK signal is expressed by the following quotation:

$$s(t) = Aa_n \cos 2\pi f_c t = A \cos (2\omega_c t + \phi_n) \text{ for } nT \leq t < (n + 1) \ T \qquad (3.22)$$

In Fig. 3.13 (a) a diagram is illustrated of how, theoretically, the phase of the carrier changes instantaneously by 180° when the baseband signal switches from 0 to 1, while Fig. 3.14 (a) represents the two states of the carrier by two vectors with a phase difference of 180°. In this sense, where $\phi = 0°$ and $\phi = 180°$ when the baseband signal is 0 and 1, respectively and where A = +V and A = −V when the baseband signal is 0 and 1, respectively. The TDM/BPSK scheme is used by many satellite communication and broadcasting systems for processing such as Forwarding Signaling/Assignment Channels, while Return Request Channels use Aloha BPSK.

3.3.7.2 Quadrature PSK (QPSK)

For M-ary, PSK is selected from M signals like: exp $[j2\pi (m - 1)/M]$, where m = 1, 2,..., M and the resulting signal $s(t)$ is written as follows:

$$s(t) = A \cos [2\pi f_c t + 2\pi/M (m - 1)] \text{ for } m = 1, 2, \ldots, M \qquad (3.23)$$

A slightly more complex form PSK is QPSK or 4-PSK, for which ϕ_n is a set of 0, $\pi/2$, π and $3/2\pi$. Then, the signal $s(t)$ is given by:

$$s(t) = A/\sqrt{2} \ (a^I_n \cos (2\pi f_c t + \pi/4) + A/\sqrt{2} \ (a^Q_n \cos (2\pi f_c t + \pi/4) \qquad (3.24)$$

where a^I_n and a^Q_n are the ±1 value data, which are converted from input data sequence a_n into the in-phase channel (I channel) and quadrature channel

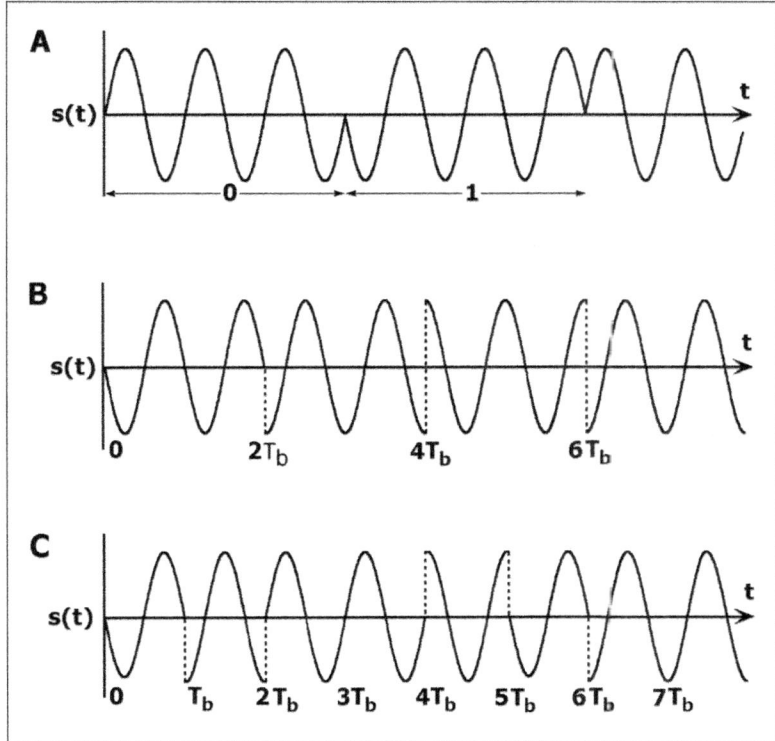

Fig. 3.13 Hybrid PSK modulations: (**a**) BPSK; (**b**) QPSK and (**c**) O-QPSK (Courtesy of Book: by Ohmori)

(Q channel), respectively. However, the relation (a^I_n, a^Q_n) is $(1, 1)$ for $\phi_n = 0$; $(1, -1)$ for $\phi_n = \pi/2$; $(-1, -1)$ for $\phi n = \pi$ and $(-1, 1)$ for $\phi_n = 3\pi/2$.

In Fig. 3.13 (b) the shape of QPSK modulated signals is presented and a QPSK scheme in the I-Q plane is shown in Fig. 3.14 (b).

Only two binary digits are needed to describe four possible states as follows: the ϕ values of 45°, 135°, 225° and 315° correspond to 00, 01, 11 and 10, respectively. In fact, each state of the signal carries two bits of information. Thus, a combination of two bits or more, which corresponds to a discrete state of a signal, is called a symbol, in the case of QPSK; the symbol rate is half the bit rate. Namely, a bit rate of 120 Mb/s corresponds, with QPSK, to a symbol rate of 6C Mb/s or Mbauds.

The QPSK schemes of modulators are basically dual channel BPSK modulator and demodulator equipment. At this point, one channel processes the u_i (I) bits and uses the reference carrier and the other process the u_q (Q) bits using a 90° phase shifted version of the reference.

The demodulation signals are passing out via Low Power Filter (LPF) and Logic circuit. The bits values u_i and u_q are selected alternately from the input bit stream.

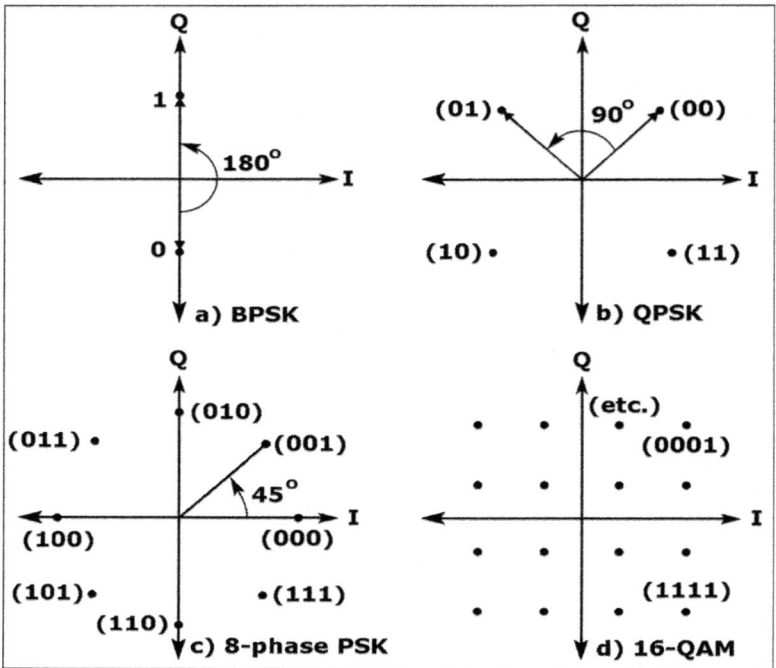

Fig. 3.14 Representation of PSK signals in I-Q plane (Courtesy of Book: by Richharia)

For example, u_i may represent the odd numbers bits and u_q the even. In this case, one binary data channel enters the QPSK modulator and the outgoing symbol rate is equal to half of the incoming bit rate.

The octal phase modulation known as 8-phase PSK is a constant amplitude scheme with a higher bandwidth efficiency of 3 b/s/Hz, shown in Fig. 3.14 (c). Thus, the demands of high bit-rate applications are related to images, video, TV and High Definition Television (HDTV) transmissions.

3.3.7.3 Offset QPSK (O-QPSK)

The O-QPSK scheme delays the quadrature bit stream by T sec relative to the in-phase bit stream to restrict the phase transition to phase changes of 0 or $\pi/2$ every T sec. In such a way, using a^I_n and a^Q_n, the equivalent low-pass and resulting signals are expressed by the following equations:

$$u(t) = \sum_{n=0}^{\infty} a^I_n p(t) \, (t - 2nT) - j a^Q_n p(t) \, (t - 2nT - T) \, e^{j\pi/4}$$

$$s(t) = \left[\sum_n a^I_n p(t)(t - 2nT)\right] \cos \, (2\pi f_c t + \pi/4) + \qquad (3.25)$$

$$\left[\sum_n a^Q_n p(t)(t - 2nT - T)\right] \sin \, (2\pi f_c t + \pi/4)$$

The data transmission occurs in the conventional QPSK at the same time in both I and Q channels. This scheme has larger phase changes than O-QPSK, which has phase changes of at most $\pm\pi/2$ data transmission and large envelope fluctuations do not occur as they do with Π-phase changes in QPSK. Figure 3.13 (c) shows the O-QPSK modulated signal's shape.

The Aloha O-QPSK (1/2-FEC) serves in return request channels by a transmission speed of 24 Kb/s, while 16 Kb/s or 9.6 Kb/s uses APC O-QPSK scheme in voice channels.

3.3.7.4 Differential PSK (DPSK)

The received signal phase is generally an $180°$ ambiguity sign, which cannot be resolved unless some known reference signal is transmitted and a comparison made. In the worst case, if such case left unresolved, the received signal could end up being the complement of the transmitted signal. When the phase condition is constant for 2 T seconds, the DPSK demodulator can obtain the optimum a posteriori probability. The DPSK scheme can also be used to remove sign ambiguity at the receiver. Differentially encoding data prior to modulation occurs when a binary 1 is used to indicate that the current message bit and prior code bit are of the same polarity and 0 to represent the condition when the two pulses are of opposite polarity. The equivalent low-pass signal in the interval of m is given by:

$$v_m = \alpha e^{j(\phi_m - \phi)} u_m(t) + n_m(t) \qquad (3.26)$$

where α = loss factor; ϕ_m = phase difference in m interval; $u_m(t)$ = signal in m interval and $n_m(t)$ = Gaussian noise in m interval. The noncoherent detection technique is useful when the carrier phase is difficult to estimate at the receiver. Thus, in satellite channels, the transmitting oscillator that generates the carrier cannot be completely stabilized and a satellite link has a low S/N ratio. It channel conditions are more severe than additive white Gaussian noise channels due to multipath fading, shadowing and Doppler effects. These problems can be solved by using differentially coherent detection of encoded signals.

Besides, for coherent detection, the bandwidth of the carrier-tracking loop at the Rx must be decreased in the proper manner to increase the S/N and obtain a good phase reference. Since propagation characteristics such as multipath fading and the Doppler effect cause rapid phase variation, the carrier phase of the Rx signal does not remain fixed long enough to be estimated. The problem can be solved with noncoherent detection, which is equivalent low-pass signal in relation to random phase value (ϕ_n) and Gaussian noise is:

$$v_m = \alpha e^{-j\phi_n} u_n(t) + n(t) \tag{3.27}$$

Therefore, noncoherent detection results in the degradation of the bit error performance with respect to coherent detection. In this case, the BER probability (P_b) for binary DPSK is determined with the following relation:

$$P_b = \tfrac{1}{2} \exp\left(-\gamma_b\right) \tag{3.28}$$

where $\gamma_b = E_b/N_0$ is the value of the S/N density ratio per bit signal. In this sense, when considering the use of DPSK schemes, a trade-off between simplified receiver complexities against reduced performance characteristics in the presence of noise, particularly when employing higher order modulation techniques, needs to be made.

3.3.7.5 π/4-QPSK

Recently, π/4-QPSK, or π/4 shift QPSK, has become very popular for mobile and other satellite networks as well as for cellular systems because it has a compact spectrum with small spectrum restoration due to nonlinear amplification and can perform differential detection. The phase point of this scheme always shifts its phase over successive time intervals by ±π/4 or ±3π/4. The spectrum of this scheme is the same as that of a QPSK that undergoes an instantaneous ±π/24 or ±π phase transition.

In such a way, the π/4-QPSK signal can reduce the envelope fluctuation due to band-limited filtering or nonlinear amplification more than can QPSK. In a more general sense, this is because differential detection can be used since π/4-QPSK is not an offset scheme. For mobile channels, a strong line-of-sight signal can be expected even in the Rician fading channel, so coherent demodulation is desirable for improved power efficiency. The bit error performance of coherently detected π/4-QPSK is the same as that of QPSK. On the other hand, differential detection is also desirable for simple hardware implementation. The bit error probability for differential coherer detected π/4-QPSK is given by:

$$P_b = e^{-2\gamma b} \sum_{k=0}^{\infty} \left(\sqrt{2} - 1\right)^k I_k\left(\sqrt{2}\gamma_b\right) - 1/2\, I_0\left(\sqrt{2}\gamma_b\right) e^{-2\gamma b} \tag{3.29}$$

where $\gamma_b = E_b/N_0$ and I_k = modified Bessel function of the first kind in k interval order. Thus, the differential detection is about 2–3 dB inferior to the coherent detection in AWGN and fading channels and noncoherent detection is also applicable. Because of these advantages, π/4-QPSK has been chosen as the standard modulation technique for several systems of cellular and satellite-based mobile communications.

Some experimental results show that in fully saturated amplifier systems, π/4-QPSK still has significant spectral restoration. In effect, to reduce this restoration, π/4-controlled transition QPSK (CTPSK) uses both sinusoidal shaping pulses and timing offsets of the phase transition between I and Q channels.

The π/4-DQPSK (Differential Quadrature PSK) variant is a differential format where the bits for a given symbol are determined by the phase change from the previous symbol. π/4 adds an π/4 offset to the phase changes compared with the phase changes in plain DPSK. This means there are a total of 8 ideal state positions (compared to the 4 for DQPSK).

3.4   Channel Coding and Decoding

Voice, video, data and telex information are transmitted in digital form through a channel that can cause degradation of these transmission signals. The noise, interference, fading and other obstacle factors experienced during transmission could increase the probability of bit error at the receiver. Differently to say, the data signal may be encoded in such a way as to reduce the likelihood of bit error.

Anyway, the coding process uses redundant bits, which contain no information to assist in the detection and correction of errors. The subject of coding emerged following the fundamental concepts of information theory laid down by Shannon in 1948, which is the relationship between communication channel and the rate at which information can be transmitted over it. Basically, the theorems laying down the fundamental limits on the amount of information flow through a channel are given.

3.4.1   Channel Processing

Channel processing is composed of special activities, which can improve the transmission techniques throughout satellite channels in connection with gain, errors, noise, interference, concentration and authenticity.

3.4.1.1   Channel Encoding

The two fundamental problems related to reliable transmission of information via channels were identified by C.E. Shannon as follows:

1. The use of minimal numbers of bits to represent the information given by a source in accordance with a fidelity criterion. In reality, this issue is usually identified as a problem of inefficient channels, to which the source coding provides most practical solutions.
2. The recovery as exactly as possible of the information after its transmission through a communication channel in the presence of noise and other interference. This is a problem of unreliable satellite systems, to which channel (error) coding is the basic solution.

Shannon proved that by proper encoding these two objectives can always be achieved, provided that the transmission rate (R_b) verifies the fundamental expression $H < R_b < C$, where H = source entropy and C = channel capacity. The BER of a digital system may be improved either by increasing E_b/N_0 or by detecting and correcting some of the errors in the received data. For the Additive White Gaussian Noise (AWGN) channel, the Shannon Hartley law states that capacity of a channel is given by the following relation:

$$C = B \log (1 + C/N) \ [b/s] \tag{3.30}$$

where B = channel bandwidth in Hz and S/N = signal-to-noise ratio at the receiver. Thus, the channel capacity is the measure rate of the maximum information quantity that two parties can communicate without error via a probabilistically modeled channel. Namely, this chain of channels is composed of information data on input rate (R_b), channel encoder with redundancy data (r) and encoded data symbols on output rate (R_c). A reverse channel model contains input encoded data symbols, channel decoder and output information data symbols. According to Shannon, if information is provided at rate R, which is less than the capacity of the channel, then a means of coding can be applied such that the probability of error of the received signal is arbitrarily small.

If this rate is greater than the channel capacity, then it will not be possible to improve the link quality by means of coding techniques. Indeed, its application could have a detrimental effect on the link. Rearranging the above equation in terms of energy-per-bit and information rate, where the information rate is equal to the channel capacity, results in the following:

$$C/B = \log (1 + E_b C/N_0 B) \tag{3.31}$$

where E_b = information bit rate; E_b = related to the carrier power and the information bit rate and C/N_0 = Carrier-to-Noise Density Ratio. Moreover, the above expression can be utilized to derive the Shannon limit, the minimum value of Eb/N_0 below which there can be no error-free transmission of information. As C/B tends to zero, this can be shown to be equal to -1.59 dB ($1/\log_2 e$). The code rate and input rate can be defined as:

$$c = n/n + r \text{ and } R_c = R_b/n \text{ [b/s]} \qquad (3.32)$$

The capacity of the channel is independent of the coding/modulation scheme used. Hence Shannon's channel coding theorem exactly stated that, for a given carrier-to-noise ratio, the error probability could be made as small as desirable, provided that the information rate (R_b) is less than the capacity (C) and a suitable coding is used.

In any case, channel coding is especially interesting because of the severe power, bandwidth and propagation limitations. Moreover, the considerable progresses in multiple access modulation schemes, resource assignment algorithms, signal processing techniques and advanced error control coding provide the most efficient means to realize highly reliable information transmission.

3.4.1.2 Digital Compression

Digital transmission in general uses compression techniques for data and video signals. The effective data transfer can be significantly increased by using data compression software. Essential results were provided on PC by the PKZIP/PKUNZIP program developed by US-based PKWARE, which in a fraction of a second gives a 2–3 times reduction in size of ASCII files and 1.5 times for many types of binary files. The ARJ compression software from Robert K. Jung is slower but more effective than PKZIP. It can also be recommended for the compression of data files containing graphic information. Thus, the real-time data compression incorporated into the most advanced modems can also increase the effective data rate of ASCII files transmission but for transmission of already compressed files with information it is better not to use the compression in the modem.

Their use in compressed video systems, where a TV Receive Only (TVRO) can also receive many channels of video from one transponder (about 6–8), has become very widespread, first in the USA and then in Europe. The compression system that has now become standard refers to an MPEG standard, formed under the auspices of the ISO and the International Electrotechnical Commission (IEC). In such a system a number of digitized videos are combined into a single bit stream in a source coder. That bit stream is then sent to a channel coder for FEC and then to a QPSK modulator, an up-converter, amplifier and an antenna for up linking to the satellite transponder. Since only one signal is present in the transponder at any on time, there is no need for back-off and full transponder power is used. At the reverse side, the compressed video downlink comprises a line in the chain of an antenna, tuner including down-converter and QPSK demodulator, then FEC detector, demultiplexer and MPEG decoder. The MPEG is determined to provide standard compression that allows video and accompanying audio signals to be compressed in channel width. The packetizer function is to enter a suitable code in the bit stream for the individual digitized TV program so that it can be separated in the receiving chain, allowing the enabled user to select the desired program. Thus, the BER is determined from the (E_b/N_0) obtained for a combination of whatever transponder EIRP, FEC coding, transmission symbol rate and receiver system are used.

If a plot of that FEC module system is not available, then the Viterbi mode FEC coding performance could be used for a good estimate of results. This type of compression has effects like: MPEG-2 compression results in the removal of most audio and video redundancy; the FES utilization scheme resulting in a rapid BER increase and the resultant (E_b/N_0) should be high enough to achieve a BER of 10^{-6} for a TVRO.

3.4.2 Coding

As is known, satellite communication systems are generally limited by the available power and bandwidth. Thus, it is of interest if the signal power can be reduced while maintaining the same grade of service (BER). As mentioned, this can be achieved by adding extra or redundant bits to the information content by using a channel coder. Otherwise, excepting several main classes of channel coder, the three most widely used variants in satellite communications are block, cyclic and convolutional encoders.

3.4.2.1 Block Codes

Binary linear block codes are expressed in the (n, k) form, where (k) is the information bits number that is converted into (n) code word bits. There are (n, k) party bits in each encoded block, where the difference between (n) and (k) bits are added by the coder as a number of redundancy bits (r). In the other words, a coded block comprising (n) bits consists in (k) information and (r) redundant bits expressed as follows:

$$n = k + r \tag{3.33}$$

Such a code is designated as a (n, k) code, where the code rate or code efficiency is given by the ratio of (k/n). Mapping between message sequences and code words can be achieved using look-up tables; although as the size of the code block increases such an approach becomes impractical. This is not such a problem as linear code words can be generated using some form of linear transformation of the message sequence. Thus, a code sequence (c) comprising of the row vector elements $(c_1, c_2, \ldots, c_n)$ is generated from a message sequence (m), comprising the row vector elements $(m_1, m_2, \ldots, m_k)$ by a linear operation:

$$c = m \, G \qquad\qquad (3.34)$$

where G = generator matrix. Thus, in general, all (c) code bits are generated from linear combinations of the (k) message bits.

A special category known as a systematic code occurs when the first (k) digits of the code are the same as the first (k) message bits, namely if input message bits appear as part of the output code bits. The remaining n-k code bits are then generated from the (k) message bits using a form of linear combination, and they are termed the party data bits. The generator matrix for a linear block code is one of the bases of the vector space of valid codewords. It defines the length of each codeword (n), the number of information bits (k) in each codeword and the type of redundancy that is added; the code is completely defined by its generator matrix. The generator matrix is a $(k \cdot n)$ matrix that is the row space of V_k.

Thus, one possible generator matrix for a typical (7, 4) linear block code has to be presented in four rows as blocks: G = 1,101,000/0110100/1110010/1010001. The distance between two coded words (for example, first 2 and second 2 digits) in a block is defined as the number of bits in which the words differ and is called the Hamming distance (d_h). The Hamming distance has the capability to detect all coded words having errors (e_d), where $e_d < (d_h - 1)$; to detect and correct (e_{dc}) bits, where $e_{dc} = (d_h - 1)/2$ and to correct t and detect (e) errors, where the Hamming distance as a minimum space between two coded blocks is given by the following equation:

In the detection process, two coded words separated by (d_h) are most likely to be mistaken for each other. The extended Golay code offers superior performance to Hamming codes but at a cost of increased receiver complexity. In practice, code words are conveniently generated using a series of simple shift registers and modulo-2 adders.

In Fig. 3.15 (Left) is shown the concept of block codes and rate, which operate on groups of bits organized as blocks, namely information bits are assembled as blocks before coding.

3.4.2.2 Cyclic Codes

The cyclic code methods are a subclass of linear codes, which code word is generated simply by performing a cyclic shift of its predecessor, shown in Fig. 3.15 (Right). In other words, each bit in a code sequence generation is shifted by one place to the right and the end bit is fed back to the start of the sequence, hence the term cyclic.

In fact, both the linear Hamming and extended Golay codes have equivalent cyclic code generators. On the other hand, non-systematic type of cyclic codes are generated using a unique generator polynomial g(p) and message polynomials in the forms presented by the following relations:

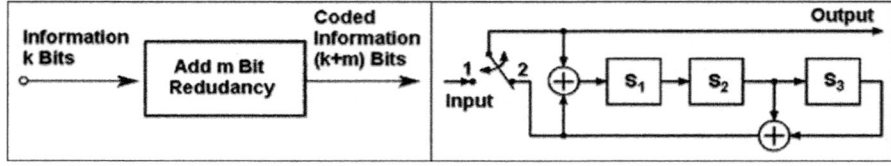

Fig. 3.15 Block and cyclic coders (Courtesy of Books: by Calcutt)

$$g(p) = p^{n-k} + g_{n-k}p^{k-1} + \ldots g_1p + 1 \qquad (3.35)$$

$$m(p) = m_{k-1}p^{k-1} + m_{k-2}p^{k-2} + \ldots + m_1p + m_0 \qquad (3.36)$$

where the generator polynomial is a factor of p^{n+1} and the value $(m_{k-1} \ldots m_0)$. When this is multiplied by the generator polynomial, it results in the generation of a code word by the following equation:

$$c(p) = \left(m_{k-1}p^{k-1} + m_{k-2}p^{k-2} + \ldots + m_1p + m_0\right)g(p) \qquad (3.37)$$

Thus, an alternative to this approach is to generate systematic cyclic codes, which can be generated in three steps, involving the use of feedback shift register:

(a) The message polynomial is multiplied by p^{n-k}, which is equivalent to shifting the message sequence by $(n - k)$ bits. This is necessary to make space for the insertion of the party bits.
(b) The product of step 1, $p^{n-k}m(p)$ is divided by the generator polynomial, $g(p)$.
(c) The remainder from step 2 is the party bit sequence, which is then added to the message sequence prior to transmissions.

The cyclic codes scheme has two methods used in satellite communication systems, such as Bose-Chadhury-Hocquenghem (BCH) and Reed-Solomon (RS).

1. **BCH Codes** – The BCH codes are the most powerful of all cyclic codes with a large range of block length, code rates, alphabets and error correction capability. These codes have been found to be superior in performance to all other codes of similar block length and code rate. Most commonly used BCH codes have a code word block length as $n = 2^m - 1$, where $(m = 3, 4 \ldots)$. For instance, the earlier satellite standards used 57 bits plus 6 party bits encoded with BCH (63, 57 code in TDM channels and for the return request channel burst employs Aloha BPSK (BCH) 4800 b/s.

2. **RS Codes** – The RS codes are a subset of the BCH codes specially suited for correcting the effect of the burst errors. The latter consideration is particularly important in the context of the satellite channels and hence, RS codes are usually incorporated into the system design. This set of codes has the largest possible code minimum distance of any linear code with the same encoder input and

output block length. At this point, the RS codes are specified using the convention RS (n, k), where n = number of code symbols word length per block; k = data symbols encoded and the difference between (n) and (k) is the number of parity symbols added to the data. The code minimum distance is given by relation:

$$d_{min} = n - k + 1 \qquad\qquad (3.38)$$

The code is capable of correcting errors such as: $e = 1/2 \, (d_{min} - 1)$ and $e = (n - k)/2$, or to use an alphabet of 2^m symbols with: $n = 2^m - 1$ and $k = 2^m - 1 - 2e$, where $m = 2,3 \ldots$ and so on. The advantage of RS codes is the reduction in the number of words (n) symbols, which are code words, producing a possibly large value of minimum distance (d_{min}).

3.4.2.3   Convolutional Codes

The second family of commonly used codes is known as convolution codes. Unlike block codes, which operate on each block independently, these codes retain several previous bits in memory, which are all used in the coding process. They are generated by a typed-shift register and two or more modulo-2 adders connected to particular stage of the register. The number of bits stored in the shift register is termed the constraint length (K). Bits within the register are shifted by (k) input bits. Each new input generates (n) output bits, which are obtained by sampling the outputs of the modulo-2 adders. The ratio of (k) to (n) is known as the code rate. These codes are usually classified according to the following convention: (n, k, K), for example (2, 1, 7), refers to a half-rate encoder of constraint length 7. It is important to know what sequence of output code bits will be generated for a particular input stream. There are several techniques available to assist with this question, the most popular being connection pictorial, state diagram, tree diagram and trellis diagram.

However, to illustrate how these methods are applied, the simple example of half-rate (1/2) encoder will be considered with constraint length k = 3. The system has two modulo-2 adders, so that the code rate is 1/2. The input bit (m) placed into the first of the shift register causes the bits in the register to be moved one place to the right. The output switch samples the output of each modulo-2 adder, one after the other, to form a bit pair for the bit just entered. The connections from the register to the adders could be one, two or three interfaces for either adder. The choice depends on the requirement to produce a code with good distance properties. A similar encoder used by the earlier satellite standard produced a half-rate convolutional encoder. Therefore, in terms of connections to the modulo-2, adders can be defined using generator polynomials in the encoder configuration.

Thus, convolutional codes are forming in convolutional coder by convolving information bits R with the impulse response of a shift register encoder, which

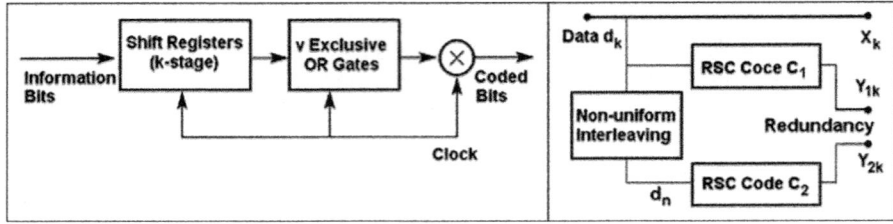

Fig. 3.16 Convolutional and turbo coders (Courtesy of Book: by Richharia)

block diagram is shown in Fig. 3.16 (Left). These types of codes use previous information bits in memory (v) and continuously produce coded bits. The constraint length of convolutional code defines the number of information bits, which influence the encoder output.

In the proper manner, the constraint length is decided by the number of shift registers or code memory. The error correcting property of the convolutional code depends on the constraint length and its value improves as code memory is increased, and in such a way decoding complexity increases.

The polynomial for the generating arm (n) of the encoder $g_n(p)$ and the generator polynomials representing encoder $g_1(p)$ can have the following relations:

$$g_n(p) = g_0(p) + g_1 p^1 + \ldots g_n p^n \text{ and} \tag{3.39}$$

$$g_1(p) = 1 + p + p^2 = 1 + p^2 \tag{3.40}$$

where the value of g_1 takes on the value of 0 or 1 and a 1 is used to indicate that there is a connection between a particular element of the shift register and the modulo-2 adder. Thus, to provide a simple representation of the encoder, generator polynomials are used to predict the output coded message sequences for a given input sequences. For instance, the input sequence 10,110 can be represented by the polynomial relation:

$$m(p) = 1 + p^2 + p^3 \tag{3.41}$$

Combining this with the respective generator polynomials and using the rules of module-2 arithmetic results in the following:

$$m(p)g_1(p) = \left(1 + p^2 + p^3\right)\left(1 + p + p^2\right) = 1 + p + p^5 \text{ and} \tag{3.42}$$

$$m(p)g_2(p) = \left(1 + p^2 + p^3\right)\left(1 + p^2\right) = 1 + p^3 + p^4 + p^5 \tag{3.43}$$

The output code sequence c(p) obtains by interleaving the above two products with the following relations:

$$c(p) = [1, 1]p^0 + [1, 0]p^1 + [0, 0]p^2 + [0, 1]p^3 + [0, 1]p^4 + [1, 1]p^5 \qquad (3.44)$$

Here, the number between brackets represents the output code sequence.

The analog earlier analog satellite transmissions used HSD channel encoding configuration for the information data stream at 56 Kb/s. The scrambling sequence on the input data stream shall be provided by the scrambler before the convolutional encoder described in CCITT Recommendation V.35 scheme. The data stream then passes differential encoder state stage 1 followed by 1/2 (half) convolutional encoding with constant length k = 7. The half (1/2) rate convolutional encoder can provide two data streams to the QPSK modulator using two generator polynomials rates as follows: $G_1 = 1 + x^2 + x^3 + x^5 + x^6$ and $G_2 = 1 + x + x^2 + x^3 + x^6$. The encoder provides two parallel data streams to the modulator: I and Q, while (Q) should lag (I) by 90° in the modulator.

On the other hand, the digital satellite standards for transmission and out-of-band signaling channels are using digital modulation and FEC mode in order to efficiently utilize satellite power and bandwidth. The basic modulation and coding techniques are filtered by 60% roll-off O-QPSK and 40% roll-off BPSK, both with convolutional coding at either rate: 1/2 or 3/4 FEC and 8-level soft decision Viterbi decoding (constraint length = 7).

In spite of that, the punctured coding is used to derive 3/4 and 1/2 rates. All BPSK channels are differentially encoded outside the FEC for all transmissions, with the exception of those fields carrying digitally coded voice, employs FEC with convolutional encoding of constraint length k = 7 and 8-level soft decision Viterbi decoding. There are two generator polynomials rates: G_1 (133 octal) and G_2 (171 octal). The transmitted bit is nominated by 1 and deleted by 0. However, the first bit in each transmission frame is the output from the G_1 polynomial and all bits are transmitted at the rate of 1/2 code. Finally, the output data from the bit selector are punctured coded data of 3/4 rate.

3.4.2.4   Concatenated Codes

These codes were originally developed for deep space communications and occur when two separate coding techniques are combined to form a large code. The inner decoder is used to correct most of the errors introduced by the channel, the output of which is then fed into the outer decoder, which further reduces the BER to the target level. That is to say, a typical concatenated coding scheme would employ half-rate convolutional encoding of constraint length 7 (2, 1, 7) – Viterbi decoding as the inner scheme and RS (255, 223) block encoding and decoding as the outer scheme. Interleaving between the inner and outer coders can be used to further improve the performance.

3.4.2.5   Turbo Codes

In information theory, turbo codes (originally in French Turbocodes) are a special class of high-performance FEC codes developed around 1990–1991 and published in 1993), along with a practical decoding algorithm. As first codes that closely approach the channel capacity, a theoretical maximum for the code rate at which reliable communication is still possible given a specific noise level. In Fig. 3.16 (Right) is shown a basic rate 1/3 turbo coder, which feeds data stream d_k directly into a Recursive Systematic Convolution (RCS) coder c_1 and after interleaving into another RCS coder C_2, which is not necessarily identical to C_1. The transmitting bit stream compresses symbol X_k and redundancies Y_{1k} and Y_{2k}, and is therefore a rate 1/3 code or it may be punctured to give higher code rate. The major importance of these codes is that they enable reliable transmission with power efficiencies close to the theoretical limit predicted by Claude Shannon.

Since their introduction, turbo codes have been proposed for low-power applications such as deep space and satellite communications, as well as for interference limited applications such as third generation cellular/personal communication services. Due to the use of a pseudo-random interleaver, turbo codes appear randomly to the channel, yet possess enough structure so that decoding can be physically realized. Developed for deep space and satellite communication applications, turbo codes offer a performance significantly better than concatenated codes. For instance, they are generated using two or more recursive systematic convolutional code generators concatenated in parallel. Here, the term recursive implies that some of the output bits of the convolutional encoder are fed back and applied to the input bit sequence, and in sense, the first encoder takes the information bits as input. The key to the turbo code generation is the presence of a permuter, which performs a function similar to an interleaver; with the only difference that here the output sequence is pseudo-random. The permuter takes a block of information bits, which should be large to increase performance, for example more than 1000 bits and produces a random, delayed sequence of output bits, which is then fed into the second encoder. Thus, the outputs of the two encoders are partly bits transmitted along with the original information bits. In order to reduce the number of transmitted bits, the party bits are punctured prior to transmission.

From various simulation results it is recognized that turbo codes are capable of achieving an arbitrarily low BER of 10–5 at an Eb/No ratio of just 0.7 dB. For instance, in order to achieve this level of performance, large block sizes of 65,532 data bits are required. Because of this prohibitively enlarged block size, an original turbo code is not well suited for real time Tel communication systems such as IS-95 CDMA cellular standard. For that reason, the work on this problem has focused on the design of short block length codes, compatible with IS-95 standard.

3.4.3 Decoding

The definition of decoding is the process of translating or converting code into plain text or any format that is useful for subsequent processes. In fact, decoding process is the reverse of encoding, which is designed to convert encoded data communication transmissions and files to their original states. The complete transmission loop requires any type of encoder followed by modulation and transmitter via transmission channel to receiver, namely to demodulator and decoder. In such a manner, decoding is the reverse method of coding and every type of decoding on the transmit side needs the same convenient decoding method on the receive side. For the reliable data transmission over satellite noisy channels, block and convolutional are widely used in most digital communication systems. Much more efficient algorithms have been found for using channel measurement information in the decoding of convolutional codes than in the decoding of block codes.

3.4.3.1 Block Decoding

The simplest means of decoding block codes is by a method of correlation whereby the decoder makes a comparison between the received code word and all permissible code words, and however it is selecting that word that gives the nearest match. This decoding will also depend on whether error detection or error correction is required, so decoders generally cannot use soft decision outputs from the demodulator, unlike the decoders for convolutional codes.

3.4.3.2 Convolutional Decoding

The effect of the transmission channel on the signal and the probability of detection of a 1 or 0 in the presence of Gaussian noise are important factors during detection. In such a manner, an output from the demodulator can be configured to give a correct decision regarding whether the incoming signal is 1 or 0. The process of decoding then depends on the two state inputs it receives.

An alternative demodulator configuration allows quantization of the predicted level which gives the decoder more necessary information regarding the probable state of the demodulator output. For example, if 3 bit ($2^3 = 8$ levels) quantization occurs then 0 0 0 would suggest a firm valuation of the level received as a 0. On the other hand, an 0 0 1 scheme suggests the 0 is received close to the threshold and this valuation as a 0 is made with less certainly. The reason for quantization is to provide the convolutional decoder with more information in order to correctly recover the transmitted information with better error performance probability.

3.4.3.3 Turbo Decoding

The turbo decoder operates by performing an interactive decoding algorithm, resulting in the partial transfer of an a priori likelihood estimate of the decoded bit sequence between the constituent decoders. Initially, the received information bits, which may be in some error due to the influence of the channel, are used to perform a priori likelihood estimates by the respective decoders. In a more precise sense, the decoders employ so called Maximum a Posteriori (MAP) algorithm to perform converge on the likely sequence of data transfer, after which the interaction between decoders ceases and the output sequence is obtained from one of the decoders.

An interleaver can be placed between the output of Decoder 1 and the input of Decoder 2, to provide an additional weighted decision input into Decoder 2; similarly, a de-interleaver is placed at the output of Decoder 2, to provide feedback to Decoder 1. The decoding time is proportional to the number of interactions between decoders.

3.4.3.4 Sequential Decoding

A sequential decoder may be used for convolutional decoding and it operates in a similar manner to the Viterbi decoder. On receipt of the incoming code word sequence this decoder will penetrate into the tree according to a decision made regarding the best path to follow. For that reason, using a trial and error technique, the decoder will progress as long as the chosen path appears correct, otherwise it will backtrack to try a different route. At this point, either soft decision or hard decision decoding is possible with the sequential decoder, although soft decision would considerably increase the computational time and storage space required.

A major advantage of sequential decoding is that the number of states examined is independent of constraint length, allowing the use of large constraint lengths and low error probability. A disadvantage is the need to store input sequences while the decoder searches for its preferred route through the tree. If the average decodes rate falls below that of the average symbol arrival rate, there is a danger that the decoder cannot cope, causing a loss of input information.

3.4.3.5 Viterbi Decoding

Viterbi decoding was developed by Andrew J. Viterbi and published in April 1967. A Viterbi decoder uses the Viterbi algorithm for decoding a bitstream that has been encoded using convolutional or trellis code. The Viterbi algorithm is the most resource consuming, but it does the maximum likelihood decoding, mostly used for decoding convolutional codes with constraint lengths $k < =10$, but values up to $k = 15$ are used in practice.

Viterbi maximum likelihood decoding of convolutional codes provides the best possible results in the presence of random errors. Thus, in an attempt to match the output sequence received by the decoder, Viterbi's algorithm models the possible state transition through a trellis identical to that used by the encoder.

Accordingly, the Viterbi decoding algorithm is a maximum likelihood path algorithm that takes advantage of the remaining path structure of convolutional codes. This method works by modeling the possible state transitions of the encoder and finding the output sequence that matches most closely to that received by the decoder. Its task is to realize that not all paths through the encoder states can contribute to the final decoded output and that many paths can be rejected after each frame is received, which keeps the problem to manageable proportions.

If the encoder remembers (v) bits, then there are 2^v possible memory states to be modeled by the decoder. Hence, this term dominates expressions for speed, complexity and cost of the decoder and currently imposes an upper limit of 8–10 on constraint length. By path maximum likelihood decoding means that of all the possible paths through the trellis, a Viterbi decoder chooses the convenient path, most likely in the probabilistic sense to have been transmitted. Viterbi decoders easily make use of either hard or soft decision making. This decoding can incorporate soft decisions very simply, which will almost double the error correction power of the code and this can provide an additional gain of up to 3 dB. Otherwise, the procedure for choosing the best Viterbi scheme is to maximize constraint length within the limits of cost and speed, to find a nonsymmetrical code with the best value of d_∞ and to use soft decisions. For instance, some mobile satellite standards utilize an 8-level soft decision Viterbi decoding in their channels (constraint length = 7).

3.4.4   Error Correction

Error Correcting Code (ECC) or FEC is a method that involves adding parity data bits to the message. These parity bits will be read by the receiver to determine whether an error happened during transmission or storage. There are several methods (such as ADPCM) that reduce the number of redundant bits in speech, audio and visual signals in order to make more economic use of bandwidth, thus it is necessary to consider methods that require the deliberate addition of redundant bits to messages. The added bits are very carefully chosen and error correction systems make it possible to achieve large savings in the power required to realize low BER. At the receiver the additional bits are used to detect any errors introduced by channels. To achieve this technique in MSC are employed: FEC, ARQ and Pseudo-noise and Interleaving. Thus, it is also possible to combine FEC and ARQ in an integration form known as a Hybrid Error Correction (HEC) transmitting scheme. At this point, however, the HEC method is used to reduce BER and the number of retranslated blocks. Such an arrangement could also be used to provide feedback information to the transmitter regarding slow variations, such as a fading.

3.4.4.1   Forward Error Correction (FEC)

The FEC is a technique where errors are detected and corrected at the receiver. Thus, this scheme requires only a one-way transmission link, since the message contains parity bits used for detection and correction of errors. In such a way, it is working only on receiving Tlx mode in radio and satellite one-way transmissions. The basic FEC technique used in MSC can be classified into two major (already explained) categories such as Convolutional and Block codes. The FEC coding as a result of convolutional coding is used in Inmarsat standards for some voice, telex and signaling channels. For example, Inmarsat standard-B uses convolutional encoder of constraint length 7- and 8-level soft decision Viterbi decoder. The coding rate is either 3/4 or 1/2, while for voice channel the code rate 3/4 is used and is derived by puncturing the rate 1/2 with k = 7 convolutional code.

On the other hand, the association of both basic coding techniques results in an even more powerful FES scheme known as the concatenated coding system. This powerful FEC scheme has been introduced in recent years, for a considerable increase of the service quality without appreciable expansion of bandwidth. While the inner code, with Viterbi decoding, can correct a large part of the random errors and very short error bursts, the residual errors at the outputs of the Viterbi decoder tend to be grouped in bursts. Thus, using a properly chosen interleaving that cuts the error bursts into shorter ones, a high rate Reed-Solomon code can be used as the outer code in order to correct most of these dispersed errors bursts to achieve a very low BER. Thus, the introduction of concatenated coding and trellis-coded modulation into MSC is the most remarkable event in the domain. Table 3.3 shows a list of FEC techniques along with their performance.

An FEC scheme can improve the quality of a digital transmission link by the following two aspects: **(a)** An BER reduction, closely related to the service quality criterion and **(b)** a saving in the E_b/N_0 or C/N_0 to be considered in the link budget. The E_b/N_0 or C/N_0 saving is often called the coding gain, expressed in dB as a difference at certain BER values, of the coded system and the reference noncoded one. Thus, in the comparison between different transmission schemes, E_b/N_0 is usually used because it is independent of the coding scheme, where the gain is given as follows:

$$G = (E_b/N_0) \text{ ref} - (E_b/N_0) \text{ cod } [\text{dB}] \tag{3.45}$$

The merit of a coding system can also be appreciated in terms of the savings in C/N_0 and C/N, considering information rate (R_b) and information transmission bandwidth, so values for Carrier-to-Noise Density Ratio (C/N_0) and the Carrier-to-Noise Ratio (C/N) are:

$$(C/N_0) = (E_b/N_0) + \log R_b \text{ and } (C/N) = (E_b/N_0) + \log R_b - 10 \log W \text{ [dB]} \tag{3.46}$$

Table 3.3 Performances of FEC techniques

Code	Decoding mode	Gain (BER = 10^{-5})	Gain (BER = 10^{-8})	Bit rate	Complexity
Convolutional	Threshold	1.5–3.0	2.5–4.0	Very High	Low
Convolutional	Viterbi (soft decision)	4.0–5.0	5.0–6.5	High	High
Convolutional	Sequential (hard decision)	4.0–5.0	6.0–7.0	High	Low
Convolutional	Sequential (soft decision)	6.0–7.0	8.0–9.0	Medium	Low
Concatenated (convol./RS)	Viterbi inner and algebraic outer	6.5–7.5	8.5–9.5	High	Medium
Concatenated (short block/ RS)	Soft inner and algebraic outer	4.5–5.5	6.5–7.5	Medium	High
Short block linear	Soft decision	5.0–6.0	6.5–7.5	Medium	High
Block (BSH/RS)	Algebraic (hard decision)	3.0–4.0	4.5–5.5	High	Medium

A coding gain in E_b/N_0 means in general a gain in C/N but the coding in C/N depends on the bandwidth expansion with respect to the reference system. It is however possible to have a coding gain without bandwidth expansions using Trellis Coded Modulation (TCM).

3.4.4.2 Automatic Request Repeat (ARQ)

The Automatic Repeat ReQuest (ARQ), also known as Automatic Repeat Query, is an error-control method for data transmission that uses acknowledgements (messages sent by the receiver indicating that it has correctly received a data frame or packet) and timeouts (specified periods of time allowed to elapse before an acknowledgment is to be received) to achieve reliable data transmission over an unreliable service. If the sender does not receive an acknowledgment before the timeout the message will be retransmitted until is received.

The ARQ is a transmission technique with which a high degree of data integrity is required and latency is not a significant factor. In reality, the ARQ scheme, based on error detection coding and a retransmission protocol, is well adapted to the situation where a two-way channel is available. Typical examples of such systems can be encountered in a computer data network using satellite links. It is worthy of notice that the ARQ and improved ARQ, as well as HEC techniques, are widely used in modern digital communications and storage systems. The ARQ method requires a two-way link, since a receiver, detecting an error, does not attempt to correct it but simply requests the transmitter to retransmit the message. Thus, the ARQ scheme basically works with the following modus:

1. **Stop and Wait ARQ** – After each message block is sent via satellite link, the transmitter waits for acknowledgement. If the message block received is in error, the transmitter will retransmit that block but if this message is correctly received, the next message block is transmitted. A half-duplex link is required, transmission on the link is possible in both directions but not at the same time.
2. **Continuous ARQ with Repeat** – The transmitter sends and the receiver acknowledges message blocks continuously. Hence, any message block not correctly received causes the transmitter to return to the block in question (incorrect received block) and recommence continuous retransmission from there. A full-duplex link is necessary for transmission in both directions simultaneously.
3. **Continuous ARQ with Selective Repeat** – In this ARQ arrangement only the block received in error is retransmitted and the transmitter continues from where it left off at the last block, instead of repeating all the subsequent even correctly received messages. Thus, full-duplex link is also necessary for transmission in both directions simultaneously.

A major advantage of ARQ compared with FEC is that decoding equipment for error corrections can be simpler and the redundancy in the total message stream is less. The ARQ efficiency is good for low error ratios but for high ratios requiring retransmission of a large number of message blocks, the system becomes inefficient. A disadvantage of ARQ is the variability of the delays experienced from end-to-end of the link and so, the possible requirement for large data stores of incoming data blocks.

Satellite transmission systems transmit packets of data that contains a 16-bit checksum field. Receiver completes an expected checksum for each packet and compares this with the actual packet received in order to verify that the packet has been correctly received.

3.4.4.3   Pseudo Noise (PN)

The PN generator will produce a set of cyclic codes with good distance properties. Thus, the name of the sequence is given, because the sequence, although deterministic, appears to have the properties of sampled white noise. Furthermore, a PN sequence is easily generated using shift registers, and has a correlation function that is high packet for zero delay and approximates to zero for other delays. The PN sequence, being deterministic, is usually for synchronization purposes between a transmitter and receiver.

Some Inmarsat standards use a scrambler circuit before FEC encoding and a descrambler at the receive end following FEC decoding. For Inmarsat standard-B and M, for instance, the scrambler/descrambler circuits are PN generators using 15 stages.

The scrambler/descrambler circuits are clocked at the rate of one shift per information bit. The first bit into the scrambler at the beginning of a frame is modulo-2, added with the output of the scrambler shift generator, corresponding to the initial state-scrambling vector. The initial state of the shift register is located at the beginning of a burst and a frame.

Considering some satellite standards, the initial state of the scrambler shift registers at the GES terminal, for SCPC channel operating in voice mode, and is sent to the GES by the FES or MES at the start of a call as part of a call set-up sequence. The FES or MES chooses any initial state (except all zeros) on a random basis for each call and signals this scrambling vector message (8D in hexadecimal form or 10,001,101) for implementation at the GES with the Least Significant Bit (LSB) in shift register No 1 and the Most Significant Bit (MSB) in shift register No 15 of the scrambler. For instance, MES terminal simultaneously sets the descrambler shift register with the same scrambling vector. Thus, for MES-to-GES channels, a fixed initial state default value of 6959 in hexadecimal form or 110,100,101,011,001, is used in MES scramblers and GES descramblers.

3.4.4.4 Interleaving

As is well known, the MSC transmission channel introduces errors of a bursting nature. Hence, in the short term, the errors introduced by the channel cannot be considered to be statistically independent or memory less, the criterion upon which most coders (block and convolutional) optimally operate. In order to mimic a statistically independent channel, a technique known as interleaving is incorporated into the transmitter chain after the output of the encoder and from the interleaver the input signal passes via the modulator. In reverse mode the output signal goes through the demodulator, deinterleaver and channel decoder. This circle presents the interleaver/deinterleaver segment within the transmission/reception chain in the satellite link. Hence, the role of the interleaver is to re-order the transmission sequence of the bits that make up the code words in some predetermined fashion, such that the effect of an error burst is minimized.

Interleaving can be performed for both block and convolutional codes. Block interleaving is achieved by firstly storing the output code words of the encoder into a two-dimensional array. Consider the case of an $(m \cdot n)$ array, where (m) is the number of code words to be interleaved and (n) is the number of code word bits. Thus, each row of the array comprises a generated code word. Once the array is full, the contents are then output to the transmitter but in this case, data is read out on a column-by-column basis. Generally speaking, the transmission of each symbol of a particular code word will be non-sequential. Namely, the input signal goes via the input sequence into the interleaver block and after processing the output sequence would correspond to the chain starting with C_{11}, C_{21}, C_{31}, C_{41}, C_{51}, C_{61}, C_{71}, C_{12}-C_{72}, ... until C_{18}-C_{78}. At this point, the effect of any error bursts will have been dispersed in time throughout the transmitted code words.

Convolutional interleavers work along similar lines, achieving performance characteristics similar to block interleaving.

At the receiver, the inverse of the interleaving function is performed by a deinterleaver and the original code words are reconstituted prior to feeding into the encoder. Namely, a burst of error affecting the transmitted bits indicated by the chain coming from interleaver block would be dispersed among the code words at the receiver.

3.5 Multiple Access Technique

Generally in telecommunication transmission systems and computer networks a channel access or multiple access method allows several terminal connected to the same multi-point transmission medium to transmit over it and to share its capacity. Examples of shared physical media are wireless networks (radio and satellite solutions), bus networks and point-to-point operating systems operating in half-duplex mode.

A transmission channel-access scheme is based on a multiplexing access method, that will allow several data streams or signals to share the same communication channel or physical medium, so in this context, multiplexing is provided by the physical layer. In addition, a channel-access scheme is also based on a multiple access protocol and control mechanism, also known as Media Access Control (MAC). In effect, media access control deals with issues such as addressing, assigning multiplex channels to different users, and avoiding collisions. Thus, media access control is a sub-layer in Layer 2 (data link layer) or of the Open Systems Interconnection (OSI) model and a component of the Link Layer of the Transmission Control Protocol/Internet Protocol (TCP/IT) model.

In particular, the OSI model is a conceptual model that characterizes and standardizes the communication functions of a telecommunication or computing system without regard to their underlying internal structure and technology. Its goal is the interoperability of diverse communication systems with standard protocols. The model partitions a communication system into abstraction layers. The original version of the model defined seven layers.

In computer and other networks the link layer is the lowest layer in the IP suite, commonly known as TCP/IP, the networking architecture of the Internet. The link layer is the group of methods and communication protocols that only operate on the link that a host is physically connected to. The link is the physical and logical network component used to interconnect hosts or nodes or in the network and a link protocol is a suite of methods and standards that operate only between adjacent network nodes of an LAN or WAN connections.

Satellite communication systems offer a number of advantages over traditional terrestrial point-to-point and computer networks, such as: (1) Wide geographic coverage including interconnection of remote terrestrial networks ("islands"); (2) Bandwidth on demand, or Demand Assignment Multiple Access (DAMA) capabilities; (3) An alternative to damaged fiber-optic ground networks for disaster recovery options; and (4) Multipoint-to-multipoint communications facilitated by the Internet and broadcasting capability of satellites.

As stated in Chap. 2, in the context of spaceflight, a satellite is an object that has been placed into orbit by human endeavor. Such objects are sometimes called artificial satellites to distinguish them from natural satellites such as the Moon.

In satellite communication, signal transferring between the sender and receiver is done with the help of satellite and its transponders. In this process, the signal that is basically a beam of modulated microwaves is sent towards the satellite. Then the satellite amplifies the signal and sent it back to the receiver's antenna present on the Earth's surface. At this point, all the signal transferring is happening in space and this type of communication is known as space communication.

Thus, multiple access technique in satellite terms involves running communication streams between multiple satellite conduits or terminals at the same time. Normally, in simple traffic a terminal only handles one stream at a time. This approach doesn't work when a satellite's owner needs it to function managing thousands of points simultaneously. As a result, satellite technology today works with three different systems that offer multiple access ability.

Therefore, in fixed and mobile satellite communication systems, as a rule, many users are active at the same time. In general, the problem of simultaneous communications between many single or multipoint satellite users, however, can be solved by using different Multiple Access (MA) techniques. Since the resources of the systems such as the transmitting power and the bandwidth are limited, it is advisable to use the channels with complete charge and to create a different MA to the channel.

This generates a problem of summation and separation of signals in the transmission and reception parts, respectively. Deciding this problem consists in the development of orthogonal channels of transmission in order to divide signals from various users unambiguously on the reception part. There are five the following principal forms of MA techniques:

1. Frequency Division Multiple Access (FDMA) is a scheme where each concerned GES or MES is assigned its own different working carrier radio frequency inside the spacecraft transponder bandwidth.
2. Time Division Multiple Access (TDMA) is a scheme where all concerned Earth stations or MES use the same carrier frequency and bandwidth with time sharing, non-overlapping intervals.
3. Code Division Multiple Access (CDMA) is a scheme where all concerned Earth stations simultaneously share the same bandwidth and recognize the signals by various processes, such as code identification. Actually, they share the resources of both frequency and time using a set of mutually orthogonal codes, such as a Pseudorandom Noise (PN) sequence.
4. Space Division Multiple Access (SDMA) is a scheme where all concerned Earth stations can use the same frequency at the same time within a separate space available for each link.

5. Random (Packet) Division Multiple Access (RDMA) is a scheme where a large number of satellite users share asynchronously the same transponder by randomly transmitting short burst or packet divisions.

Currently, these methods of multiple accesses are widely in use with many advantages and disadvantages, together with their combination of hybrid schemes or with other types of modulations. Hence, multiple access technique assignment strategy can be classified into three methods as follows: (1) Preassignment or fixed assignment; (2) Demand Assignment (DA) and (3) Random Access (RA); the bits that make up the code words in some predetermined fashion, such that the effect of an error burst is minimized.

In the preassignment method satellite channel plans are previously determined for chairing the system resources, regardless of traffic fluctuations. This is suitable for communication links with a large amount of steady traffic. Since most mobile users in MSC do not communicate continuously, the preassignment method is wasteful of the satellite resources.

In Demand Assignment Multiple Access (DAMA) satellite channels are dynamically assigned to users according to the traffic requirements. Due to high efficiency and system flexibility, DAMA schemes are suited to MSC systems. In RA a large number of mobile users use the satellite resources in bursts, with long inactive intervals. In effect, to increase the system throughout, several mobile Aloha methods have been proposed. The MA techniques permit more than two Earth stations to use the same satellite network for interchanging information. Several transponders in the satellite payload share the frequency bands in use and each transponder will act independently of the others to filter out its own allocated frequency and further process that signal for transmission.

This feature allows any GES located in the corresponding coverage area to receive carriers originating from several MES and vice versa and carriers transmitted by one MES can be received by any GES. This enables a transmitting Earth station to group several signals into a single, multi-destination carrier. Access to a transponder may be limited to single carrier or many carriers may exist simultaneously. The baseband information to be transmitted is impressed on the carrier by the single process of multi-channel modulation.

Depending on how the available bandwidth is allocated to the users these techniques can be classified as narrowband and wideband systems.

The term narrowband technique is used to relate the bandwidth of the single channel to the expected coherence bandwidth of the channel. The available spectrum is divided in to a large number of narrowband channels, which are operated by frequency division duplexing (FDD). In narrow band FDMA, a user is assigned a particular channel, which is not shared by other users in the vicinity and if FDD is used then the system is called FDMA/FDD. The narrow band TDMA allows users to use the same channel but allocated a unique time slot to each user on the channel, thus separating a small number of users in time on a single channel. However, for narrow band TDMA, there generally are a large number of channels allocated using

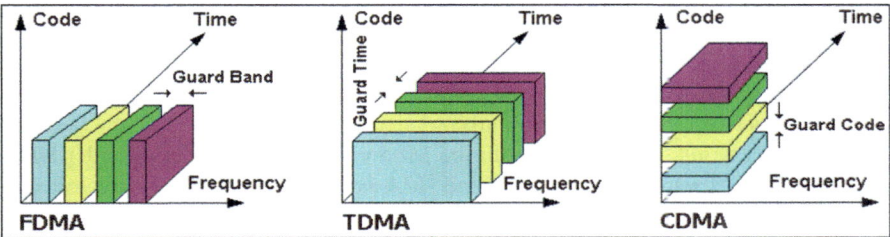

Fig. 3.17 Multiple access techniques (Courtesy of Web: by Google)

either FDD or TDD, each channel is shared using TDMA. Such systems are called TDMA/FDD and TDMA/TDD access systems.

In wideband systems, the transmission bandwidth technique of a single channel is much larger than the coherence bandwidth of the channel. Thus, multipath fading doesn't greatly affect the received signal within a wideband channel, and frequency selective fades occur only in a small fraction of the signal bandwidth.

3.5.1   Frequency Division Multiple Access (FDMA)

The most common and first employed MA scheme for satellite communication systems is FDMA concept shown in Fig. 3.17 (**FDMA**), where transmitting signals occupy non-overlapping frequency bands with guard bands between signals to avoid interchannel interference. The bandwidth of a repeater channel is therefore divided into many sub-bands each assigned to the carrier transmitted by an Earth station. The MES transmit continuously and the channel transmits several carriers simultaneously at a series of different frequency bands. Because of interchannel interference, it has to be provided guard intervals between each band occupied by a carrier to allow for the imperfections of oscillators and filters. The downlink Rx selects the required carrier in accordance with the appropriate frequency.

When the satellite transponder is operating close to its saturation, nonlinear amplification produces Intermodulation (IM) products, which may cause interference in the signals of other users. In order to reduce IM, it is necessary to operate the transponder by reducing the total input power according to input back off and that the IF amplifier provides adequate filtering.

Therefore, FDMA allocates a single satellite channel to one user at once. In fact, if the transmission path deteriorates, the controller switches the system to another channel. Although technically simple to implement, FDMA is wasteful of bandwidth because the voice channel is assigned to a single conversation, whether or not somebody is speaking. Moreover, this technique cannot handle alternate forms of data, only voice transmissions. In such a way, this system's advantages are that it is simple technique using equipment proven over decades to be reliable and it will remain very commonly in use because of its simplicity and flexibility.

It does have some disadvantages however:

1. A FDMA method is the relatively inflexible system and if there are changes in the required capacity, then the frequency plan has to change and thus, involve many Gateways and Earth terminals.
2. Multiple carriers cause IM in both the MES HPA and in the transponder HPA. Reducing IM requires back off of the HPA power, so it cannot be exploited at full capacity.
3. As the number of carriers increase, the IM products between carriers also increase and more HPA back off is needed to optimize the system. The throughput decreases relatively rapidly with the number of transmission carriers, therefore for 25 carriers it is about 40% less than with 1 carrier.
4. The FM system can suffer from what is known as a capture effect, where if two received signals are very close in frequency but of different strengths, the stronger one tends to suppress the weaker one. For this reason the carrier power has to be controlled carefully.

Therefore, with the FDMA technique, the signals from the various users are amplified by the satellite transponder in a given allocated bandwidth at the same time but at different frequencies. Depending on the multiplexing and modulation techniques employed, several transmission hybrid schemes can be considered and in general may be divided into two categories, based on the traffic demands of Earth stations on MCPC and SCPC.

3.5.1.1 Multiple Channels Per Carrier (MCPC)

The main elements of the MCPC mode are multiplexer, modulator and transmitter using a satellite uplink, when GES multiplexes baseband data is received from a terrestrial network and destined for various FES or MES terminals.

Moreover, the multiplexed data are modulated and transmitted to the allocated frequency segment, when the bandwidth of the transponder is shared among several MES, each with different traffic requirements. The transponder bandwidth is divided into several fixed segments, with several time frequency divisions allocated to these MES terminals. Namely, between each band segment is a guard band, which reduces the bandwidth utilization efficiency and the loss is directly related to the number of accessing MES in the network as shown in Fig. 3.17 (**FDMA**). Depending on the number of receiving MES, a total number of carriers will pass through the satellite transponder.

On the other hand, the signals received from different MES extract the carrier containing traffic addressed to GES by using an appropriate RF filter, demodulator, baseband filter and demultiplexer. The output of the demodulator consists in multiplexed telephone channels for a few MES together with the channels addressed to them. A baseband filter is used to filter out the desired baseband frequency segment and finally, a demultiplexer retrieves individual telephone channels and feeds them into the terrestrial network for onward transmission.

Each baseband filter of GES receive stations in this scheme corresponds to a specific one in the GES transmitting station. However, any change in channel capacity requires the return of this filter, which is difficult to implement. Thus, many schemes may be categorized according to the type of baseband signal.

3.5.1.2 Single Channel Per Carrier (SCPC)

For certain applications, such as the provision of MSC service to remote areas or individual MES, traffic requirements are low. In reality, assigning multiple channels to each MES is wasteful of bandwidth because most channels remain unutilized for a significant part of the day. For this type of application the SCPC type of FDMA is used. In the SCPC system each carrier is modulated by only one voice or by low to medium bit rate data channel. Some old analog systems use Companded FM but most new systems are digital PSK modulated. In the SCPC scheme, each carrier transmits a single carrier.

The assignment of transponder channels to each MES may be fixed Pre-Assigned Multiple Access (PAMA) or variable Demand Assignment Multiple Access (DAMA), the channel slots of the transponder are assigned to different MES units according to their instantaneous needs. In the case of PAMA, a few SCPC channels, about 5 to 10, are permanently assigned to each MES. In case of DAMA, a pool of frequency is shared by many MES terminals. When necessary, each MES requests a channel from frequency management of the Network Control Station (NCS), which may always attempt to choose the best available channel or a lower quality one until an unoccupied channel has been found. The allocation is then announced on a signaling channel known as a broadcast channel, which is received by the calling and called MES terminal, which then tune to the allocated channel. The communication takes place on the allocated channel and the end of call is announced by a signaling message, following which the NCS returns the channel to the common pool.

The SCPS solution requires an Automatic Frequency Control (AFC) pilot to maintain the spectrum centering on a channel-by-channel basis. This is usually achieved by transmitting a pilot tone in the centre of the transponder bandwidth. It is transmitted by designated reference GES and all users use this reference to correct their transmission frequency. A receiving user uses the pilot tone to produce a local AFC system which is able to control the frequency of the individual carriers by controlling the frequency of the LO.

Hence, drift in MSC translation frequency and frequency variations caused by the Doppler effect and the carriers retain their designated frequencies relative to each other. This feature is essential, because if uncorrected, the sum of the total frequency error can cause carrier overlapping, as carrier bandwidths are small. Thus, a stable receive frequency permits the GES demodulator design to be simplified. Centrally controlled networks, such as Inmarsat are simple to manage missions, because they provide a higher usage of channels and can use simple demand-assignment equipment. The SCPS scheme is cost-effective for networks consisting in a significant number of GES with a small number of channels.

3.5.2 Forms of FDMA Operations

There are several hybrid schemes of multiplexed FDMA in combination with SCPS, PSK, TDM and TDMA techniques.

3.5.2.1 SCPC/FM/FDMA

The baseband signals from the network or users each modulate a carrier directly, in either analog or digital form according to the nature of SCPC signal in question. Therefore, each carrier accesses the satellite on its particular frequency at the same time as other carriers on the different frequencies from the same or other station terminals. Information routing is thus, performed according to the principle of one carrier per link. The Inmarsat-A standard use SCPS, utilizing analog transmission with FM for telephone channels.

Thus, in calculating the channel capacity of the SCPC/FM system it is necessary to ensure that the noise level does not exceed specified defined values. The CCIR Recommendations for an analog channel state that the noise power at a point of zero, the relative level should not exceed 10,000 WOP with a 50 dB test tone, namely the noise ratio. In this way, it is assumed that the minimum required carrier-to-noise ratio per channel is at least 10 dB.

3.5.2.2 SCPC/PSK/FDMA

In this arrangement, each voice or data channel is modulated onto its own RF carrier. The only multiplexing occurs in the transponder bandwidth, where frequency division produces individual channels within the bandwidth, so Inmarsat is using various channel types of this multiplex scheme. The satellite transponder carrier frequencies may be PAMA or DAMA. For PAMA carriers the RF is assigned to a channel unit and the PSK modem requires a fixed-frequency LO input. For DAMA, the channels may be connected according to the availability of particular carrier frequencies within the transponder RF bandwidth. For this arrangement, the SCPC channel frequency required is produced by a frequency synthesizer.

The forward link assigned by TDM in shore-to-ship direction uses the SCPC/DA/FDMA solution for Inmarsat standard-B voice/data transmission. This standard in the return link for channel request employs Aloha O-QPSK and for low speed data/telex uses the TDMA scheme in ship-to-shore direction. The Inmarsat-Aero in forward ground-to-aircraft direction uses packet mode TDM for network broadcasting, signaling and data and the circuit mode of SCPS/DA/FDMA with distribution channel management for service communication links. Thus, the request for channel assignment, signaling and data in the return aircraft-to-ground direction the Slotted Aloha BPSK (1/2 – FES) of 600 b/s is employed and consequently, the TDMA scheme is reserved for data messages.

3.5.2.3 TDM/FDMA

This arrangement allows the use of TDM groups to be assembled at the satellite in FDMA, while the PSK is used as a modulation process at the Earth station. Systems such as this are compatible with FDM/FDMA carriers sharing the same transponders and the terminal requirements are simple and easily incorporated. The Inmarsat standard-B system for telex low speed data uses this scheme in the shore-to-ship direction only and in the ship-to-shore direction uses TDMA/FDMA. The CES TDM and SES TDMA carrier frequencies are pre-allocated by Inmarsat. Each CES is allocated at least one forward CES TDM carrier frequency and a return SES TDMA frequency. Thus, additional allocations can be made depending on the traffic requirements. The channel unit associated with the CES TDM channel for transmission consists in a multiplexer, different encoder, frame transmits synchronizer and modulator. So at the SES, the receive path of the channel has the corresponding functions to the transmitted end. The CES TDM channels use BPSK with differential coding, which is used for phase ambiguity resolution at the receive end.

3.5.2.4 TDMA/FDMA

As previously stated, however, the TDMA signals could occupy the complete transponder bandwidth. In fact, a better variation of this is where the TDMA signals are transmitted as a sub-band of transponder bandwidth, the remainder of which being available for example for SCPC/FDMA signals. Thus, the use of a narrowband TDMA arrangement is well suited for a system requiring only a few channels and has the all advantages of satellite digital transmission but can suffer from intermodulation with the adjacent FDMA satellite channels. Accordingly, the practical example of this scheme is the Tlx service of the Inmarsat-B system in ship-to-shore direction, which, depending on the transmission traffic, offers a flexible allocation of capacity for satellite communication and signaling slots.

3.5.3 Time Division Multiple Access (TDMA)

The TDMA application is a digital MA technique that permits individual Earth station transmissions to be received by the satellite in separate, non-overlapping time slots, called bursts, which contain buffered information. The satellite transponder receives these bursts sequentially, without overlapping interference and is then able to retransmit them to the MES terminal. Synchronization is necessary and is achieved using a reference station from which burst position and timing information can be used as a reference by all other stations. Each MES terminal must determine the satellite system in use, time and range so that the transmitted signal bursts, typically QPSK modulated, are timed to arrive at the satellite in the proper time slots.

The offset QPSK modulation is used by Inmarsat-B SES. So as to ensure the timing of the bursts from multiple MES, TDMA systems use a frame structure arrangement to support telex in the ship-to-shore direction. Therefore, a reference burst is transmitted periodically by a reference station to indicate the start of each frame to control the transmission timing of all data bursts. A second reference burst may also follow the first in order to provide a means of redundancy. In the proper manner, to improve the imperfect timing of TDMA bursts, several synchronization methods of random access, open-loop and closed-loop have been proposed.

In Fig. 3.17 **(TDMA)** a concept of TDMA is illustrated, where each mobile terminal transmits a data burst with a guard time to avoid overlaps. Since only one TDMA burst occupies the full bandwidth of the satellite transponder at a time, input back off, which is needed to reduce IM interference in FDMA, is not necessary in TDMA. At any instant in time, the transponder receives and amplifies only a single carrier. Thus, there can be no IM, which permits the satellite amplifier to be operated in full HPA saturation and the transmitter carrier power need not be controlled. Because all MES transmit and receive at the same frequency, tuning is simplified. This results in a significant increase in channel capacity. Another advantage over FDMA is its flexibility and time-slot assignments are easier to adjust than frequency channel assignments. The transmission rate of TDMA bursts is about 4800 b/s, while the frame length is about 1.74 s and the optimal guard time is approximately 40 ms, using the open-loop burst synchronization method.

There are some disadvantages because TDMA is more complex than FDMA:

1. Two reference stations are needed and complex computer procedures, for automated synchronizations between MES terminals.
2. Peak power and bandwidth of individual MES terminals need to be larger than with FDMA, owing to high burst bit rate.

Therefore, in the TDMA scheme, the transmission signals from various users are amplified at different times but at the same nominal frequency, being spread by the modulation in a given bandwidth. Depending on the multiplexing techniques employed, two transmission hybrid schemes can be introduced for use in MSS.

3.5.3.1   TDM/TDMA

The Inmarsat analog standard-A uses the TDM/TDMA arrangement for telex transmission. Each SES has at least one TDM carrier and each of the carriers has

20 telex channels of 50 bauds and a signaling channel. Moreover, there is also a common TDM carrier continuously transmitted on the selected idle listening frequency by the NCS for out-of-band signaling. The SES remains tuned to the common TDM carrier to receive signaling messages when the ship is idle or engaged in a telephone call. When an SES is involved in a telex forward call it is tuned to the TDM/TDMA frequency pair associated with the corresponding CES to send messages in shore-to-ship direction. Telex transmissions in the return ship-to-shore direction form a TDMA assembly at the satellite transponder. Each frame of the return TDMA telex carrier has 22 time slots, while each of these slots is paired with a slot on the TDM carrier. The allocation of a pair of time slots to complete the link is received by the SES on receipt of a request for a telex call. Otherwise, the Inmarsat-A uses for forward signaling a telex mode, while all other MSS Inmarsat standards for forward signaling and assignment channels use the TDM BPSK scheme.

The new generation Inmarsat digital standard-B (inheritor of standard-A) uses the same modulation TDM/TDMA technique but instead of Aloha BPSK (BCH) at a data rate of 4800 b/s for the return request channel used by Inmarsat-A, new standard-B is using Aloha O-QPSK (1/2 – FEC) at a data rate of 24 Kb/s. This MA technique is also useful for the small satellite terminal onboard maritime, land and aeronautical applications. In this case, the forward signaling and sending of messages in ground-to-mobile direction use a fixed assigned TDM carrier. The return signaling channel uses hybrid, slotted Aloha BPSK (1/2 FEC) with a provision for receiving some capacity and the return message channels in the mobile-to-ground direction are modulated by the TDMA system at a data rate of 600 b/s.

3.5.3.2   FDMA/TDMA

The Iridium system employs a hybrid FDMA/TDMA access scheme, which is achieved by dividing the available 10.5 MHz bandwidth into 150 channels introduced into the FDMA components. Each channel accommodates a TDMA frame comprising eight time slots, four for transmission and four for reception. Each slot lasts about 11.25 ms, during which time data are transmitted in a 50 Kb/s burst. Each frame lasts 90 ms and a satellite is able to support 840 channels. Thus, a user is allocated a channel occupied for a short period of time, during which transmissions occur.

3.5.4   Code Division Multiple Access (CDMA)

The CDMA solution is based on the use the modulation technique also known as Spread Spectrum Multiple Access (SSMA), which means that it spreads the information contained in a particular signal of interest over a much greater bandwidth than the original signal.

In this MA scheme the resources of both frequency bandwidth and time are shared by all users employing orthogonal codes, shown in Fig. 3.17 **(CDMA)**. Therefore, the CDMA is achieved by a PN (Pseudo-Noise) sequence generated by irreducible polynomials, which is the most popular CDMA method. In this way, a SSMA method using low-rate error correcting codes, including orthogonal codes with Hadamard or waveform transformation has also been proposed.

Concerning the specific encoding process, each user is actually assigned a signature sequence, with its own characteristic code, chosen from a set of codes assigned individually to the various users of the system. This code is mixed, as a supplementary modulation, with the useful information signal. On reception, from all the signals that are received, a given user is able to select and recognize, by its own code, the signal, which is intended for it, and then to extract useful information. The other received signal can be intended for other users but they can also originate from unwanted emissions, which gives CDMA a certain anti-jamming capability. For this operation, where it is necessary to identify one CDMA transmission signal among several others sharing the same band at the same time, correlation techniques are generally employed. From a commercial and military perspective this MA is still new and has significant advantages. Interference from adjacent satellite systems including jammers is better solved than with other systems.

This scheme is simple to operate as it requires no synchronization of the transmitter and is more suited for a military MES. Small mobile antennas can be very useful in these applications, without the interference caused by wide antenna bandwidths. Using multibeam satellites, frequency reuse with CDMA is very effective and allows good flexibility in the management of traffic and the orbit/spectrum resources. The Power Flux Density (PFD) of the CDMA signal received in the service area is automatically limited, with no need for any other dispersal processes. It also provides a low probability of intercept of the users and some kind of privacy, due to individual characteristic codes. The main disadvantage of CDMA by satellite is that the bandwidth required for the space segment of the spread carrier is very large, compared to that of a single unspread carrier, so the throughput is somewhat lower than with other systems.

Therefore, in the CDMA scheme, the signals from various users operate simultaneously, at the same nominal frequency but are spread in the given allocated bandwidth by a special encoding process.

Depending on the multiplexing techniques employed the bandwidth may extend to the entire capacity of the transponder but is often restricted to its own part, so CDMA can possibly be combined in the hybrid scheme with FDMA and/or TDMA. The SSMA technique can be classified into two methods: Direct Sequence (DS) and Frequency Hopping (FH). A combined system of DC and FH is called a hybrid CDMA system and the processing gain can be improved without increases of chip rate. The hybrid system has been used in the military Joint Tactical Information Distribution System (JTIDS) and OmniTRACS, which is Ku-band MSS, developed by the Qualcomm Company. In a more precise sense, the CDMA technique was developed and implemented by experts of the Qualcomm Company in 1987.

At present, the CDMA system advantages are practically effective in new satellite systems, such as Globalstar, also developed by Qualcomm, which is devoted to MSS handheld terminals and Skybridge, involved in FSS. This type of MA is therefore attractive for handheld and portable MSS equipment with a wide antenna pattern.

Antennas with large beam widths can otherwise create or be subject to interference with adjacent satellites. This MA technique is very attractive for commercial, military and even TT&C communications because some Russian satellites use CDMA for command and telemetry purposes.

The Synchronous-CDMA (S-CDMA) scheme proves efficiently to eliminate interference arising from other users sharing the same carrier and the same spot beam. Interference from other spot beams that overlap the coverage of the intended spot is still considerable. This process to ensure orthogonality between all links requires signalling to adjust transmission in time and frequency domains for every user independently.

3.5.4.1 Direct Sequence CDMA (DS-CDMA)

This dominant DS-CDMA technique is also called Pseudo-Noise (PN) modulation, where the modulated signal is multiplied by a PN code generator, which generates a pseudo-random binary sequence of length (N) at a chip rate (R_c), much larger than information bit rate (R_b), with a relation as follows:

$$R_c = N \cdot R_b \qquad (3.47)$$

This sequence is combined with the information signal cut into small chip rates (R_c), thus, speeding the combined signal in a much larger bandwidth (W ~ R_c), namely, the resulting signal has wider frequency bandwidth than the original modulated signal. The transmitting signal can be expressed in the following way:

$$s(t) = m(t) \, p(t) \, \cos \, (2\pi f_c t) = m(t) \, p(t) \, \cos \, \omega_c t \qquad (3.48)$$

where $m(t)$ = binary message to be transmitted and $p(t)$ = spreading NP binary sequence. At the receiver the signal is coherently demodulated by multiplying the received signal by a replica of the carrier.

Neglecting thermal noise, the receiving signal at the input of the detector of Low-Pass Filter (LPF) is given by:

$$r(t) = m(t) \, p(t) \, \cos \, \omega_c t \, (2 \, \cos \, \omega_c t) = m(t) \, p(t) + m(t) \, p(t) \, \cos \, 2\omega_c t \quad (3.49)$$

The detector LLF eliminates the HF components and retains only the LW components, such as $u(t) = m(t) \, p(t)$. This component is then multiplied by the local code [$p(t)$] in phase with the received code, where the product $p(t)^2 = 1$. At the output of the multiplier this gives:

$$x(t) = m(t)\ p(t)\ p(t) = m(t)\ p(t)^2 = m(t)\ [V] \tag{3.50}$$

The signal is then integrated over one bit period to filter the noise. The transmitted message is recovered at the integrator output, so in fact, only the same PN code can achieve the despreading of the received signal bandwidth. In this process, the interference or jamming spectrum is spread by the PN codes, while other user's signals, spared by different PN codes, are not despread.

Interference or jamming power density in the bandwidth of the received signal decreases from their original power. Otherwise, the most widely accepted measure of interference rejection is the processing gain (G_p), which is given by the ratio R_c/R_b and value of $G_p = 20$–60 dB. The input and output signal-to-noise ratios are related as follows:

$$(S/N)_{Output} = G_p(S/N)_{Input} \tag{3.51}$$

Therefore, for DS-CDMA a transmitter will assemble a bit stream consisting of message and code information as a function of time.

In the forward link, the GES (Hub Station) transmits the spread spectrum signals, which are spread with synchronized PN sequence to different MSC users. Since orthogonal codes can be used, the mutual interference in the network is negligible and the channel capacity is close to that of FDMA. Conversely, in the return link, the signals transmitted from different MES users are not synchronized and they are not orthogonal. The first case is referred to as synchronous and the second case as asynchronous SSMA.

However, the nonorthogonality causes interference due to the transmission of other users in the satellite network and as the number of simultaneously accessing users increases, the communication quality gradually degrades in a process called Graceful Degradation.

3.5.4.2 Frequency Hopping CDMA (FH-CDMA)

The FH-CDMA system works similarly to the DS system, since a correlation process of de-hopping is also performed at the receiver. The difference is that here the pseudo-random sequence is used to control a frequency synthesizer, which results in the transmission of each information bit rate in the form of (N) multiple pulses at different frequencies in an extended bandwidth. The transmitted and received signals of FH-CDMA system have the following forms:

$$\begin{aligned} s(t) &= m(t)\ \cos\ \omega_c(t)\ t \\ r(t) &= m(t)\ \cos\ \omega_c(t)\ t \cdot 2\ \cos\ \omega_c(t)\ t = m(t) + m(t)\ \cos\ 2\omega_c(t)\ t \end{aligned} \tag{3.52}$$

Thus, at the receiver the carrier is multiplied by an unmodulated carrier generated under the same conditions as at the transmitter. The second term in the receiver is eliminated by the LPF of the demodulator. The relation of processing gain for FH is:

$$G_p = W/\Delta f \qquad (3.53)$$

where W = frequency bandwidth and Δf = bandwidth of the original modulated signal. At this point, coherent demodulation is difficult to implement in FH receivers because it is a problem to maintain phase relation between the frequency steps. Due to the relatively slow operation of the frequency synthesizer, DS schemes permit higher code rates than FH radio.

3.5.5 Space Division Multiple Access (SDMA)

Previous systems have used frequency, time and code domains to achieve multiple access schemes, while SDMA uses spatial separation.

The significant factor in the performance of MA in a satellite communications system is interference caused by different factors and other users. In the other words, the most usual types of interference are co-channel and adjacent channel interference. The co-channel interference can be caused by transmissions from non-adjacent cells or spot beams using the same set of frequencies, where there is minimal physical separation from neighboring cells using the same frequencies, while the adjacent channel interference is caused by RF leakage on the subscriber's channel from a neighboring cell using an adjacent frequency.

This can occur when the user's signal is much weaker than that of the adjacent channel user. Signal-to-Interference Ratio (SIR) is an important indicator of call quality; it is a measure of the ratio between the mobile phone signal (the carrier signal) and an interfering signal. A higher SIR ratio means increasing overall system capacity.

Taking into account that within the systems of satellite communications, every user has their own unique spatial position, this fact may be used for the separation of channels in space and as a consequence, to increase he SIR ratio by using SDMA. In effect, this method is physically making the separation of paths available for each satellite link.

Terrestrial telecommunication networks can use separate cables or radio links but on a single satellite, independent transmission paths are required. Thus, this MA control radiates energy into space and transmission can be on the same frequency: such as TDMA or CDMA and on different frequencies, such as FDMA, which Frequency Reuse in SDMA is illustrated in Fig. 3.18.

The SDMA scheme in satellite communications is implementing mode that optimizes the use of radio spectrum and minimizes system cost by taking advantage of the directional properties of dish antennas. In SDMA, satellite dish antennas transmit signals to numerous zones on the earth's surface. The users antennas are

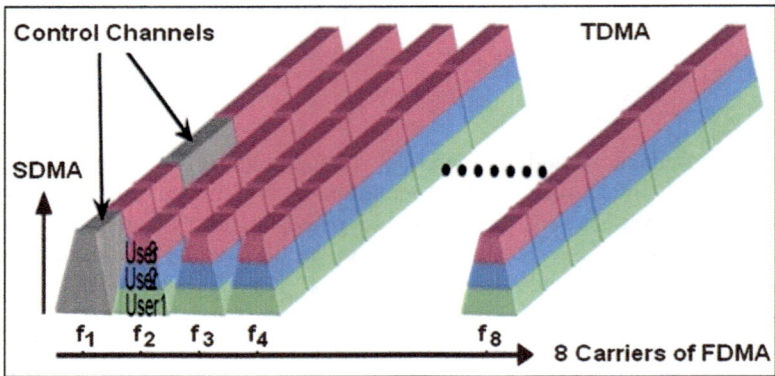

Fig. 3.18 Frequency reuse in SDMA (Courtesy of Web: by Google)

highly directional, allowing duplicate frequencies to be used for multiple surface zones.

Consider a scenario in which signals must be transmitted simultaneously by one satellite to mobile or fixed receivers in 20 different surface zones. In a conventional satellite system, 20 channels and 20 antennas would be necessary to maintain channel separation. In SDMA, there can be far fewer channels than zones. Thus, if duplicate-channel zones are sufficiently separated, the 20 signals can be transmitted to Earth stations using four or five channels. The narrow signal beams from the satellite antennas ensure that interference will not occur between zones using the same frequency. The SDMA technique requires careful choice of zones for each transmitter, and also requires precise antenna alignment. A small error can result in failure of one or more channels, interference among channels, and/or confusion between surface coverage zones.

3.5.5.1 Special Effects of SDMA in Satellite Systems

The SDMA technology has been successfully used in satellite communications for several years. As stated, the SDMA technique can also be integrated with all the different MA techniques in use, such as FDMA, TDMA and CDMA, and therefore can be applied to any fixed or mobile communication system. However, we shall see that the ways in which the SDMA technique can be introduced and the advantages it provides differ depending on the system under consideration. As aforesaid described, modifications required to realize the SDMA technique are limited to the satellite array, and thus do not involve fixed or mobile units. However, this allows introducing this technique in existing fixed or mobile satellite systems, with no need to modify their characteristics.

The ability to reject jammer of adaptive arrays can be ensured for performing SDMA mode where several fixed or mobile users are allowed to share the same classical access in a cell for cellular systems or in a channel for satellite systems, leading to a capacity increase. In such a way, this sharing can be realized with possibility through the use of adaptive beam forming and interference rejection on the satellite uplink and downlink communication for users, which are located at different angular sectors.

In using SDMA, either FDMA or TDMA are needed to allow GES to roam in the same satellite beam or for polarization to enter the repeater. Thus, the frequency reuse technique of same frequency is effectively a form of SDMA scheme, which depends upon achieving adequate beam-to-beam and polarization isolation. Using this system reverse line means that interference may be a problem and the capacity of the battery is limited.

On the other hand, a single satellite may achieve spatial separation by using beams with horizontal and vertical polarization or left-hand and right-hand circular polarization. This could allow two satellite beams to cover the same Earth surface area, being separated by the polarization. Thus, the satellite could also have multiple beams using separate antennas or using a single antenna with multiple feeds.

In case of multiple satellite communication systems, spatial separation can be achieved with orbital longitude or latitude and for intersatellite links, by using different planes. Except for frequency reuse, this system provides on-board switching techniques, which, in turn, enhance channel capacity. Additionally, the use of narrow beams from the satellite allows the Earth station to operate with smaller antennas and so produce a higher power density per unit area for a given transmitter power.

Therefore, through the careful use of polarization, beams (SDMA) or orthogonal (CDMA), the same frequency spectrum may be reused several times, with limited interference among fixed or mobile users.

The more detailed benefits of an SDMA system include the following:

1. The number of cells required to cover a given area can be substantially reduced.
2. Interference from other systems and from users in other cells is significantly reduced.
3. The destructive effects of multipath signals, copies of the desired signal that have arrived at the antenna after bouncing from objects between the signal source and the antenna can often be mitigated.
4. Channel reuse patterns of the systems can be significantly tighter because the average interference resulting from co-channel signals in other cells is markedly reduced.
5. Separate spatial channels can be created in each cell on the same conventional channel. In other words, intra-cell reuse of conventional channels is possible.
6. The SDMA station radiates much less total power than a conventional station. One result is a reduction in network-wide RF pollution. Another is a reduction in power amplifier size.
7. The direction of each spatial channel is known and can be used to accurately establish the position of the signal source.
8. The SDMA technique is compatible with almost any modulation method, bandwidth, or frequency band including GSM, PHP, DECT, IS-54, IS-95 and

other formats. The SDMA solution can be implemented with a broad range of array geometry and antenna types.

Another perspective of the realization of SDMA systems is the application of smart antenna arrays with different levels of intelligence consisting in the antenna array and digital processor. Since the frequency of transmission for satellite communications is high enough (mostly 6 or 14 GHz), that the dimensions of an array placed in orbit is commensurable with the dimensions of the parabolic antenna, is a necessary condition to put such systems into orbit.

Thus, the SDMA scheme mostly responds to the demands of LEO and MEO constellations, when the signals of users achieve the satellite antenna under different angles ($\pm 22°$ for the MEO). In this instance, ground level may be split into the number of zones of service coverage determined by switched multiple beam pattern lobes in different satellite detections, or by adaptive antenna separations, which scenario is illustrated in Fig. 3.19 (Left). For instance, there are two different beamforming approaches in different SDMA satellite communications: (1) The multiple spot beam antennas are the fundamental way of applying SDMA in large fixed and mobile satellite systems and (2) Adaptive array antennas dynamically adapt to the number of users.

3.5.5.2 Switched Spot Beam Antenna

Switched Multi-Beam Antennas are designed to track each subscriber of a given cell with an individual beam pattern as the target subscriber moves within the cell (spot). Therefore, it is possible to use array antennas and to create a group of overlapping beams that together result in omnidirectional coverage. This is the simplest technique comprising only a basic switching function between separate directive antennas or predefined beams of an array. Beam-switching algorithms and RF signal-processing software are incorporated into smart antenna designs. For each call, software algorithms determine the beams that maintain the highest quality signal and the system continuously updates beam selection, ensuring that customers get optimal quality for the duration of their call. One might design overlapping beam patterns pointing in slightly different directions, similar to the ones shown in Fig. 3.19 (Left). Every so often, the system scans the outputs of each beam and selects the beam with the largest output power. The black cells reuse the frequencies currently assigned to the mobile terminals, so they are potential sources of interference. In fact, the use of a narrow beam reduces the number of interfering sources seen at the base station. Namely, as the mobile moves, the smart antenna system continuously monitors the signal quality to determine when a particular beam should be selected.

Switched-beam antennas are normally used only for the reception of signals, since there can be ambiguity in the system's perception of the location of the received signal. In fact, these antennas give the best performance, usually in terms of received power but they also suppress interference arriving from directions

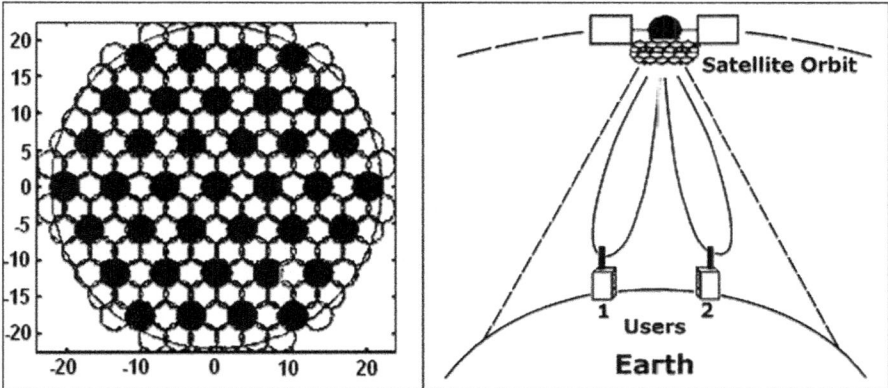

Fig. 3.19 Beam patterns and adaptive antenna for SDMA (Courtesy of Paper: by Zaharov)

away from the active antenna beam's centre, because of the higher directivity, compared to a conventional antenna, some gain is achieved. In high-interference areas, switched-beam antennas are further limited since their pattern is fixed and they lack the ability to adaptively reject interference. Such an antenna will be easier to implement in existing cell structures than the more sophisticated adaptive arrays but it gives only limited improvement.

3.5.5.3 Adaptive Array Antenna Systems

Adaptive Array Antenna Systems select one beam pattern for each user out of a number of preset fixed beam patterns, depending on the location of the subscribers. At all events, these systems continually monitor their coverage areas, attempting to adapt to their changing radio environment, which consists in (often mobile) users and interferers. Thus, in the simplest scenario, that of a single user and no interferers, the system adapts to the user's motion by providing an effective antenna system pattern that follows the mobile user, always providing maximum gain in the user's direction.

The principle of SDMA with adaptive antenna system application is quite different from the beam-forming approaches described in Fig. 3.19 (Right).

The events processed in SDMA adaptive array antenna systems are as follows:

1. A "Snapshot", or sample, is taken of the transmission signals coming from all of the antenna elements, converted into digital form and stored in memory.
2. The SDMA digital processor analyzes the sample to estimate the radio environment at this point, identifying users and interferers and their locations.
3. The processor calculates the combining strategy for the antenna signals that optimally recovers the user's signals. With this strategy, each user's signal is received with as much gain as possible and with the other users/interferers signals rejected as much as possible.

4. An analogous calculation is done to allow spatially selective transmission from the array. Each user's signal is now effectively delivered through a separate spatial channel.

5. The system now has the ability to both transmit and receive information on each of the spatial channels, making them two-way channels.

As a result, the SDMA adaptive array antenna system can create a number of two-way spatial channels on a single conventional channel, be it frequency, time, or code. Of course, each of these spatial channels enjoys the full gain and interference rejection capabilities of the antenna array. In theory, an antenna array with (n) elements can support (n) spatial channels per conventional channel. In practice, the number is somewhat less because the received multipath signals, which can be combined to direct received signals, takes place. In addition, by using special algorithms and space diversity techniques, the radiation pattern can be adapted to receive multipath signals, which can be combined. Hence, these techniques will maximize the SIR or Signal-to-Interference and Noise Ratio (SINR).

3.5.5.4 SDMA/FDMA

This modulation arrangement uses filters and fixed links within the satellite transceiver to route an incoming uplink frequency to a particular downlink transmission antenna, shown in Fig. 3.20 (Left). A basic configuration of fixed links may be set up using a switch that is selected only occasionally. Thus, an alternative solution allows the filter to be switched using a switch matrix, which is controlled by a command link. Because of the term SS (Switching Satellite) this scheme would be classified as SDMA/SS/FDMA, which block diagram is shown in Fig. 3.20 (Right).

The satellite switches are changed only rarely, only when it is desired to reconfigure the satellite, to take account of possible traffic changes. The main disadvantage of this solution is the need for filters, which increase the mass of the payload.

3.5.5.5 SDMA/TDMA

This solution is similar to the previously explained SDMA/SS/FDMA in that a switch system allows a TDMA receiver to be connected to a single beam. Switching again is only carried out when it is required to reconfigure the satellite. Under normal conditions, a link between beam pairs is maintained and operated under TDMA conditions.

The utilization of time slots may be arranged on an organized or contention basis and switching is achieved by using the RF signal. Thus, on board processing is likely to be used in the future, allowing switching to take place by the utilization of baseband signals. The signal could be restored in quality and even stored to allow transmission in a new time slot in the outgoing TDMA frame.

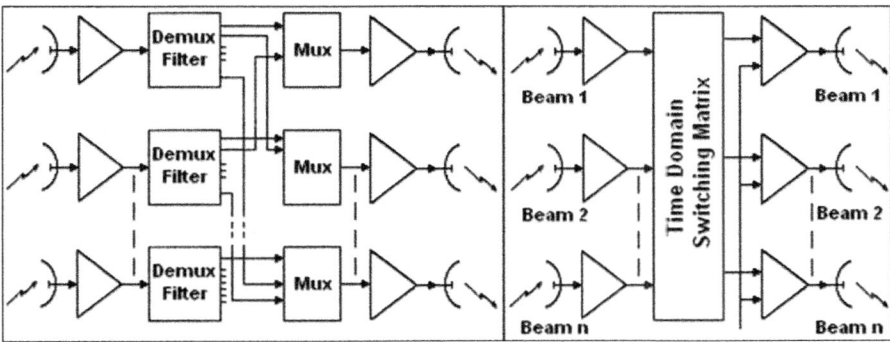

Fig. 3.20 Block diagram of SDMA/FDMA and SDMA/SS/FDMA-SDMA/SS/TDMA (Courtesy of Book: by Calcutt)

This scheme is providing up and downlinks for the later Intelsat VI spacecraft, known as SDMA/SS/TDMA, which block diagram is shown in Fig. 3.20 (Right). This MA is using to allow TDMA traffic from the uplink beams to be switched to downlink beams during the course of TDMA frame. At this point, the connection exists at a specific time for the burst duration within the frame time before the next connection is made, and so on.

3.5.5.6 SDMA/CDMA

This arrangement allows access to a common frequency band and may be used to provide MA to the satellite, when each satellite stream is decoded in order to obtain the destination address. Onboard circuitry must be capable of determining different destination addresses, which may arrive simultaneously, while also denying invalid users access to the downlink. Onboard processors allow the CDMA bit stream to be retimed, regenerated and stored on the satellite. Because of this possibility the downlink CDMA configurations need not be the same as for uplink and the Earth link may thus, be optimized.

3.5.6 Random Division Multiple Access (RDMA)

For data transmission, a bit stream may be sent continuously over an established channel without the need to provide addresses or unique words if the channel is not charred. In fact, where charring is implemented, data are sent in bursts, which thus, requires unique words or synchronization signals to enable time-sharing with other users, to be affected in the division of channels. Each burst may consist in one or more packets comprising data from one or more sources that have been assembled over time, processed and made ready for transmission. However, this type of multiplex scheme is also known as Packet MA. Packet access can be used in special RDMA solutions, such as Aloha, where retransmission of blocked packets may be required.

Random access can be achieved to the satellite link by contention and for that reason is called a contention access scheme. This type of access is well-suited to satellite networks containing a large number of stations, such as MES, where each station is required to transmit short randomly-generated messages with long dead times between messages.

The principle of RDMA is to permit the transmission of messages almost without restriction, in the form of limited duration bursts, which occupy all the bandwidth of the transmission channel. Therefore, in other words, this is MA with time division and random transmission and an attribute for the synonym Random Division Multiple Access is quite assessable. A user transmits a message irrespective of the fact that there may be other users equally in connection. The probability of collisions between bursts at the satellite is accepted, causing the data to be blocked from receipt by the Earth station. In case of collision, the destination Earth station receiver will be confronted with interference noise, which can compromise message identification and retransmission after a random delay period. The retransmissions can occur as many times as probably are carried out, using random time delays.

Therefore, such a scheme implies that the transmitter vies for satellite resources on a per-demand basis. It will provide that no other transmitter is attempting to access the same resources during the transmission burst period, when an error-free transmission can occur. The types of random protocols are distinguished by the means provided to overcome this disadvantage. The performance of these protocols is measured in terms of the throughput and the mean transmission delay. Throughput is the ratio of the volume of traffic delivered at the destination to the maximum capacity of the transmission channels. The transmission time, i.e., delay is a random variable. Its mean value indicates the mean time between the generation of a message and its correct reception by the destination station.

3.5.6.1 Aloha

The most widely used contention access scheme is Aloha and its associated derivatives. This solution was developed in the late 1960s by the University of Hawaii and allows usage of small and inexpensive Earth stations (including MES) to communicate with a minimum of protocols and no network supervision. This is the simplest mode of operation, which time-shares a single RF divided among multiple users and consists in stations randomly accessing a particular resource that is used to transmit packets. When an Aloha station has something to transmit, it immediately sends a burst of data pulses and can detect whether its transmission has been correctly received at the satellite by either monitoring the retransmission from the satellite or by receiving an acknowledgement message from the receiving party. Should a collision with another transmitting station occur, resulting in the incorrect reception of a packet at the satellite, the transmitting station waits for a random period of time, prior to retransmitting the packet.

Otherwise, a remote station (MES or FES terminals) uses Aloha to get a Hub station (GES) terminal's attention. Namely, the MES terminal sends a brief burst requesting a frequency or time slot assignment for the main transmission. Thus, once the assignment for MES is made, there is no further need for the Aloha channel, which becomes available for other stations to use. After that, the main transmissions are then made on the assigned channels. At the end, the Aloha channel might be used again to drop the main channel assignments after the transmission is completed. The advantages of Aloha are the lack of any centralized control, giving simple, low-cost stations and the ability to transmit at any time, without having to consider other users.

In the case where the user population is homogenous, so that the packet duration and message generation rate are constant, it can be shown that the satellite traffic carried S (packet correctly interpreted by the receiver), as a function of total traffic G (original and retransmitted message) is given by the relation:

$$S = G \exp(-2G) \text{ [packet/time slot]} \tag{3.54}$$

where (S = transmission throughput) and (G) are expressed as a number of packets per time slot equal to the common packet duration. The Aloha protocol cannot exceed a throughput of 18% and the mean transmission time increases very rapidly as the traffic increases due to an increasing number of collisions and packet retransmissions.

The Aloha mode is relatively inefficient with a maximum throughput of only 18.4% (1/2). However, this has to be counter-weight against the gains in simple network complexity, since no-coordination or complex timing properties are required at the transmitting MES.

3.5.6.2   Slotted Aloha

This form of Aloha or S-Aloha, where the time domain is divided into slots equivalent to a single packet burst time; there will be no overlap, as is the case with ordinary Aloha. In Fig. 3.21 is shown a simplified arrangement that presents this point. The transmissions from different stations are now synchronized in such a way that packets are located at the satellite in time slots defined by the network clocks and equal to the common packet duration. There cannot be partial collisions, because every collision arises from complete superposition of packets. The time-scale of collision is reduced to the duration of a packet, whereas with the Aloha protocol, this timescale is equal to the duration of both packets. This situation divides the probability of collision by two and the throughput becomes:

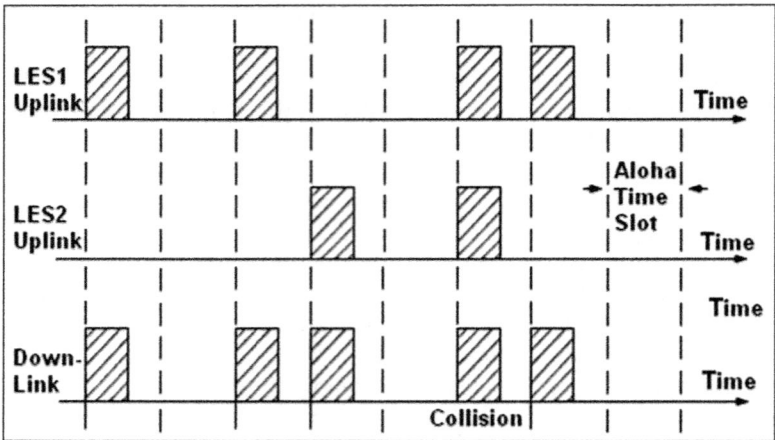

Fig. 3.21 Time slot organization for Slotted Aloha (Courtesy of Book: by Calcutt)

$$S = G \exp(-G) \ [\text{packet/time slot}] \tag{3.55}$$

This protocol enables collisions between new messages and retransmission to be avoided and increases the throughput of S-Aloha in the order of 50–60% by introducing a frame structure, which permits the numbering of time slots. Each packet incorporates additional information indicating the slot number reserved for retransmission in case of collision.

For the same value of utilization as basic Aloha, the time delay and probability of packet loss are both improved. The major disadvantages of S-Aloha are that more complex equipment in the Earth station is necessary, because of the timing requirement and because there are fixed time slots, customers with a small transmission requirement are wasting capacity by not using the time slot to its full availability.

3.5.6.3 Slot Reservation Aloha

This solution of an extension for the slotted-Aloha scheme allows time slots to be reserved for transmission by an Earth station. In general terms this mode of operation is termed a Packed Reserved Multiple Access (PRMA). Slot reservation basically takes the following two forms:

1. **Implicit** – When a station acquires a slot and successfully transmits, the slot is reserved for that station for as long as it takes the station to complete its transmission. The network controller then informs all stations on the network that the slot is available for contention once more. There is only the problem that a station with much data to transmit could block the system to other users.

2. **Explicit** – Every user station may send a request for the reservation of a time slot prior to transmission of data. A record of all time slot occupation and reservation requests is kept. Actually, a free time slot could be allocated on a priority basis. Some kind of control for the reservation of slots is necessary and this could be accomplished by a single or all stations being informed of slot occupancy and reservation requests.

3.6   Satellite Broadband and Internet Protocols

Under the third and fourth-generation IMT-2000 services, high-speed and large-capacity multimedia data and Internet will be delivered through the satellite and radio networks and a wide variety of transmission services, such as global roaming to be provided at the same time. In such a way, many solutions and protocols will be employed in terrestrial radio and telecommunication systems, including ISDN, IP, Transmission Control Protocol (TCP), IP/ATM, Unspecified Bit Rate (UBR) and Available Bit Rate (ABR) broadband and Internet solutions and protocols. With its new advanced service menu, the IMT-2000 network is required to provide not only mobility management for individual terminals but also various service control functions. With capabilities such as QoS, in which the desired communications quality is individually established for each session, the IP/ATM switching system provides for a sophisticated and economic satellite network configuration.

Usage of different satellite fixed and mobile solutions has increased dramatically in past years serving all satellite applications. Along with that trend there has been an increasing demand for higher transmission rates. Hence, to cope with this demand, new technologies have emerged onto the radio and telecommunications stage. The IP/ATM over satellite is one fine example, which is creating waves in the global telecommunication world today and is ideal for a wide range of satellite applications, including data transfer, voice, imaging, full motion video and multimedia. Its characteristics include scalability, which make it ideal for the traffic demands of broadband multimedia expected on data networks of the future. Just as the Internet revolutionized worldwide communications, ATM brings new meaning to high-speed satellite networking.

The development of the Internet technologies is the concern of the Internet Engineering Task Force (IETF), which publishes its recommendations on the website under a series of Request for Comments (RFC) documents, with a specific identification number for each RFC. These documents are available for downloading from the IETF website in the form of information FYI RFC documents or those that specify Internet Standards (STD RFC).

3.6.1   Satellite Internet Protocol (IP)

The drive towards the establishment of an information society will bring together the two most successful solutions of all of the technological advances of the latter quarter of the twentieth century. In fact, this is the integration of Internet with satellite fixed and mobile communications and in particular to deliver Mobile IP to all mobile customers. In this respect, fourth generation (4G) mobile networks will be based upon an IP-environment.

Significant effort around the world is now underway towards standardizing such a mobile environment through such organizations as the Mobile Wireless Internet Forum (MWIF), IETF, 3GPP and 3GPP2. Recently, two different approaches to the network architecture by 3GPP and 3GPP2 have been proposed. First, the 3GPP solution for the W-CDMA radio interface, is based on an evolution of the personal satellite network, with enhancements to the call control functionalities obtained though the introduction of a new network element, Called State Control Function (CSCF), to allow the provision of Data over IP (DoIP) and Voice over IP (VoIP) satellite services. Second, the 3GPP2 solution for the CDMA2000 radio interface system has adopted a forthcoming packet network architecture incorporating Mobile IP functionalities in support of packet data mobility. Eventually, it is hoped that a single, harmonized solution will emerge from the two distinct approaches. In the future fixed and mobile network, where different satellite applications will operate alongside terrestrial networks, various categories of service will co-exist, where "best effort", which is presently available over the Internet, will operate with guaranteed QoS classes.

The move towards an all IP network should facilitate the networking between satellite and terrestrial mobile networks, since the problem of providing mobile link connectivity across the different networks reduces to the level of attaining the appropriate route to direct the packets of information to the appropriate terminal. In the IP environment, the fixed and mobile terminal has a temporary IP address, known as the Care-of-Address (CoA), which is associated with a correspondent node of a particular access network. The CoA is in addition to the IP address that is permanently assigned to the mobile terminal, which is its home address, which is stored at its home sub-network. Thus, the mobile node's CoA is made known to the home agent. Packets addressed to the mobile node are thus, tunneled by the home agent to the correspondent node, identified by the CoA. Namely, when a mobile node moves from one access network to another, a new CoA will be assigned by the visiting network operator corresponding to the new correspondent node.

The fixed and mobile node then makes the home agent aware of the change in the CoA, through the transmission of a binding update message. The future fixed and mobile networks for all satellite applications will use whatever wireless networking is available at reasonable cost. The LEO, MEO, HEO or GEO satellite constellation may be used, as well as wireless VHF Data Link (VDL), all kind of future Stratospheric Platform Systems (SPS) and perhaps next-generation of cellular networks, if in the future survives the competition systems previously mentioned.

Unique, cost-effective satellite techniques will also be deployed using GEO satellite direct broadcasting, known as Digital Video Broadcasting (DVB). Digital TV satellites with data embedded in the Moving Picture Expert Group 2 (MPEG-2), transport streams to transmit data to the fixed and mobile VSAT terminals. The return path would be an inexpensive, low-bandwidth, duplex channel.

3.6.1.1   IP Security Protocol (IPSec)

The IPSec protocol is a framework of open standards developed by the IETF, which can provide security for the transmission of sensitive information over unprotected networks such as the Internet. In this sense, the IPSec protocol acts at the network layer, protecting and authenticating IP packets between participating IPSec devices, such as Cisco or other IP routers. With IPSec, data can be transmitted across a public network without fear of observation, modification, or spoofing. This mode enables fixed applications such as Virtual Private Networks (VPN), including Intranets, Extranets and remote user access.

Fixed and Mobile users will be able to establish a secure connection back to their office, similarly to the encryption method. For example, the user can establish an IPSec known as a "tunnel" with a corporate firewall, requesting authentication services in order to gain access to the corporate network; all of the traffic between the user and the firewall will then be authenticated. The user can then establish an additional IPSec tunnel requesting data privacy services with an internal router or end system.

Therefore, the IPSec method is a framework of open IP standards that provides data confidentiality, data integrity and data authentication between participating peers. IPSec provides these security services at the IP layer; it uses Internet Key Exchange (IKE) to handle the negotiation of protocols and algorithms based on local policy and to generate the encryption and authentication keys to be used by IPSec. In fact, IPSec can be used to protect one or more data flows between a pair of hosts, between a pair of security Gateways, or between a security Gateway and a host.

3.6.1.2   Mobile Transmissions Over IP (MToIP)

The IP data link can be used for transmissions such as regularly made long-distance phone calls, videoconferencing connections or sending Fax messages. In such a way, IP telephony is known as a Voice over IP (VoIP), whilst similar to this, Videophone can also be established over IP (VPoIP) and Fax over IP (FoIP). Actually, these services are the transmission of telephone, video or Fax calls over a data network like one of the many terrestrial networks that make up the Internet or will be possible through Mobile Satellite Internet (MSI) as future MSC standards.

On the other hand, a VoIP application meets the challenges of combining legacy voice networks and packet networks by allowing both voice and signaling information to be transported over the packet network. A FoIP application enables the networking of standard Fax machines with packet networks. It accomplishes this by extracting the Fax image from an analog signal and carrying it as digital data over the packet network. The VPoIP mode is a similar application, which can enable the mobile and semi fixed networking of standard video cameras, videoconferencing or videophones with mobile packet network or via mobile ISDN links.

However, the TDM over IP (TDMoIP) is a transport technology that extends T1, E1, T3, E3, serial data or analog voice circuits transparently across IP or Ethernet networks. When used for voice mode, these TDM over IP circuits are transparent to signaling and provide superior voice quality with much lower latency than VoIP. The TDM scheme over IP supports all PBX features and all modem and fax rates. In other words, TDMoIP is not limited to voice and can extend to circuits carrying any protocol over IP including Frame Relay, ATM, ISDN, SS7, SNA, HDLC, Asynchronic, Synchronic, X.25, as well as H.320 and H.324 video over IP (VPoIP). The TDM scheme over IP was recently developed by the RAD Company to provide a simple, inexpensive migration strategy to IP-based networks.

3.6.1.3   Mobile IP Version 6 (MIPv6)

The IP was introduced in the Arpanet in the mid 1970s; thus the version in common used today is IP version 4 (IPv4). Although several protocol suites (including Open System Interconnection) have been proposed over the years to replace IPv4, none have succeeded because of IPv4's large and continually growing, installed base.

Nevertheless, IPv4 was never intended for the current Internet, either in terms of the number of hosts, types of applications, or security concerns. However, in the early 1990s, IETF recognized that the only way to cope with these changes was to design a new version of IP as a successor to IPv4. In such a way, the IETF formed the next IP generation (IPgen) Working Group to define this transitional protocol to ensure long-term compatibility between the current and new IP versions and support for current and emerging IP-based applications.

The result of this effort was IP version 6 (IPv6), described in RFC 1883–1886; these four RFC were officially entered into the Internet Standard Track in December 1995. IPv6 is designed as an evolution from IPv4 rather than as a radical change. Useful features of IPv4 were carried over in IPv6 and less useful features were dropped. Mobile IPv6 is an IP-layer mobility protocol for the IPv6 Internet, being standardized by the IETF. Thus, the idea is that when mobility, like any other functionality, is implemented in the network layer, it needs to be implemented only once and will then be transparently available for all higher layer protocols. It remains to be seen how well this promise is fulfilled in practice. There are, however, some applications such as mobile VPN access, for which Mobile IP is clearly a good solution.

The mobile network is configured in the way that the first half of an IPv6 address indicates the subnet to which the address belongs and it is used for routing IP packets across the Internet. When a mobile Internet host, known as a Mobile Node

(MN) in the Mobile IPv6 terminology, moves to a different place in the network topology, its subnet and IP address necessarily change. This creates two kinds of problems: existing connections (e.g., TCP connections and IPSec security associations) between the mobile and other hosts known as a Correspondent Nodes (CN) become invalid and the mobile is no longer reachable at its old address for new connections. The former problem is important in stateful protocols and has little effect on stateless protocols, such as Hyper-Text Transfer Protocol (HTTP). The latter problem typically concerns servers and not client computers.

Mobile IPv6 protocol has two basic goals: all transport-layer and higher layer connections and security associations between the mobile and its correspondents should survive the address change and the mobile host should be reachable as long as it is connected to the Internet somewhere in the world. Mobile IP makes some quite strong assumptions about the environments in which it is used. First, that all mobile hosts have a home network and a Home Address (HoA) on that network. This is a reflection from a time when mobility was an exception: few Internet nodes would be mobile and even they would for most of the time remains stationary at home.

In any case, Mobile IP solves the reachability problem by ensuring that the mobile is always able to receive packets sent to its home address. The IP address of a stationary IP node normally serves two purposes: it is both an identifier for the node and an address that is used for routing messages to the node. Mobile IP preserves this dual use of home addresses, which are an identifier for the mobile, as well as an address to which correspondents can send packets.

The mobile's current location, called Care-of Address (CoA), on the other hand, is a pure address and serves no identification purposes. Any IPv6 address can be or become mobile and there is no way of distinguishing a mobile and stationary host by just looking at its address. This is because the Mobile IP protocol was originally designed to be transparent to the mobile's correspondents and the correspondent did not need to know that the mobile, in fact, was a mobile.

An additional technical advantage in the adoption of the MIPv6 IP solution as a final stage of standardization should be determined to provide the opportunity to perform seamless communications handover between satellite and TTN. Recent trials involving handover between TTN using MIPv6 have demonstrated the feasibility of such an approach. At any rate, researchers are now addressing the needs of integrated space/terrestrial mobile communication networks, based upon packet-oriented service delivery.

3.6.2   Transmission Control Protocol (TCP)

The transmission of Internet packets in Tx-Rx direction is primarily achieved using the TCP solution. The TCP mode is a connection-oriented transport protocol that sends data as an unstructured stream of bytes and provides the functionality to ensure that the transmission rate of data over the network is appropriate for the capabilities of the Rx device, as well as the devices that are used to route the data from the Tx to the Rx. Thus, by using sequence numbers and acknowledgment messages, TCP can provide a sending node with delivery information about packets transmitted to a destination node. Where data has been lost in transit from source to destination, TCP can retransmit the data until either a timeout condition is reached or until successful delivery has been achieved or can also recognize duplicate messages and will discard them appropriately.

If the sending PC is transmitting too fast for the receiving computer, TCP can employ flow control mechanisms to slow data transfer. The TCP method can also communicate delivery information to the upper-layer protocols and the applications it supports and is responsible for ensuring that the network's resources are divided in an equitable manner among all users of the network. Applications such as File Transfer Protocol (FTP) and HTTP (the language of the Web) rely on TCP to transport their data over the network as quickly as the network will allow.

3.6.2.1   TCP/IP Over Satellite

As far a TCP/IP connection is concerned, a mobile satellite network should be viewed as any other network connection. Given the way that TCP operates, the transmission of IP packets over the MSC link poses several problems that need to be overcome if services are to be delivered efficiently. In such a manner, the major difficulty is due to the latency of the satellite link, which when combined with a burst error channel and the characteristics of the TCP protocol itself, can result in an inefficient means of transmission. This is because TCP operates using a conservative congestion control mechanism, whereby new data can be transmitted only when an ACK (acknowledgement) from a previous transmission has been received.

Therefore, with this in mind, the need for a high quality link between the satellite and MES is re-emphasized, since packets in error are presently deemed to be due to congestion on the satellite network. Hence, in this case, TCP responds by reducing its transmission rate accordingly. It can be seen that TCP operates on the basis of "best effort", founded on the available resources of the network.

When starting transmission, TCP enters the network in a restrained manner, whereby the initial rate of transmission is carefully controlled to avoid overloading the network with traffic. The TCP method achieves this by employing a congestion control mechanism such as slow start, congestion avoidance, fast retransmit and recovery.

Slow start is used, as its name suggests, at the start of transmission or after congestion of the network has been detected and the data transmission is reduced. On the other hand, congestion control is used to gradually increase the transmission rate once the initial data rate has been ramped-up using the slow start algorithms. In this sense, the fast transmit and fast recovery algorithms are used to speed up the recovery of the transmission rate after significant congestion in the communication network has been detected.

3.6.2.2 TCP Intertwined Algorithms

Modern implementations of TCP contain four intertwined algorithms that have never been fully documented as Internet control mechanism standards, such as: Slow Start, Congestion Avoidance, Fast Retransmit (FRet) and Fast Recovery (FRec).

1. **Slow Start** – The old TCP mode would start a connection with the sender injecting multiple segments into the network, up to the window size advertised by the receiver. While this works when the two hosts are on the same LAN, if there are routers and slower links between the sender and the receiver, problems can arise. Some intermediate router must queue the packets and it is possible for that router to run out of space. Hence, the algorithm to avoid this is called slow start. It operates by observing that the rate at which new packets should be injected into the network is the rate at which the acknowledgments are returned by the other end.

   Slow start adds a window to the sender's TCP known as Congestion Window (CWnd), which is initialized to one segment when a new connection is established with a host on another network. Each time an ACK is received, the CWnd is increased by one segment. The Receiver Advertised Window (RWnd) is the maximum amount of data that can be buffered at the receiver, which is determined by the minimum of CWnd and RWnd and ensuring that the Rx is not overloaded with data. The sender can transmit up to the minimum of the CWnd and the RWnd. The CWnd is flow control imposed by the sender, while the RWnd is flow control imposed by the Rx. The former is based on the sender's assessment of perceived network congestion; the latter is related to the amount of available buffer space at the Rx for this connection. The sender starts by transmitting one segment and waiting for its ACK. When that ACK is received, the CWnd is incremented from one to two and two segments can be sent. When each of those two segments is acknowledged, the CWnd is increased to four. This provides an exponential growth, although it is not exactly exponential because the receiver may delay its ACK, typically sending one ACK for every two segments that it receives. At some point the capacity of the Internet may be reached and an intermediate routers will start discarding packets. This tells the sender that its congestion window has become too large. However, it can be seen that after transmitting the first segment, the data sender remains idle until an ACK message has been received. For a GEO satellite, this would take in the region of roughly 500–570 ms.

2. **Congestion Avoidance** – Congestion can occur when data arrives on a big pipe (a fast LAN) and gets sent out via a smaller pipe (a slower WAN). It can also occur when multiple input streams arrive at a router whose output capacity is less than the sum of the inputs. Therefore, congestion avoidance is a way to deal with lost packets of data. The assumption of the algorithm is that packet loss caused by damage is very small (much less than 1%); therefore the loss of a packet signals congestion somewhere in the network between the source and

destination. There are two indications of packet loss: a timeout occurring and the receipt of duplicate ACK.

The congestion avoidance and slow start are independent algorithms with different objectives. But when congestion occurs TCP must slow down its transmission rate of packets into the network and then invoke slow start to get things going again. Hence, in practice they are implemented together. They require that two variables be maintained for each connection: a CWnd and a Slow Start Threshold (SSThres) size. If CWnd is less than or equal to SSThres, TCP is in slow start, while TCP is performing congestion avoidance. Slow start continues until TCP is halfway to where it was when congestion occurred and then congestion avoidance takes over. Slow start has CWnd begin at one segment and be incremented by one segment every time an ACK is received.

3. **Fast Retransmit** – Before describing the change, realize that TCP may generate an immediate ACK (a duplicate ACK) when an out-of-order segment is received, with a note that one reason for doing so was for the experimental fast retransmit algorithm. Thus, this duplicate ACK should not be delayed, the purpose is to let the other end know that a segment was received out of order and to tell it what sequence number is expected. Since TCP mode does not know whether a duplicate ACK is caused by a lost segment or just a reordering of segments, it waits for a small number of duplicate ACK to be received. It is assumed that if there is just a reordering of the segments, there will be only one or two duplicate ACK before the reordered segment is processed, which will then generate a new ACK. If three or more duplicate ACK are received in a row, it is a strong indication that a segment has been lost. The TCP mode then performs a retransmission of what appears to be the missing segment, without waiting for a retransmission timer to expire.

4. **Fast Recovery** – After fast retransmit sends what appears to be the missing segment, the congestion avoidance but not slow start is performed. That is to say, this is practically the fast recovery algorithm. It is an improvement that allows high throughput under moderate congestion, especially for large windows. The reason for not performing slow start in this case is that the receipt of the duplicate ACK tells TCP that more than just a packet has been lost. Since the receiver can only generate the duplicate ACK when another segment is received, that segment has left the network and is in the receiver's buffer. That is, there is still data flowing between the two ends and TCP does not want to reduce the flow abruptly by going into slow start.

3.6.3   Mobile Asynchronous Transfer Mode (ATM)

The ATM solution is a network switching technology used by Broadband Integrated Service Digital Network (BISDN). It uses a technique called cell switching, breaking all data into cells, or packets and transmits them from one location on the network to another, connected by switches. So, the small, constant cell size allows

ATM equipment to transmit audio, video and computer data over the same network and assure that no single type of data hogs the line.

The latest implementations of ATM support data transfer rates from 25 to 622 Mb/s, which compares to a maximum of 100 Mb/s for Ethernet. It is a protocol designed to handle isochronous (time critical) data such as video and telephony (audio), in addition to more conventional data communications between computers. These protocols are capable of providing a homogeneous network for all traffic types.

The same protocols are used regardless of whether the application is to carry conventional telephony (VoIP), Fax over IP (FoIP), entertainment video over IP (VPoIP) or computer network traffic over LAN, MAN, WAN or satellite networks.

Small and constant packet size allows switching to be implemented in hardware, rather than have routing done in the software. This makes ATM switches sufficiently rapid that multiple isochronous data can be statistically multiplexed together. The significance of this is that ATM protocols provide bandwidth on demand. Network management software will allow small amounts of bandwidth to be set aside for simple transactions, such as E-mails, while allowing more bandwidth for resource intensive multimedia applications. This sort of technology means that a level of bandwidth can be guaranteed and that a connection is not delayed or interrupted by network traffic, which in ATM parlance is known as QoS.

The ATM transport protocol is universal since it can be used for all kinds of networks be it physical networks, (twisted pair, coax and fibre optics) or virtual networks, such as radio and satellite systems. Today's emphasis on multimedia presentations, videoconferencing, remote lecturing, etc., has made ATM more attractive to network administrators. Simply throwing more bandwidth onto the network is not the solution to the problem, since not all demands on the network are equal. The ATM's ability to provide the bandwidth and QoS guarantees makes it the obvious upgrade choice. Other areas where ATM is used include Digital TV, HDTV and video vending. A public ATM network would offer some entirely new services to people in their homes. The most significant type of new services will be the possible video services. What will be so unique about these services is the opportunity for a viewer to interact more with the programs they receive. At this point, the use of the ATM solution would reduce costs considerably, making ATM a very attractive technology for mobile applications, such as oil companies, shipping and airlines. Many designers of satellite systems are thinking about implementing the ATM protocol, which transmits data that have been placed in cells of a constant length (53 bytes).

Thus, the ATM guarantees data transmission at a rate ranging between 2 Mb/s and 2.4 Gb/s. The protocol acts on the principle that a virtual channel should be set up between two points whenever such a need appears. This is what makes the ATM protocol different from the TCP/IP protocol, in which messages are transmitted in packet form, where each packet may reach the recipient via a different route. The ATM protocol enables data transmission through various media. Taking into account the header of the cell (cell-tax), which takes 5 bytes, the application of the ATM protocol may not appear to be so cost-effective when the rate of transmission is low and the capacity of the link (e.g., in two-way modem channels) becomes a basic limitation.

As noted, the ATM technology is expected to provide QoS-based networks that support voice, video and data applications. Initially, the ATM protocol was originally designed for fiber-based terrestrial networks that exhibit low latencies and error rates. With the increasing demand for electronic connectivity across the world, satellite networks play an indispensable role in the deployment of global networks. The Ka-band communication link using the GHz frequency satellites spectrum can reach user terminals across most of the populated world. Thus, the ATM-based satellite networks can effectively provide real time as well as non-real time communications services to remote areas.

There are four different types of ATM service.

1. **Constant Bit Rate (CBR)** – The CBR ATM service specifies a fixed bit rate so that data is sent in a steady stream. This is analogous to a leased line.
2. **Variable Bit Rate (VBR)** – This ATM service provides a specified throughput capacity but data is not sent evenly. This is a popular choice for voice and videoconferencing data.
3. **Unspecified/Undefined Bit Rate (UBR)** – It does not guarantee any throughput levels. This is used for applications, such as file transfer, that can tolerate delays.
4. **Available Bit Rate (ABR)** – It provides a guaranteed minimum capacity but allows data to be burst at higher capacities when the network is free.

However, there is opinion that ATM holds the answer to the Internet bandwidth problem but others are skeptical. Thus, ATM creates a fixed channel, or route, between two points whenever a data transfer begins. This differs from TCP/IP, in which messages are divided into packets and each packet can take a different route from source to destination. This difference makes it easier to track and bill data usage across an ATM.

3.6.3.1 IP/ATM Over Satellite

As mentioned, the ATM terrestrial technology can be successfully used in satellite communication networks to enhance the overall performance of the network. This network as usual, can comprise corresponding satellite configuration and several ground stations communicating via satellite on the one hand and ATM Switch TTN infrastructure on the other, interconnected with the main component of the network, known as ATM Satellite Internetworking Unit (ASIU). The ASIU is responsible for management and control of system resources as well as the overall system administrative functions, like real-time bandwidth allocation, network access control, system timing and traffic control. This fixed satellite configuration can be very easily implemented in MSC with connecting MES terminals via satellite with GES and ASIU to the ATM Switching TTN.

The internal architecture of the ASIU in the satellite direction stage is to extract the ATM cells from the various transport methods. After the ATM cells are extracted, they are error corrected and then buffered according to their type. After that, the cells are encoded and transmitted through the modem to the satellite. Hence, looking at the various components in a bit more detail the first stage will be the extraction of ATM cells from the transport protocols, using the following methods:

1. **Plesiochronous Digital Hierarchy (PDH)** – The PDH protocol was developed to carry digitized voice efficiently. At any rate, it mainly operates by multiplexing various rates of bit streams at the highest allowed clock speed. When necessary, it adds a stuffing bit which the demultiplexer can later remove and therefore, the stuffing procedure is inefficient. Also, rerouting signals after network failures and managing remote network elements are extremely difficult.
2. **Physical Layer Convergence Protocol (PLCP)** – On the other hand, PLCP mode works differently. The PLCP combines 12 ATM cells in one frame format with a header in front. Consequently, this method is not very good for satellite transmission as it is very sensitive to burst errors, which may result in the loss of a whole frame and loss of synchronization of the PLCP device.
3. **Synchronous Optical Network (SONET) or Synchronous Digital Hierarchy (SDH)** – For the reasons already explained above PDH and PLCP are not suitable to be used with ATM and engineers have been turning to SDH as the protocol for ATM. The SONET, similar to the SDH protocol, uses pointer bytes to indicate the location of the first byte in the payload of the SDH frame. Thus, the SDH protocol incorporates a cell delineation mechanism for the acquisition and synchronization of the ATM cells on the receiver side of the network and has some considerable advantages over PDH/PLCP, the main being: flexibility of service, allowing the network operator to respond quickly to customers' requirements; improved quality and supervision, permitting the operator to increase the quality of the offered services;

Furthermore, the SDH protocol reduces operations costs, by using efficient network management technology, which makes the remote control of the network possible mostly without on-site activities. Then the higher rates of transfer are well defined and direct multiplexing is possible without an intermediate multiplexing stage. In fact, despite these advantages SDH has an important problem in that incorrect pointer detection may produce an incorrect payload extraction and an error block. As a result, all received blocks may be incorrect and severely errored. To overcome this problem, an efficient error mechanism, error detection and recovery must be used in the next stage of the ASIU architecture.

However, satellite systems have several inherent constraints. The resources of the satellite network, especially the satellite and the Earth station, are expensive and typically have low redundancy; these must be robust and be used efficiently. Thus, the large delays in GEO and delay variations in LEO systems affect both real-time and non-real time applications. In an acknowledgment and timeout-based congestion control mechanism (like TCP), performance is inherently related to the delay-bandwidth product of the connection. Moreover, TCP Round-Trip Time (RTT) measurements are sensitive to delay variations that may cause false timeouts and retransmissions. As a result, the congestion control issues for broadband satellite networks are somewhat different from those of low-latency terrestrial networks. Both interoperability issues as well as performance issues need to be addressed before a transport-layer protocol like the TCP model can satisfactorily work over long latency satellite ATM networks.

Satellite ATM networks can be used to provide broadband access to remote locations, as well as to serve as an alternative to fiber-based backbone networks. In either case, a single satellite is designed to support thousands of Earth terminals. The Earth terminals set up Virtual Channels (VC) through the on board satellite switches to transfer ATM cells among one another.

Because of the limited capacity of a satellite switch, each MES has a limited number of VC it can use for TCP/IP data transport. In backbone networks, these MES are IP/ATM edge devices that terminate ATM connections and route IP traffic in and out of the ATM network. Namely, these high-capacity backbone routers must handle thousands of simultaneous IP flows. As a result, the routers must be able to aggregate multiple IP flows onto individual VC. Flow classification may be done by means of a QoS that can use IP source–destination address pairs, as well as transport-layer port numbers. Therefore, the QoS manager can further classify IP packets into flows based on the differentiated services code points in the TOS byte of the IP header.

In addition to flow and VC management, all MES terminals also provide a means for congestion control between the IP and ATM networks. The on-board ATM switches must perform traffic management at the cell and VC levels. Hence, TCP hosts implement various TCP flow and congestion control mechanisms for effective network bandwidth utilization. The enhancements that perform intelligent buffer management policies at the switches can be developed for UBR service to improve transport layer throughput and fairness. A policy for selective cell drop based on per-VC accounting can be used to improve fairness.

Providing a minimum Guaranteed Rate (GR) to UBR traffic has been discussed as possible candidate to improve TCP performance over UBR. The goal of providing GR is to protect the UBR service category from total bandwidth starvation and provide a continuous minimum bandwidth guarantee. It has been shown that in the presence of high loads of higher priority Constant Bit Rate (CBR), Variable Bit Rate (VBR) and Available Bit Rate (ABR) traffic, TCP congestion control mechanisms benefit from a minimum GR.

Moreover, Guaranteed Frame Rate (GFR) has recently been proposed in the ATM Forum as an enhancement to the UBR service category. GFR will provide a minimum rate guarantee to VC at the frame level.

The GFR service also allows for the fair usage of any extra network bandwidth. GFR is likely to be used by applications that can neither specify the traffic parameters needed for a VBR VC, nor have the capability for ABR (for rate-based feedback control). Current internetworking applications fall into this category and are not designed to run over QoS-based networks. Routers separated by satellite ATM networks can use the GFR service to establish VC between one another. GFR can be implemented using per-VC queuing or buffer management.

3.6.3.2   UBR Over Satellite

The Unspecified Bit Rate (UBR) service class is intended for delay-tolerant or non-real time applications, that is, those that do not require tightly constrained delay and delay variation, such as traditional computer communications applications. Sources are expected to transmit non-continuous bursts of cells. Namely, the UBR service supports a high degree of statistical multiplexing among sources and includes no notion of a per-VC allocated bandwidth resource. Thus, transport of cells in UBR service is not necessarily guaranteed by mechanisms operating at the cell level. However, it is expected that resources will be provisioned for UBR service in such a way as to make it usable for some sets of applications. The UBR service may be considered as an interpretation of the common term "best effort service". The UBR service is typically used for data transmission applications such as file transfer and E-mail. Neither an ATM-attached router nor an ATM switch provides traffic or QoS guarantees to a UBR virtual circuit.

As a result, UBR VC can experience a large number of cell drops or a high cell transfer delay as cells move from the source to the destination device. In its simplest form, an ATM switch implements a tail drop policy for the UBR service category. Hence, if cells are dropped, the TCP source loses time waiting for the retransmission timeout. Even though TCP congestion mechanisms effectively recover from loss, the link efficiency can be very low, especially for large delay-bandwidth networks. In general, link efficiency typically increases with increasing buffer size. In fact, the performance of TCP over UBR can be improved by using buffer management policies. In addition, TCP performance is also affected by TCP congestion control mechanisms and TCP parameters such as segment size, timer granularity, receiver window size, slow start threshold and initial window size.

The TCP Reno implements the fast retransmit and recovery algorithms that enable the connection to quickly recover from isolated segment losses. However, fast retransmit and recovery cannot efficiently recover from multiple packet losses within the same window. A modification to Reno TCP is proposed, so that the sender can recover from multiple packet losses without having to timeout. The TCP solutions with Selective Acknowledgment (SACK) are designed to efficiently recover from multiple segment losses. With SACK, the sender can recover from multiple dropped segments in about one round trip. The studies show that in low-delay networks, the effect of network-based buffer management policies is very important and can dominate the effect of SACK. The throughput improvement provided by SACK is very significant for long latency connections. When the propagation delay is large, timeout results in the loss of a significant amount of time during slow start from a window of one segment.

Reno TCP (with fast retransmit and recovery) results in the worst performance (for multiple packet losses) because timeout occurs at a much lower window than in Vanilla TCP. With SACK TCP, a timeout is avoided most of the time and recovery is complete within a small number of round trips. The NewReno modification to the fast retransmit and fast recovery algorithms (TCP Reno) has been therefore proposed to counteract multiple packet drops where the SACK option is not available. For lower delay satellite networks (LEO), both NewReno and SACK TCP provide high throughput but as the latency increases, SACK significantly outperforms NewReno, Reno and Vanilla.

3.6.3.3   ABR Over Satellite

The ABR service category is another option to implement TCP/IP over ATM. This service category is specified by a Peak Cell Rate (PCR) and a Minimum Cell Rate (MCR), which is guaranteed by the network. ABR connections use a rate-based closed-loop end-to-end feedback control mechanism for congestion control. The network tries to maintain a low cell loss ratio by changing the Allowed Cell Rate (ACR) at which a source can send. Switches can also use the Virtual Source/Virtual Destination (VS/VD) feature to segment the ABR control loop into smaller loops. Studies have indicated that ABR with VS/VD can effectively reduce the buffer requirement for TCP over ATM, especially for long delay paths. In addition to network-based drop policies, end-to-end flow control and congestion control policies can be effective in improving TCP performance over UBR. The fast retransmit and recovery mechanism can be used in addition to slow start and congestion avoidance to quickly recover from isolated segment losses. Otherwise, the SACK option has been proposed to recover quickly from multiple segment losses.

The TCP performance over ABR service is important for the satellite IP network. Namely, a key ABR feature, VS/VD, highlights its relevance to long delay paths. Most of the consideration assumes that the switches implement a rate-based switch algorithm like ERICA+. Credit-based congestion control for satellite networks has also been suggested, in long-latency satellite configurations, therefore the feedback delay is the dominant factor in determining the maximum queue length. A feedback delay of 10 ms corresponds to about 3670 cells (at OC-3) of queue for TCP over ERICA+, while a feedback delay of 550 ms corresponds to 201,850 cells. This indicates that satellite switches need to provide at least one feedback delay of buffering to avoid loss on these high delay paths. A point to consider is that these large queues should not be seen in downstream workgroup or WAN switches, because they will not provide as much buffering. Satellite switches can isolate downstream switches from such large queues by implementing the VS/VD option, while VS/VD can effectively isolate nodes in different VS/VD loops. As a result, the buffer requirements of a node are bound by the feedback delay-bandwidth product of the upstream VS/VD loop. At this point, VS/VD helps to reduce the buffer requirements of terrestrial switches connected to Gateway terminals. The feedback delay-bandwidth products of the satellite hop are about 160,000 cells and dominates the feedback delay-bandwidth product of the terrestrial hop (about 3000 cells). Without VS/VD, terrestrial switch S, a bottleneck, must buffer cells up to the feedback delay-bandwidth product of the entire control loop (including the satellite hop). With a VS/VD loop between the satellite and the terrestrial switch, the queue accumulation due to the satellite feedback delay is confined to the satellite switch. The terrestrial switch only buffers cells that are accumulated due to the feedback delay of the terrestrial link to the satellite switch.

3.6.4 Fixed Digital Video Broadcasting-Return Channel Via Satellite (DVB-RCS)

According to a recent Satellite Industry Association's report, over 130 M direct satellite TV subscribers are spread around the world. The key enablers attributed to this success include modern satellite systems capable of powerful transmissions over vast expanses, efficient transmission, highly integrated, low-cost receiver systems and a vast variety of rich content at an affordable price. Current audio and video compression techniques dwell specifically on the MPEG multimedia standards, which constitute the industry's standards for compression and transport. At this point, a majority of direct broadcast satellite systems beaming Standard-Definition TV (SDTV) either use or can support the MPEG-2 standard, while HDTV and multimedia broadcast systems rely firmly on the standard's upgraded version, MPEG-4 mode. However, in last two decades is implemented the new standard of DVB-RCS system.

To encourage mass-scale acceptance, regulatory bodies promote standardization of the satellite broadcast service. At this point, ITU recognizes at least six Direct to Home (DTH) broadcast systems that encompass television, data and multimedia. Modern satellite and handheld receiver technologies enable television and radio broadcasts to handheld and portable personal devices. In the meantime, numerous commercial systems that offer sound and television services to the individual and groups have emerged in the last decade. The final developments of this technique offers the new personal and mobile solutions that will be in the future recognized as Digital Video Broadcast-Satellite Personal (DVB-SP) and Digital Video Broadcast-Satellite Mobile (DVB-SM) standards, both of which support commercial mobile television services.

The major development in satellite broadcasting technology was the standardization of the DVB-RCS, which allows the users within direct broadcast terrestrial network (DVB-T) to communicate directly with the broadcast satellite network (DVB-S) through an assigned return channel. The DVB-T cell can be comprised by UMTS/GPRS, ISDN, B-ISDN, and ATM broadband networks and the DVB-S cells may include rural, Consumer broadcasting SoHo/SME LAN. Corporate WAN and multicasting networks. This greatly simplifies the overall network architecture and associated network management procedures, in that now all kind of communication solutions takes place over the same access network. Thus, the DVB-RCS system has been specified for indoor use only, so can be envisaged that very soon, research and standardization efforts will be directed towards establishing a suitable mobile standards. Will be enough to develop corresponding mobile satellite antenna, and this could open up significant new opportunities in the mobile satellite sector.

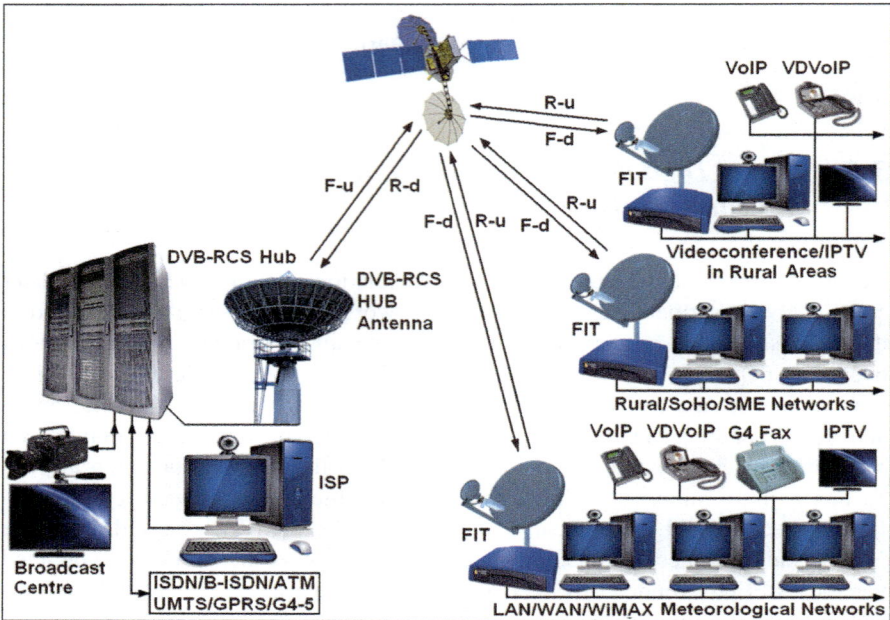

Fig. 3.22 Mixed DVB-RCS solutions. *Acronyms*: *F-d* forward downlink, *F-u* forward uplink, *R-d* return downlink, *R-u* return uplink, *ISP* internet service provider, *FIT* fixed interactive terminal, *SME* small/medium enterprise, *SoHo* small office/home office, *VoIP* voice over IP, *VDVoIP* voice, data, and video over IP (Courtesy of Manual: "Global Mobile CNS" by Ilcev)

A sample of fixed DVB-RCS system is depicted in Fig. 3.22. The DVB-RCS network enables via Hub as a GES with C, Ku or Ka-band antenna to interface the TTN (DVB-T) cell via corresponding GEO satellite connections (C, Ku or Ka-band GEO) to the Fixed Interactive Terminals (FIT) or Remote (DVB-S) cells for the following services:

1. Regenerate rural communications: VoIP, IPTV, Internet access and interactive TV/radio two separate way broadcasting (Telephony/Broadband/Broadcast);
2. Broadband access: Asymmetric Digital Subscriber Line (ADSL) anywhere and anytime, Internet access/E-mail, Consumer, SoHo/SME LAN, Corporate WAN, Intranet/VPN, FTP and HTTP; The FTP scheme is service for moving and copying an electronic file of any type from one computer to another over the Internet. For instance, it can be used both for downloads and uploads;
3. Teleservice of all E-solutions, which include E-medicine and E-education and others: Videoconference, Image/Video/Audio transfer and Interactive distance learning; and
4. Multicasting: Web casting, Video streaming, Satellite newsgathering and Push/Pull data delivery and Voice Data and Video over IP (VDVoIP).

The Hub terminal supports existing DVB-RCS compliant Forward Link System (FLS) and provides all the ground required interfaces and management functions necessary to set up a DVB-RCS service over 100 Mb/s downstreams. The Hub operator can interface DVB-RCS terminals to a terrestrial network or service provider, and manage all the operational aspects of the system. The Hub station design optimized for robustness and stability in operations and can be delivered in a number of different configurations to suit customers precise applications. Its architecture is also designed to accommodate upgrades and expansions of DVB-RCS network and offer to DVB-RCS providers a choice of C, Ku or Ka-band antenna and RF equipment, so one ordinary DVB-RCS configuration can handless 1 to 10 forward link transmitters and from one to several hundred return link receivers.

The DVB-RCS FIT for rack mounting is composed by Outdoor Unit (ODU) of antenna and Indoor Satellite Unit (ISU) of Remote transceiver to provide the most attractive and reliable DVB solutions and ensures full compatibility with any DVB-RCS system or VSAT modem, ETSI/CE approval. The DVB-RCS Remote is desktop terminals and router that provide two-way IP communications via satellite at C, Ku and Ka-Band frequencies.

These terminals for corporations, institutions, home offices and householders offer an open-interface for high-capacity broadband access that bypasses the "last mile" bottleneck associated with terrestrial infrastructure. Therefore, DVB-RCS system offers broadband access to core IP networks using standard technologies such as DVB-RCS, DVB-S, latest DVB-S2 and IP interfacing DVB-T with fixed and mobile user terminals.

In such a way, the broadband network delivers two-way IP connectivity both between user terminals in the satellite system and between user terminals and the terrestrial network coming in a scalable choice of performance with a range of data IP throughputs from 4 to 30 Mb/s.

The broadcast MPEG2/DVB-S service is available in one-way (unidirectional) from the Hub to the users terminals. With very low cost per bit, service prices become comparable to those offered by terrestrial networks and can be delivered where other technologies cannot reach. This network can provide solutions in vas variety of user segments, such as: Enterprises and Private Networks, Broadcasting and Content Distribution (BCD), Satellite News Gathering, Satellite Emergency and Security Management, Defense Information Management, and so on.

3.6.5   *Mobile Digital Video Broadcasting-Return Channel Over Satellite (DVB-RCS)*

The convergence of MSC and Internet technique has opened many opportunities to deliver new multimedia service over hybrid satellite systems to MES. The interactive nature of the Internet has paved the way for new generation Mobile Satellite Systems (MSS) to support interactivity. Apart from the convergence between

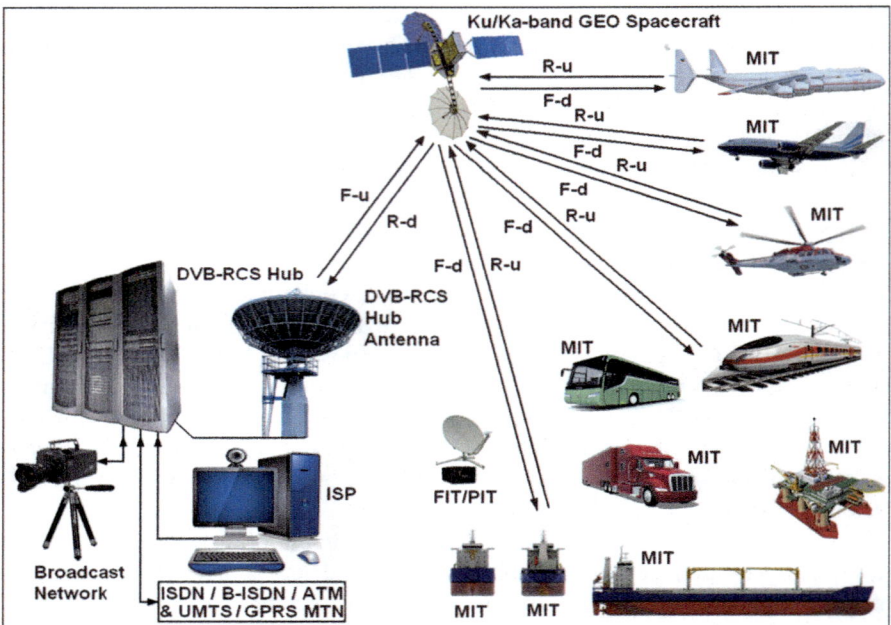

Fig. 3.23 Mobile DVB-RCS solutions. *Acronyms*: *ISP* internet service provider, *IPTV* IP television, *VoIP* voice over IP, *VDVoIP* voice, data, and video over IP, *PIT* portable interactive terminal, *MIT* mobile interactive terminal, *FIT* fixed interactive terminal, *F-d* forward downlink, *F-u* forward uplink, *R-d* return downlink, *R-u* return uplink (Courtesy of Manual: by Ilcev)

mobile and Internet technologies, the other major technical driver is the convergence between mobile and fixed technologies.

By supplementing satellite broadcasting systems with a narrowband uplink, new interactive services can be facilitated in DVB and DAB solutions. This is foreseen for fixed network operation and could equally be adopted onto a mobile network such as the UMTS, thus for demonstrating the convergence concept of satellite personal and mobile communications, Internet and broadcasting technology.

A configuration of mobile DVB-RCS system shown in Fig. 3.23 is designed by author in 2000. The DVB-RCS mobile network includes an Hub as a GES with C, Ku or Ka-band antenna to interface the TTN (DVB-T) cell via corresponding satellite connections (C, Ku or Ka-band GEO) to the Mobile Interactive Terminals (MIT) or Remote (DVB-S) cells for the following mobile services: Maritime and Offshore Installations, Land (Road and Rail) solutions and Aeronautical service. This infrastructure is the best solution for establishment a network for satellite meteorological gathering onboard ships, vehicles aircraft and even SPS stations. In any case, however, the modern satellite meteorological network needs the most reliable communication link between satellites and mobiles for weather data transfer.

3.7 MPEG Multimedia Standards

The encoding of both high-fidelity (non-speech) audio signals and video signals today is dominated by the standards defined by the International Standards Organization (ISO) MPEG, subsequently adopted by both the ISO/IEC and the ITU. The first standard, MPEG-1, "Coding of moving pictures and associated audio for digital storage media at up to about 1.5 Mb/s, issued in 1992, and its successor MPEG-2, "Generic coding of moving pictures and associated audio information", issued in 1994, define generic coding for both moving pictures (i.e. video) and associated audio information.

Work on an MPEG-3 standard was discontinued. More recently, MPEG-4 primarily focuses on new functionality, but also includes improved compression algorithms, and for the first time includes speech compression algorithms, such as CELP. Each standard comprises a number of parts, for example, MPEG-4 has 23 parts, each of which focuses on a particular aspect of the encoding (video, audio, data encapsulation, etc.). Significantly, MPEG standards only define in detail the source decoder, providing a toolbox of algorithms, and standardized bit stream formats. Detailed encoder architecture is typically not defined (although example implementations are given), thereby allowing equipment manufacturers to differentiate their products and ensure compatibility of user's equipment.

3.7.1 Audio Broadcasting

With reference to the audio broadcasting in general is necessary to be turn attention to the general audio waveform, and in particular to the high fidelity audio, including stereo music, multichannel surround-sound content and speech encoding.

3.7.1.1 MPEG-2 Audio Layer II (MP2) Encoding

MPEG-1 provided for three different types of audio encoding with sampling rates of 32 kHz (kSamples/s), 44.1 and 48 kHz for monophonic (mono), dual mono and stereophonic (stereo) channels. MPEG-2 added support for up to five audio channels and sampling frequencies down to 16 kHz. Subsequently, MPEG-2 Advanced Audio Encoding (AAC) added sampling from 8 to 96 kHz with up to 48 audio channels (and 15 embedded data streams). MPEG-2 audio layer I is the simplest encoding scheme and is suited for encoded bit rates above 128 kb/s per channel (Pan 1995). Audio layer II, known as MP2, has an intermediate complexity and is targeted at bit rates around 128 kb/s. MP2, also known as Musicam, forms the basis of the Digital Audio Broadcasting (DAB) system and is incorporated into the DVB standard.

3.7.1.2   MPEG-2 Audio Layer III (MP3) Encoding

MPEG-2 audio layer III, known as MP3, provides increased compression (or, alternatively, improved quality at the same data rate) compared with MP2 at the expense of slightly increased complexity and computational effort. Today, MP3 is widely used for the storage and distribution of music on personal computers via the Internet and for digital satellite broadcasting by 1 worldspace (see Chap. 10). The MP3 psychoacoustic model uses a finer-frequency resolution than the MP2 poly-phase quadrature filter band provides (Pan 1995). MP3 divides the audio spectrum into 576 frequency bands and processes each band separately. It does this in two stages. First, the spectrum is divided into the normal 32 bands, using a polyphase quadrature filter, in order to ensure compatibility with audio layers I and II. In MP3, however, each band is further divided into 18 sub-bands using a Modified Discrete Cosine Transform (MDCT). The MP3 MDCT is a variant of the discrete cosine transform that reuses a fraction of the data from one sample to the next. We have learnt in Chap. 8, that following the transformation into the frequency domain, component frequencies can be allocated bits according to their audibility using the masking levels in each filter. We also noted that MP3 exploits inter-channel redundancies, for example in situations when the same information is transmitted on both stereo channels. Typically, MP3 permits compression of CD-quality sound by a factor of -12.

3.7.2   Video Broadcasting

As with audio encoding, the dominant video encoding standards are currently the MPEG standards, notably MPEG-2 for standard-definition video and, increasingly, MPEG-4 for high-definition video. The MPEG-2 video decoders contain a 'tool-box' of compression algorithms grouped into a number of subsets called 'profiles'. Furthermore, profiles support a number of levels (combinations of image size, frame rate, etc.), and decoders may implement a subset of profiles.

3.7.2.1   MPEG-2 Video Encoding

The MPEG video encoding algorithms aim to transmit differences between frames where possible and use a DCT (a form of Fourier transform) to encode the difference information into the frequency domain, discarding insignificant high spatial frequency information that would not normally be noticeable. For this reason, they are sometimes referred to as hybrid block interframe Differential PCM (DPCM)/Discrete Cosine Transform (DCT) algorithms. A video buffer is used to ensure a constant bit stream on the user side. Typical bit rates for MPEG-2 encoded standard-definition video are in the region 3–15 Mɔ/s, which is around 10–50:1 compression is achieved over the raw PCM bit rate.

3.7.2.2   High-Definition TV and MPEG-4

Even greater compression ratios are required for the transmission of high-definition TV video. HDTV frame resolutions of up to 1920×1080 pixels are in use (at 25 Hz) – with up to 5 times as many pixels per frame compared with SDTV. While MPEG-2 High Level can support resolutions of up to 1920×1080 (sometimes referred to as full-HD), HDTV generally requires the increased compression available in MPEG-4 to be viable.

The MPEG-4 Advanced Video Coding (MPEG-4 part 10 AVC), jointly developed with the ITU Video Coding Experts Group (as H.264), employs additional techniques to achieve compression ratios greater than those for MPEG-2, and is used by several satellite services for broadcasting HDTV. MPEG-4 AVC is utilized in BluRay videodiscs. Specifically, MPEG-4 AVC utilizes a number of features to achieve higher compression than MPEG-2. Some of the more significant features of MPEG-4 AVC are:

1. Up to 32 reference pictures, however may be used for motion compensation, rather than just 1 (I-Pictures) or 2 (B-Pictures);
2. The macroblock size used for motion compensation may be varied from 16×16 to 4×4 pixels with subpixel precision, and new 4×4 and 16×16 pixel block transforms; and
3. Improved non-linear quantization size control and improved entropy encoding.

3.7.2.3   Multiplexing and Transporting

In Fig. 3.24 is illustrated packetization of encoded video and audio elementary streams to produce a Packetized Elementary Stream (PES). The PES packets are next combined with system-level information to form Transport Streams (TS) or Programme Streams (PS).

The systems part of the MPEG-2 specification (ISO/IEC 13818–1:2000(E)) specifies all syntactical and semantic rules to combine MPEG-2 video and audio elementary streams (output of an encoder), including other types of data content, into a single or multiple streams to enable storage or transmission.

The programme stream consists of one or more streams of PES packets of common time base into a single stream. The stream is useful for operation in relatively error-free environments such as interactive multimedia applications.

The transport stream consists of one or more independent programmes into a single stream. This type of stream is useful in error-prone environments as satellite broadcasts. Packets are 188 bytes in length. Transport stream rates may be fixed or variable. Values and locations of Programme Clock Reference (PCR) fields define the stream rate. The transport stream design is such as to facilitate: retrieval and

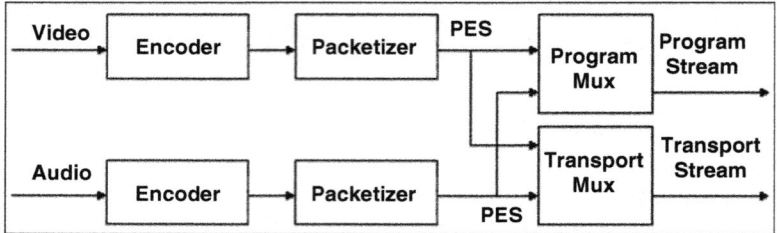

Fig. 3.24 MPEG programme and transport stream (Courtesy of Book: by Richharia)

decoding of a single programme within the transport stream, extraction of only one more programme or contents from a transport stream and conversion of a programme into a transport stream including recovery at Rx.

Figure 3.25 illustrates the concept of demultiplexing and decoding of a single programme from a received transport scheme containing one or more programme streams. The input stream to the demultiplexer/decoder includes a system layer wrapped about a compression layer. The system layer facilitates the operation of the demultiplexer block, and the compression layer assists in video and audio decoding. Note that, although audio and video decoders are illustrated, these may not always be necessary when the transport stream carries other classes of data.

3.8   Direct-to-Home Broadcast System

Recognizing the advantages of digital television transmissions, and their potential to transport HDTV efficiently, considerable effort was directed in the 1980s and 1990s (notably in the United States, Europe and Japan) towards the development of digital transmission systems, resulting in the design of several terrestrial and satellite systems. The convergence of computing, telecommunications and broadcast disciplines led the developers to adapt a generic architecture that would offer a variety of enabling services in addition to SDTV and HDTV.

Enabling technology and services include a digital video recorder to facilitate recording of programme directly in a digital format, interactivity (e.g. multichannel display or multiple camera angle displays), receiver-enabled home networking and reception onboard mobiles.

Broadcast Satellite Service (BSS) radio frequency plan of ITU provides a useful framework with guaranteed availability of spectrum in Ku-band to each member country, allowing high-power radio transmissions amenable for reception at home via small non-obtrusive antenna on low-cost receivers. It is necessary to highlight the technology of the DVB-S2 satellite system, as it incorporates a wide range of recent technology advances to provide a highly flexible and efficient medium, and moreover it is uniquely identified by the ITU as a broadcast system for digital satellite broadcasting system with flexible configuration.

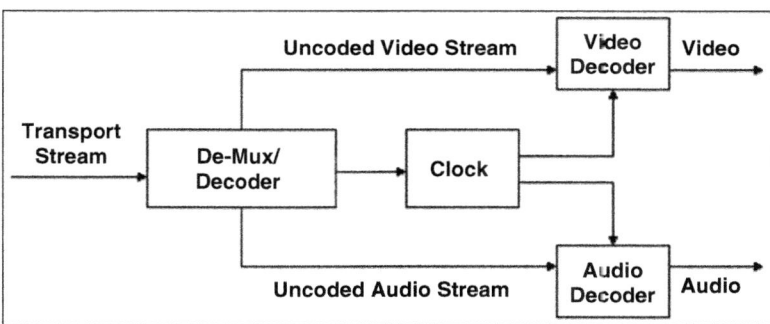

Fig. 3.25 Demultiplexing and decoding of MPEG (Courtesy of Book: by Richharia)

3.8.1   Transmission System Architecture

A typical direct-to-home system comprises incoming signals to the uplink facility from one or more transmission sources, such as a studio, local terrestrial broadcast signal feed, pre-recorded material, etc. Occasionally, additional material such as advertisements may be added locally at pre-agreed points of the incoming programme. The incoming signals are monitored, routed within the facility and if necessary, readjusted and synchronized. The prerecorded material is checked for quality, edited, when necessary, and read into video file servers to be played at the broadcast time.

Commercial systems incorporate a facility known as conditional access to facilitate reception solely by the authorized users. Other functionalities include additions of Service Information (SI) and Electronic Programme Guide (EPG), analogue-to-digital conversion, compression, multiplexing to create a suitable transport stream, error control and modulation. The SI and EPG equipment creates signals for displaying programmes related information, for example programmes title, start/end time, etc.

The compression equipment is typically MPEG-2 standard, although migration to MPEG-4 is endemic because of its tighter encoding. At this point, several channels are multiplexed to provide a single high-rate channel for transmission. The stream is FEC mode, modulated and transmitted. The ITU BSS plan recommends uplink transmissions in the 17.5 GHz band and downlink transmissions to the user community in the 12.5 GHz band, but their transmission powers are restricted by the FSS regulations. The BSS satellites are placed in geostationary orbits. The satellites use transparent transponders capable of transmitting very high powers through spot, and often, shaped beams to be able to provide the high signal quality necessary for reception on small DTH receivers.

The DTH receivers typically use a 45–60 cm dish at Ku-band, depending on satellite EIRP. The relatively high gain of the receive antenna in conjunction with high-power satellite transmissions provides sufficient link margins to counter rain fades common in this band.

3.8.2 *Generic Reference Integrated Receiver Decoder (IRD) Model*

The indoor unit housing the electronics is known as the Integrated Receiver Decoder (IRD). The core functions of all the direct-to-home television systems are nominally identical, and hence a generic model of the IRD is feasible. Operators may tailor the remaining functions around these core functional entities. The ITU proposes reference architecture of the IRD on this premise. The model provides a structured definition of functionalities to facilitate a generic receiver design. As observed in the preceding section, the core functions relate generically to a transmission system. The additional essential functions relate to the service provision, operation of the system and additional or complementary features, which may differ depending on implementation. These functions and units include: a satellite tuner, output interfaces, an operating system and applications, EPG, Service/system Information (SI), Conditional Access (CA), a display, a remote control, Read Only Memory (ROM), Random Access Memory (RAM), FLASH memory, an interactive module, a microcontroller and units to support teletext, subtitling, etc.

Software-reconfigurable receivers are common as they simplify an upgrade to the receiver. The upgrade may, for example, become necessary to repair a software anomaly, add a new functionality or reconfigure receiver subsystems when a new satellite transponder is deployed. Modern receivers typically include a digital video recorder permitting users to record a programme directly in a digital format, or to store content, ready for an 'instant' video-on-demand display, thereby avoiding the interactive delay. Often L-band IF, signals from a single dish/LNB flow to one or more receivers in the customer's home. Home networking systems permit interworking between receivers, thereby permitting programmes recorded on one set to be viewed by other sets elsewhere in the house, and, in addition, support features such as security and quality of service management.

3.9 Transmission Standards

The development of digital television has evolved independently around the world, and hence several types of transmission system are in use. The majority operates in the 11/12 GHz downlink band. Being digital, the systems can support numerous applications and services efficiently be it television, multimedia, data services or audio. Thus, to assist in the selection of an appropriate system, ITU recommends four systems – system A, system B, system C and system D.

3.9.1 Digital Video Broadcast-Second Generation (DVB-S2) Standard

The DVB transmission project, initiated in Europe but subsequently extended world-wide, defines digital broadcast standards by consensus among the participants representing the manufacturing industry, broadcasters, programme providers, network operators, satellite operators and regulatory bodies. The DVB standards embrace broadcast transmission technology across all media of cable, terrestrial and satellite. The DVB-S specifications were standardized in 1993. A second-enervations specification, DVB-S2, was produced in 2003 in response to a growing demand for more capacity, which was standardized by the European Telecommunication Standard Institute (ETSI) as EN 302307 (ETSI Online). In August 2006 the ITU's study group on satellite delivery issued a recommendation (BO.1784) that DVB-S2 be the preferred option for a Digital Satellite Broadcasting System with Flexible Configuration (Television, Sound and Data), entitled system E.

The DVB-S standards support up to 12 categories of transmission medium encompassing a plethora of media contents, among others, standard- and high-definition television, radio and data with or without user interactivity. The standard includes specifications for Internet Protocol (IP) data, software downloads and transmissions to handheld devices. The standards, namely DVB-Satellite (DVB-S), DVB-Return Channel Satellite (DVB-RCS) and DVB-Second Generation (DVB-S2) apply to fixed user terminals. The DVB-Satellite Handheld (DVB-SH) standard, discussed later in the chapter, applies to handheld terminals.

Many parts of the DVB-S specifications are shared between various transmission media. The source coding of video and audio and formation of the transport stream comprises MPEG-2 tailored for satellite systems (MPEG specifications are too generic). DVB-S also supports H.264/AVC video and MPEG-4 high-efficiency AAC audio, and additionally, audio formats such as Dolby AC-3 and DTS coded audio. Guidelines are also available for transporting IP content. The specifications support teletext and other data transmitted during the vertical blanking period, subtitles, graphical elements, service information, etc. DVB-S's enhanced version, DVB-S2, is based on three key concepts: (1) Best transmission performance; (2) Total flexibility; and (3) Reasonable receiver complexity.

The specifications enable delivery of services that could never have been delivered by DVB-S. According to the developers, the DVB-S2 standard is not intended to replace DVB-S in the short term for conventional TV broadcasting applications but is rather aimed at new applications such as the delivery of HDTV and IP-based services, fly-away small DSNG stations, low-cost Internet access to rural areas and in developing countries, etc. The DVB-S2 specifications, in conjunction with recent advances in video compression, have enabled the widespread commercial launch of HDTV services. The supported applications are as follows: (1) Standard and high-definition television broadcasts; (2) Interactive services, including Internet access for consumer applications; (3) Professional applications: Digital TV Contribution (DTVC) facilitating point-to-multipoint transmissions, and Digital Satellite News Gathering (DSNG); (4) Content distribution; and (5) Internet trunking.

In addition to MPEG-2 video and audio coding, DVB-S2 is designed to handle a variety of advanced audio and video formats, which the DVB Project is currently defining. In fact, DVB-S2 accommodates most common digital input stream format, including single or multiple MPEG transport streams, continuous bit streams, IP as well as ATM packets. It is 30% spectrally more efficient than its predecessor, employing an adaptable modulation scheme consisting of QPSK and 8-PSK for broadcast applications and also 16-APSK and 32-APSK for professional applications such as newsgathering and interactive services.

In addition, the modulation and coding schemes may be dynamically adapted to variable channel condition on a frame-by-frame basis scheme. The coding arrangement consists of a Bose-Chaudhuri-Hocquenghem (BCH) outer code scheme and a Low-Density Parity Check (LDPC) inner code. The communication performance lies within 0.7 dB of the theoretical limit. The flexibility offered by variable coding and modulation provides different levels of protection to services as needed.

The specifications support operation on any type of satellite transponder characteristics with a large variety of spectrum efficiencies and associated C/N requirements. Thus, the DVB-S2 broadcast services comprise a backward compatible mode and a more optimized version, which is not backward compatible. The system allows interactive services by return channels established either via satellite or another medium incorporating the added flexibility of Adaptive Coding and Modulation (ACM) to the forward channel through feedback. Structured as a toolkit, DVB-S2 attempts to optimize transmission efficiency and flexibility, keeping receiver costs at an acceptable level.

3.9.2 DVB-S2 Architecture

Figure 3.26 shows a block schematic of functional components of a DVB-S2 transmission system. There are two levels of framing structures, one at Baseband (BBFRAME) and the other at Physical Layer (PLFRAME).

Thus, BBFRAME includes signaling to configure the receiver for the given specification and service. The PLFRAME comprises a regular sequence of frames, each coded and modulated for the given channel condition and application, containing a few signaling bits for synchronization and physical layer signaling.

Referring to BBFRAME shown in Fig. 3.26, the mode and stream adaptation block interfaces with the incoming stream, provides synchronization and supports the adaptive coding-modulation schemes. It merges multiple streams and slices them into blocks, each of which are modulated and coded homogeneously. A header of 80 bits containing transmission characteristics is next appended to the baseband data block to assist reception. The information within the header, for example, informs the receiver as to whether the transmission stream is single or multiple, the type of coding modulation schemes, signal format, etc. When the

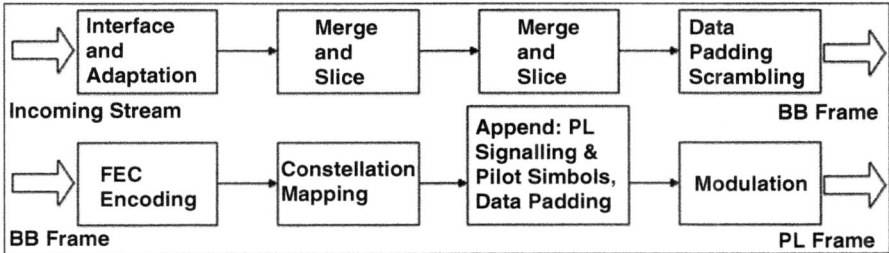

Fig. 3.26 DVB-S2 transmission system (Courtesy of Book: by Richharia)

volume of data is insufficient to fill the frame, data padding is introduced, and finally the frame is scrambled and passed over for coding.

The FEC coding-modulation schemes are instrumental in high transmission efficiency. The FEC code comprises a Low-Density Parity Check (LDPC) code that exhibits a low distance from the Shannon limit for the specified decoder complexity (equating to 14 mm^2 of silicon 0.13 μm technologies).

The LDPC codes are suitable for iterative decoding at reasonable complexity because of their easily parallelizable decoding algorithm, which can use simple operations resulting in reduced complexity and storage. To avoid error floors at low error rates, a BCH outer code of the same block length as the LPDC code is concatenated. The coded payload modulates the carrier with a QPSK, 8-PSK, 16-APSK or 32-APSK scheme, as required by the given application data rate and link conditions. The corresponding spectral efficiency ranges from 0.5 to 4.5 b/symbol. The QPSK and 8-PSK schemes are applied for broadcast applications through nonlinear transponders, whereas 16 and 32-APSK schemes are better suited for applications operating through transponders driven in the quasi-linear region. The DVB-S2 provides backward compatibility with DVB-S through a hierarchical modulation scheme.

The Physical Layer (PL) signals are composed of a regular sequence of frames, within each of which the modulation and coding schemes remain unchanged, that is a homogeneous operation. A header of 90 bits preceding the payload assists in synchronization at the receiver, detection of modulation and coding parameters, etc. This header must be made particularly robust, as the LPDC/BCH scheme is not applied here and the packet will not be detectable unless the header is detected correctly. An interleaved first-order Reed-Muller block code in tandem with the π/2-BPSK modulation scheme was found to be suitable.

3.9.3   Digital Video Broadcast-Third Generation (DVB-S3) Standard

1. **DVB-S3 Standard** – An Israeli company Novelsat proposed a more efficient successor to the DVB-S2 transmission system, known as DVB-S3. In 2010 its engineers have tested their signals on Eutelsat's W3A satellite as well as on AsiaSat 5, Amos-3 and an Intelsat spacecraft. It gained an average improvement of 28% over today's DVB-S2 standard. The DVB-S2 transmission operates very close to the "Shannon Theorem", which suggests there is a limit on the amount of transmitted data that can be handled within a certain bandwidth without being overwhelmed by noise. The Israeli group looked again at the mathematical algorithms needed to compress digital TV signals and incorporate fresh algorithms into new modulators and modems, but this Novelsat's solution improves the bit-rate carried by an average of 28%. This gain results from reducing the "guard band" at the end of each channel from 20% of the channel's spectral width to just 5%. The first generation of DVB-2 transmissions had a guard band of 35% of a channel's width. At this point, by exploiting the greater processing power available now compared with the time over a decade ago when DVB-S2 standard was developed, it has been able to reduce the roll off band to just 5% at each end of a channel, which accounts for much of the throughput improvement.

   After launching its new DVB-S3 (NS3) modulation technology in April 2011, positioned to be substantially more efficient than DVB-S2, start-up Novelsat Company says that it has already established 32 live satellite trials. The system will first be used in a series of VSAT delivery satellites, using a simple software upgrade, but it is really targeted at becoming a replacement technology for DVB-S2 delivering video from the sky to consumer homes and mobiles. This Company is already in talks to begin a standardization process with the DVB Project, and that a 72 MHz transponder channel, which uses NS3, is anything from 28% to 90% more efficient than DVB-S2. Novelsat claims that its new system will deliver 358 Mb/s in a 72 MHz transponder channel, which it says is roughly 20% more than DVB-S2 could possibly manage and more than 28% better than most real world situations. Thus, many DVB-S2 transponders today deliver only 168 Mb/s, which makes it a substantial improvement of more than double.

2. **DVB-S2X Standard** – The DVB-S2 standard was designed to boost provision of new services, primarily focused on HDTV and DTH (Direct-To-Home) services. The DVB Project (DVB), the consortium of industry and regulatory bodies behind DVB-S2, says the specification is so good that it shouldn't need to develop a DVB-S3 specification.

However, soon after is developed new DVB-S2X standard as an extension of the DVB-S2 specification that provides additional technologies and features. This standard has been published as (ETSI EN 302307 Part 2 and with DVB-S2 being Part 1. It offers improved performances for the core applications of DVB-S2, including DTH, contribution, VSAT and Digital Satellite News Gathering (DSNG). The specification also provides an extended operational range to cover emerging markets such as mobile applications.

Among the rest, DVB-S2 has been specified near 15 years ago with a strong focus on DTH. Since then, new requirements have come up and DVB-S2X offers the necessary technical specifications. This work was done by the DVB Technical Module sub-group on satellite chaired by Dr. Alberto Morello (at RAI), who also led the work on DVB-S2. This standard is supporting significantly higher spectral efficiency for the C/N typical for professional applications such as contribution links or IP-trunking. It also supports very low C/N down to -10 dB for mobile applications (e.g. maritime, aeronautical, trains, etc.).

Chapter 4
Atmospheric Electromagnetic Radiation

This chapter is introducing the practical and theoretical fundamentals of Earth atmospheric electromagnetic (EM) radiation as an integration sequence of satellite meteorological and weather observation. Some radiation of the EM spectrum is absorbed by the atmosphere, but some is passing through the atmosphere and reaches the Earth's surface, such as visible light, infrared, ultraviolet and microwave radiations. At this point, EM radiation is energy that is propagated through free space or through a material medium like atmosphere in the form of EM waves.

In a more proper sense, space and atmospheric EM radiation is a form of energy that is produced by oscillating electric and magnetic disturbances or by the free movements of electrically charged particles traveling through a vacuum or matter. In fact, the electric and magnetic fields come at right angles to each other and combined wave moves perpendicular to both magnetic and electric oscillating fields thus the disturbance. In addition, electron radiation is released as photons, which are bundles of light energy that travel at the speed of light as quantized harmonic waves. This energy is then grouped into categories based on its wavelength into the EM spectrum. The electric and magnetic waves travel perpendicular to each other and have certain characteristics, including amplitude, wavelength and frequency. Therefore, time varying electric and magnetic fields are mutually linked with each other at right angles and perpendicular to the direction of motion.

4.1 Fundamentals of Atmospheric Radiative Transfer

All kind of information received by a satellite about the Earth and its atmosphere comes in the form of EM radiation. It is necessary, therefore, to understand the mechanisms by which this EM radiation is generated and how it interacts with the atmosphere. The context about EM radiation is included in the following sequences, but all are concentrated on aspects that are essential for satellite meteorology. Radiative transfer in atmosphere is the specific physical phenomenon of

© Springer International Publishing AG 2018

S.D. Ilčev, *Global Satellite Meteorological Observation (GSMO) Theory*,
DOI 10.1007/978-3-319-67119-2_4

energy transfer in the form of EM radiation. At this point, the propagation of radiation through a medium is affected by absorption, emission and scattering processes.

The atmosphere is filtering the energy received from the Sun and from the Earth. Radiative transfer describes the interaction between radiation and matter in the atmosphere, such as gases, aerosols, cloud droplets, etc. The three key processes to be taken into account are:

1. **Emission** – The emission spectrum of a chemical element or chemical compound is the spectrum of frequencies of electromagnetic radiation emitted due to an atom or molecule making a transition from a high energy state to a lower energy state;
2. **Absorption** – Absorption of electromagnetic radiation is the way in which the energy of a photon is taken up by matter, typically the electrons of an atom. In such a way, the EM radiation energy is transformed into internal energy of the absorber, for example thermal energy. A material's absorption spectrum is the fraction of incident radiation absorbed by the atmospheric matter over a range of frequencies; and
3. **Scattering** – Scattering of an incident radiation by the atmospheric matter is the process by which small particles suspended in a medium of a different index of refraction diffuse a portion of the incident radiation in all directions. Thus, with scattering, there is no energy transformation, but a change in the spatial distribution of the energy. Scattering, along with absorption, causes attenuation problems with radar and other measuring devices.

The interaction between atmospheric matter issues on the one hand and solar and terrestrial radiation on the other hand plays a leading role for life conditions at the Earth's surface:

- Stratospheric ozone filters the solar ultraviolet radiation;
- Absorption by a few gas-phase species, such as water, methane or carbon dioxide, of the terrestrial radiation defines the so-called greenhouse effect of the Earth, which results in a surface temperature greater than 273 K ($-0.15\ °C$, otherwise, the Earth would be a "white planet"; and
- More generally, the state of the atmosphere is determined by its energy budget, which is comprising the radiative fluxes (solar and terrestrial radiation) and the latent and sensible heat fluxes.

The radiative properties and the concentrations of the atmospheric trace species determine the general behavior of the atmosphere components. For example, a few species, emitted by anthropogenic activities, play a decisive role by increasing the greenhouse effect or by decreasing the filtration of ultraviolet radiation, by taking critical part in the consumption of stratospheric ozone.

4.1.1 Nature of Radiation

Energy transfer from one place to another is accomplished by any one of three processes. Conduction is the transfer of kinetic energy of atoms or molecules (heat) by contact among molecules traveling at varying speeds. Convection is the physical displacement of matter in gases or liquids. Radiation is the process whereby energy is transferred across space or atmosphere without the necessity of a transfer medium, in contrast with conduction and convection.

The observation of a target in space by a device separated by some distance is the act of remote sensing, for example ears sensing acoustic waves are remote sensors. Thus, remote sensing with satellites for meteorological research has been largely confined to passive detection of radiation emanating from the Earth/atmosphere system. All satellite remote sensing systems involve the measurement of electromagnetic radiation. In such a way EM radiation has the properties of both waves and discrete particles, although the two are never manifest simultaneously.

All forms of EM radiation travel in a vacuum at the same velocity, which is approximately 3×10^{10} cm/s and is denoted by the letter c. Thus, EM radiation is usually quantified according to its wave-like properties, which include intensity and wavelength. For many applications it is sufficient to consider electromagnetic waves as being a continuous train of sinusoidal shapes.

If radiation has only one color, it is said to be monochromatic. The color of any particular kind of radiation is designated by its frequency, which is the number of waves passing a given point in 1 s and is represented by the letter f, with units of cycles/s or Hertz. The length of a single wave, i.e., the distance between two successive maxima is called the wavelength and is denoted by λ, namely with units of microns to meters.

For monochromatic radiation, the number of waves passing a fixed point in 1 s multiplied by the length of each wave is the distance the wave train traveled in 1 s. But that distance is equal in magnitude to the velocity of light c, which is presented the following equation:

$$c = f \lambda = 3 \times 10^{10} \ \mathrm{cm/s} \tag{4.1}$$

Since the frequency of a wave f is equal to the number of wavelengths in a distance of 3×10^{10} cm, which it is usually a large number. Therefore, it is often much more convenient, (especially in the infrared region of the spectrum) to consider the number of waves in 1 cm, called the wavenumber. In fact, this new unit the wavenumber is designated by v (cm^{-1}) and the following relation:

$$v = 1/\lambda (\mathrm{cm}^{-1}) \tag{4.2}$$

However, wavenumber becomes a relatively small unit in the microwave region, in such a way is using frequency in GHz units in that region. Similarly, because centimeters (cm) are usually too large for wavelength units, so it can be used

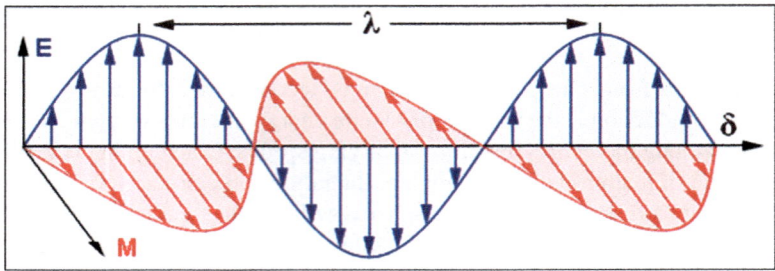

Fig. 4.1 Diagram of electromagnetic wave radiation (Courtesy of Manual: by Ilcev)

micrometers (μm) for most of the spectrum and for the very short wavelength can be used Angstrom units (Å).

4.1.1.1 Basic of Radiation

The atmospheric EM radiation consists of alternating electric and magnetic fields, which schematic diagram is illustrated in Fig. 4.1. The electric field vector E is perpendicular to the magnetic field vector M and the direction of propagation δ is perpendicular to both. Radiation is often specified by its wavelength, which is the distance between crests of the electric or magnetic field. In Fig. 4.2 is depicted the EM spectrum of waves radiation. In fact, a broad range of wavelengths from the ultraviolet to the microwave region is useful in satellite meteorology.

An alternate way to describe EM radiation in the atmosphere is to give its frequency, which is the rate at which the electric or magnetic field oscillates when observed at a point. The fundamental unit of frequency is the hertz (Hz), or one cycle per second. The frequency v is related to the wavelength λ by the following equation:

$$v = c/\lambda \tag{4.3}$$

where c is the speed with which EM radiation travels in direction of propagation δ and is known as the speed of light. In a vacuum the speed of light is 2.99792458×10^8 m s^{-1}. In the atmosphere, it travels slightly more slowly, due to interaction with air molecules. The index of refraction n of a substance is the ratio of the speed of light in vacuum to the speed with which electromagnetic radiation travels in that substance.

At sea level, the index of refraction of air is approximately 1.0003. For most purposes, therefore, the speed of light in a vacuum can be safely used even in the atmosphere. However, strong vertical gradients of atmospheric density and humidity result in strong vertical gradients of n, see more about in section Scattering. These cause bending of EM rays and can cause slight mislocation of satellite scan spots. One also sees radiation specified by wavenumber k, which is the reciprocal of

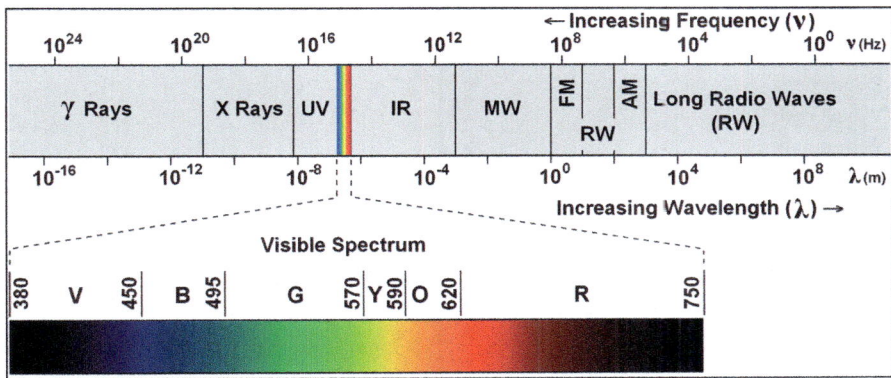

Fig. 4.2 Diagram of electromagnetic spectrum of waves radiation (Courtesy of Manual: by Ilcev)

the wavelength. Traditionally, wavenumber is expressed in inverse centimeters (cm). Radiation with a 15-μm wavelength has a 667-cm^{-1} wavenumber, for example. Since wavenumber is inversely proportional to wavelength, it is directly proportional to frequency. Therefore, a fundamental property of electromagnetic radiation is that it can transport energy. Many of the units used to quantify electromagnetic radiation are based on energy. These units are summarized in Table 4.1.

The basic unit of radiant energy is the joule (J). Radiant flux value is radiant energy per unit time, measured in watts [W; joules per second (J s^{-1})]. Radiant flux depends on area, which is often inconvenient; it is usually normalized by surface area. Radiant flux density is radiant flux crossing a unit area. It is measured, of course, in watts per square meter (W m^{-2}). Radiant flux density is so frequently used that it is subdivided to indicate which way the energy is traveling. Radiant exitance M is radiant flux density emerging from an area, and irradiance E is radiant flux density incident on an area.

As a conclusion it can be said that radiation is emission and propagation of energy in the form of rays or waves. The energy is radiated or transmitted in the form of rays, waves or particles. A stream of particles or EM waves is emitted by the atoms and molecules of a radioactive substance as a result of nuclear decay. There are two possible viewpoints for describing EM radiations. A radiation is composed of particles as photons. Similarly, it can be viewed as a wave propagating at the speed of light with c ~ 3 × 108 ms^{-1} in a vacuum, and with a similar value in air. It is then characterized by its frequency v in s − 1 or in Hertz (Hz) or equivalently by its wavelength λ = c/v (usually expressed in nm or in μm).

4.1.1.2 Solid Angle

In nature, EM radiation is a function of direction in space or in atmosphere (matter). The directional dependence is taken into account by employing the solid angle.

Table 4.1 Radiation symbols and units

Quantity	Recommended symbols	SI units
Frequency	ν	Hz
Wavelength	λ	m
Wavenumber	k	m^{-1}
Radiant energy	Q	J
Radiant exposure	H	$J\,m^{-2}$
Radiant flux	Φ	W
Radiant flux density	M, E	Wm^{-2}
Radiant Exitance	M	Wm^{-2}
Irradiance	E	Wm^{-2}
Radiance	L	$Wm^{-2}\,sr^{-1}$
Emittance	ε	Unitless
Absorbance	α	Unitless
Reflectance	ρ	Unitless
Transmittance	τ	Unitless
Absorption coefficient	σ_a	m^{-1}
Scattering coefficient	σ_s	m^{-1}
Extinction coefficient	σ_e	m^{-1}
Single scatter albedo	$\tilde{\omega}$	Unitless
Absorption number	$\tilde{\alpha}$	Unitless
Vertical optical depth	δ	Unitless
Slant path optical depth	δ_{sl}	Unitless
Scattering angle	ψ	rad
Scattering phase function	$p(\psi_s)$	sr^{-1}
Bidirectional reflectance	γ	sr^{-1}
Anisotropic reflectance factor	ξ	Unitless
Albedo	A	Unitless

Namely if one draws lines from the center of the unit sphere to every point on the surface of an object, the area of the projection on the unit sphere is the solid angle, which is depicted in Fig. 4.3. The solid angle of an object that completely surrounds a point is 2π steradians (sr), the area of the unit sphere.

The concept of solid angle is used for quantifying the solar radiation received by a surface. Let σ be a surface element on a sphere of radius r centered at point O. Thus, a solid angle Ω is then defined as the ratio of σ to the square of r: $\Omega = \sigma/r^2$. In spherical coordinates, the differential surface element $d\sigma$ is generated by the variations of the zenithal angle $d\theta$ and of the azimuthal angle $d\varphi$, which definition is illustrated in Fig. 4.4. In the left part of the figure, the surface elements dA and dS are represented by their projection in a vertical plane. The differential solid angle is then given by the following equation:

Fig. 4.3 Illustration of a solid angle (Courtesy of Book: by Kidder)

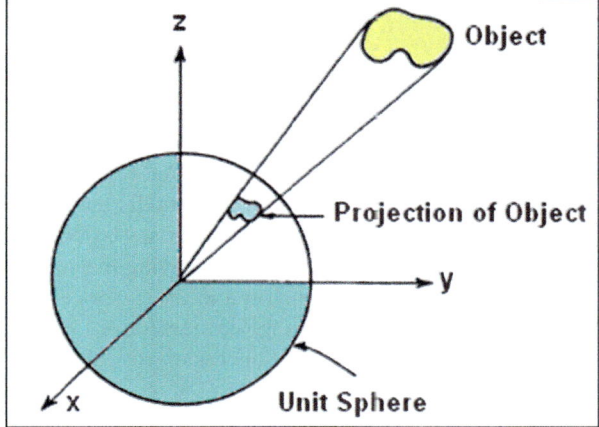

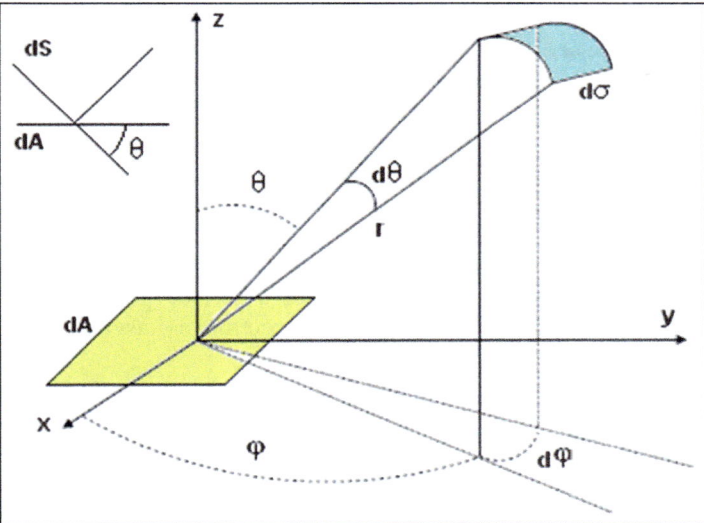

Fig. 4.4 Definition of a solid angle (Courtesy of Book: by Sportisse)

$$d\Omega = d\sigma/r^2 \qquad (4.4)$$

The solid angle subtended by an infinite plane is (2π sr). For an object with cross-sectional area (Ac) at a distance r from a point (Ac « r^2), the solid angle is Ac/r^2. Thus, solid angle is traditionally represented by the symbol Ω. If θ represents the zenith angle, the angle that is measured from the vertical or from the normal to a surface, and φ represents the azimuth angle, then a differential element of solid angle can be expressed mathematically. Namely, since $d\sigma = r\sin\theta d\varphi \times rd\theta$, this yields as follows:

$$d\Omega = \sin\theta \, d\theta \, d\phi = -d\mu \, d\phi \qquad (4.5)$$

$$d\Omega = -d\mu \, d\phi \qquad (4.6)$$

where $\mu = \cos\phi$. A solid angle is measured in *steradian* (sr). For a sphere, as $\theta \sim [0, \pi]$ and $\phi \sim [0, 2\pi]$, it can be obtained upon integration $\Omega = 4\pi$.

Radiant flux density per unit solid angle is known as radiance and is preferably assigned the symbol L. Suppose that a small element of surface is emitting radiation with radiance L. Thus, a question that arises is: What is the radiant exitance, that is, what is the total amount of radiation leaving the surface? This question is answered by integrating the radiance over the 2π sr above the surface. However, radiance represents the radiation leaving (or incident on) an area perpendicular to the beam. For other directions, it will be necessary to weight the radiance by $\cos\theta$.

4.1.1.3 Radiance and Irradiance

In radiometry, radiance is the radiant flux emitted, reflected, transmitted or received by a surface, per unit solid angle per unit projected area, and spectral radiance is the radiance of a surface per unit frequency or wavelength, depending on whether the spectrum is taken as a function of frequency or of wavelength. In fact, radiance is the power of EM radiation per unit area (radiative flux) incident on a surface. Radiant emittance or radiant exitance is the power per unit area radiated by a surface.

However, irradiance is the radiant flux (power) received by a surface per unit area. Spectral irradiance is the irradiance of a surface per unit frequency or wavelength, which depends on whether the spectrum is taken as a function of frequency or of wavelength.

In Fig. 4.4 is considered a differential surface area dA, so let dE_λ be the radiative energy.

(expressed in joules) intercepted by dA for incident photons of wavelength in $[\lambda, \lambda + d\lambda]$, during a time interval dt, in the solid angle $d\Omega$. In such a way, the monochromatic radiance is defined as the energy that is propagated through the surface dS and generated by dA in a direction perpendicular to the incident direction ($dS = \cos\theta \times dA$), namely it is presented by the following equation:

$$dI_\lambda = dE_\lambda/dS \, d\lambda \, dt \, d\Omega \qquad (4.7)$$

The radiance is usually expressed in $Wm^{-2} \, nm^{-1} \, sr^{-1}$ (with $1\,W = 1\,Js^{-1}$).

Monochromatic irradiance is defined as the component of the monochromatic radiance that is normal to dA, upon integration over the whole solid angle, which normal direction is given by the vertical axis in Fig. 4.4, by the following equation:

$$F_\lambda = \int \Omega \cos \theta \, I_\lambda d\Omega \qquad (4.8)$$

Upon integration over all the wavelengths, the irradiance is defined by relation:

$$F = \int_\lambda F_\lambda d\lambda \qquad (4.9)$$

It is expressed as a (received) power per unit area of surface (in Wm^{-2}).

4.1.1.4 Energy Transitions

Energy transition is generally defined as a long-term structural change in energy systems, which have occurred in the past and still is occurring. In fact, energy is transferred between the Earth's surface and the atmosphere via conduction, convection and radiation. Quantum mechanics describes the energy levels of a given molecule. The key point is that the energy levels are given by a discrete sequence, let us say $(E_n)_n$, specific of the spectroscopic properties of the molecule.

The simplest illustration is detailed in the discrete energy levels exercise with the case of a particle supposed to be "trapped in a well". Thus, the well is actually defined by energy potential, which is considered as a particle "trapped in a well". For convenience, we suppose that the well corresponds to a one-dimensional interval, let us say [0, 1], and let x be the spatial variable.

The particle location is introduced by a probability density function, p(x), which can be computed from the wavefunction f (x) as $p(x) = |f(x)|^2$. The wavefunction is governed by the following Schrödinger equation:

$$-h^2/2m \, d^2f/dx^2 + V(x)f = Ef \qquad (4.10)$$

where m = particle mass, V (x) = energy potential that defines the well, E = corresponds to the particle energy and h = 6.63×10^{-34} J, which de facto is Planck's constant. The particle motion is free inside the well (V = 0) but the particle is "trapped" (its probability density function is null at the boundaries). Here it is necessary to calculate the possible levels of energy (E). The solution for governing equation for f is as follows:

$$d^2f/dx^2 = f\left(2mE/h^2\right) \qquad f(0) = f(1) = 0 \qquad (4.11)$$

The solutions are in the form $f(x) \sim x \, (\sin(\sqrt{2} \, mE/h^2)$, where relation $\sqrt{2} \, mE/h^2 = n\pi$, with n as a positive integer. In fact, this results in a discrete spectrum of possible values for the energy level, presented by the following relation:

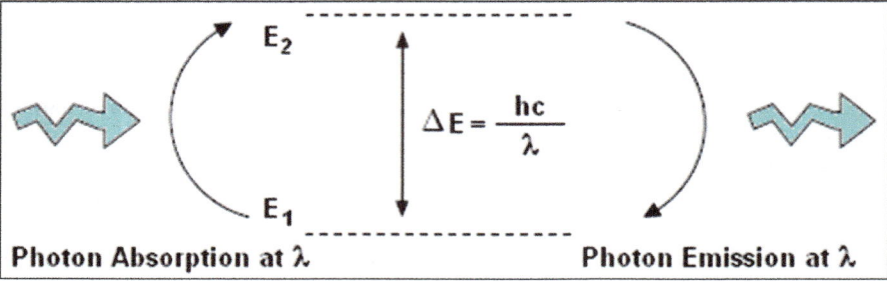

Fig. 4.5 Diagram for emission and absorption of a Photon (Courtesy of Book: by Sportisse)

$$En = n^2 \left(h^2/2m \right) \pi^2 \tag{4.12}$$

Considering a molecule with a given energy level, let us say E1, the emission of a photon by this molecule corresponds to a transition from E1 to a lower energy level, let us say E2 < E1. On the opposite, the absorption of a photon by this molecule implies the transition from $E1$ to a higher energy level, let us say E2 > E1. The wavelength of the photon is fixed by the energy transition. In Fig. 4.5 is shown transition between two energy levels E1 and E2 ($E_1 < E_2$), so wavelength is given by Planck's law: $\lambda = hc/(E2 - E1)$, states that:

$$\Delta E = h\nu = hc/\lambda \tag{4.13}$$

Thus, a photon can therefore be absorbed or emitted only if its wavelength corresponds to a possible transition. As a result, for a given molecule, absorption and emission are possible only in specific parts of the radiation spectrum (determined by the spectroscopic properties of the molecule). If λ has a small value, the energy gap is large: the shortwave radiations (e.g. ultraviolet radiation) are the most energy-containing ones. On the other hand, if the wavelength has a large value, the energy gap is low and the longwave radiations (e.g. infrared radiation) do not contain a lot of radiative energy.

4.2 Energy Emissions

Energy levels associated with molecules, atoms and nuclei clusters are in general discrete, quantized energy levels and mutual transitions between those levels typically involve the absorption or emission of photons. Electron energy levels have been used as the example here, but quantized energy levels for molecular vibration and rotation also exist. In such a way, transitions between vibrational quantum states typically occur in the infrared and transitions between rotational quantum states are typically in the microwave region of the EM spectrum.

The emission spectrum of a chemical element or chemical compound is the spectrum of frequencies of EM radiation emitted due to an atom or molecule making a transition from a high energy state to a lower energy state. The photon energy of the emitted photon is equal to the energy difference between the two states. Thus, there are many possible samples of electron transitions for each atom and each transition has a specific energy difference. This collection of different transitions, leading to different radiated wavelengths, will make up an emission spectrum. In fact, each element's emission spectrum is unique. so spectroscopy can be implemented to identify the elements in matter of unknown composition. Similarly, the emission spectra of molecules can be used in chemical analysis of substances.

4.2.1   Blackbody Emission

All material above absolute zero in temperature emits radiation. Explaining the nature of this radiation was one of the chief problems facing physicists in the nineteenth century. As usual, though, nature guards her secrets. If one looks at two different kinds of material, each at the same temperature, one finds that the radiation being emitted by them is de facto different. This led physicists to invent the perfect emitter, known as a blackbody, which emits the maximum amount of radiation at each wavelength. Although some materials come very close to being perfect emitters in some wavelength ranges, no real material is a perfect blackbody. Fortunately, the radiation inside a cavity whose walls are thick enough to prevent any radiation from passing directly through them can be shown to be the radiation that would be emitted by a blackbody. For instance, by observing the radiation inside cavities (through small holes) physicists knew, by the late nineteenth century, the empirical relationship between blackbody radiation and the two variables on which it depends: temperature and wavelength.

Planck's law for emission distribution describes the spectral density of EM radiation that is emitted by a black body in thermal equilibrium at a given temperature T. The law is named after Max Planck, who proposed it in 1900. It is a pioneering result of modern physics and quantum theory. In such a way, the radiative energy emitted by a "body" at equilibrium depends on its temperature. Intuitively, it is expected that the higher the temperature is, the higher the emission is. The maximum of radiative energy that can be emitted per unit area of surface and per time unit defines the so-called blackbody emission. For a body at certain temperature T, the maximum of emitted radiance at wavelength λ is given by the so-called Planck distribution:

$$B\lambda(T) = \left(2hc^2/\lambda^5\right) \left[1/\exp\left(hc/\lambda k_B T\right) - 1\right] \qquad (4.14)$$

where $k_B = 1.38 \times 10^{-23}\,\text{JK}^{-1}$ is Boltzmann's constant, and the unit is $\text{Wm}^{-2}\,\text{nm}^{-1}$.

For instance, in considering a blackbody, defined as a substance that absorbs all incident radiation, the blackbody is supposed to be inside a box, such that the walls do not emit nor absorb radiation. The reflections on the walls are therefore supposed to be perfect, so the blackbody receives an incident radiation I, that is exactly the emitted radiation.

Namely, let E1 < E2 be two energy levels of the blackbody. At equilibrium, the probability of having Ei (i = 1, 2) is given by the Maxwell-Boltzmann distribution, which is equal to exp.$(-E_i/k_B T)$, up to a normalization factor. The probability $P1 \to 2$ that the absorption of radiation leads to a transition from E1 to a higher value E2 is proportional to the number of molecules at state E1, and is expressed by the following equation:

$$P_{1\to 2} = \alpha I \exp(-E_1/k_B T) \tag{4.15}$$

where α is multiplying factor. The transition from E2 to a lower value E1 is driven by two processes: first, the spontaneous emission of a photon and, second, the emission induced by the absorption of the incident radiation I. Thus, the transition probability $P_{2\to 1}$ is therefore composed of two terms: the first one is proportional to the number of molecules at state E2 while the second one is proportional to I.

$$P_{2\to 1} = \beta I \exp(-E_2/k_B T) + \gamma \exp(-E_2/k_B T) \tag{4.16}$$

where first relation before + is inducted emission and second relation after + is spontaneous emission, with β and γ two multiplying factors. In such a way, at equilibrium, $P_{1\to 2} = P_{2\to 1}$, so that the number of molecules at a given energy level is constant. The incident radiation is exactly the emitted radiation and satisfies as follows:

$$I = \gamma/\alpha \exp[(-\Delta E/kBT) - \beta] \tag{4.17}$$

with $\Delta E = E_2 - E_1$, connected to the wavelength γ by $\Delta E = hc/\gamma$. This justifies Planck's law expressed in Eq. (4.14), if $\alpha = \beta$, namely if the probability of absorption is equal to the probability of induced emission.

Another important aspect of the Planck function is its integral over wavelength. The total radiant exitance from a blackbody is expressed with the following equation:

$$M_{BB} = \int_0^{\infty} \pi B_{\lambda}(T) \, d\lambda = \pi^5/15 \left(c_1 c_2^{-4} T^4 \right) = \sigma T^4 \tag{4.18}$$

where value σ is specified as Stefan-Boltzmann constant, and Eq. (4.18) is called the Stefan-Boltzmann law.

4.2.2 Surface Emission

Surface emissivity is critical for observation of surface skin temperature and infrared cloud properties when the observed radiance is influenced by the surface radiation. The surface emissivity of a material is its effectiveness in emitting energy as thermal radiation. In fact, thermal emission from a land or se) surface is dependent on both the spectral emissivity, an intrinsic property of the surface determined by its composition and texture and the surface temperature. Thus, usually neither of these properties is well known a priori so that remote sensing of a surface has the problem of separating emissivity and temperature.

Due to compositional and structural effects the emission from a real solid may significantly deviate from that of a blackbody at the same temperature. To quantify these deviations, a measure of radiating efficiency called the directional spectral emissivity is defined as the ratio of the emitted thermal radiance of a body to the radiance expected from a blackbody at the same temperature. Therefore, at any given wavelength λ, emissivity is defined as the ratio of the actual emitted radiance R_λ, to that from an ideal blackbody B_λ, such as:

$$\varepsilon_\lambda(\theta, \Phi) = R_\lambda(\theta, \Phi, T)/B_\lambda(T) \tag{4.19}$$

where sensor zenith θ and azimuth Φ must also be considered. Emissivity ranges between zero and one for all real substances. Water emissivity stays above 0.97 for all infrared spectral regions; land surface emissivity usually are above 0.90 but go as low as 0.75 near 4 microns for bare soil and desert surfaces. In Fig. 4.6 are shown surface emissivities at nadir viewing for some common surfaces, such as leaves (green or 1), snow (gray or 2), water (blue or 3), savanna (turquoise or 4), and desert (tan or 5) for wavelengths between 3.5 and 13.5 microns. Savanna and desert show a significant decrease in emissivity around 4 microns; desert also shows a significant decrease around 9 microns.

4.2.3 Medium Emissivity

Emissivity is the measure of an object's or medium ability to emit infrared energy. Emitted energy indicates the temperature of the object. In fact, emissivity can have a value from 0 of shiny mirror to 1.0 of blackbody). However, most organic, painted or oxidized surfaces have emissivity values close to 0.95. Infrared (thermal) energy, when incident upon matter, be it solid, liquid or gas, will exhibit the certain properties, such as absorption, reflection, and transmission to varying degrees.

Absorption is already explained in Sect. 4.1, however reflection is the change in direction of a wavefront at an interface between two different media so that the wavefront returns into the medium from which it originated. Common examples include the reflection of light, sound and water waves. Transmission is the degree to

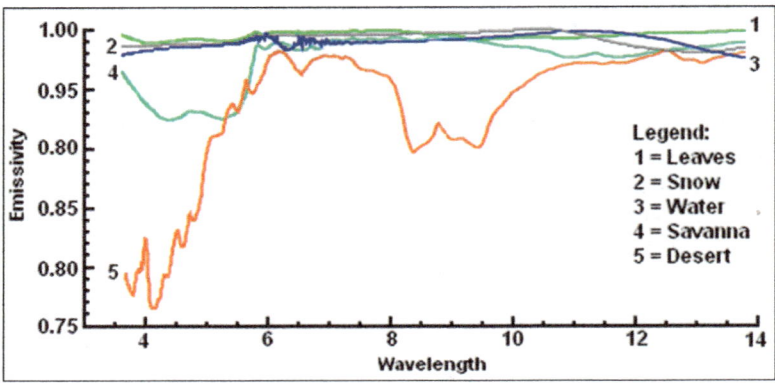

Fig. 4.6 Surface emissivities (Courtesy of Book: by Menzel)

which thermal energy passes through a material. There are few materials that transmit energy efficiently in the infrared region between 7 and 14 μm. Germanium is one of the few good transmitters of infrared energy and thus it is used frequently as lens material in thermal imaging systems.

In more particular sense, emissivity at a given wavelength is the ratio of infrared energy radiated by an object at a given temperature to that emitted by a blackbody at the same temperature The emissivity of a blackbody is unity at all wavelengths. In Fig. 4.7 is depicted emissivity diagram of ambient radiated by the heat surface provides reflected, transmitted and emitted energy, which can be measured by special sensors.

However, a realistic medium is not just a blackbody. The radiative energy that is actually emitted by a medium at temperature T, for a given wavelength λ, can be expressed as:

$$E\lambda(T) = \varepsilon_\lambda(T)B_\lambda(T) \tag{4.20}$$

where ε_λ (T) = so-called emissivity at wavelength λ and at temperature T. By definition $\varepsilon_\lambda \leq 1$ (unitless). For example, in the case of infrared radiation, the radiative behavior of fresh snow ("whitest" snow) is similar to a blackbody, with value $\varepsilon_\lambda \approx 1$, which typical values of the emissivity in the infrared region, for different surface types are presented in Table 4.2. Thus, on the opposite, the urban environment has a low emissivity, which will strongly impact the urban climate.

4.2.4   Earth and Sun Applications

The Sun can be considered as a blackbody with a temperature of 5800 K or 5526.85 C°. Applying Wien's law justifies that the Sun's emission peaks in the visible region, which is maximum at 500 nm). The Earth can be considered as a blackbody

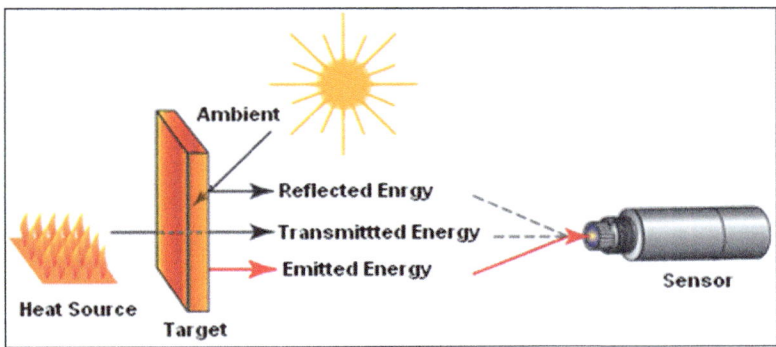

Fig. 4.7 Emissivity of ambient (Courtesy of Manual: by Internet)

Table 4.2 Typical values of the emissivity

Surface	εIR	Surface	εIR
Sea	0.95–1	Grass	0.90–0.95
Fresh snow	0.99	Desret	0.85–0.90
"old" snow	0.80	Forest	0.95
Liquid water clouds	0.25–1	Concrete	0.70–0.90
Cloud cirrus	0.10–0.90	Urban	0.85

at 288 K or 14.85 C°. Thus, the maximum is then in the infrared region (10 μm). Upon a direct application of the Stefan-Boltzmann law, the Sun emits about 204 as much as the Earth (5800/288 ≈ 20). The key point is that both emitted radiation spectra can be split, which scenario is shown in Fig. 4.8.

As illustrated in Fig. 4.8, normalized emission spectrum for the Earth (blackbody at T = 288 K) and for the Sun (blackbody at T = 5800 K).

On the other hand, the atmosphere will have a different behavior with respect to the solar and terrestrial radiation, namely in a first approximation, the infrared terrestrial radiation is absorbed while the atmosphere is transparent to the solar visible radiation.

4.3 Radiative Properties of Matter

The emissivity of a surface at a specified wavelength may vary as temperature changes since the spectral distribution of emitted radiation changes with temperature, in comparison to a blackbody at the same temperature. The emissivity of a surface varies between zero and one. The emissivity of a real surface varies as a function of the surface temperature, the wavelength and the direction of the emitted radiation. The emissivity of a surface at a specified wavelength may vary as temperature changes since the spectral distribution of emitted radiation changes with temperature. The total hemispherical emissivity is defined in terms of the

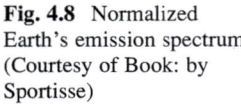

Fig. 4.8 Normalized Earth's emission spectrum (Courtesy of Book: by Sportisse)

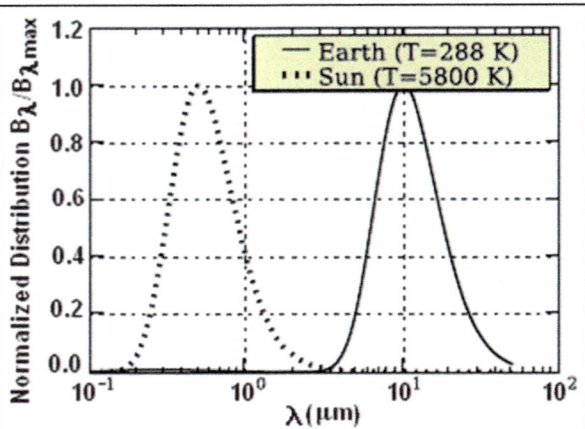

radiation energy emitted over all wavelengths in all directions. Radiation is a complex phenomenon, so the dependability of its properties in wavelength and direction makes it even more complicated. Thus, the gray and diffuse approximation methods are commonly used to perform radiation calculations.

All radioactive particles and waves, from the entire EM spectrum to alpha, beta and gamma particles, possess the ability to eject electrons from atoms and molecules and to create ions. Matter in bulk comprises particles that, compared to radiation, may be said to be at rest, but the motion of the molecules that compose matter, which is attributable to its temperature, is equivalent to travel at the speed rate of hundreds of metres per second. Although matter is commonly considered to exist in three forms, such as solid, liquid and gas, a review of the effects of radiation on matter must also include mention of the interactions of radiation with gasses, attenuated (low-pressure) gases, plasmas and matter in states of extraordinarily high density. In bulk, matter can exist in several different forms or states of aggregation, known as phases, depending on ambient pressure, temperature and volume. A phase is a form of matter that has a relatively uniform chemical composition and physical properties (such as density, specific heat, refractive index and so forth).

4.3.1 Electromagnetic Spectrum and Waves

As stated above, the EM spectrum is a continuum of all EM waves arranged according to frequency and wavelength. The Sun, Earth and other bodies radiate EM energy of varying wavelengths. Thus, EM energy passes through space at the speed of light in the form of sinusoidal waves, and EM waves form a continuous spectrum, which diagram of EM waves radiation is shown in Fig. 4.2.

The electromagnetic spectrum is the collective term for all known frequencies and their linked wavelengths of the known photons (EM radiation). The "EM spectrum" of an object has a different meaning, and is instead the characteristic distribution of EM radiation emitted or absorbed by that particular object. At this point, the EM spectrum extends from below the low frequencies used for modern radio communication to gamma radiation at the short-wavelength (HF) end, thereby covering wavelengths from thousands of kilometers down to a fraction of the size of an atom. Visible light lies toward the shorter end, with wavelengths from 400 to 700 nanometers. The limit for long wavelengths is the size of the universe itself, while it is thought that the short wavelength limit is in the vicinity of the Planck length. Until the middle of the twentieth century it was believed by most physicists that this spectrum was infinite and continuous.

The different types of radiation are defined by the amount of energy found in the photons. Radio waves have photons with low energies, microwave photons have a little more energy than radio waves, infrared photons have still more, then visible, ultraviolet, X-rays, and, the most energetic of all, gamma-rays. Gamma-rays have the shortest wavelength of any type of radiation. The EM spectrum contains microwaves, radio waves, infrared waves, optical waves, ultraviolet rays, X-rays and gamma rays.

Visible light is just a small part of the EM spectrum that ranges from low frequency radio waves to high frequency gamma rays. The visible spectrum starts just above the infrared region and ends just below the ultraviolet region. White light can be split up into many colours by using a prism. In fact, this visible light is just part of the whole spectrum of EM radiation. However, not all types of EM radiation are visible. Each type has a different wavelength and a different use in everyday life, for example EM radiation can be used for wireless communications.

4.3.2 Absorption and Emission of Radiation

Emissivity is a measure of how strongly a body radiates at a given wavelength, which value ranges between zero and one for all real substances. However, a gray body is defined as a substance whose emissivity is independent of wavelength. In the atmosphere, clouds and gases have emissivities that vary rapidly with wavelength. The ocean surface has near unit emissivity in the visible regions.

For a body in local thermodynamic equilibrium the amount of thermal energy emitted must be equal to the energy absorbed; otherwise the body would heat up or cool down in time, contrary to the assumption of equilibrium. In an absorbing and emitting medium in which I_λ is the incident spectral radiance, the emitted spectral radiance R_λ is given by:

$$R_\lambda = \varepsilon_\lambda B_\lambda = a_\lambda I_\lambda \qquad (4.21)$$

where a_λ represents the absorptance at a given wavelength.

If the source of radiation is in thermal equilibrium with the absorbing medium, then:

Fig. 4.9 Kirchhoff's law of radiation (Courtesy of Paper: by Robinson)

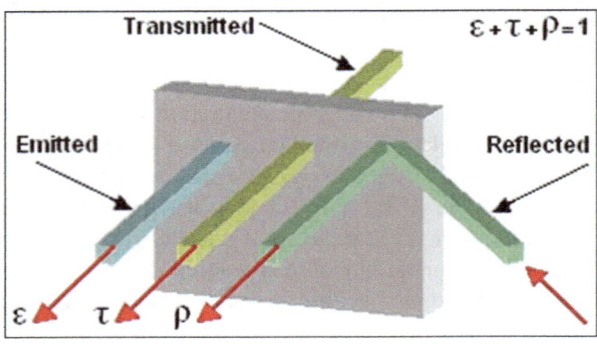

$$I_\lambda = B_\lambda \quad \text{so that :} \quad \varepsilon_\lambda = a_\lambda \tag{4.22}$$

This is often referred to as Kirchhoff's Law. In qualitative terms, it states that materials that are strong absorbers at a given wavelength are also strong emitters at that wavelength, and so, similarly weak absorbers are weak emitters. According to the image shown in Fig. 4.9, Kirchhoff's Law also states that at any temperature, the ratio of emissive power e_λ of a body to its absorptive power a_λ, for a particular wavelength, is always constant and is equal to the emissive power of perfect black body for that wavelength. This implies the ratio between e_λ and a_λ for any body is a constant quantity ($=E_\lambda$).

Here will be shortly introduced the following terms:

1. **Convection** – It **is** the mode of transmission of heat, through fluids, in which the particle of fluids acquire heat from one region and deliver the same to the other regions by leaving their mean positions and moving from one point to another.
2. **Radiation** – It is that process heat transmission in which heat travels from one point to another in straight lines, with velocity of light, without heating the intervening medium.
3. **Absorptive Power** – It is power of the substance ($a = Q_1/Q$) defined as the ratio between amounts of heat absorbed by it to the total amount of heat incident upon it.
4. **Reflecting Power** – It is power of a substance ($r = Q_2/Q$) defined as the ratio between amounts of heat reflected by the substance to the total amount of heat incident upon it.
5. **Transmitting Power** – It is power of a substance ($t = Q_3/Q$) defined as the ratio between amounts of heat transmitted by the body to the total amount of heat incident upon it.

The mutual relation between three above powers is:

$$a + r + t = Q_1/Q + Q_2/Q + Q_3/Q = [Q_1 + Q_2 + Q_3]/Q = Q/Q = 1 \tag{4.23}$$

Radiant Emittance Radiant emittance of a body (E) at a temperature (T) is defined as the total amount of energy, for all wavelengths, radiated per unit time, per unit area by the body, is as follows:

$$E = \int_0^\infty \varepsilon_\lambda d\lambda \tag{4.24}$$

Energy Density Total energy density (U) at any point is defined as the radiant energy per unit volume, around that point, for wavelengths taken together, such as:

$$E = \int_0^\infty u_\lambda d\lambda \tag{4.25}$$

A fraction of the incident radiation is absorbed along the path of propagation in a medium, (such as here presented atmosphere). Namely, the Beer-Lambert law, which is also referred to as the Beer-Lambert-Bouguer law, governs the reduction in the radiation intensity $I\lambda$ at wavelength λ. In Fig. 4.10 is illustrated absorption of an incident radiation traversing a medium (Gray Box), so if S stands for the medium thickness (oriented in the direction of propagation), the evolution of the radiation intensity is as follows:

$$dI_\lambda/dS = -a_\lambda(S)I_\lambda \tag{4.26}$$

where $a_\lambda(S)$ = absorption coefficient at wavelength λ (what is depending on the medium). The unit of a_λ is, for instance, m^{-1} or cm^{-1}. Assuming that the medium is homogeneous, then a_λ has a constant value in the following relation:

$$I_\lambda(S) = I_\lambda(0)\exp(-Sa_\lambda) \tag{4.27}$$

Consider a medium composed of p absorbing species, with densities n_i ($i = 1, ..., p$), expressed in molecule cm^{-3}.

A way to define the absorption cross section is to consider an incident flux of energy per surface, F (in Wcm^{-2}). The resulting absorbed energy is then $F_a = \sigma_a F$ (expressed in W).

Another classical concept is the so-called optical depth τ_λ (unitless), which is defined for a monochromatic radiation by:

$$d\tau_\lambda = a_\lambda(S) ds \tag{4.28}$$

Rewriting the Beer-Lambert law yields as follows:

$$dI_\lambda/d\tau_\lambda = I_\lambda \tag{4.29}$$

The absorbing coefficient is then obtained by summing over all species. The contribution for a given species depends on the density and on the so-called

Fig. 4.10 Gray box
(Courtesy of Book: by
Sportisse)

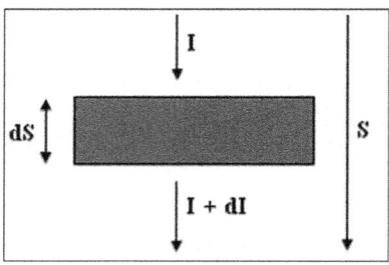

absorption crossection (the effective cross section resulting in absorption), $\sigma^a_i\,(\lambda, S)$, usually expressed in cm^2, such as:

$$a_\lambda(S) = \sum_{I=1}^{p}\; n_i(S)\, \overset{a}{\underset{i}{\sigma}}\,(\lambda, S) \tag{4.30}$$

4.3.3   Atmospheric Scattering of Radiation

Radiation scattered from a particle is a function of several things: particle shape, size and index of refraction, then wavelength of radiation and viewing geometry. Thus, in 1908 Mie applied Maxwell's1 equations, which describe electromagnetic radiation, to the case of a plane electromagnetic wave incident on a sphere. The far-field radiation (that observed at many radii from the sphere) is the scattered radiation. Mie showed that for a spherical scattered, the scattered radiation is a function of only viewing angle, index of refraction and the size parameter defined as:

$$X = 2\pi r/\lambda \tag{4.31}$$

where r = radius of the sphere. Namely, the size parameter can be used to divide scattering into three regimes, which scenario of scattering regimes is shorn in Fig. 4.11.

Scattering is a physical process by which a particle in the path of an EM wave continuously abstracts energy from the incident wave and re-radiates that energy in all directions. The particle may be thought of as a point source of the scattered energy. In the atmosphere, the particles responsible for scattering cover the sizes from gas molecules (~10–8 cm) to large raindrops and hail particles (~1 cm). The relative intensity of the scattering pattern depends strongly on the ratio of particle size to wavelength of the incident wave. If scattering is isotropic, the scattering pattern is symmetric about the direction of the incident wave. Thus, a small anisotropic particle tends to scatter light equally into the forward and backward

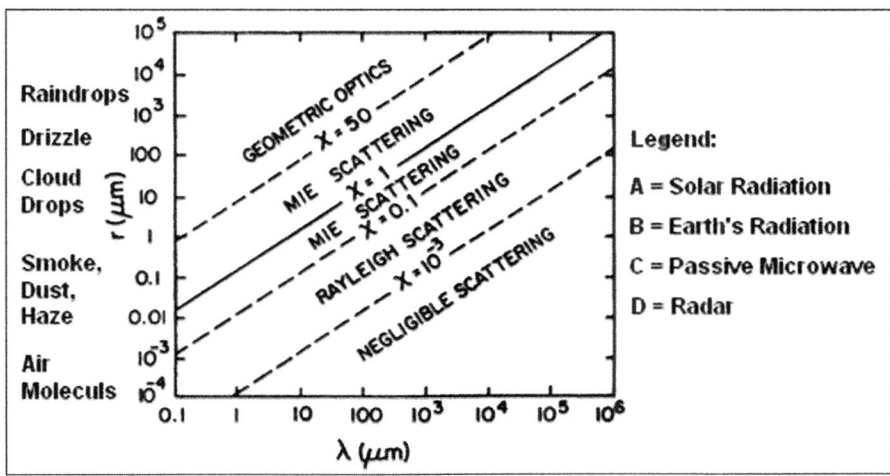

Fig. 4.11   Scattering regimes (Courtesy of Book: by Kidder)

directions. However, when the particle becomes larger, the scattered energy is increasingly concentrated in the forward directions. Distribution of the scattered energy involving spherical and certain symmetrical particles may be quantitatively determined by means of the EM wave theory. When particles are much smaller than the incident wavelength, the scattering is called Rayleigh scattering. Beside, for particles whose sizes are comparable to or larger than the wavelength, the scattering is customarily referred to as Mie scattering.

Let consider a gaseous molecule or a particle, aerosol or cloud drop, with a characteristic size. The incident radiation is also scattered in all directions, and the shape of the scattered intensity strongly depends on the characteristic size.

The scattering of the incident EM wave by a gas-phase molecule or by a particle mainly depends on the comparison between the wavelength (λ) and the characteristic size (d). It can be recalled that $d \sim 0.1$ nm for a gas-phase molecule, d $\sim$ [10 nm, 10 μm] for an aerosol and d $\sim$ [10, 100] μm for a liquid water drop. The wide range covered by the body size will induce different behaviors. Three scattering regimes are usually distinguished: the Rayleigh scattering (typically for gases), the scattering represented by the optical geometry's laws (typically for liquid water drops) and the so-called Mie scattering (for aerosols).

1. **Rayleigh Scattering** – If d $<<$ λ (the case for gases), the EM field can be assumed to become homogeneous at the level of the scattering body. In fact, this defines the so-called Rayleigh Scattering (also referred to as molecular scattering). The scattered intensity in a direction with an angle θ to the incident direction, at the distance r from the scattering body, which scenario Scattering of an Incident Radiation (I_O) is shown in Fig. 4.12. Thus, for a media of mass concentration C, composed of spheres of diameter d and of density ρ, is then given by the following equation:

Fig. 4.12 Scattering of an
incident radiation (I_O)
(Courtesy of Book: by
Sportisse)

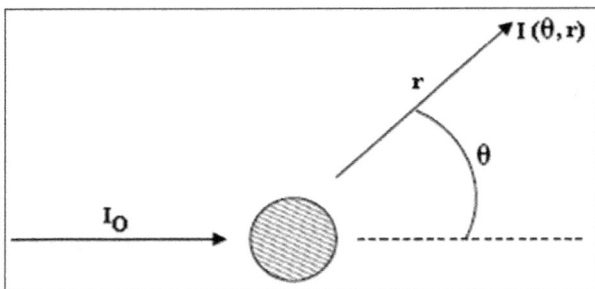

Fig. 4.12 Scattering of an incident radiation (I_O) (Courtesy of Book: by Sportisse)

$$I(\theta, r) = I_0 \left[(8\pi^4/r^2\lambda^4) \, (\rho^2 d^6/C^2) \right] (m^2 - 1/m^2 + 1)^2 (1 + \cos^2\theta) \qquad (4.32)$$

The incident intensity is I_0. m is the complex refractive index, specific to the scattered body, it is defined as the ratio of the speed of light in the vacuum to that in the body, and depends on the chemical composition for aerosols (e.g. $m = 1.34$ for water at $\lambda = 450$ nm.

This formula is inversely proportional to λ^4, and scattering is therefore much stronger for the shortwave radiations, which is devoted to the sky colors. Thus, as a result, the terrestrial longwave radiations are weakly scattered, and that the Rayleigh scattering is an increasing function of the size (d) and is a decreasing function of the distance (r). Moreover, Rayleigh scattering is symmetric between the backward and forward directions:

$$I(\theta, r) = I(\pi - \theta, r) \qquad (4.33)$$

2. **Optical Geometry** – If $d \gg \lambda$ (this is the case of liquid water drops with respect to the solar radiation), the laws of optical geometry can be applied, leading to the understanding of many physical phenomena (e.g. rainbow formation). The scattering weakly depends on the wavelength.
3. **Mie Scattering** – If $d \approx \lambda$ (the case for most of atmospheric aerosols), the simplifications used above are no longer valid. A detailed calculation of the interaction between the EM field and the scattering body is required by the Mie theory. The intensity of the scattered radiation in the direction with an angle θ to the incident direction at a distance r is:

$$I(\theta, r) = I_0 \left[\lambda^2 (i_1 + i_2)/4\pi^2 r^2 \right] \qquad (4.34)$$

where i_1 and i_2 are the intensity Mie parameters, given as complicated functions of d/λ, θ and m. Thus, the parameters i_1 and i_2 are characterized by a set of maxima as a function of the angle θ.

4.3.4 Surface Reflection

A small part of the Sun's energy is directly absorbed, particularly by certain gases such as ozone and water vapor. Some of the Sun's energy is reflected back to space by clouds and the Earth's surface. Reflected radiation, particularly reflected solar radiation and reflected MW radiation, is very important to satellite meteorology. Several quantities are used to describe reflected radiation. Thus, the most basic is the bidirectional reflectance γ_r, which is related to the fraction of radiation from incident direction (θ_i, Φ_i) that is reflected in the direction (θ_r, Φ_r). Perhaps the best way to define this question it is to write the formula for the radiance reflected from a small element of surface:

$$L_r(\theta_r, \Phi_r) = \int_0^{2\pi} \int_0^{\pi/2} L_i(\theta_i, \Phi_i)\gamma_r(\theta_r, \Phi_r; \theta_i, \Phi_i) \, \cos\theta_i \sin\theta_i \, d\theta_i d\Phi_i \qquad (4.35)$$

Basically, radiation from incident direction (θ_i, Φ_i) illuminates a small element of surface. Taking into account the effect of incident angle $L_i \cos\theta_i$ is available to be reflected. Thus, a fraction γ_r $(\theta_i, \Phi_i; \theta_r, \Phi_r)$ is reflected into direction (θ_r, Φ_r). Integrating over all incident solid angles gives the reflected radiance in direction (θ_r, Φ_r). An important property of the bidirectional reflectance is known as the Helmboltz reciprocity principle. It states that the bidirectional reflectance is invariant if the directions of incoming and outgoing radiation are interchanged:

$$\gamma_r(\theta_r, \Phi_r; \theta_i, \Phi_i) = \gamma_r(\theta_i, \Phi_i; \theta_r, \Phi_r) \qquad (4.36)$$

This equation is used in the construction of tables of bidirectional reflectance from satellite observations because not all incident radiation angles can be observed.

4.3.5 Solar and Terrestrial Radiation

The solar radiation reaching the Earth originates (for our purposes) from a layer of the Sun called the photosphere, which coincides with the visible disk of the Sun and has a radius of 6.96×105 km. The radiation leaving the photosphere is very nearly that of a 6000-K or a 5727 C$^\circ$ blackbody. Calculating the absorption spectrum of the atmospheric compounds (namely of the absorption cross sections σ_i^a (λ) is the objective of spectroscopy. This issue is based on the possible energy transitions for a given molecule.

1. **Absorption of Solar Radiation** – The X-ray region of the electromagnetic spectrum (the most energetic radiation) is filtered in the ionosphere through the ionization process. Let A be a molecule or an atom, then the ionization process reads:

$$A + hv\ (\lambda) - \ \rightarrow A^+ + e^- \tag{4.37}$$

Ionization requires high energies, defined by the concept of ionization potential, which potential corresponds to the maximum of the wavelengths (thus to the minimum of the energies) for which ionization occurs. Ionization takes place in the upper atmosphere. Once the X-ray region is filtered, the remaining part of the spectrum is not energetic enough so that ionization is no longer possible. As expected, the electron density is an increasing function of altitude. For example, high values are responsible for the so-called *black out* that affects the communications of a space shuttle in the reentry phase (at an altitude of about 90–100 km).

2. **Ultraviolet and Visible Radiation** – The comparison between the spectrums of solar radiation at the top of the atmosphere and that at sea level is shown in Fig. 4.13. The difference between the two curves corresponds to the absorption of solar radiation in the atmosphere. Thus, for the ultraviolet solar radiation, the absorption is strong for molecular oxygen (O_2), ozone (O_3), water vapor (H_2O) and carbon dioxide (CO_2). These species, especially stratospheric ozone, filter the ultraviolet radiation, which anneals its adverse effects to health and vegetation. Here can be noted that the splitting between the ultraviolet and visible radiation:

   – The shortwave solar radiations are absorbed in the ionosphere (X-ray region), so at this point, in the mesosphere (Schumann-Runge continuum for O_2, $\lambda \sim [150, 200]$ nm) and in the stratosphere (Hartley continuum for O_3, $\lambda \sim [200, 300]$ nm);
   – The atmosphere is transparent for the visible solar radiations, and therefore, this property defines the so-called atmospheric window. This is a key point since it makes it possible to heat and to light the Earth's surface.

3. **Absorption of the Terrestrial Radiation** – The longwave infrared radiations, which are corresponding to terrestrial and atmospheric emissions, can be absorbed by water vapor (H_2O), CH_4, CO_2 and O_3. These gases are characterized by their strong absorption of the infrared radiations, such as green house gase*s*, which main absorption bands related to the greenhouse gases. Here can be noted that the main absorption bands are related to the greenhouse gases (GHG). The altitude at which the absorbing and then emitting gases are located can be obtained by comparing the spectrum with the Planck distribution. In such a way, the corresponding emission temperature results in an altitude (by using the vertical distribution of temperature).

   In Fig. 4.14 is illustrated the radiance spectrum in the infrared region, as it would be measured by a sensor at the altitude of 70 kilometers, above a region with a temperature of 305 K. The spectrum is computed by a numerical model that solves the radiative transfer equation (MODTRAN) for a standard atmosphere (USA 1976, clear sky). The Planck's distributions for the blackbody emissions are plotted for a few temperatures, such as values at 220, 240,

Fig. 4.13 Radiance spectrum at upper atmosphere and sea level (Courtesy of Book: by Sportisse)

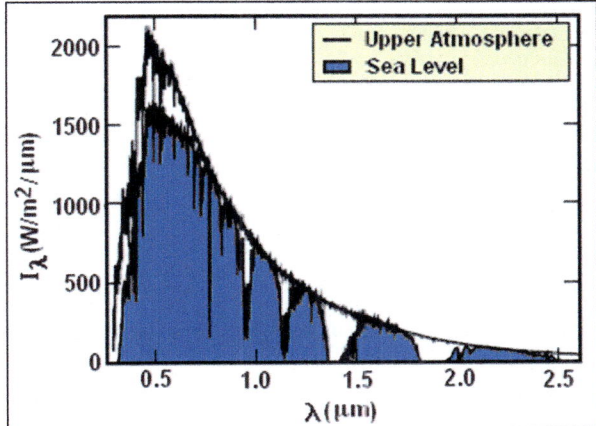

Fig. 4.14 Radiance spectrum at upper atmosphere and sea level (Courtesy of Book: by Sportisse)

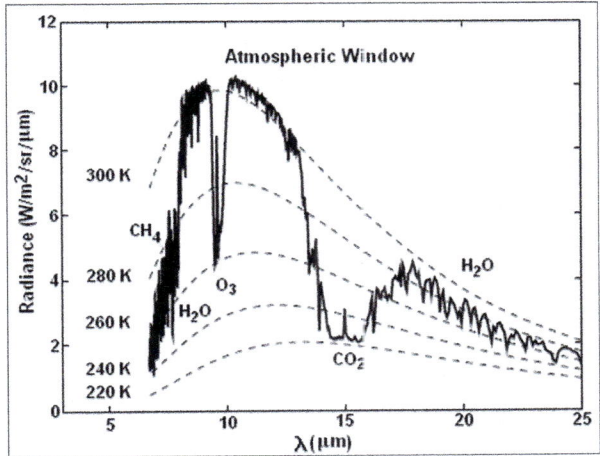

260, 280 and 300 K. However, the corresponding GHG are indicated near the absorption peaks.

4.3.6 Conservation of Energy

Most of the energy our Planet and atmosphere get comes as light from the Sun and most of this light is visible. Some is higher energy (i.e., more energy per photon) ultraviolet light and some of it is lower energy infrared radiation. Most of it bounces off us right back out into the blackness of space and some of it gets absorbed in a

Fig. 4.15 Schematic for
reflectance, absorptance and
transmittance in layers
(Courtesy of Book: by
Menzel)

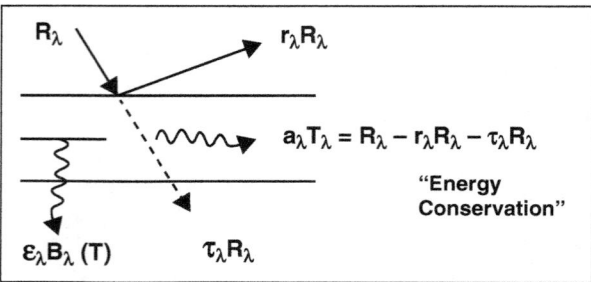

Fig. 4.15 Schematic for reflectance, absorptance and transmittance in layers (Courtesy of Book: by Menzel)

number of ways. Some of the light's energy goes into increasing the kinetic energy of individual atoms that get hit. The overall kinetic energy of atoms is calling heat and much of this heat involves the warming of the atmosphere.

Consider an atmospheric layers of gases absorbing medium where only part of the total incident radiation I_λ is absorbed, and the remainder is either transmitted through the layer or just reflected from it, which scenario in a layer of gases is illustrated in Fig. 4.15. In other words, if a_λ, r_λ, and τ_λ represent the fractional absorptance, reflectance, and transmittance, respectively, then the absorbed part of the radiation must be equal to the total radiation minus the losses due to reflections away from the layer and transmissions through it. Hence, this relation is expressed by the following equation:

$$a\lambda I\lambda = I\lambda - r\lambda I\lambda - \tau_\lambda I\lambda \quad \text{or} \quad a_\lambda + r_\lambda + \tau_\lambda = 1 \qquad (4.38)$$

which says that the processes of absorption, reflection, and transmission account for all the incident radiation in any particular situation. This is simply conservation of energy, so for a blackbody $a_\lambda = 1$, so it follows that $r_\lambda = 0$ and $\tau_\lambda = 0$ for blackbody radiation. Thus, in any window region $\tau_\lambda = 1$, $a_\lambda = 0$ and $r_\lambda = 0$. Radiation incident upon any opaque surface, $\tau_\lambda = 0$, is either absorbed or reflected, so at this point that relation is shown such as:

$$a_\lambda + r_\lambda = 1 \qquad (4.39)$$

At any wavelength, strong reflectors are weak absorbers (i.e., snow at visible wavelengths), and weak reflectors are strong absorbers (i.e., asphalt at visible wavelengths).

From Kirchhoff's Law is possible also to write as follows:

$$\varepsilon_\lambda + r_\lambda + \tau_\lambda = 1 \qquad (4.40)$$

which is giving emission, reflection and transmission account for all the incident radiation for media in thermodynamical equilibrium.

4.3.7   Selective Absorption and Emission

The atmosphere of the Earth exhibits absorptance that varies drastically with wavelength. Atmospheric absorptance is small in the visible part of the spectrum, while it is large in the infrared. This has a profound effect on the equilibrium temperature at the surface of the Earth. Thus, the following problem illustrates this point. Assume the Earth behaves like a blackbody and the atmosphere has an absorptance a_S for incoming solar radiation and a_L for outgoing longwave radiation. In such a way, E is the solar irradiance absorbed by the Earth atmosphere system, a known function of Albedo. The implied irradiance emitted from the Earth's surface Y_s and the irradiance emitted by the atmosphere Y_a (for both upward and downward) requires the following radiative equilibrium at the surface:

$$(1 - a_S) E - Y_s + Y_a = 0 \tag{4.41}$$

and at the top of the atmosphere as follows:

$$E - (1 - a_L) Y_s - Y_a = 0 \tag{4.42}$$

Solving for Ys yields as follows:

$$Y_s = [(2 - a_S)/(2 - a_L)] E \tag{4.43}$$

and then is getting the following equations:

$$Y_s = [(2 - a_L) - (1 - a_L) (2 - a_S)/(2 - a_L)] E \tag{4.44}$$

Since $a_L > a_S$, the irradiance and hence the radiative equilibrium temperature at the Earth surface is increased by the presence of the atmosphere. With $a_L = 0.8$ and $a_S = 0.1$ and $E = 241$ Wm^{-2}, Stefans Law yields a blackbody temperature at the surface of 286 K, in contrast to the 255 K it would be if the atmospheric absorptance was independent of wavelength ($a_S = a_L$). The atmospheric gray body temperature in this example is 245 K.

4.3.8   Composition of the Earth's Atmosphere

To describe the interaction of the Earth's atmosphere with solar radiation, the atmosphere's composition must be understood. In fact, the atmosphere is composed of a group of nearly permanent gases and a group of gases with variable concentration. In addition, the Earth's atmosphere also contains various solid and liquid particles such as aerosols, water drops, and ice crystals, which are highly variable in space and time.

The Earth's atmosphere is composed of the following molecules: nitrogen (78%), oxygen (21%) and argon (1%), what is total account for more than 99.99% of the permanent gases.

Then trace amounts of carbon dioxide, neon, helium, methane, krypton, hydrogen, nitrous oxide, xenon, ozone, iodine, carbon monoxide and ammonia. However, lower altitudes also have quantities of water vapor.

These gases have virtually constant volume ratios up to an altitude of about 60 km in the atmosphere. It should be noted that although carbon dioxide is listed here as a permanent constituent, it's concentration varies as a result of the combustion of fossil fuels, absorption and release by the ocean, and photosynthesis. Water vapour concentration varies greatly both in space and time depending upon the atmospheric condition. In fact, its variation is extremely important in the radiative absorption and emission processes. On the other hand, ozone concentration also changes with respect to time and space, and it occurs principally at altitudes from about 15 to about 30 km, where it is both produced and destroyed by photochemical reactions. Most of the ultraviolet radiation is absorbed by ozone, preventing this harmful radiation from reaching the earth's surface.

4.3.9 Different Atmospheric Absorptions and Emissions

1. **Absorption and Emission of Solar Radiation in the Atmosphere** – This phenomenon is accomplished by molecular storing of the electromagnetic radiation energy. Molecules can store energy in various ways. Any moving particle has kinetic energy as a result of its motion in space. This is known as translational energy, which averaged translational kinetic energy of a single molecule in the x, y and z directions is found to be equal to $kT/2$, where k is the Boltzmann constant and T is the absolute temperature. Thus, a molecule which is composed of atoms can rotate, or revolve, about an axis through its centre of gravity and, therefore, has rotational energy. The atoms of the molecule are bounded by certain forces in which the individual atoms can vibrate about their equilibrium positions relative to one another. The molecule, therefore, will have vibrational energy.

   These three molecular energy types (translational, rotational and vibrational) are based on a rather mechanical model of the molecule that ignores the detailed structure of the molecule in terms of nuclei and electrons. It is possible, however, for the energy of a molecule to change due to a change in the energy state of the electrons of which it is composed. Thus, the molecule also has electronic energy. The energy levels are quantized and take discrete values only. As we have pointed out, absorption and emission of radiation takes place when the atoms or molecules undergo transitions from one energy state to another. In general, these transitions are governed by selection rules. Atoms exhibit line spectra associated with electronic energy levels. Molecules, however, also have rotational and vibrational energy levels that lead to complex band systems.

Solar radiation is mainly absorbed in the atmosphere by O_2, O_3, N_2, CO_2, H_2O, O, and N, although NO, N_2O, CO and CH_4, which may occur in very small quantities, also exhibit absorption spectra. Absorption spectra due to electronic transitions of molecular and atomic oxygen and nitrogen, and ozone occur chiefly in the ultraviolet (UV) region, while those due to the vibrational and rotational transitions of triatomic molecules such as H_2O, O_3 and CO_2 lie in the infrared region. There is very little absorption in the visible region of the solar spectrum.

2. **Atmospheric Absorption and Emission of Thermal Radiation** – The Earth also emits electromagnetic radiation covering all frequencies. However, the global mean temperature of the earth-atmosphere system is only about 250 K, a temperature that is obviously much lower than that of the Sun's photosphere. As a consequence of Planck's law and Wien's displacement law discussed earlier, the radiance (intensity) peak of the Planck function is smaller for the earth's radiation field and the wavelength for the radiance (intensity) peak of the earth's radiation field is longer. The energy emitted from the earth-atmosphere system is called thermal infrared (or terrestrial) radiation. In Fig. 4.16 is shown the Earth radiance to space measured by the Infrared Interferometer Spectrometer Instrument (IRIS) onboard Nimbus IV spacecraft as well as Planck emission and line spectra are indicated, plus the spectral distribution of radiance emitted by a blackbody source at various temperatures in the terrestrial range (in terms of wavenumber). In some spectral regions the envelope of the emission spectrum is very close to the spectrum emitted from a blackbody with a temperature of about 300 K, which is about the temperature of the surface.

In these spectral regions, the atmosphere is transparent to that radiation. In other spectral regions the emission spectrum is close to the spectrum emitted from a blackbody with a temperature of about 220 K, which is about the temperature at the tropopause. This occurs in spectral regions where the atmosphere is opaque or absorbing to that radiation and the atmosphere is said to be trapping the radiation.

In Fig. 4.17 is shown measurements from an airborne interferometer in certain portions of the infrared spectrum and radiation trapped by various gases in the atmosphere. Thus, among these gases, carbon dioxide (CO_2), water vapour (H_2O), and ozone (O_3) are the most important absorbers. Some minor constituents not shown in Fig. 4.17), such as carbon monoxide (CO), nitrous oxide (N_2O), methane (CH_4) and nitric oxide (NO) are relatively insignificant absorbers insofar as the heat budget of the Earth-atmosphere is concerned.

The global distribution of CO_2 is fairly uniform, although there has been observational evidence indicating a continuous global increase over the past century owing to increased combustion of fossil fuels. This leads to the question of the earth's climate and possible climatic changes due to the increasing CO_2 concentration. Unlike CO_2, however, H_2O and O_3 are highly variable both with respect to

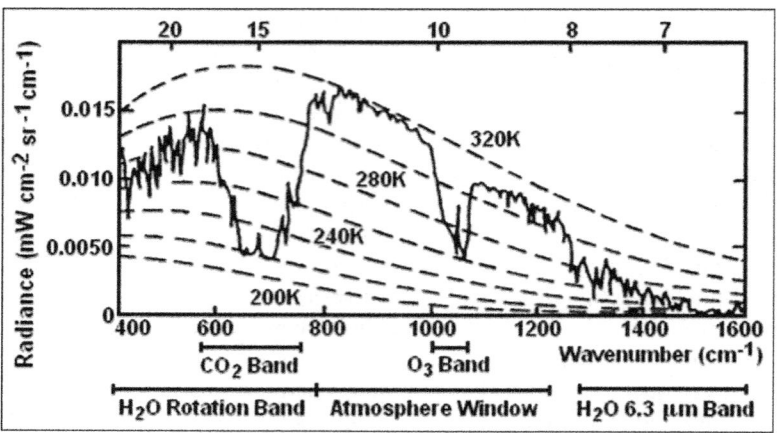

Fig. 4.16 Infrared portion of the Earth-atmosphere emitted radiation to space observed from satellite Nimbus IV (Courtesy of Book: by Menzel)

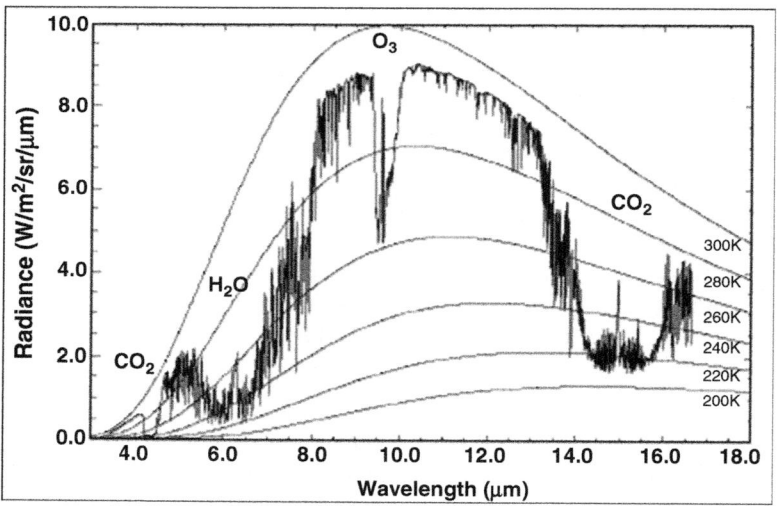

Fig. 4.17 Earth-atmosphere emitted radiances overlaid on Planck function envelopes (Courtesy of Book: by Menzel)

time and the geographical location. These variations are vital to the radiation budget of the earth-atmosphere system and to long-term climatic changes.

4.4   Additional Applications to the Earth's Atmosphere

The thermal energy can be transferred between the Earth's surface and the atmosphere via conduction, convection and radiation. In previous context is already stated about radiation and convection as basic resources of thermal energy in atmosphere and on the Earth surface and underground. Conduction is the process by which heat energy is transmitted through contact with neighboring molecules.

The additional applications important to the Earth's atmosphere in Satellite meteorology, which will be included in this sector together with heat energy transfer, are specific facts for aerosols and Earth and atmosphere Albedo.

4.4.1   Transfer of the Heat Energy

The heat source for our planet is the Sun. Energy from the sun is transferred through space and through the Earth's atmosphere to the Earth's surface. Since this energy warms the earth's surface and atmosphere, some of it is or becomes heat energy. There are three ways heat is transferred into and through the atmosphere such as: radiation (the Sun heats the ground), convection (the warm air raises) and conduction (the ground heats the air), which scenario is depicted in Fig. 4.18.

Conduction and convection are the two common modes of heat transfer in the Earth and in the atmosphere. Conduction arises from interactions between neighboring atoms, whereas convection transports heat by fluid movement resulting from the interaction between gravity and thermally induced density changes. First, consider conduction. In anisotropic medium, conductive heat flow is:

$$q = k\ VT \tag{4.45}$$

where q = heat flux (calories cm/sec^2), k = thermal conductivity of the medium and VT = temperature gradient.

Since conductivity and temperature gradient can vary individually with depth, each must be specified for a given location. Near the surface of the Earth, the gradient is dT/dz. and z is increasing with depth. Thus, the rate at which heat is conducted out ward through the Earth depends in a simple way on how temperature and thermal conductivity vary with depth. However, to predict the variation of temperature with depth requires knowing the vertical distribution of heat sources (radioactivity) and thermal diffusivity, so a measurement of the relative ability of a material to conduct and to retain heat k/pC, in which p is density of the rock-fluid system and C is heat capacity per unit mass. Because the vertical distribution of heat sources is not routinely determined, most emphasis is given to determining their mal diffusivity. The wide range of thermal diffusivities in rocks and minerals results from variation in composite on and porosity (above). Water, which responds

Fig. 4.18 Heat energy transfer (Courtesy of Manual: by NOAA)

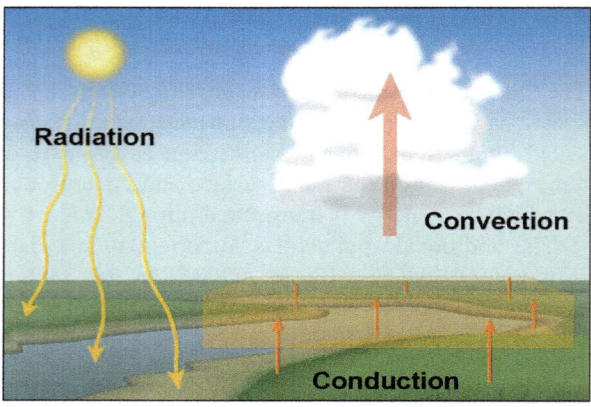

only sluggishly to thermal changes in its environment, has a thermal diffusivity an order of magnitude less than that of typical rocks and minerals.

4.4.2   Feedbacks in the Atmosphere

Understanding how rising GHG levels may affect Earth's energy balance is complicated because of feedbacks in Earth's climate system. Thus, feedbacks are interactions between climate variables such as temperature, precipitation, and vegetation and elements that control the greenhouse effect, such as clouds and Albedo. Positive feedbacks amplify temperature change by making the greenhouse effect stronger or by reducing Albedo, so they make the climate system more sensitive to the properties that trigger them. Negative feedbacks have a dampening effect on temperature change, making the climate system less sensitive to the factors that trigger them.

Feedbacks can be very complex process and may take place over short or long time spans. Important feedbacks in Earth's atmosphere include:

1. **Water Vapor Feedback (Positive)** – The atmosphere can hold increasing amounts of water vapor as the temperature rises, because the pressure of water vapor in equilibrium with liquid water increases exponentially with temperature. The presence of more water vapors temperature raises increases the greenhouse effect, as well as the absorption of solar radiation, which further raises temperature. Thus, this is the strongest and best understood feedback mechanism in the atmosphere, because it is based on the straightforward fact that warm air can hold more water vapor than cool air.

2. **Cloud Feedback on Terrestrial Radiation (Positive)** – Because warmer temperatures increase water vapor amounts, they can increase cloudiness and further raise temperature. This is a very strong feedback that is not well understood. It is hard to know whether or how much cloudiness will increase as temperature does,

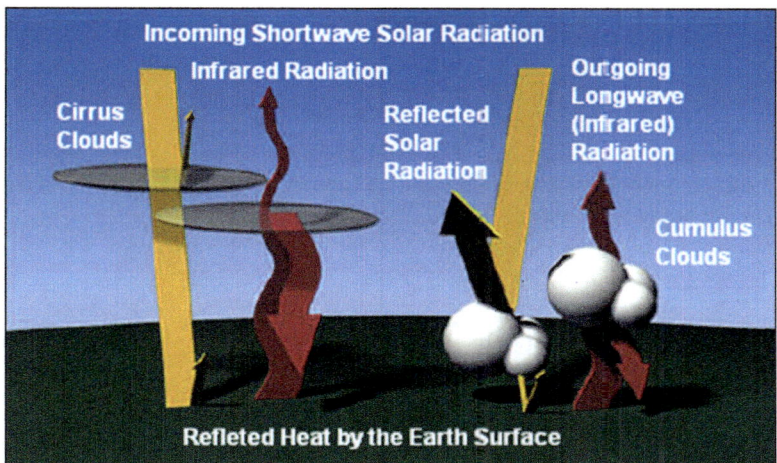

Fig. 4.19 Effects of Cirrus and Cumulus clouds on Earth's energy balance (Courtesy of Manual: by NASA)

because cloudiness depends more on upward air motion than on temperature or water vapor levels directly.

3. **Cloud Feedback on Solar Radiation (Negative)** – As temperature and water vapor levels rise, cloudiness may increase. Greater cloudiness raises Earth's Albedo, reflecting an increasing fraction of solar radiation back into space and decreasing temperature, although some cloud types are more reflective than others, shown in Fig. 4.19. This is another very strong feedback that is not well understood because it is hard to know whether or how much cloudiness will increase with temperature. Also, as noted in the previous example, clouds can also absorb infrared radiation, raising temperatures.

4. **Vegetation Feedback on Solar Radiation (Negative).** As temperatures rise, desert may expand with increasing Earth's Albedo and decreasing temperature. This is a very complex feedback. In such a way, it is uncertain whether deserts will expand, or conversely, whether higher CO levels might stimulate higher plant growth levels and increase vegetation instead of reducing it.

5. **Ice-Albedo Feedback on Solar Radiation (Positive).** Rising temperatures cause polar glaciers and floating ice sheets to recede and decreasing Earth's Albedo and increasing air temperatures. This feedback is very strong at times when polar ice has expanded widely, such as at the peak of ice ages. It can work in both directions, helping ice sheets to advance as Earth cools and accelerating the retreat of ice sheets during warming periods. There is relatively little polar ice on land today, so this feedback is not likely to play a major role in near-term climate change. Temperature increases large enough to melt most or all of the floating ice in the Arctic could sharply accelerate global climate change, because ocean water absorbs almost all of the incident solar radiation whereas ice reflects most sunlight.

4.4.3   Specific Facts for Aerosols

The extinction properties of a particle are determined by its extinction efficiency, defined as the ratio of the scattering cross section to the interception surface. At this point, for a particle of diameter d, the interception surface is $A = \pi(d/2)^2$. The extinction comprises a part associated to absorption and a part associated to scattering, namely such as:

$$Q^{ext}{}_\lambda = Q^a{}_\lambda = Q^d{}_\lambda = \sigma^a{}_\lambda + \sigma^d{}_\lambda / \pi(d/2)^2 \tag{4.46}$$

The values of absorption and scattering efficiencies, $Q^a{}_\lambda$ and $Q^d{}_\lambda$ respectively, are functions of the following size parameter:

$$\alpha_\lambda = \pi d / \lambda \tag{4.47}$$

where d = particle diameter (the particle is supposed to be a sphere); the properties of the medium defined by the particle (due to its chemical composition), described by its complex refractive index as follows:

$$m_\lambda = n_\lambda + jk_\lambda \quad \text{where } j2 = -1 \tag{4.48}$$

The real part n_λ is related to scattering, while the imaginary part k_λ is related to absorption. Actually, m_λ is normalized with respect to the "ambient" medium, while here is air, whose refractive index is about 1 for visible radiation.

The refractive index in the visible region of the electromagnetic spectrum is given in Table 4.3 for different particle types (with different chemical composition).

The extinction coefficient for a particle can be deduced from the extinction efficiency. For a particle density n (expressed in number of particles per volume of air), is obtaining:

$$b^{ext}{}_\lambda = \sigma^{ext}{}_\lambda \cdot n = \left(\pi d^2 / 4\right) \left(Q^{ext}{}_\lambda \cdot n\right) \tag{4.49}$$

The extinction efficiency is a function of the refractive index m_λ and of the size parameter α_λ. For large values of the size parameter ($d/\lambda >> 1$, typically for cloud drops with visible radiation), Q^{ext} is about 2, which value is shown in Fig. 4.20, and from which is yielding the extinction coefficient as follows:

$$b^{ext}{}_\lambda \approx \pi d^2 / 2 \cdot n \tag{4.50}$$

Table 4.3 Typical values of aerosol

Aerosol type	n_λ	k_λ
Water	1.34	0.
Ammonium	1.53	-5×10^{-3}
Sulfate	1.43	0.
Sea salt	1.5	0.
Soot	1.75	-0.45
Mineral aerosol	1.53	-8.5×10^{-3}
Organic aerosol	1.53	-8.5×10^{-3}

Fig. 4.20 Evolution of the extinction efficiency as a function of the size (Courtesy of Book: by Sportisse)

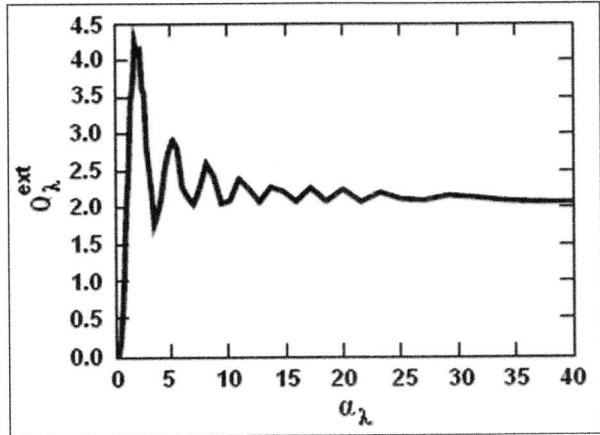

4.4.4   Earth and Atmosphere Albedo

Albedo is the fraction of solar energy of shortwave radiation reflected from the Earth back into space. It is a measure of the reflectivity of the Earth's surface and atmosphere. In fact, probably the most frequently used reflection quantity is the A = Albedo (A), which is the ratio of M = radiant exitance (M, due to reflection) to E = irradiance, giving as flows:

$$E = \int_o^{2\pi} \int_o^{\pi/2} L_i(\theta_i, \Phi_i) \ \cos\theta_i \ \sin\theta i \ d\theta_i d\Phi_i \tag{4.51}$$

$$M = \int_o^{2\pi} \int_o^{\pi/2} L_r(\theta_r, \Phi_r) \ \cos\theta_r \ \sin\theta r \ d\theta_r d\Phi_r \tag{4.52}$$

$$A = M/E \tag{4.53}$$

Albedo is a unitless ratio between zero and one. Thus, as defined in Eq. (4.53) it is a function of neither the incoming nor the outgoing angles; however, this does not mean that the Albedo is constant. If the incoming radiation changes, the Albedo will change.

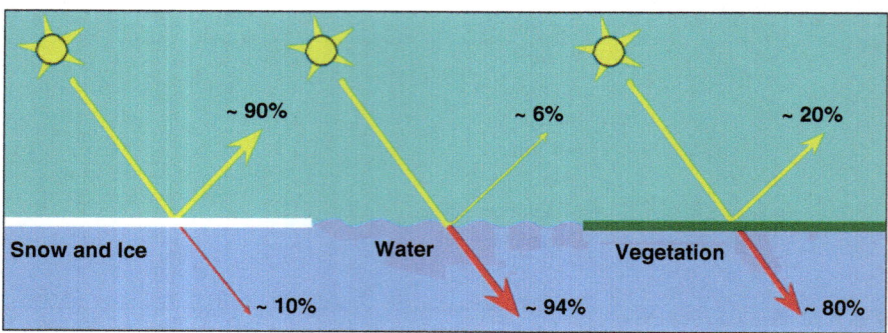

Fig. 4.21 Albedo from snow-ice, water and vegetation (Courtesy of Presentation: by Carana)

The word Albedo is used to describe the percentage of sunrays reflected from the Earth or atmosphere back into space. The more radiation that is reflected, the higher the Albedo and the less global warming occur. Lighter colored surfaces have a higher Albedo, so in such a way, when sunrays hit snow and ice, there is a high Albedo, when most sunlight hitting the surface bounces back towards space. When the sunrays hit water, more solar radiation is absorbed by the ocean, which scenario is increasing the temperature. The slightly better situation is the case with all types of vegetations. Last two cases have low Albedo and contribute to global warming.

For a given wavelength, the Albedo of a surface (by extension of an atmospheric layer) is defined as the fraction of the incident radiation that is scattered backward. Thus, as shown in Fig. 4.21 the values of Albedo varies according to the Earth surface type. Albedo is not important at high latitudes in winter, because there is hardly any incoming sunlight. It becomes important in spring and summer when the radiation entering through leads can greatly increase the melt rate of the sea ice. Therefore, Albedo is defined as the reflective quality of a surface. It is expressed as a percentage of reflected insolation to incoming insolation and zero percent is total absorption while 100% is total reflection. In terms of visible colors, darker colors have a lower Albedo, that is, they absorb more insolation, and lighter colors have high Albedo or higher rates of reflection. The angle of the sun also impacts Albedo value and lower sun angles create greater reflection because the energy coming from a low sun angle is not as strong as that arriving from a high sun angle. Additionally, smooth surfaces have a higher Albedo while rough surfaces reduce it.

4.5 Radiative Transfer Equations (RTE)

Radiative transfer is defined as the process of transmission of the EM radiation through the atmosphere and the influence of the atmosphere or it is a mechanism for exchanging energy between the atmosphere and the underlying surface and between their different layers. Infrared radiation emitted by the atmosphere and

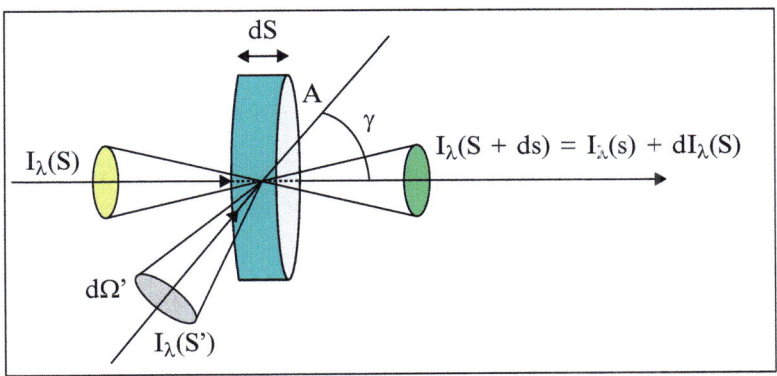

Fig. 4.22 Radiative heat transfer (Courtesy of Presentation: by Liew)

measured by satellite sensors is the basis for remote sensing of atmospheric temperature structure. In fact, the atmospheric effect is classified into multiplicative effects and additive effects. Accordingly, the multiplicative effect comes from the extinction by which incident energy from the earth to a sensor will reduce due to the influence of absorption and scattering. The additive effect comes from the emission produced by thermal radiation from the atmosphere and atmospheric scattering, which is incident energy on a sensor from sources other than the object being measured.

The three processes of radiation transfer, such as heat emission, absorption and scattering, are actually coupled, which radiative heat transfer is depicted in Fig. 4.22. The radiative transfer equation reads as follows:

$$dI_\lambda/ds = -[a_\lambda(S) + d_\lambda(S)]I_\lambda(S) + a_\lambda(S) B_\lambda(T)$$

$$+ d_\lambda/4\pi \int P(\Omega, \Omega')I_\lambda(\Omega') d\Omega' \tag{4.54}$$

The sum $a_\lambda + d_\lambda$ defines the extinction coefficient, usually written as b^{ext}_λ and thus where value I = incident radiation (radiance), dI = increment of radiance, ds = path length, $a_\lambda(S)$ = absorption coefficient at wavelength λ, $d_\lambda(S)$ = scattering coefficient, Ω = solid angle of incident, Ω' = solid angle of scattering, T = temperature in K and P = phase function.

The optical depth determines the opacity of the medium and is defined in the similar way as Eq. (4.28) such as:

$$d\tau_\lambda = [a_\lambda(S) + d_\lambda(S)]ds \tag{4.55}$$

In such a way it is possible to get the following relation:

$$dI_\lambda/d\tau = -I_\lambda(\tau) + \omega_a B_\lambda[T(\tau)] + \omega_d/4\pi \int P(\Omega, \Omega')I_\lambda(\Omega')\, d\Omega' \qquad (4.56)$$

where values $\omega_a = a_\lambda/(a_\lambda + d_\lambda)$ and $\omega_d = d_\lambda/(a_\lambda + d_\lambda)$ are the absorption and scattering Albedo, respectively.

In fact, here are investigated two simplified cases, such as the case of infrared radiation (only absorption and emission are taken into account) and the case of visible radiation (only scattering is described).

In the following context will be introduced the additional radiative transfer equations, such as definition of extinction coefficient K, values of absorber or scattering density ρ and emission coefficient j, are expressed by the following equations:

$$dI = \rho \cdot K \cdot I \cdot ds = \rho \cdot j \cdot ds \quad \text{than:} \quad j = \rho \cdot B(T) \qquad (4.57)$$

Concerning values of extinction coefficient K and Albedo for single scattering, it is possible to express the following equation:

$$j = \omega_0(K/4\pi)\rho \int_\Omega P(\Omega, \Omega')I(\Omega')\, d\Omega' \qquad (4.58)$$

4.5.1 Primers of Radiations

As stated before, the atmosphere filters the energy received from the Sun and also from the Earth. Here will be introduced only two important primers of radiations, such as infrared and visible radiations.

4.5.1.1 Infrared Radiation

Infrared radiation is a type of EM radiation, as are radio waves, ultraviolet radiation, X-rays and microwaves. Infrared (IR) light is the part of the EM spectrum that people encounter most in everyday life, although much of it goes unnoticed. It is invisible to human eyes, but people can feel it as heat. Infrared radiation is emitted or absorbed by molecules when they change their rotational-vibrational movements. So, it excites vibration modes in a molecule through a change in the dipole moment, making it a useful frequency range for study of these energy states for molecules of the proper symmetry. The IR spectroscopy examines absorption and transmission of photons in the infrared range.

Not all gas molecules are able to absorb IR radiation. For example, nitrogen (N_2) and oxygen (O_2), which make up more than 90% of Earth's atmosphere, do not absorb infrared photons. However, other significant greenhouse gases include water vapor (H_2O), methane (CH_4), nitrous oxide (N_2O) and ozone (O_3).

Scattering can be neglected for the infrared radiation. Moreover, the temperature T is a function of the altitude, so it can be written $T(\tau)$ with τ an increasing function of the altitude (the terrestrial radiation propagates from the bottom to the top of the atmosphere). In such a way, integrating equation (4.56) yields as follows:

$$I_\lambda/(\tau) = I_\lambda(0)e^{-\tau} + \int_0^\tau B_\lambda\left[T\left(\tau'\right)\right] e^{(\tau'-\tau)}d\tau' \tag{4.59}$$

The first term is a pure extinction term while the second term describes the emission from the atmosphere, while $I_\lambda(0)$ is the emitted radiation at the Earth's surface.

4.5.1.2   Visible Radiation

Visible radiation is EM radiation that causes the sensation of sight or light. Namely, the visible spectrum is the portion of the EM spectrum that is visible to the human eye. Thus, EM radiation in this range of wavelength is called visible light or simply light. A typical human eye will respond to wavelengths from about 380 to 780 nanometres (nm). In terms of frequency, this corresponds to a band in the vicinity of 430–770 THz.

For the visible region of the EM spectrum, it can neglect both absorption and emission in the atmosphere, thus the following equation is giving as follows:

$$dI_\lambda(\tau)/d\tau = -I_\lambda(\tau) + 1/4\pi \int P(\Omega,\Omega')I_\lambda(\Omega') \, d\Omega' \tag{4.60}$$

Thus, the optical depth τ is a decreasing function of the altitude (the radiation propagates from the top to the bottom of the atmosphere. The boundary condition I_λ (0) corresponds to the solar radiation received at the top of the atmosphere. There does not exist any analytical solution in the general case. Solving this equation can be performed with the method of successive orders. The solution is built by solving successively the systems such as:

$$I_\lambda^{n+1}(\tau)/d\tau = -I_\lambda^{n+1}(\tau) + 1/4\pi \int P(\Omega,\Omega')I_\lambda^n(\Omega') \, d\Omega' \tag{4.61}$$

Scattering is then applied to the radiation computed in the previous iteration. In such a way, it can then apply a superposition approach since the radiative transfer equation is linear, yielding the following equation:

$$I_\lambda = \sum_{n=0}^{\infty} I_\lambda{}^n \qquad (4.62)$$

Exact analytical solutions for the radiative transfer equation presented in Eq. (4.54) are usually excessively difficult to obtain and explicit solutions are generally impossible to express for all situations except the very simplest cases. Therefore, research on radiative heat transfer in participating media has generally proceeded in two different directions: exact (analytical and numerical) solutions of highly idealized configurations and as well as approximate solutions methods for more complex realistic scenarios.

4.5.2   Radiative Budget for the Earth Atmosphere System

Incoming solar radiation is absorbed by the Earth's surface, water vapor, gases and aerosols in the atmosphere. This incoming solar radiation is also reflected by the Earth's surface, by clouds and by the atmosphere. Energy that is absorbed is emitted by the Earth-atmosphere system as longwave radiation.

4.5.2.1   Solar Constant and Emission Effective Temperature

The solar constant is defined as the annual average solar radiation received outside the Earth's atmosphere on a plane normal to the incident radiation at the mean Sun to Earth distance and has a value close to $2 \, cal - cm^{-2} \, min^{-1}$ or $1370 \, Wm^{-2}$. Thus, the actual solar irradiance varies by 3–4% of this value during the year due to the eccentricity of earth's orbit about the Sun.

If is considered the incoming solar radiation at the top of the atmosphere as made of 100 units, 30 units are reflected back to space (6 by the atmosphere, 20 by clouds and 4 by the earth's surface), then at this point 19 units are absorbed by the atmosphere (16 by gases and 3 by clouds). The remaining 51 units are absorbed by the earth's surface.

Out of these 51 units, 6 are lost to space directly and 45 are returned upwards and absorbed by the atmosphere and clouds (7 by convection and conduction, 23 by evaporation as latent heat and 15 by longwave radiation). The atmosphere and clouds have already absorbed 19 units from the solar radiation, making a total of 64 units, which are returned to space as longwave radiation. The budget is thus balanced at the top of the atmosphere.

At the Earth's surface and at any level in the atmosphere, the net radiation is the balance of four radiative fluxes, downward solar radiation and longwave radiation, and upward solar and longwave radiation. These can be measured with special instruments installed on the ground or flown on balloons as radiometer sondes. The prime factors involved in the radiation budget of the earth-atmosphere system are the Albedo or reflectance properties of land, ocean and cloud tops, scattering

Fig. 4.23 Radiative flux
received by the Earth
(Courtesy of Book: by
Sportisse)

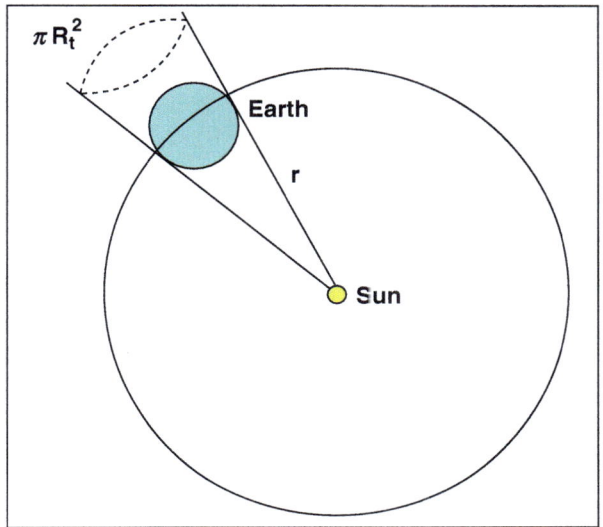

properties of aerosols, dust and particulate matter in the atmosphere, and vertical
profiles of temperature and concentration of gases which absorb longwave radiation
(water vapour, CO_2, ozone). Data on these variables if available can be used to
compute the radiation budget components indirectly.

The solar constant of a given planet is defined by the solar radiation flux per unit
area of the planet surface is depicted in Fig. 4.23, with the detail of the main
notations. This is radiative flux received by the Earth of radius R_t and at a distance r
from the Sun. Thus, the solar flux per unit area of surface (expressed in Wm^{-2})
received by a sphere of radius r and centered at the Sun, is giving the following
relation:

$$S = 4\pi R_S^2 \sigma \, T_S^4 / 4\pi^2 \tag{4.63}$$

where RS and TS are the radius and the temperature of the Sun, respectively.
However the interception surface defined by the Earth is πR_t^2, resulting in a
received flux $\pi R_t^2 \times S$ and in flux per unit area of the Earth's surface is as follows:

$$S/4 = \pi R_t^2 \cdot S/\pi R_t^2 \tag{4.64}$$

With $r = 1.5 \times 10^8$ km is distance between the Sun and the Earth, it will be
possible to obtain the solar constant for the Earth, such as $S \approx 1368 \, Wm^{-2}$. For a
given planet, the emission effective temperature is defined as the emission temperature of a blackbody in radiative balance with the radiative fluxes received by the
planet. Thus, the radiative budget for the Earth/atmosphere system is calculated as
shown in Fig. 4.23, where the solar radiation is intercepted by a surface πR_t^2 (with
R_t the Earth's radius), the fraction A is reflected back to space (with A the global

Albedo of the Earth/atmosphere system, A $\approx$ 0.3). Finally, upon division by the Earth's surface πR_t^2, this gives the following radiative budget (in Wm^{-2}), it is possible to get the following relation:

$$\sigma T_S^4 = \left[\pi R_t^2 S(1 - A)\right]/\pi R_t^2 = S(1 - A)/4 \qquad (4.65)$$

With T_e the emission effective temperature of the Earth/atmosphere system will be possible to provide rearranging that yields:

$$T_e^4 = S(1 - A)/4\sigma \qquad (4.66)$$

The emission effective temperature is therefore a function of the Albedo and of the distance to the Sun (which defines the solar constant). With A $\approx$ 0.3 can be calculated $T_e \approx$ 255 K, to be compared with the mean temperature at the Earth's surface, which value is about 288 K. The difference 33 K corresponds to the greenhouse effect (which makes it possible to have a surface temperature greater than -18 °C) and results from energy redistribution from the atmosphere to the ground. Thus, the available flux at the Earth's surface is S/4 $\approx$ 342 Wm^{-2}, which is usually written as F_S.

4.5.2.2 Energy Budget for the Earth/Atmosphere System

The temperature of the Earth/atmosphere system is mainly fixed by the radiative properties of the Earth and of the atmospheric compounds (gases, clouds and aerosols). The global energy budget for the Earth/atmosphere system is shown in Fig. 4.24, which indicative values of fluxes are expressed in Wm^{-2}. Therefore, here are considered the received solar energy, the energy fluxes for the Earth and the energy fluxes for the atmosphere.

In fact, the resulting budget is a global budget and does not take into account the seasons, the diurnal cycles (day/night) and the spatial location. Except for the solar flux, the relative uncertainties are at least of 10%. Thus, the fluxes are only crude estimations. Here will be introduced received solar energy, Earth's energy budget and atmosphere's energy budget.

1. **Received Solar Energy** – The received solar energy value is the only energy source for the Earth/atmosphere system. The incident solar radiation represents 342 Wm^{-2}, to be split as follows:

   - 77 Wm^{-2} is reflected back to space by clouds and aerosols (namely 22%);
   - 67 Wm^{-2} is absorbed by gases and clouds (namely 20%);
   - 168 Wm^{-2} is scattered and then absorbed by the Earth (namely 49%); and
   - 30 Wm^{-2} is reflected by the Earth and then scattered to space (namely 9%).

   Thus, the planetary Albedo is therefore about 0.31, which is the reflected energy 107 Wm^{-2}, to be compared with the received energy, 342 Wm^{-2}.

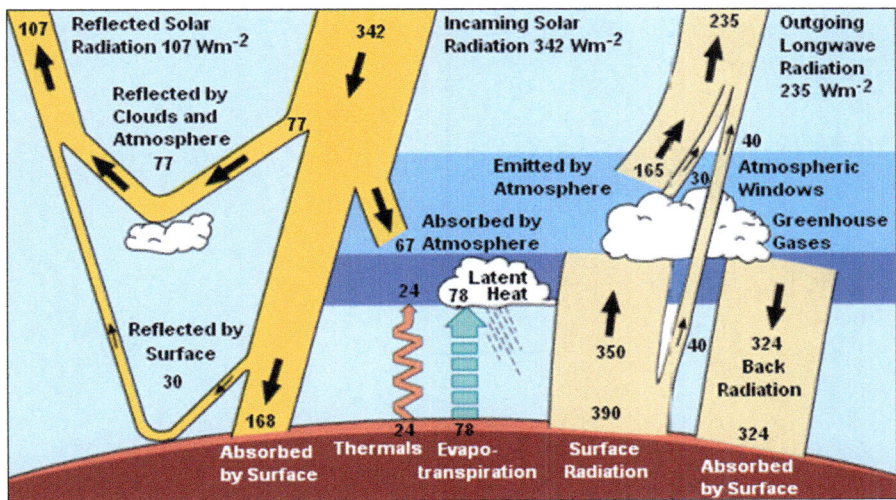

Fig. 4.24 Global energy budget for the Earth/atmosphere system (Courtesy of Manual: by Kiehl)

2. **Earth's Energy Budget** – The energy fluxes for the Earth are related to:

   – Radiative energy;
   – Sensible heat connected to the vertical turbulent motions;
   – Latent heat, produced by the water cycle; and
   – Heat conduction in the soil.

   With the same units as above, the Earth absorbs 168 Wm^{-2} in the shortwave solar radiation. It emits 390 Wm^{-2} in the longwave radiation into the atmosphere. Thus, as seen before, the emitted radiation is strongly absorbed by the atmosphere and actually, up to 350 Wm^{-2} is absorbed (namely 90%) and 40 Wm^{-2} (namely 10%) is transmitted to space through the atmospheric window.

   Moreover, the Earth surface receives 324 Wm^{-2} from the longwave radiation emitted by the atmosphere. The whole part is supposed to be absorbed at the Earth's surface. The radiative budget is then positive for the Earth, so in such a way, there is a gain of radiative energy for the Earth ($168 + 324{-}390 = 102$ Wm^{-2}). However, at equilibrium, the energy budget for the non-radiative part is therefore negative, the Earth has an energy loss (-102 Wm^{-2}) by latent heat (water evaporation) and sensible heat (vertical turbulent motion).

3. **Atmosphere's Energy Budget** The atmosphere absorbs 67 Wm^{-2} in the shortwave solar radiations and 350 Wm^{-2} in the longwave terrestrial radiations. It emits 519 Wm^{-2} in the longwave radiations (including 324 to the Earth and 195 to space). The radiative budget for the atmosphere is therefore negative ($350 + 67{-}519 = -102$ Wm^{-2}), so the atmosphere has a loss of radiative energy. At equilibrium, the energy gain is provided by latent and sensible heat fluxes (102 Wm^{-2}, coming from the Earth).

Chapter 5
Satellite Meteorological Parameters

Satellite meteorological parameters help to provide the initial conditions for Numerical Weather Prediction (NWP). Desired fields include those of temperature, moisture, winds, clouds, fog, aerosols, precipitation, surface properties, heating and other parameters. Thus, for high horizontal resolution and global coverage, weather satellite data are an unrivaled source of NWP information. The basic form of this information is satellite sensor-incident, wavelength-dependent radiance or equivalently and brightness temperature. The process of retrieving meteorological data from the observed radiances involves solving an inverse problem, which often is ill posed or poorly conditioned. In fact, a variety of methods have been used to make the inverse problem more tractable, such as in essence, for which some a priori information is required. The inverse problem in meteorology is the opposite of the calculation of radiances, given the relevant meteorological profiles, surface conditions and sensor geometry. This so-called forward problem is an integral part of physical retrieval methods that are now generally favored over conventional statistical retrieval methods.

Physical retrieval methods, in practice, actually rely to a certain degree on a priori statistics. Temperature retrievals, especially in the southern hemisphere, are of proven importance in NWP involving global scale models. In fact, vertical temperature profiles are obtained from observed radiances in a set of emission bands with varying optical depths. In comparison to conventional radiosonde data, current temperature retrievals have coarse vertical resolution. Beside, there is also some evidence that retrieved synoptic features are somewhat smoothed horizontally in spite of the high horizontal resolution of the temperature retrievals. The usefulness of other geophysical parameters for NWP is not well established and is largely untested, although a great many retrieval methods have been proposed or developed for a variety of potentially interesting parameters. The reason for this is twofold: these data are expected to have the most impact on mesoscale modeling, but mesoscale analysis systems are still relatively primitive. Moisture variables, for instance, specific humidity, clouds, and precipitation, are retrievable and are potentially very useful. It is theoretically possible to retrieve specific humidity profiles by

© Springer International Publishing AG 2018
S.D. Ilčev, *Global Satellite Meteorological Observation (GSMO) Theory*,
DOI 10.1007/978-3-319-67119-2_5

using methods analogous to those used to retrieve temperatures. However, provided results to date with available different sensors have not been wholly satisfactory. Other moisture parameters are somewhat easier to retrieve, for example, cloud cover may be obtained by a variety of techniques from imagery in visible and/or infrared spectral regions, and vertically integrated water vapor is well correlated with emissions at microwave frequencies, especially over the ocean.

Unfortunately, these parameters are not readily assimilated by a global NWP model, and special methods are required. A variety of other meteorological parameters, which may be used for NWP are retrievable. These parameters include winds at the surface from microwave sensors and cloud drift winds aloft, as well as other surface properties, such as soil moisture, Albedo, snow cover, temperature, and fluxes. In summary, remote sensing of meteorological parameters is important for NWP. The potential for a much greater impact exists. Greater accuracy and vertical resolution, better retrieval methods, and better and more flexible analysis methods are desirable.

5.1   Introduction to Satellite Weather Observation

Weather observation and forecast system is an initial value problem indicating the need for accurate information and initial condition that is achieved through different meteorological data assimilation. The traditional meteorological data coverage is spatially and temporally limited, however satellite meteorological data provides much better coverage in both space and time. At this point, about 90% of the meteorological data that goes into the assimilation of any analysis-forecast system comprise of data from weather satellites and rest from the conventional site platforms at sea, on the ground and in the air. In the other words, satellite meteorological services involve receiving satellite data and images from international and national satellites, their processing for generation of images in all channels, derivation of operational products, their archiving in the memory drivers and their real time utilization for weather forecasting. The important satellite weather products include weather imagery and atmospheric properties via different meteorological instruments.

Therefore, the most broadly used products from satellites are weather observations that enable forecasts. Since satellite images have become readily available, no tropical cyclone (hurricane or typhoon) has gone undetected, which provides affected coastal areas with advance warning and crucial time to prepare. In addition, monitoring of tropical cyclones via satellite and determining their position helps ships and aircraft to avoid accidents. This exemplifies are not showing only how satellite observations have transformed the Earth sciences, but also how the improved predictability of Earth processes can provide direct societal benefits. Satellite weather forecasts more than a few hours into the future are made with the aid of NWP models. In fact, by assimilating satellite observations, which yield dramatically improved and continually updated knowledge of the state of the atmosphere, meteorologists can devise models that project the weather into the future with much improved accuracy compared to presatellite forecasts. Consequently, 7-day forecasts have improved significantly in accuracy over the past decades, particularly for the relatively data-sparse southern hemisphere. Needless to say, these improvements in forecast skills are saving countless human lives and have an enormous economic value and saving the energy sector alone huge amounts of investments.

5.1.1 Satellite Meteorological Instruments for Observation and Monitoring

Various instruments are used for Earth surface observation and measuring different weather phenomena and parameters, while space or satellite meteorological instruments and sensors make comprehensive and large-scale observations of different meteorological elements at the ground level as well in the upper layers of the atmosphere.

At the global scale, meteorological observations and monitoring are recorded at three main levels, such as surface observatories, upper air observatories and space-based or satellite observation platforms. The World Meteorological Organization (WMO), as a specialized agency of the United Nations (UN), coordinates these observations such as:

1. **Surface Meteorological Observations** – This system has instruments for measuring and recording weather elements like temperature, air pressure, humidity, clouds, precipitations and wind. Specialized observatories also record elements like radiation, ozone atmospheric trace gases, pollution and atmospheric electricity. So, these observations are taken all over the globe at fixed times of the day as decided by the WMO and the use of instruments are made conforming to international standards, which make observations globally compatible.

    In general, surface meteorological observations are normally classified into five categories depending upon their instruments and the number of daily observations taken. The highest category is Class-I. Typical instrumental facility available in a Class-I observatory consists of the following: Maximum and minimum thermometers, Anemometer and wind vane, Dry and Wet bulb thermometer, Rain gauge and Barometer. Thus, all observations are taken in these observatories normally at 00, 03, 06, 09,12,15,18, 21 h (GMT) around the globe.

    However, for some logistic reasons, some of the observatories take limited number of daily observations upper air observation during daytime only.

    The different surface weather stations are a facility for observation and monitoring, either on land or sea, with instruments and equipment for measuring atmospheric conditions to provide information for weather forecasts and to study the weather and climate factors. The weather measurements include all above mentioned as manual observations are taken at least once daily, while automated measurements are taken at least once an hour. Weather conditions out at sea are taken by ships and buoys, which measure slightly different meteorological quantities such as Sea Surface Temperature (SST), wave height, and wave period. Drifting weather buoys outnumber their moored versions by a significant amount.

The different meteorological instruments used in surface weather stations can be installed in the Personal Weather Stations (PWS), dedicated Ships Weather Stations (SWS), at sea Buoys Weather Stations (BWS) and Synoptic Meteorological Stations (SMS). Except for those instruments requiring direct exposure to the elements (anemometer, rain gauge), the instruments should be sheltered in a vented box, usually a Stevenson Screen, to keep direct sunlight off the thermometer and wind off the hygrometer. The instrumentation may be specialized to allow for periodic recording otherwise significant manual labour is required for record keeping. Automatic transmission of meteorological data can be performed in the different format, such as aeronautical METAR, maritime Weather Warnings (WX) and also all meteorological data can be transmitted via satellite equipment and collected by the Direct Readout Ground Earth Stations (DRGES).

2. **Upper Air Meteorological Observations** – Different helium filled balloon carrying a radiosonde is released twice daily at the Upper Air Observatory (UAO). Weather data, such as wind direction and velocity, temperature and relative humidity of the atmosphere daily measured on height of up to 100,000 ft. or about 30 km are recorded and transmitted to the ground station for further processing. However, in the similar way these weather data can be transmitted via special very small satellite transmitters via Inmarsat or Iridium satellite communication networks to the DRGES infrastructures. In the similar way, the special UAO radiosondes can be also involved in the implementation of new observing systems such as wind profilers and commercial maritime, aircraft or any observations. Otherwise, the radiosonde balloons can be also filled with hydrogen for flight in upper atmosphere and can be tracked by the special theodolite set up on the building roof to track the balloon. For nighttime operation, a lantern can be hung from the balloon to facilitate tracking. Based on the balloon's elevation, azimuth and rate of ascent, winds at different heights above ground were calculated. Since such measurements relied entirely on visual observation, they could not be carried out when low clouds were present.

3. **Satellite or Space Meteorological Observations** – Weather satellites or space systems make comprehensive and large-scale observations of different meteorological elements at the ground level as well in the upper layers of the atmosphere. The GEO satellites provide space-based observations about different weather conditions, such as valuable observations of temperature, cloud cover, wind and associated weather phenomena.

Satellite meteorological instruments for observation and monitoring are more important and modern tools than surface and upper air meteorological observatories, because they have many more functions and abilities to ensure a different observation and monitoring of a large number of meteorological data. Here will be mentioned just important instruments and all in particular will be introduced in the next chapter.

For instance, there are 11 instruments onboard Chinese PEO meteorological satellite FY-3 A/B, including three imaging remote sensors Visible and Infra-Red Radiometer (VIRR), Medium Resolution Spectral Imager (MERSI) and Micro Wave Radiation Imager (MWRI), three satellite sounding instruments Infrared Atmospheric Sounder (IRAS), Micro Wave Temperature Sounder (MWTS) and Micro Wave Humidity Sounder (MWHS), two ozone instruments Solar Backscatter Ultraviolet Sounder (SBUS), Total Ozone mapping Unit (TOU), two Earth radiation budget instruments Earth Radiation Monitor (ERM), Solar Irritation Monitor (SIM) and one space environment monitoring instrument Space Environment Monitor (SEM). This satellite has nearly 100 remote-sensing channels in the spectral range, from ultraviolet to Visible (VIS), Infrared (IR) and microwave,

However, FY-3C is equipped with all 11 payloads, but MWTS is upgraded to MWTS-II, MWHS to MWHS-II, and a new payload, GNSS Occultation Sounder (GNOS), is on board FY-3C. MWTS-II will increase the channels from 4 to 13, and MWHS-II will increase the channels from 5 to 15. The FY Chinese satellite series provide the following instruments: (1) Imaging Instruments; (2) Sounding Instruments; (3) Ozone Instruments; (4) Earth Radiation Instruments; and (5) Space Environment Monitoring Instruments. A new method of sensor cross-calibration is using the Chinese Multi-channel Visible Infrared Scanning radiometer (MVIRS), which has six channels equivalent to the US Advanced Very High Resolution Radiometer (AVHRR) and Moderate Resolution Imaging Spectrometer (MODIS).

5.1.2   Satellite Weather Imagery

In the early 60s satellite global images started becoming available to the meteorological observations centres around the world, thus soon after the skills of satellite image data and interpretation started with developments very rapidly. Until then, knowledge about clouds had been documented into cloud atlases on the basis of what had been observed from the ground or aircrafts. The view from satellite altitudes was, however, completely different and clouds seen by satellites had to be interpreted in an altogether different manner.

As is known, in the early days of the meteorological satellite era, the picture technology was primitive. However, on 21 December 1963 was launched the US meteorological PEO satellite TIROS-8 first equipped with the Automatic Picture Transmission (APT) capability that made satellite pictures available the world over to meteorological agencies in real time as the satellite passed over their region. The ground units APT pictures received directly from satellites flying overhead were produced on facsimile chart paper, mostly of inferior quality and liable to fading in a short time. The images many times lacked clarity and hence they had to be examined with great care. Thus, the technology improved really fast, even including the photographic paper, when interactive computer image processing systems like Man computer Interactive Data Access System (McIDAS) became common. Since 1973, McIDAS software performs many functions with satellite imagery, observational reports, numerical forecasts and other data. Those functions include displaying, analyzing, interpreting, acquiring and managing the data. Now of course, there are web sites on the Internet on which anyone can access satellite imagery in near real time.

5.1.3   Characteristics of Satellite Imagery

Most of the early interpretation weather schemes used six characteristics to identify clouds in satellite images, which were introduced by Conover in 1962, are as follows:

1. Cloud brightness relating particularly to the depth and composition;
2. Texture – whether smooth, fibrous, opaque or mottled;
3. Form of elements, which whether are regular or irregular;
4. Pattern of elements, which are associated to the values of topography, air flow, vertical and horizontal wind shear;
5. Size, both of the patterns and the individual elements; and
6. Vertical structure, for example shadows thrown below.

In addition, cloud classification schemes were developed on the basis of cloud patterns and introduced by Hopkins in 1967, such as follows:

1. Vertical features: (a) Circular or spiral bands, (b) Crescent-shaped or comma-shaped cloud masses, (c) Quasi-circular cloud masses, and (d) Curved or linear bands;
2. Major cloud bands, in which the length is much greater than the width; and
3. General features: (a) Minor bands like cloud streets, jet stream bands, lee waves and etc., (b) Cumuliform features, polygonal cells, and (c) Stratiform features, fog areas and etc.

Although nowadays most of the satellite weather picture interpretation works and products are carried out on image processing computers, and various interpretation aids are available to the analysts, it is necessary to have a basic knowledge of the fundamental characteristics mentioned above and their importance.

It is also important to include that interpretation of individual satellite images should never be done in isolation, as may happen for example, when one sees a satellite image flashed on a television channel. In addition, the day and time of the image, the geographical area covered, the gray scale and different enhancement used, must all be known and given due consideration. Other available observational data and synoptic weather charts should also always be referred to. Continuity in time is essential and earlier images should be compared for detecting changes or confirming the interpretation. Images from all available channels should also be compared. The information derived from the VIS, IR, Water Vapour (WV) and Micro Wave (MW) channels imagery is not redundant but complementary, and when all channels are considered together, they help to remove uncertainties or ambiguities in the interpretation process and to identify clouds and surface features uniquely.

5.1.3.1   Visible Channel Imagery

Clouds that appear brighter in Visible (VIS) imagery are those, which have a large Albedo because of great depth, high cloud water and ice content and small cloud droplet size. Thus, clouds that appear gray are those having shallow depth, low cloud water/ice content and large cloud droplet size. Vertically grown cumulonimbus clouds appear most prominently in VIS images whereas thin cirrus cannot be clearly seen. Small cumulus clouds usually take the form of open hexagonal cells while stratocumulus clouds appear as closed cells, while low level stratus clouds or areas covered by fog can be identified by their uniform brightness and sharp boundaries. Tall cumulonimbus cloud tops some times throw shadows on lower cloud layers. Since convection is the main source of rainfall in the global tropics, in the early years of satellite weather observation, the distribution of Highly Reflective Clouds (HRC) was considered important and extensively documented.

A visible satellite image uses a channel recorded with visible light, so it is similar to what we would see from space only it is depicted in black and white. Bright areas (high amount of reflected light) show clouds and snow cover, while dark areas show ground and ocean. In Fig. 5.1 is illustrated visible satellite image at 11:00 AM NZST on 17 April 2015, courtesy of Japanese multipurpose GEO satellite MTSAT. The thicker the cloud, the more visible light is reflected back to the satellite, and the brighter it will appear on the image. Thinner clouds do not reflect as much so will appear a lighter gray.

5.1.3.2   Infra Red Channel Imagery

The most important advantage of Infra Red (IR) satellite imagery is that it is available at any time, unlike visible imagery, which is available only in daylight hours. Secondly, IR radiances are a measure of the temperature of the radiating surface. However, clouds that appear white in IR imagery are those, which have cold cloud top temperatures such as cumulonimbus and cirrus clouds. A mature cumulonimbus cloud can be further recognized by its sharp edge on the windward side and a fuzzy edge on the other side resulting from the cirrus plumvisiblee. Thus, the plume may get blown downwind over several hundred kilometers and give an indirect indication of the upper level wind speed and direction. Clouds in IR imagery do not have the kind of texture that is seen in VIS imagery. In such a way, IR images cannot discriminate between low clouds and the sea surface or

Fig. 5.1   VIS Image (Courtesy of Manual: by MTSAT)

between fog and land surface, because of the absence of sufficient temperature difference.

An IR satellite image uses a channel recorded from infrared energy, which is typically the 10–11 μm wavelength used for these images. This is a longer wavelength, invisible to the human eye, but can tell us a lot about the temperature of an object using a similar principle for night vision goggles that detect thermal energy. Infrared images have the advantage that we can "see" clouds on them throughout the night, as opposed to visible images that can only show reflected visible light during the day.

Sometimes it is possible to "enhance" an IR image to highlight certain temperatures with certain colours. In the following example, the image has been enhanced to show colder cloud tops (higher clouds) as red and pink, while the warmer (lower) cloud tops are a lighter gray. In Fig. 5.2 is illustrated IR satellite image at 11:00 AM NZST on 17 April 2015, courtesy of MTSAT.

5.1.3.3   Water Vapour Channel Imagery

A Water Vapour image (WV) is also taken with infrared energy, but at a slightly different wavelength. Water likes to absorb and emit energy at a certain infrared

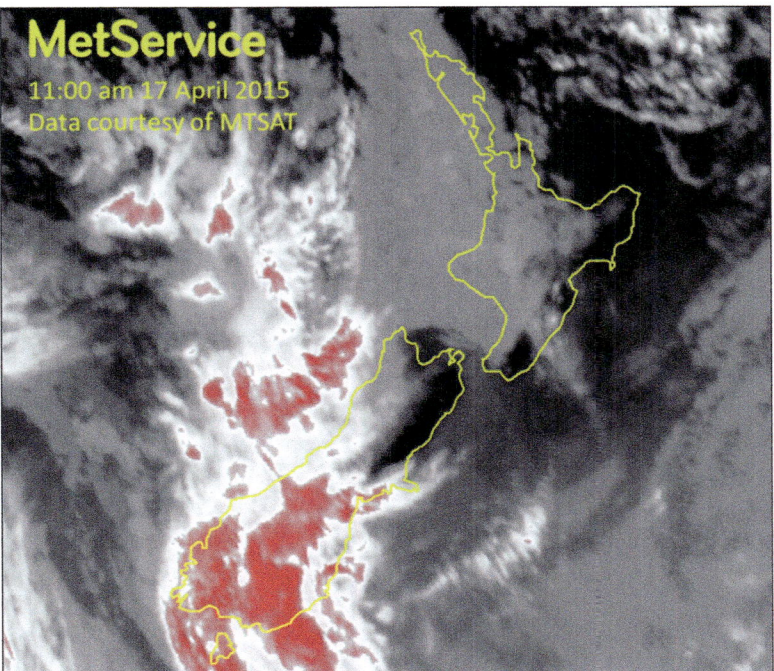

Fig. 5.2 IR Image (Courtesy of Manual: by MTSAT)

wavelength of 6.7–7.3 μm, and it can take advantage of this by measuring that wavelength of energy, so that it can detect areas of high and low moisture in the mid and upper levels of the atmosphere.

Thus, water vapor imagery is used to analyze the presence and movement of water vapor moisture in the upper and middle levels of the atmosphere. Above the troposphere there is very little moisture. However, water vapor imagery does not detect moisture in the lower atmosphere and near the surface. Only the top most level of moisture is detected using these wavelengths. Namely, if there is a thick layer of clouds in the upper levels then it will not be able to see the clouds that are closer to the surface. The same applies to water vapor. When upper level moisture is present, this layer of moisture will prevent the detection of water vapor that is below this layer. If the upper levels are dry then water vapor can be detected in the middle layers of the atmosphere. There is usually though enough moisture in the middle and upper levels that moisture in the lower levels is not detected.

As stated above, water vapor absorbs radiation in the wavelength from 6.7 to 7.3 microns. Suppose there is a thick layer of moisture in the upper levels. This layer of moisture will absorb the IR radiation in the 6.7–7.3 micron wavelengths that are being emitted from the lower troposphere up to this layer. Thus, the satellite receives less radiation in the 6.7–7.3 micron spectrum where there is a higher concentration of upper level moisture.

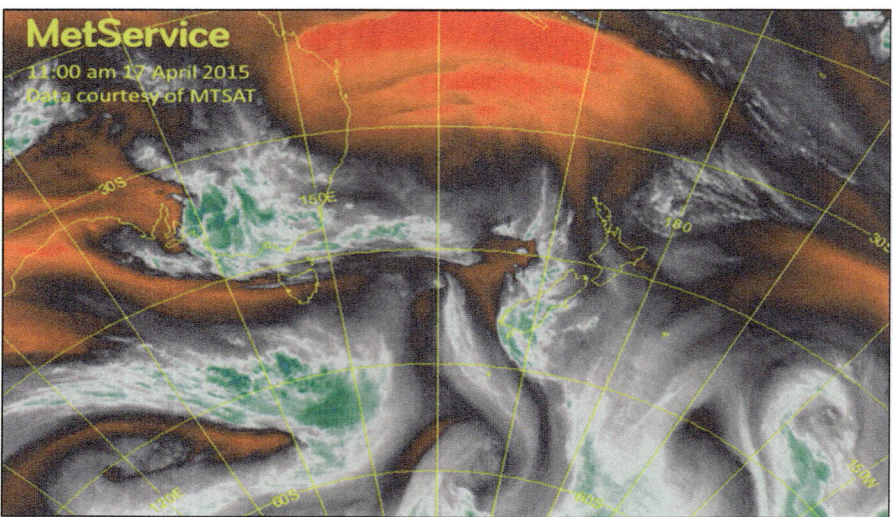

Fig. 5.3 WV Image (Courtesy of Manual: by MTSAT)

The weather satellite will interpret this greater absorption as a colder temperature and a higher concentration of water vapor. If the upper levels are dry then less radiation is absorbed by moisture in the 6.7–7.3 micron spectrum in the upper levels. The satellite will interpret this lesser absorption as a warmer temperature and a lesser concentration of moisture. A high emission in the 6.7–7.3 micron spectrum indicates the emission is coming closer to the lower levels of the troposphere where it is warmer and more moisture is present.

A dark color or warm color indicates a relative lack of upper level moisture. It does not mean though there is a lack of moisture in the lower levels or at the surface. It could be very moist at the surface or it could be fairly dry. A white or cold color indicates a high concentration of water vapor. This layer of water vapor is absorbing radiation in the 6.7–7.3 micron range.

Water vapour satellite image at 11:00 AM NZST on 17 April 2015 is depicted in Fig. 5.3, by courtesy of MTSAT, as an example water vapor image that is colorized. Clouds and other areas of high mid and upper level moisture are depicted as white and green, while dry and wormer areas aloft are shown as orange and red.

Below are presented some uses of water vapor imagery, such as follows:

1. Water vapor imagery can be used to detect dry slots. Within the dry slot precipitation intensity diminishes and precipitation chance is reduced. Dry slots can happen when a mid-latitude cyclone or tropical system advects air from a dry air source into its circulation.
2. Water vapor imagery can be used to detect moisture advecting in from the tropics. This moisture-laden air can produce big precipitation events when it is advected into a storm system.

3. Dry air in the middle and upper atmosphere indicates there are no significant upper level dynamics aloft that will cause the air to rise over those areas. Very moist air aloft though indicates there could be significant dynamic lifting the air such as jet stream divergence and positive vorticity advection.
4. Water vapor imagery allows the forecaster to see a complete atmospheric motion that is not just where the clouds are. This circulation can be used to point out troughs and ridges and where vertical motions are rising and sinking.
5. Water vapor imagery can be used to pick out the exact position of upper level lows.

5.1.3.4   True Colour RGB Channel Imagery

The amount of imager data from the world's weather satellites is impressive and will increase dramatically when GEO and PEO satellites come online. But it poses a challenge: figuring out how to extracts, distill and package the data into products that are easy for forecasters to interpret and use. If somebody ever hears of a "false colour" or "natural colour" Red, Green and Blue (RGB) image, this just means that meteorologists are only using one visible channel instead of two, and the near-infrared channel doubles up as both red and green. They do this when the satellite doesn't record as many visible channels, so this type of image looks pretty realistic as well, but it's just a matter of not needing to use as much actual visible light to create the image.

Therefore, combination of RGB processing different channels offers a simple yet powerful solution. In fact, it consolidates the information from different spectral channels into single products that provide more information than any one image can provide. Since different objects absorb/emit energy at different wavelengths (see the water vapour section above as one example), one of the most powerful ways we can use satellite images is by combining the different channels so that we can tell the difference between, say, clouds and snow or land and sea. One simple way of doing this is channel differencing, where we subtract one channel from another and then display the resulting image. Good examples of how this can be useful include the detection of fog and volcanic ash.

Another common way of combining images is the creation of an RGB image in which can be assigned different channels to be displayed as different colour schemes. For example, if somebody assigns one near-infrared channel and two visible channels to red, green, and blue colours respectively, it is possible to get a "true colour" RGB image that looks almost as if observers were viewing the earth from space with our own eyes. Thus, in Fig. 5.4 is illustrated true colour RGB image at 10:30 AM NZST on 17 April 2015, by courtesy of NASA MODIS.

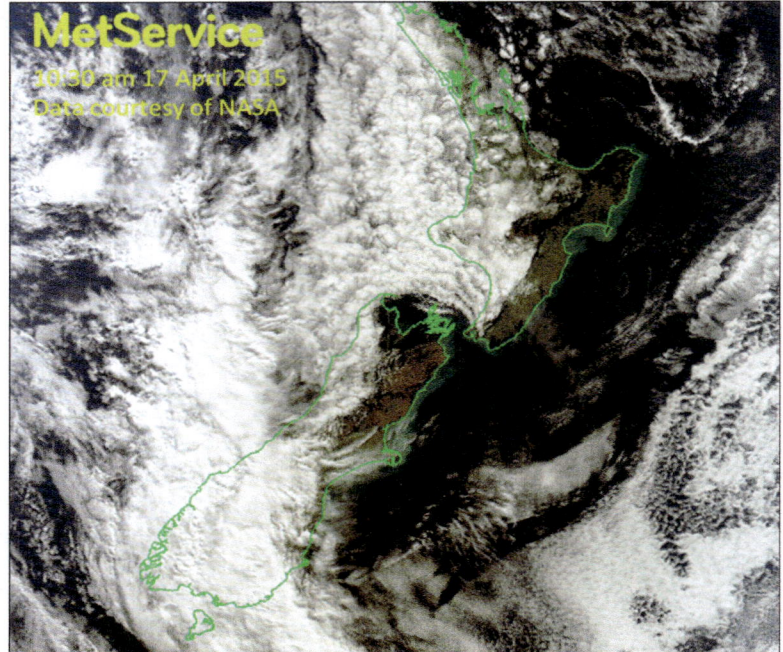

Fig. 5.4 True RGB Image (Courtesy of Manual: by NASA)

5.1.4   Weather and Atmospheric Properties

Weather is the conditions of the atmosphere experienced by humans at any given moment. Weather is used to refer to the atmospheric conditions that exist for a short duration during hours or days. These conditions fluctuate and vary temporally over a given place. Often people confuse this with the term, climate. Thus, climate refers to the average atmospheric conditions over a long period of time during a month or season of a place. Those conditions are at times when people feel our atmosphere that is very hot or cold, sultry and humid or dry, rainy, stormy etc. However, all these experiences are due to atmospheric conditions such as, temperature, humidity, wind speed and its direction, rainfall, etc. These conditions vary both spatially and temporally.

5.1.4.1   Atmospheric Temperature Measurements

Temperature is the measure of thermal or internal energy of the molecules within an object or gas. Meteorologist can measure temperature of an object using either direct contact or remote sensing. Temperature of air is closely related to other atmospheric properties, such as pressure, volume and density. Temperature is an index of sensible heat, which does not measure the amount of heat energy. It indicates the degree of molecular activity or kinetic energy of molecules. As it describes the kinetic energy, it specifies the speed of the movement of molecules. In case of gas (air), molecules move and change their location; but in case of solid, the molecules do not change their location but only vibrate it their place. The speed of this vibration is described by temperature.

When a body has higher temperature, heat will flow from it to another body with lower temperature. Thus, that is heat flows from a body with high temperature to a body with low temperature. Temperature is measured using different types of thermometers. Temperature is reported in any of the three scales, such as Celsius scale, Fahrenheit scale, and Absolute (Kelvin) scale. Of all the above, Celsius scale is used throughout the world to report the atmospheric temperature. It is named after the Swedish Astronomer Andres Celsius, which 0 °C is the triple point temperature at which the gaseous, liquid and sold states of water are at equilibrium under standard pressure. The boiling point of water under standard pressure conditions is at 100 °C. Kelvin scale is based on absolute zero. At absolute zero molecular activity ceases theoretically. Absolute zero in Kelvin scale is equal to -273.16 °C. Each degree on Kelvin scale is equal to that of Celsius scale. Accordingly, the value of 273 is added to Celsius value for converting it into Kelvin. Fahrenheit scale is an old scale used formerly, which is still used in United States of America.

The temperatures of the atmosphere at various altitude level as well as sea and land surface temperatures can be inferred from satellite measurements. In fact, these measurements can be used to locate weather fronts, determine the strength of tropical cyclones, study urban heat islands and monitor the global climate. Wildfires, volcanoes and industrial hot spots can also be found via thermal imaging from weather satellites.

Satellite temperature measurements have been obtained for troposphere since 1978. Thus, by comparison, the usable balloon or radiosonde record begins in 1958. Weather satellites do not measure temperature directly. They measure radiance in various wavelength bands. Since 1978 Microwave (MW) Sounding Units (MSU) onboard the US PEO satellites have measured the intensity of upwelling microwave radiation from atmospheric oxygen, which is related to the temperature of broad vertical layers of the atmosphere.

Measurements of IR radiation pertaining to sea surface temperature have been collected since 1967. Satellite datasets are presenting that over the past four decades the troposphere has warmed and the stratosphere has cooled. In fact, both of these trends are consistent with the influence of increasing atmospheric concentrations of greenhouse gases.

The term "upper air" refers to the part of the atmosphere well above Earth's surface. The temperature of the atmosphere in this region is a fundamental climate variable. However, recent changes in atmospheric temperature have been attributed to human-induced climate change. Continued monitoring of atmospheric temperature is critical to advancing human understanding of the sensitivity of Earth's climate to changing atmospheric composition. There are several methods available for the measurement of the upper air as follows:

1. **Radiosondes** – Radiosondes are commonly called weather balloons as small instruments lifted aloft by helium-filled balloons. All radiosonde measurements made by temperature, pressure, and humidity sensors are radioed back to the surface.
2. **Microwave (MW) Sounders** – These MW sounders are satellite-borne instruments that measure the radiance of Earth at microwave frequencies, which allows scientists to deduce the temperature of thick atmospheric layers.
3. **Infrared (IR) Sounders** – These are satellite-borne instruments as well that measure the radiance of Earth at infrared frequencies, which allows scientists to retrieve the temperature of thick atmospheric layers using inversion algorithms.
4. **GPS Radio Occultation** – The approach uses satellite-borne GPS receivers to measure the refraction of the GPS signals by the Earth's atmosphere. This allows for the retrieval of vertical temperature and moisture profiles.

Therefore, each satellite product measures the mean temperature of the atmosphere in the thick layer. Moreover, this atmosphere brightness temperature T_B measured by the satellite can be described as an integral over the height above Earth's surface Z of the atmospheric temperature T_A weighted by a weighting function W(Z), plus a small contribution due to emission by Earth's Surface $\tau \varepsilon T_S$ gives the following equation:

$$T_B \approx \int_Z T_A(Z) W(Z) dZ + \tau \varepsilon T_S \qquad (5.1)$$

The exact form of the weighting function depends on the temperature, humidity and liquid water content of the atmospheric column being measured. Thus, representative weighting functions based on the mean state of the atmosphere are sometimes useful.

5.1.4.2 Atmospheric Density Measurements

The density of air ρ (known as and air density) is the mass per unit volume of Earth's atmosphere. In general, density measures the heaviness of an object or how closely packed the substance is. Density is related to both the type of material that an object is made of and how closely packed the material is. Air density, like air pressure, decreases with increasing altitude. It also changes with variation in temperature and humidity. At sea level and at 15 °C air has a density of approximately 1.225 kg/m^3 according to International Standard Atmosphere (ISA). The density of dry air (ρ in kg/m^3), can be calculated using the ideal gas law, expressed as a function of absolute pressure (p_A), specific gas constant for dry air (R_S in J/kg · K) and absolute temperature (T_A in K):

$$\rho = p_A / RS \cdot TA \qquad (5.2)$$

5.1.4.3   Atmospheric Pressure Measurements

In general sense, pressure is the force exerted over a given area or object, either because of gravity pulling on it or other motion the object has Molecules in the air produce pressure through both their weight and movement, and in such a way this pressure is connected to other properties of the atmosphere. However, atmospheric pressure, sometimes also called barometric pressure, is the pressure within the atmosphere of Earth. In most circumstances Earth atmospheric pressure is closely approximated by the hydrostatic pressure hydrostatic pressure caused by the weight of air above the measurement point. As elevation increases, there is less overlying atmospheric mass, so that Earth atmospheric pressure decreases with increasing elevation.

Pressure measures force per unit area, with SI units of Pascals ($1~Pa = 1~N/m^2$). On average, a column of air one cm^2 (0.16 sq. in) in cross-section, measured from sea level to the top of the Earth's atmosphere, has a mass of about 1.03 kg (2.3 lb) and weight of about 10.1 Newtons. That weight across one cm^2 is a pressure of $10.1~N/cm^2$ or $101~kN/m^2$ (kPa). A column 1 sq in. (6.5 cm^2) in cross-section would have a weight of about 14.7 lb. (6.7 kg) or about 65.4 N. At low altitudes above the sea level, the pressure decreases by about 1.2 kPa for every 100 m. For higher altitudes within the troposphere, the following equation (the barometric formula) relates atmospheric pressure p to altitude h can be presented by the following equation:

$$p = p_0 \cdot \exp\left(-g \cdot M \cdot h / R_0 \cdot T_0\right) \tag{5.3}$$

where p_0 = sea level standard atmospheric pressure (value is 101,325 Pa), g = Earth-surface gravitational acceleration (with value of 9.80665 m/s^2), M = molar mass of dry air (value is 0.0289644 kg/mol), R_0 = universal gas constant (8.31447 J/(mol · K) and T_0 = sea level standard temperature (288.15 K).

5.1.4.4   Atmospheric Humidity Measurements

Humidity is a measure of the amount of moisture in the air. It tells how comfortable it is to be outside and if there is enough moisture to create clouds and rain. At any temperature, water vapor will be present in the air. There is a limit to the amount of water vapor, which can exist in equilibrium with the air (the saturation pressure). The mass of water vapor in a unit volume of moist air is known as the absolute humidity (measured in g/m^3), which is presented by the following relation:

$$H_A = m_W / V_{WAV} \tag{5.4}$$

where H_A = absolute humidity (mass/volume), m_w = mass of water and V_{WAV} = total wet air volume.

The relative humidity H_R is the value commonly mentioned in weather forecasts and it is ratio between the actual and saturation partial pressures of water vapor (measured in %):

$$H_R = p_w/p_s \times 100 \qquad (5.5)$$

where p_w = partial vapor pressure of water and p_s = saturation vapor pressure of water.

5.2 Temperature and Trace Gases

The most important application of satellite measurements is atmospheric sounding, that is, retrieving vertical profiles of temperature and trace-gas concentrations, especially those of water vapor and ozone. Here will be discussed the theory, practice and accuracy of such retrievals, and methods of assimilating these data into numerical models. In these retrievals, radiance from the surface is considered to be noise for which corrections must be made. It will be also discussed the opposite problem, that is, sensing surface parameters, such as land and sea surface temperature, in which the atmospheric contribution to the radiance is noise that must be eliminated.

5.2.1 Sounding Theory

There are two basic types of passive atmospheric sounding: vertical sounding, in which the sounding instrument senses radiation coming from the atmosphere and the Earth's surface, and limb sounding, in which only the limb of the atmosphere is sensed. So, both sounding methods utilize Schwarzchild's equation. Active atmospheric sounding includes the use of lidar and radar. In this chapter we focus on passive techniques, which are in use today.

5.2.1.1 Vertical Sounding Theory

Vertical sounding involves a satellite-based sounding instrument sensing radiation coming from the atmosphere and the Earth's surface, which is illustrated by the schematic shown in Fig. 5.5. In such a way, vertical sounding theory involves implementation of integrated form of Schwarzchild's Equation written such as:

Fig. 5.5 Vertical (Nadir) Scanning Geometry (Courtesy of Manual: by SSC)

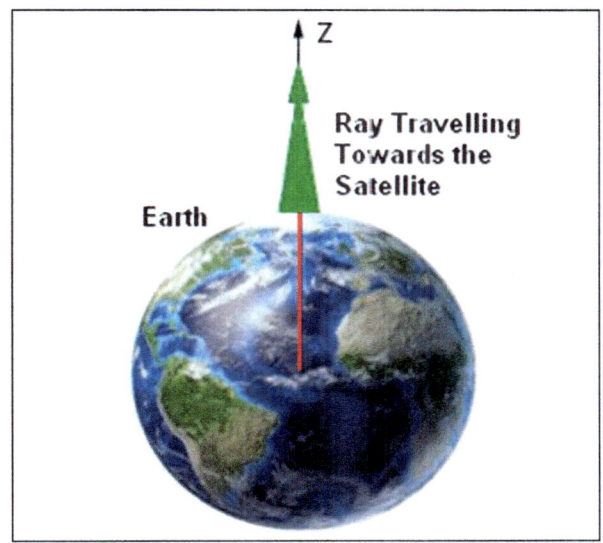

$$L_\lambda = L_o \exp\left(-\delta_o/\mu\right) + \int_0^{\delta_o} \exp\left[-(\delta_o - \delta_\lambda)/\mu\right] B_\lambda(T)\, d\delta_\lambda/\mu \qquad (5.6)$$

where value L_λ = radiance detected by the satellite, which applies to an atmosphere free of scattering and L_o = radiance emerging from the Earth's surface. Although optical depth δ is a physically meaningful quantity, it is not necessarily optimal for retrieving soundings. In fact, instead to substitute h, a generalized height coordinate can be replaced with height z, pressure p, ln (p), optical depth, transmittance or any other function, which is monotonic in height, h_o and h_T represent the Earth's surface and the top of the atmosphere, respectively. Recalling that τ_λ (δ_1, δ_2), the vertical transmittance between levels δ_1 and δ_2, is given by the following equation:

$$\tau_\lambda(\delta_1, \delta_2) \equiv \exp\left(-|\delta_1 - \delta_2|\right) \qquad (5.7)$$

After some manipulation can be presented the following two equations:

$$L_\lambda = L_o \tau_o^{1/\mu} + \int_{h_o}^{h_T} B_\lambda(T)\, W_\lambda(h, \mu)\, dh \quad \text{where} \quad W_\lambda(h, \mu) = d/dh\left(\tau_\lambda^{1/\mu}\right) \qquad (5.8)$$

where τ_o = vertical transmittance from the surface to the satellite, and $\tau_\lambda^{1/\mu}$ is the slant-path transmittance from the surface level h to the satellite. While, W_λ (h, μ) is called a weighting function, because it weights value $B_\lambda(T)$ in the atmospheric term. Of course, if the scene is overcast, an infrared sounder will see only down to the cloud top, in which case L_o and τ_o should be replaced with the cloud top values L_c and τ_c, respectively, and the atmospheric term should be integrated from h_c to h_T.

The blackbody radiance term can be retrieved as a function of value h if the atmosphere is observed at a number of wavelengths whose weighting functions sample the atmosphere (i.e. weighting functions that peak at different atmospheric heights). The temperature itself can then be obtained. Note that slight adjustments have to make to the above equations for soundings made over water. Thus, this is because of the contribution made from radiation reflected from the water.

In the microwave portion of the spectrum, the weighting function must be modified slightly for soundings made over water. For infrared observations and for microwave observations over dry ground, the emittance of the surface is approximately one, and there is very little reflection. The surface term is usually approximated as $B_\lambda(T_o)\ \tau_\lambda^{1/\mu}$, where T_o is the skin temperature, that is, the temperature of the ground itself. Water, however, reflects as much as 60% of the microwave radiation incident on it. In this case, the surface radiance L_o is composed of two terms, the radiation emitted by the surface and the reflected sky radiance:

$$L_o = \varepsilon_o B_\lambda(T_o) + (1 - \varepsilon_o)\, L_{sky} \tag{5.9}$$

where the water surface has been approximated as a specular reflector with reflectance $1 - \varepsilon_o$ and the sky brightness may be determined by applying Schwarzchild's equation:

$$L_{sky} = B_\lambda\left(T_{space}\right)\tau_o^{1/\mu} + \int_{h_o}^{h_T} B_\lambda(T)\, d/dh \left[-(\tau_o/\tau_\lambda)^{1/\mu}\right] dh \tag{5.10}$$

where value $(\tau_o/\tau_\lambda)^{1/\mu}$ is the slant-path transmittance from level h to the Earth's surface, while Tspace is the space temperature (approximately 2.7 K) caused by radiation left over from the Big Bang. Combination of 5.8 and 5.10 radiance measured from a satellite is:

$$L_\lambda = \varepsilon_o B_\lambda(T_o)\tau_o^{1/\mu} + (1 - \varepsilon_o)\, B_\lambda\left(T_{space}\right)\tau_o^{2/\mu} + \int_{h_o}^{h_T} B_\lambda(T)\, W_\lambda(h, \mu)\, dh \tag{5.11}$$

Following the above equation is presenting as follows:

$$W_\lambda(h, \mu) = \left[1 + (1 - \varepsilon_o)\, (\tau_o/\tau_\lambda)^{2/\mu}\right] d\tau_\lambda^{1/\mu}/dh \tag{5.12}$$

This is a general weighting function that approaches W_λ (h, μ) in Eq. (5.8) as value ε_o approaches one. The differences between these equation and Eq. (5.8) arise from the fact that reflection affords the satellite the opportunity to see the atmosphere twice, once before reflection, and once after reflection.

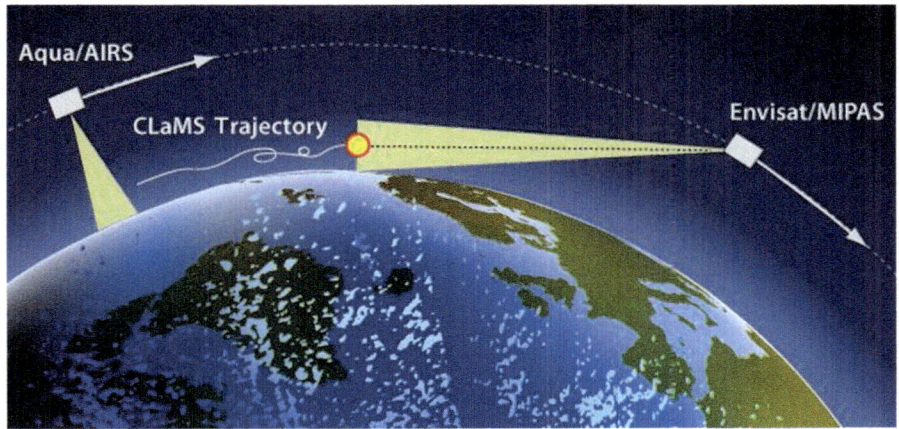

Fig. 5.6 Nadir and Limb Scanning Geometry (Courtesy of Manual: by IEEE)

5.2.1.2 Limb Sounding Theory

The observation of the atmosphere is made with a technique that is called limb sounding. Starting from the position of the instrument (spacecraft, aircraft or radiosonding balloon inside the stratosphere), the instrument collects radiation emitted from the atmosphere along the lines of view that observe it at small corners below the horizon. In Fig. 5.6 is illustrated a schematic overview of the three primary data modalities: while AIRS aboard NASA's Aqua satellite scans the atmosphere directly below the satellite (at satellite nadir), Envisat's Michelson Interferometer for Passive Atmospheric Sounding MIPAS instrument uses a so called limb geometry, i.e. it performs a tangential scan. For each MIPAS detection (yellow dot), the system provides CLaMS trajectories, which are started at the location of the detection.

The lines of vision penetrate part of the stratosphere and part of the troposphere reaches a low altitude portion and go back to the space again. The instrument measures spectrum generated by the spontaneous emission of the atmosphere along the line of sight in the millimeter wave region in the B, C and D bands. From the spectrum analysis, information on the composition of the atmosphere can be obtained. By using a set of measurements performed at different viewing angles (which correspond to different atmospheric heights), a vertical concentration profile can be determined. In addition, an important technique in satellite meteorology is limb sounding in which the sounder onboard the US GOES I-M spacecraft also scans the Earth's limb.

Consider a ray traveling in the positive x direction toward the satellite, the height above the surface of the ray at its closest approach to Earth is called the tangent height z_T. In such a way, the radiance measured by the satellite is given by the following equations:

$$L_\lambda(z_T) = \int_{-\infty}^{\infty} B_\lambda(T) \, [d\tau_\lambda(x, z_T)/dx] \, dx = \int_{-\infty}^{\infty} B_\lambda(T) \, W_\lambda(z, z_T) \, dz \qquad (5.13)$$

From where the next stage can be realized the following relation:

$$W_\lambda(z, z_T) = \{[d\tau_\lambda(x, z_T)/dx]_+ + [d\tau_\lambda(x, z_T)/dx]_-\}|dx/dz| \qquad (5.14)$$

In spice of that, + and − subscripts indicate values along the positive and negative x axes, respectively.

5.2.2   Retrieval Methods

The physical problem that has to be solved is what temperature and trace gas concentration profiles could have produced a set of observed radiances? Namely, this is called the inverse problem or retrieval problem. The opposite problem, called the forward problem, is needed to calculate outgoing radiances given temperature and trace-gas profiles.

Clouds cover approximately 50% of the Earth so about 50% of soundings is contaminated with clouds. Given clear-column radiances, the forward problem is easy to solve, but the retrieval problem is difficult because the solution is not unique. In this case, even if a noise free radiometer that measured radiances at all wavelengths could be constructed, a unique solution to the radiative transfer equation would not be guaranteed. When a finite number of wavelengths are observed and the measurements are contaminated with noise, an infinite number of solutions are possible.

The retrieval problem becomes one of finding temperature profiles that satisfy the radiative transfer equation and approximate the true profiles as closely as possible. Approaches to the retrieval problem can be classified into three general areas such as: physical retrievals, statistical retrievals, and hybrid retrievals.

5.2.2.1   Physical Retrievals

In physical retrieval schemes the ease of the forward problem is exploited in an iterative process as follows:

1. A first-guess temperature profile is chosen.
2. The weighting functions are calculated.
3. The forward problem is solved to yield estimates of the radiance in each channel of the radiometer.
4. If the computed radiances match the observed radiances within the noise level of the radiometer, the current profile is accepted as the solution.
5. If convergence has not been achieved, the current profile is adjusted (see below).

6. Steps 3 through 5 (or 2 through 5) are repeated until a solution is found.

The two most widely used methods of adjusting the temperature profile are those of both Chahine and Smith (1970). Chahine retrieves temperatures for as many levels as there are channels in the radiometer. Suppose the radiometer has J channels, and then the scheme retrieves the temperature at the levels J located at the peak of the weighting functions.

In this scheme there is a one-to-one correspondence between channel j of the radiometer and level j, where the weighting function of channel j peaks. Let $T_j^{(n)}$ be the nth estimate of the temperature at the jth level and Bj $[T_j^{(n)}]$ be the resultant Planck radiance at level j at the wavelength of channel j. In addition, Let $L_j^{(n)}$ be the nth estimate of the radiance in channel j, calculated using the $T_j^{(n)}$, and let $L_j{\sim}$ be the observed radiance in channel j. So, Chahine iterates the temperature by iterating the Planck radiance such as:

$$B_j\left[T_j^{(n+1)}\right] = B_j\left[T_j^{(n)}\right]\left[L_j \sim /L_j^{(n)}\right] \tag{5.15}$$

The iterated temperature at level j is found using the inverse Planck function. This scheme works because if, for example, the calculated radiance in channel j is greater than the observed radiance, it is reasonable to adjust downward the Planck radiance (and thus the temperature) at the level where the weighting function for channel j peaks. Since the peak of the weighting function is the strongest contributor to the radiance, using the ratio of the observed to the calculated radiance to adjust the Planck radiance is also reasonable. Smith's scheme for adjusting temperature profiles is similar to that of Chahine, but he relaxes the requirement that the temperature be retrieved at only J levels. Suppose the temperature is to be retrieved at K levels. Let $T_k^{(n)}$ be the nth estimate of the temperature at level k, and $B_jT_k^{(n)}$ the resulting Planck radiance at the wavelength of channel j. As above, let $L_j^{(n)}$ and $L_j{\sim}$ be, respectively, the nth estimate of, and the observed value of, the radiance in channel j. At each level, Smith obtains J estimates of an iterated Planck radiance such as:

$$B_j\left[T_{jk}^{(n+1)}\right] = B_j\left[T_k^{(n)}\right] + \left[L_j \sim -L_j^{(n)}\right] \tag{5.16}$$

where the J estimates of $T_k^{(n+1)}$, obtained by the inverse Planck function, are denoted by $T_{jk}^{(n+1)}$. Not all of these J estimates are equally good because each channel sees some levels better than others. Smith solves for $T_k^{(n+1)}$ as a weighted average of the $T_{jk}^{(n+1)}$ using the weighting functions as weights:

$$T_k^{(n+1)} = \sum_{j-1}^{i} W_{jk}T_{jk}^{(n+1)} / \sum_{j-1}^{i} W_{jk} \tag{5.17}$$

Smith's scheme is more flexible than that of Chahine in that it allows the user to choose the levels at which he will retrieve temperatures (consistent, of course, with the predetermined levels at which the transmittances are calculated). Since Smith's levels are not restricted to be at the peaks of the weighting functions, thus it cannot iterate the temperature at a level by using a single channel. At each level, thus, it obtains a suggested temperature change from each channel and lets the weighting function discriminate among them.

5.2.2.2   Statistical Retrievals

In statistical retrievals, the radiative transfer equation is not directly used. These methods assume that the radiometer has been designed so that its channels (weighting functions) will vertically sample the atmosphere. Thus, a set of radio-sonde soundings that are nearly coincident in real time and space with satellite soundings is compiled. This set, called the training data set, is used to calculate a statistical relationship between observed radiances and atmospheric temperatures.

These relationships are then applied to other observed radiances to retrieve temperatures. The process can be described as follows. Suppose there are N sounding pairs in the training data set. For instance, let l be the J x 1 column vector containing the/radiances observed by the radiometer in one of these soundings. Let t be the K x 1 column vector, paired with l, which contains the temperatures (and perhaps dewpoints) at K levels in the atmosphere. And finally, let the symbol (x) represent the element-by-element average of the vector x. Statistical retrieval methods require finding the K x J matrix C and the vectors (l) and (t) such that:

$$t - (t) = C[l - (l)] \qquad (5.18)$$

Here is important to note that once C, (t), and (l) are known, retrieval of temperatures from observed radiances is a very significant simple task which involves only vector subtraction, matrix multiplication, and vector addition. In fact, this simplicity makes a statistical method attractive for operational retrieval schemes in which numerous soundings methods must be processed. There are several ways to find C, (t), and (l), namely at this point the simplest is the regression solution. Let (t) and (l) be the average of all the vectors t and l, respectively, in the training data set. Let T' be the K x N matrix whose columns are the vectors t – (t), and let L' be the J x N matrix whose columns are the vectors l – (l). The matrix C, which in a least-squares sense, minimizes errors in Eq. (5.18) is giving such as:

$$C = T'L'^{T}(L'L'^{T})^{-1} \qquad (5.19)$$

where the superscript value T indicates that the transpose of the matrix is to be used and the superscript value –1 indicates the matrix inverse. This scheme has been used extensively to retrieve satellite soundings. One problem with the regression solution is that no "filtering" of noise from the input temperatures or radiances is done. As a result, the value of C matrix can be unstable, namely, small radiance errors can produce unacceptably large errors in the retrieved temperatures. At this point,

Smith and Woolf in 1976 developed a technique to filter noise using statistical eigenvectors. Their scheme was used operationally for many years. However, more recently, Thompson in 1992 employed singular value decomposition to improve statistical retrievals.

5.2.2.3 Hybrid Retrievals

Hybrid retrieval methods are in between physical and statistical retrieval methods, also known as inverse matrix methods. They appear much like purely statistical methods, but they do not require the large training data set. They use weighting functions like physical retrievals, but they do not directly involve integration of the radiative transfer equation. The radiative transfer equation is linearized about a standard temperature profile $T_s(h)$, which provides as follows:

$$L_\lambda = L_{\lambda s} + \varepsilon_o(dB_\lambda/dT)\, T'_{so}\tau_o + \int_{h_o}^{h_T} W_\lambda(h)\,(dB_\lambda/dT)\,T'_s(h)\,dh \qquad (5.20)$$

where $L_{\lambda s}$ is the radiance if $T_s(h)$ were the actual profile, and $T'_s(h)$ is the deviation from T_s at level h.

A quadrature (numerical integration) scheme is chosen, which converts Eq. (5.20) into a matrix equation such as:

$$1' = At' + e \qquad (5.21)$$

where value l' is the column vector of radiance deviations from the standard profile, t' is the vector of temperature deviations from $T_s(h)$ and e is a column vector that contains the errors of measurement of the radiances. The matrix A contains the weighting functions, the quadrature weight, and the Planck sensitivity factors dB_λ/dT, value A is calculated with a knowledge of the transmittances as in the case of physical retrievals. Suppose, temporarily, that we have a set of coincident radiosonde and satellite observations. Let the uppercase letters indicate matrices, which have N columns for the N sounding pairs in the data set:

$$L' = AT' + E \qquad (5.22)$$

Substituting Eqs. (5.22) and (5.19) is giving the following equation:

$$C = T'(AT' + E)^T\left[(AT' + E)(AT' + E)^T\right]^{-1} \qquad (5.23)$$

Expanding and using the matrix relationship $(AB)^T = B^T A^T$ is giving as follows:

$$C = T' \left(T'^T A^T + E^T\right) \left(AT'T'^T A^T + AT'E^T ET'^T A^T + EE^T\right)^{-1} \qquad (5.24)$$

Now, the system assumes that the measurement errors are uncorrelated with temperature deviations. Thus $T'E^T$ and ET'^T are negligible in comparison with the other terms. In such a way is presented so call minimum variance method such as:

$$C = S_T A^T \left(A S_T A^T + S_E\right)^{-1} \qquad (5.25)$$

where value S_T is the temperature covariance matrix $(T'T'^T/N)$ and S_E is the radiance error covariance matrix (EE^T/N).

The interesting property of Eq. (5.25) is that all of the components can be determined separately: value A can be calculated from the transmittances, S_T can be determined from a sample of radiosonde soundings and value S_E can be measured in the laboratory as part of instrument calibration. For instance, it is possible to drop the assumption that a data set of coincident radiosonde-satellite soundings has been collected. If it is further assumed that the temperature variance is independent of level and then can be given such as:

$$S_T = \sigma_T^2 I \qquad (5.26)$$

where I is the identity matrix and σ_T^2 is the variance of temperature. Similarly, if radiance error in one channel is uncorrelated with radiance error in any other channel and if the variance of radiance error is independent of channel, then is given as follows:

$$S_E = \sigma_E^2 I \qquad (5.27)$$

Finally, Eq. (5.25) reduces and then is presented such as:

$$C = A^T \left(AA^T + \gamma I\right)^{-1} \qquad (5.28)$$

where $\gamma = \sigma_E^2/\sigma_T^2$. In practice, γ is often used as a tuning parameter, that is, a number that is changed until the best retrievals are obtained. The main advantage of the hybrid methods is that they are easier to put into operation than the statistical or physical methods. Thus, they share with the physical retrieval method the disadvantage of depending on knowledge of the transmittances. Most of them share with the statistical retrieval method the advantage of including statistical knowledge of atmospheric structure.

5.2.3 Operational Retrievals

The retrieval method is only a small part of the process by which radiance from operational satellites are converted into temperature and moisture soundings. So in this section, will be discussed the processes that are used operationally to produce soundings from NOAA and GOES satellites.

5.2.3.1 TIROS Operational Vertical Sounder

The US TIROS Operational Vertical Sounder (TOVS) equipped aboard NOAA's TIROS series of PEO satellites consists of three instruments such as: the High Resolution Infrared Radiation Sounder (HIRS), the Microwave Sounding Unit (MSU) and the Stratospheric Sounding Unit (SSU). This system is onboard the NOAA 6 through NOAA 14 and TIROS-N satellites (part of the Polar Orbiting Environmental Satellite (POES) system). The MSU and SSU have been replaced with improved instruments, such as the Advanced Microwave Sounding Unit-A (AMSU-A) and Advanced Microwave Sounding Unit-B (AMSU-B), onboard the newer generations of satellites.

Surface radiation received by HIRS channel 9 is suppressed by 9.7 micron ozone band absorption in the lower stratosphere layers such that high lower stratospheric ozone values directly result in cold HIRS channel 9 radiances. An algorithm is applied to estimate the "total" column ozone based upon these suppressed HIRS 9.7 micron radiances.

Therefore, new generation of Advanced TOVS (ATOVS) consists of HIRS, the AMSU-A and AMSU-B for retrieving temperature, humidity and ozone sounding in all weather conditions. The Microwave Humidity Sounder (MHS) developed by Eumetsat, replaced AMSU-B on NOAA-19 and the European Metop satellites Metop-1 and Metop-2. Thus, currently the ATOVS is generating products from the NOAA-19, Metop-1, and Metop-2. This instrument package provides information on temperature and humidity profiles, total ozone, clouds and radiation on a global scale to the operational user community.

A sounding, for our use, is a vertical atmospheric temperature or moisture profile derived from radiance measurements. The radiance measurements are from a variety of instruments flown on weather satellites. The National Environmental Satellite, Data and Information Service (NESDIS) of the US NOAA currently generates soundings from geostationary and polar weather satellites. The Office of Satellite and Product Operations currently generates sounding products from four operational systems ATOVS, Microwave Integrated Retrieval System (MIRS), Infrared Atmospheric Sounding Interferometer (IASI) and GOES systems. Each PEO satellite provides full global coverage each day, while data from polar satellites are useful for anybody requiring global coverage. Thus, each sounding product generated by Office of Satellite and Product Operations (OSPO) has its own section in the web site.

In this sector will be introduced the following five components of TOVS:

1. **TOVS Preprocessor** – The TOVS processes all data received directly from the Satellite Operations Control Center. Then, digital counts are converted to radiances, and each scan spot is Earth-located. The Stratospheric Sounding Unit (SSU) data are further processed by the Stratospheric Mapper, in such a way, a variety of corrections are applied to the HIRS/2 and the MSU data:

   - The MSU data are corrected for antenna side-lobes.
   - A limb correction is applied to both the MSU and HIRS/2 data so that the other modules can treat the data as if they were all obtained at nadir. The limb correction consists of regression equations, formed from synthesized radiance data, which result in a correction to be added to the radiance. Thus, if L_{jm} is the radiance in channel j at scan angle m, then the correction is given by the following relation:

$$\Delta L_{jm} = \sum_{n=1}^{i} a_{jmn} L_{nm} + b_{jm} \tag{5.29}$$

   - An attempt is made to correct the window channels for water vapor absorption.
   - For daytime data, an Albedo, estimated using the 3.7-μm channel, is used to correct the 4.3-μm channels for reflected sunlight.
   - Since the MSU has coarser spatial resolution than does the HIRS/2, the MSU data are interpolated to the locations of the HIRS/2 scan spots.

   Finally, the TOVS preprocessor obtains solar zenith angles, terrain elevations, and initial guess values of skin temperature and surface Albedo. Thus, all these data are staged to disk to wait for further processing.

2. **The TOVS Atmospheric Radiance Module** – The Atmospheric Radiance Module's job is to deliver spatially averaged, clear column radiances to the Retrieval Module. Perhaps surprisingly, the Atmospheric Radiance Module consumes most of the computer time in the TOVS system. First, the data are divided into boxes of nine HIRS/2 scan spots cross-track and seven scan spots along-track. One sounding is retrieved to represent this group of 63 scan spots. In fact, the nominal resolution of the operationally retrieved TOVS soundings is approximately 250 km. Next, the 63 scan spots are tested to determine whether they are contaminated by clouds. Several tests are employed:

   - In daylight, the Albedo measured by HIRS/2 channel 20 is compared with the Albedo expected for a cloud-free scene. Clouds, of course, would increase the visible Albedo.
   - At night, channel 19 (3.7 μm) is less sensitive to clouds than channel 8 (11 μm). During the day, clouds reflect sunlight at 3.7 μm; therefore, channel 19 is much more sensitive to clouds than channel 8. Thus, a large

difference in the brightness temperatures of these two channels is an indication of clouds.

- The 11-μm brightness temperature is compared with the expected surface temperature. Too low a brightness temperature indicates cloud contamination.
- The HIRS/2 brightness temperatures are used in regression equations to estimate the MSU brightness temperatures. If clouds contaminate the scan spot, the estimated MSU brightness temperatures will be less than the observed ones.

Finally, Atmospheric Radiance Module produces clear-column radiances. If four or more of the 63 scan spots are determined to be clear, the clear-column radiances are calculated as a weighted average of the observed radiances from the clear spots.

3. **TOVS Stratospheric Mapper** – Because the SSU does not scan as far out as the HIRS/2 and the MSU, HIRS/2 scan spots at the edges of the scan have no corresponding SSU data. This problem is solved by the Stratospheric Mapper, which maintains a global map of SSU radiances on a latitude-longitude grid. The map is updated as new observations arrive. The SSU data are corrected for limb effects before mapping.

4. **The TOVS Retrieval Module** – The Retrieval Module system performs the retrievals using clear-column radiances produced by the Atmospheric Radiance Module and the Stratospheric Mapper. Until September 1988, temperatures and water vapor below 100 hPa were retrieved using statistical eigenvectors. Above 100 hPa, temperatures and total ozone were retrieved by regression. As of this writing, temperatures and water vapor are retrieved using the Minimum Variance Simultaneous (MVS) method according to Eqs. (5.23) and (5.24). The method is "simultaneous" in the sense that temperature and water vapor are retrieved using one matrix.

5. **Archived TOVS Data** – The TOVS sounding data are archived by NESDIS. Users who are interested in ordering historical data, however, should be aware that the archive tapes come in the form of layer-mean virtual temperatures for 15 layers between the surface and 0.4 hPa. Water vapor is archived in the form of layer precipitable water in the three lowest layers (surface-850, 850–700, and 700–500 hPa). The tropopause temperature and pressure, total ozone, cloud-top pressure, cloud amount and an average value of adjacent-pair or N^* are saved.

5.2.3.2   VISSR Atmospheric Sounder (VAS)

The Visible Infrared Spin-Scan Radiometer (VISSR) onboard GOES Satellites known as VISSR Atmospheric Sounder (VAS) instrument is a radiometer with three operating modes such as follows: the operational VISSR mode, the Multi-Spectral Imaging (MSI) mode and the Dwell Sounding (DS) mode. The VISSR mode is routinely used by NOAA/NESDIS for its operational products, which include an 0.9-km visible picture and an 11-micrometer Infrared (IR) picture of

6.9-km resolution, while the two other modes are VAS unique. The MSI mode can provide the operational VISSR capability plus two additional IR bands. The DS mode is used primarily for sounding, to obtain the temperature and moisture profiles.

Therefore, the VAS, when operated in the VISSR mode, provides which are compatible with all original NESDIS VISSR data users. This mode provides both visible and IR data. The IR (11 μm) data are 6.9-km resolution and encoded to 8 bits. Thus, in order to readily exercise the capabilities of the VAS mode and yet allow NESDIS to fulfill its operational commitments, a Transparent VAS Mode (TVM) configuration was implemented. This mode allowed the VAS Demonstration to obtain both MSI and DS data routinely, while NESDIS concurrently supplied operational VISSR data to its users.

The VAS soundings have been retrieved operationally at NESDIS's VAS Data Utilization Center (VDUC) since the summer of 1987. In first stage, the raw VAS data are converted to brightness (equivalent blackbody) temperatures and Earth located. As with the TOVS retrieval system, scan spots are blocked before processing. The VAS system is designed to be flexible, so the number of spots in a block is variable. Typically, however, a block of 11 x 11 scans spots is used, which results in a field of retrieved soundings with approximately 75-km resolution. Next, the data are filtered for clouds. Since visible data are not collected during dwell sounding, the cloud filter relies on three window infrared channels: 3.9, 11.2, and 12.7 μm (channels 12, 7, and 8), such as:

- If the 3.9-μm brightness temperature is significantly different (3–5 K) from the 11.2-μm brightness temperature, the scan spot is assumed to be cloudy.
- If the 11.2-μm brightness temperature (corrected for moisture effects by the so called split-window technique, see below) is significantly colder than the surface air temperature (obtained from analyzed surface observations), the scan spot is classified as cloudy.

Besides, if a majority of the scan spots appear to be clear, the clear radiances are averaged and a sounding is retrieved down to the surface. On the other hand, during the retrieval process itself, the sounding can be classified as cloudy if the algorithm cools the first guess surface temperature by more than 6 K. In addition, if most of the scan spots are cloudy, the cloudy spots are averaged and cloud-top pressure is estimated by comparing the 11.2-μm brightness temperature, which can be corrected for moisture effects with the first guess temperature profile. A sounding is retrieved down to the cloud top and values below cloud top are estimated by interpolating between the cloud-top value and the surface analysis. These cloudy soundings, however, are of questionable accuracy.

The retrieval algorithm is a hybrid scheme was described by Smith in 1986. First-guess temperature and moisture profiles are selected, usually from a model forecast such as the Nested Grid Model (NGM) or the Global Spectral Model. Thus, surface temperature and humidity are obtained from analyzed fields of the corresponding hourly observations. The first-guess estimate of the skin temperature is obtained by regression from the 11.2, 12.7, and (for noncloudy scan spots) 3.9-μ m brightness temperatures. Temperatures above the 100-hPa level are obtained from the latest analyses or from climatology. In fact using the first-guess profiles, the weighting functions and the brightness temperatures in the VAS channels are calculated. The temperature and moisture profiles are calculated by making a single correction to the first-guess profiles, which is why the scheme is known as the "simultaneous" or "one-step" method.

5.2.4   Limb Sounding Retrievals

Limb soundings utilize broadband observations in one or two channels, however at many tangent heights rather than narrowband satellite observations in many spectral intervals. Because of the sharpness of the weighting functions, temperature retrievals would seem to be straightforward. Any of the retrieval methods discussed in Sect. 5.2.2 should work nicely. The problem, as pointed out by Gille and House (1971), is that a reference level, where both the height and pressure are known, is needed. Suppose that the pressure at reference height z_0 is p_0. Moreover, using a single wideband channel, a unique temperature profile can be retrieved. A different assumption about the pressure at z_0 yields a different temperature profile. Gille and House solved this problem by utilizing a second, narrower (more opaque) channel. For each channel, a temperature profile is retrieved by assuming a pressure at reference level z_0. So, if the assumed pressure is incorrect, the two temperature profiles will not agree. Iterating p_0 between inversions yields accurate temperature profiles. However, many inversions are necessary to obtain a single temperature profile, which means that the calculations are lengthy. Fortunately, rapid, accurate retrieval schemes have been developed by Gordley and Russell in 1981; and Rodgers in 1976 and in 1984.

In the past of satellite meteorology system, three infrared limb sounders have flown on the Nimbus satellites. The Limb Radiance Inversion Radiometer (LRIR) flew on Nimbus-6, and both the Limb Infrared Monitor of the Stratosphere (LIMS) and the Stratospheric and Mesospheric Sounder (SAMS) flew on Nimbus-7.

The Atmospheric Infrared Sounder (AIRS) as modern satellite meteorological instrument is a high spectral resolution spectrometer with 2378 bands in the thermal infrared (3.7–15.4 μm) and 4 bands in the visible (0.4–1.0 μm). These ranges have been specifically selected to allow determination of atmospheric temperature with an accuracy of 1 °C in layers 1 km thick, and humidity with an accuracy of 20% in layers 2 km thick in the troposphere. In the cross-track direction, a ± 49.5° swath centered on the nadir is scanned in 2 s, followed by a rapid scan in 2/3 s taking routine calibration related data that consist of four independent Cold Space Views, one view of the Onboard Blackbody Calibrator, one view of the Onboard Spectral Reference Source, and one view of a photometric calibrator for the VIS/NIR photometer. Each scan line contains 90 IR footprints, with a resolution of 13.5 km at nadir and 41 km × 21.4 km at the scan extremes from nominal 705.3 km orbit. The Vis/NIR spatial resolution is approximately 2.3 km at nadir.

The heart of the AIRS spectrometer is the multi-aperture, echelle grating spectrometer and corresponding multiple detector arrays. The IR Spectrometer Assembly system includes a pupil-imaging, (that is, detectors are located at a pupil stop of the spectrometer optics as opposed to the detector units being at a field stop) telescope which views the Earth and calibrator assemblies through a rotating scan mirror in the scan head assembly system. While in-flight calibration measurements are made once per scan line, namely every 2.667 s, data from ten or more scan lines are combined by the ground calibration software to update calibration coefficients.

The AIRS spectrometer instrument will make measurements of the Earth's atmosphere and surface that will allow scientists to improve weather prediction and to observe changes in Earth's climate. For example, inaugurating a new generation of operational atmospheric monitors, AIRS, the Advanced Microwave Sounder (AMSU) and the Humidity Sounder Brazil (HSB) are scheduled to fly together on the second satellite platform of NASA's Earth Observing System (EOS AQUA).

5.2.5 Ozone and Other Gases

Ozone has absorption lines in all major portions of the electromagnetic spectrum and thus can be measured with a variety of techniques, such as HIRS/2, limb and new generations of other sounders.

The old generation HIRS mode is a 20-channel infrared scanning radiometer that performs operational atmospheric sounding. The HIRS/2 sounder flies on the NOAA/TIROS-N type satellites and provides the basic 20 channel IR temperature and humidity soundings of the TOVS system. All HIRS/2 channels are used to provide information on temperature and humidity profiles, surface temperature, cloud parameters and total ozone. Calibration of HIRS/2 is performed every 40 scan lines, there is a gap of three scan lines in data coverage. Thus, HIRS scans across track in a "stop and stare" mode at a scan rate of 6.4 s. There are 56 Earth Views per scan. The circular field-of-views have a diameter of 20 km at nadir, and the swath covers ± 1080 km. The operational HIRS/2 sounder has the 9.7-μm channel.

Planet et al. (1984) regressed clear-column radiances, produced during operational TOVS processing, against surface-based observations of total ozone (column integrated ozone) made with a Dobson spectrophotometer. They found that three HIRS/2 channels yielded an acceptable regression retrieval scheme. Channel 3 (14.49 μm) is sensitive to stratospheric temperature, channel 8 (11.11 μm) is sensitive to surface temperature, and channel 9 (9.71 μm) is sensitive to ozone, stratospheric temperature, and (slightly) surface temperature.

The regression retrieval scheme essentially uses channels 3 and 8 to correct channel 9 for temperature dependence. Planet et al. showed the retrieval scheme to be accurate within 38%. Susskind et al. (1984) and Reuter and Susskind (1986) have developed a physical retrieval scheme for total ozone using HIRS/2 and MSU data that also gives good results.

The new generation HIRS/4 sounder provides calibrated vertical profiles of temperature and humidity; information on cloud cover, cloud top height, cloud top temperature and cloud phase, as well as surface Albedo Thus, this is also a 20-channel infrared scanning radiometer that performs operational atmospheric sounding. HIRS has 19 infrared channels (3.8–15 μm) and one visible channel. The swath width is 2160 km, with a 10 km resolution at nadir. IR calibration of the HIRS/4 is provided by programmed views of two radiometric targets m-dash, the warm target, mounted on the instrument baseplate, and a view of deep space. Data from these views provides sensitivity calibrations for each channel at 256 s intervals, if commanded. Internally generated electronic signals provide calibration and stability monitoring of the detector amplifier and signal processing electronics. Thus, the HIRS radiometric sounder uses CO_2 absorption bands for temperature sounding (CO_2 is uniformly mixed in the atmosphere). Otherwise, the HIRS instrument also measures water vapour, ozone, N_2O and cloud and surface temperatures.

Microwave Limb Sounder (MLS) is instrument is designed to improve our understanding of ozone in Earth's stratosphere, especially how ozone is depleted by processes of chlorine chemistry. The instrument measures naturally occurring microwave thermal emission from the limb of Earth's atmosphere to remotely sense vertical profiles of selected atmospheric gases, temperature and pressure. This instrument designed by JPL flies aboard NASA's Aura spacecraft under the agency's Earth Observing System program together with three other instruments. The MLS instrument is designed to study the natural thermal radiation emitted from Earth's limb (the edge of Earth's atmosphere) to gather measurements on atmospheric gases, temperature and pressure. These data can be used to better understand the causes of ozone changes and pollution in the upper troposphere. The old generation of Limb sounders also uses the 9.6-ìμm bands to retrieve ozone. Using the emissivity growth approximation of Gordley and Russell (1981), Ember et al. (1984) retrieved ozone profiles from LIMS data and compared them with balloon-based measurements (ozonesondes) and rocket-based measurements. Ozone was retrieved from ~300 hPa to ~0.1 hPa. Calculated accuracies, based on experiment uncertainties, ranged from 15%, in the region 1–3 hPa, to about 40% at 100 hPa and 0.1 hPa. For instance, comparison with ozonesondes yielded mean differences within 10% in the region 7–50 hPa. The rms differences were ~15%. Comparison with rocket ozone measurements in the region 0.3–50 hPa yielded mean differences less than 16% and rms differences in the range 12–25%.

5.2.6 Split-Window Technique

Most of the preceding discussion concerns the estimation of atmospheric quantities. The retrieval schemes attempt to correct for any signal from the surface. Some meteorologically important quantities, such as sea and land surface temperatures are properties of the Earth surface. To retrieve them, atmospheric modification of the upwelling radiation must be corrected. An extremely important technique used in many of these retrieval schemes is the so-called split-window technique. The split-window technique has its roots in papers by Saunders (1967) and Anding and Kauth (1970). Its best statement, however, is in McMillin and Crosby (1984). Surface observations are made in atmospheric windows.

However, these windows are dirty; that is, the radiance leaving the surface is modified by the atmosphere. For infrared windows, the radiative transfer equation can be written:

$$L_\lambda = B_\lambda(T_s)\tau_o(\lambda) + B_\lambda(T_A)\left[1 - \tau_o(\lambda)\right] \tag{5.30}$$

where $\tau_o(\lambda)$ is the transmittance from the surface to the satellite at wavelength λ, while T_s is the surface brightness (equivalent blackbody) temperature and TA is a mean brightness temperature of the atmosphere given by the following equation:

$$B_\lambda(T_A) = [1/1 - \tau_o(\lambda)] \int_{\tau_0}^{1} B_\lambda(T)\, d\tau \tag{5.31}$$

Some windows, for instance, particularly the 10.5–12.5-μm windows, are wide enough to permit observations in two channels. The split-window technique uses observations at two wavelengths to eliminate the influence of T_A and solve for T_s. If U is the column-integrated water vapor (the precipitable water), then is getting as follows:

$$\tau_\lambda \approx \exp\left[-\beta_\lambda(U/\mu)\right] \tag{5.32}$$

Since the weighting functions peak at the surface, T_s, T_{B1} and T_{B2} will be close to T_A. Thus, expanding the Planck functions about T_A is giving as follows:

$$B_\lambda(T) \approx B_\lambda(T_A) + (dB_\lambda/dT)(T - T_A) \tag{5.33}$$

where the partial derivative is evaluated at relation $T = T_A$. At this point, by approximating the Planck function as locally linear, a slightly less accurate but more popular form of the split-window equation can be derived:

$$T_s = T_{B1} + \eta(T_{B1} - T_{B2}) \tag{5.34}$$

Otherwise, two important points must be made about the split-window equation. First, the split-window technique is a correction technique. The difference between observations at two wavelengths is used to correct one of the observations for atmospheric effects to yield an improved estimate of surface radiance or temperature. Second, the factor η does not depend on the amount of absorber U. Thus, since the transmittance is relatively large in atmospheric windows, Eq. (5.32) can be approximated as follows:

$$\tau_\lambda \approx 1 - \beta_\lambda U/\mu \quad \text{where}: \quad \eta \approx \beta_1/\beta_2 - \beta_1 \qquad (5.35)$$

Otherwise, the most accurate way of calculating value η is by simulating satellite-measured radiances using a variety of model atmospheres with differing absorber amounts. Thus, if AVHRR channels 4 (11 μm) and 5 (12 μm) are used for the first and second wavelengths, respectively, then $\eta \approx 3$. Here can be noted that η is positive because channel 5 is more sensitive to water vapor than is channel 4. However, in practice the split-window technique is rarely used directly. Rather, it provides theoretical justification for regressing the surface temperature of interest against measured or simulated brightness temperatures.

5.3   Winds Flow

To specify the current state of the atmosphere and to predict its future state, one must know the flow field in addition to the mass field. Here will be discussed the ways in which winds (the flow field) can be derived from satellite measurements.

5.3.1   Cloud and Vapour Tracking

The great advantage of GEO satellites is that they can make frequent images of the same area. Since the launch of ATS 1 in 1966, meteorologists have watched the movement of clouds and attempted to use them to estimate winds. The concept is very simple, because the vector difference of the location of a cloud in two successive images divided by the time interval between images is an estimate of the horizontal wind at the level of the cloud. Winds estimated by this method are called cloud-track or cloud-drift winds. Hubert (1979) reviews the techniques used in cloud tracking. The two basic methods for tracking clouds, manual tracking and automatic tracking, are discussed in the next two sections.

1. **Manual Tracking** – In manual tracking, an analyst individually locates the clouds to be tracked on each of two or more consecutive images (Hubert and Whitney 1971). This is usually done by positioning a cursor at the cloud center on a video display device. In such a way, a computer takes the cloud positions and calculates the wind vectors using navigation algorithms. Often three images are used. The wind vector at the center time is the average of the vectors before and after the center time. In fact, if the two vectors are not consistent, the analyst is asked to recheck the positions. Manual tracking has two advantages over automatic tracking. First, people are very good at choosing appropriate clouds to track. Low clouds can even be tracked through thin cirrus overcast. Second, in theory more cloud vectors can be obtained manually than with automatic

methods because individual clouds (or cloud systems) are tracked rather than areas of clouds. However, a disadvantage of manual tracking is that it is extremely tedious, which effectively limits the number of clouds that can be manually tracked.

2. **Automatic Tracking** – Automatic tracking is done primarily using the cross-correlation method1 of Leese et al. (1971) as improved by Smith and Phillips (1972) and Philips et al. (1972). In this technique, tracking of individual cloud features, which is difficult for a computer, is abandoned in favor of a computationally intensive method in which the average motion of an area of clouds is calculated. The automatic cloud tracking system was applied to METEOSAT 6.7 μm water vapor measurements to learn whether the system can track the motions of water vapor patterns. In fact, data for the midlatitudes, subtropics, and tropics were selected from a sequence of Meteosat pictures for 25 April 1978. Trackable features in the water vapor patterns were identified using a clustering technique and the features were tracked by two different methods. In flat (low contrast) water vapor fields, the automatic motion computations were not reliable, but in areas where the water vapor fields contained small scale structure (such as in the vicinity of active weather phenomena) the computations were successful. Cloud motions were computed using Meteosat satellite infrared observations (including tropical convective systems and midlatitude jet stream cirrus). For instance, the SRI International (SRI) automatic tracking system (SATS) computes cloud motion vectors from sequences of observations made by geosynchronous weather satellites. The data used previously in cloud tracking were in the visible and infrared channels as provided by the US GOES constellation.

5.3.2 Winds from Soundings

In general, the flow field is not independent of the mass field. Given temperature soundings and an estimate of the sea-level pressure, thus one can calculate the height of any pressure surface. In the extratopics, winds are related to the height field. Three relationships come immediately to mind. The geostrophic wind is related to the gradient of the geopotential:

$$V_g = 1/f\acute{\kappa} \cdot \Delta\Phi \qquad (5.36)$$

where value Φ is the geopotential (gz), f is the Coriolis parameter $= 2(d\Omega_e/dt) \sin\Theta$, $d\Omega_e/dt =$ angular velocity of rotation of Earth, $\Theta =$ latitude, and $\acute{\kappa}$ is a unit vector in the vertical direction. The gradient wind blows in the same direction as the geostrophic wind, but with magnitude is getting as follows:

$$V_{gr} = 2V_g/1 + \left[1 + 4\left(V_g/fR_T\right)\right]^{1/2} \tag{5.37}$$

where values R_T is the radius of curvature of the trajectory of an air parcel. The gradient wind is a better approximation for curved flow than the geostrophic wind. Finally, the quasi-geostrophic approximation (see Holton 1992) may be used to estimate the wind. The question is how accurate are winds calculated from satellite temperature soundings?

Namely, Hayden in 1984 used surface pressure reports and station elevations to calculate the 1000-hPa height field. He then used VAS soundings and the hypso-metric equation to calculate the heights of pressure surfaces aloft. He estimated winds from the heights using the three approximations mentioned above. For 2 days in 1982, he compared these winds with rawinsonde observations (sample size = 174). The gradient wind estimate was the best of the three. At 300 hPa the comparison between the satellite-estimated gradient wind and the rawinsonde measured wind was good. The bias was less than 1 ms^{-1}, and the rms error was under 10 m s^{-1}. The correlation coefficient was 0.87 and at 850 hPa, the results were not as good. Although the rms errors were lower than at 300 hPa, the correlation coefficient was much lower, essentially zero on one of the days.

These results make physical sense because satellite soundings are least accurate near the Surface. However, the mean temperature of a deep layer of the troposphere is accurately retrieved from radiance data. Therefore, the winds aloft ought to be accurately estimated by this technique, assuming that the gradient wind equation is capable of approximating the actual wind. Hayden's results are supported by those of Lord et al. (1984), who compared winds derived from satellite soundings with aircraft observations. It is interesting to note that winds derived from VAS data complement cloud track winds because VAS winds can only be calculated in clear areas.

5.3.3   Ocean Surface Winds

Unique properties of the ocean surface or of oceanic storms have allowed the development of several techniques to measure surface winds over the ocean.

1. **Active Microwave Winds** – Since the invention of radar, there has been an interest in the radar return from the sea surface (sea clutter). Thus, an extensive series of observations using onboard aircraft mounted radars during the 1960s indicated that there is a relationship between surface wind and radar return.

   NASA then began development of specialized radar known as a scatterometer, which very accurately measures the backscattering cross section per unit sea surface area. This is a unitless parameter called the normalized radar cross section, $\sigma°$. Scatterometers were flown on aircraft and on Skylab, first US space station launched in May 1973. The results were sufficiently good that a scatterometer was planned for Seasat. Namely, Seasat's mission was to demonstrate the feasibility of global ocean monitoring from space. Its instrument complement consisted of an VIRR); one passive microwave instrument, the Scanning Multichannel Microwave Radiometer (SMMR), identical to the one flown on Nimbus-7); and three active microwave instruments (radars): an Altimeter (ALT), Synthetic Aperture Radar (SAR) and a scatterometer, the Seasat-A Satellite Scatterometer (SASS). Since 2000 modern Earth observation and weather forecasting satellites have onboard scatterometer such as: NSCAT (NASA Scatterometer) instrument on ADEOSI (1996–97), SeaWinds instrument on QuikSCAT (2001–2009), OSCAT-2 instrument on SCATSAT-1 (launched 2016), SCAT instrument on Oceansat-2 (2009–2014), ISS-RapidSact on the International Space Station (2014–2016) and ASCAT on MetOp satellites.

2. **Passive Microwave Winds** – Microwave radiation emitted from the sea surface is a function of wind speed for two reasons. First, the same wind-generated waves that alter the radar backscatter coefficient also change the surface emittance. According to experimental emittance calculations at 19.4 GHz of a wind-roughened ocean surface as a function of incidence (viewing) angle. The emittance change is strongly dependent on polarization and viewing angle. For nadir viewing, wind-induced roughness has little effect on emittance. However, for nonnadir viewing, roughness increases emittance for horizontally polarized radiation; roughness decreases emittance for vertically polarized radiation out to 55° and increases it thereafter. Here has to be noted that the angle between the viewing azimuth and the wind direction affects the emittance slightly. The second reason that sea surface emittance is a function of wind speed is that at wind speeds above about 7 ms^{-1} foam begins to form on the sea surface. The fraction of the surface covered by foam increases with wind speed. Since foam is essentially black, emittance increases rapidly, and nearly linearly, with wind speed above 7 ms^{-1}. Wilheit (1979) has proposed a sea surface emittance model, which includes the effects of foam and roughness.

The change of sea surface emittance with wind speed is the basis for wind speed retrieval algorithms using SMMR data. Wilheit and Chang (1980) developed such algorithms by regressing simulated brightness temperatures or simple functions of brightness temperature against wind speed. Moreover, Wilheit et al. (1984) tested and improved these algorithms using SMMR instrument observations. Several problems were encountered with the real data, but the best SMMR-estimated winds were within about 2.7 ms^{-1} rms of winds measured by NOAA environmental data buoys or ships. The Special Sensor Microwave Imager (SSM/I) instrument on the US Defense Meteorological Satellite Program (DMSP) can also estimate surface wind speed over the ocean. Thus, since surface emittance is only weakly dependent on wind direction, passive MW techniques yield only wind speed.

3. **Tropical Cyclone Winds** – The strongest and most dangerous winds over the ocean are the result of tropical cyclones. One of the chief contributions of meteorological satellites is simply the detection of these storms. Since the

1960s, no tropical cyclone anywhere on Earth has gone undetected, while shipping interests are pleased to know about the location of tropical cyclones and also to know the wind speed to keep their ships at a safe distance. Not all tropical cyclones are equally dangerous in terms of wind speed and the ocean waves that result, but satellite data is necessary to estimate hurricane winds.

The most important technique for estimating winds in tropical cyclones was developed by Dvorak (1973, 1975, 1984). Each tropical cyclone goes through a life cycle, which may be classified into one of several types by its appearance in visible or infrared satellite images. In addition to typing the storm, its current strength can be determined from the satellite images. The Dvorak technique is a method using enhanced IR and/or VIS satellite imagery to quantitatively estimate the intensity of a tropical system. Thus, cloud patterns in satellite imagery normally show an indication of cyclogenesis before the storm reaches tropical storm intensity. Indications of continued development and/or weakening can also be found in the cloud features. Using these features, the pattern formed by the clouds of a tropical cyclone, expected systematic development, and a series of rules, an intensity analysis and forecast can be made. This information is then standardized into an intensity code.

The next technique for estimating winds in tropical cyclones is microwave (MW) sounding. The MW sounders are nearly unaffected by clouds, which can probe the interior of tropical cyclones. Thus, the surface pressure drop in the center of a tropical cyclone is the result of warming in the storm's core. This warming is maximum in the upper troposphere but below the tops of the clouds. It is therefore not detectable with infrared sounders. Rosenkranz et al. (1978) first noticed that the 55.45-GHz channel of the Nimbus-6 Scanning Microwave Spectrometer (SCAMS) showed a warm anomaly (difference between temperatures inside and outside the storm) above a strong western Pacific typhoon. This warm anomaly is explained by the fact that the 55.45 GHz weighting function peaks in the upper troposphere in the region of the maximum tropical cyclone temperature anomaly. On the other hand, Kidder et al. (1978) found that the 55.45-GHz brightness temperature anomaly is correlated with tropical cyclone central pressure, and Kidder et al. (1980) succeeded in relating the radial gradient of 55.45-GHz brightness temperature to surface pressure gradient and then to surface wind speed at outer radii, specifically to the radii of 15.4-ms^{-1} (30-knot) and 25.7-ms^{-1} (50-knot) winds, which are important for ship routing.

Velden and Smith (1983) extended this technique in two ways. First, they used data from the Microwave Sounding Unit (MSU) on board the NOAA satellites. The MSU has better spatial resolution than did the SCAMS, and it measures the brightness temperatures more accurately. Second, they retrieved atmospheric temperature profiles inside the tropical cyclones. The retrieved temperature anomaly at 250 hPa is correlated with both the central pressure (minimum sea-level pressure) and with intensity (maximum sustained sea-level wind speed). They found rms errors of about 6 hPa in central pressure estimates and about 6 ms^{-1} in intensity estimates. This technique has been used semi-operationally at the US National Hurricane Center.

5.3.4 Doppler Wind Measurements

The Upper Atmosphere Research Satellite (UARS) was a NASA-operated satellite orbital observatory whose mission was to study the Earth's atmosphere, particularly the protective ozone layer. ... It entered Earth orbit at an operational altitude of 600 km (370 NM), with an orbital inclination of 57°. Thus, the UARS platform has two instruments devoted to measuring horizontal winds, which are important because the dynamics and chemistry of the upper atmosphere are inextricably linked. The instruments are the High Resolution Doppler Imager (HRDI) and the Wind Imaging Interferometer (WINDII). Both view the limb of the atmosphere and measure the Doppler shift in emission and absorption lines in the visible and near-infrared portion of the spectrum.

The chemical species detected are molecular oxygen, atomic oxygen, and the OH molecule. The Doppler shift determines the relative velocity between the satellite and the volume of air being sampled. Correcting for the motion of the satellite and the rotation of the Earth yields one component of the horizontal wind. By observing the same location both as the satellite approaches and as it departs from an area, one can estimate the horizontal wind vector. The instruments are able to scan vertically to give a vertical profile of the wind with about 4-km vertical resolution. It is expected that winds will be measured to accuracies of between 5 and 15 ms^{-1}, depending on altitude. In such a way, since UARS was launched as this book was being written, the reader is referred to the literature for results of these wind measurements.

5.4 Clouds and Aerosols

Clouds cover roughly half the Earth, and aerosols are always present. They constitute major factors in determining the radiation budget of Earth and they play crucial, if as yet not completely understood, roles in modulating the climate. They also contaminate radiometric observations necessary for sounding the atmosphere and remotely sensing different surface parameters. Spacebased instruments are the only means by which the global distribution of clouds and aerosols can be adequately sampled.

5.4.1 Clouds from Sounders

Many cloud properties are of interest to the meteorologist and researchers. The ones most often retrieved are cloud-top temperature, cloud-top height (or pressure), and cloud amount (the fractional area covered by cloud). Useful also are reflectance, emittance, optical depth, phase (ice or water), liquid water content and drop size distribution. Attempts to retrieve cloud properties can be diviced into schemes that use sounding instruments, schemes that use visible and infrared imaging instruments, and schemes that use microwave instruments.

Most efforts to detect clouds using sounder data have been directed toward removing cloud contamination so that accurate, clear-column radiances can be calculated. In addition, it is also desirable, however, to retrieve the properties of the clouds. If we assume that clouds reflect little radiation at the wavelengths used by sounders, then the radiation sensed by a radiometer can be written as the sum of three terms:

$$L(\lambda) = (1 - N)\, L_{clr}(\lambda) + N\varepsilon L_{cld}(\lambda) + N(1 - \varepsilon)\, L_{clr}(\lambda) \qquad (5.38)$$

where $L(\lambda)$ is the measured radiance of a scene, $L_{clr}(\lambda)$ is the radiance from the scene if clouds were not present (the clear-column radiance), $L_{cld}(\lambda)$ is the radiance that would come from the cloud if it were black, value N is the cloud amount, ε is the cloud emittance, and $1 - \varepsilon$ is the cloud transmittance. The three terms represent, respectively, the radiation from the clear area, the radiation emitted by the cloud and the radiation transmitted through the cloud. Combining the first and third terms results as follows:

$$L(\lambda) = \left(1 - N'\right) L_{clr}(\lambda) + N' L_{cld}(\lambda) \qquad (5.39)$$

where $N' \equiv N\varepsilon$ is called the effective cloud amount. The above equation is the basis of efforts to determine clear-column radiances for sounding retrievals and to determine the cloud properties N' and p_c, the pressure of the cloud top.

5.4.2 Clouds from Imagers

Most of the work on cloud estimation has been done using window channels on imaging instruments. Thus, these instruments have the advantage of much higher spatial resolution than sounders. Perhaps the most often used technique for cloud characterization is simple manual inspection of satellite photographs or video images. These analyses are used daily for weather analysis and forecasting and in the construction of nepbanalyses, which are maps showing chiefly cloud type and amount at various atmospheric levels (Fye 1978). In this segment will be shortly introduced the following techniques:

1. **Threshold Techniques** – Thresholding technique is one of the powerful methods for satellite image segmentation, which is useful in discriminating objects from the background in many classes of scenes. The basic technique consists in thresholding the image of a spectral band in the short wavelength range (preferably a blue band). Thus, pixels whose reflectance is above the threshold are declared cloudy. This method is not very subtle and often does not detect thin clouds, it also makes many false detections. It is also possible to

check that the cloud is white, but the contribution of this verification is not really effective, because thin clouds are not perfectly white, while many bright pixels are white, in cities for example. In theory, the oldest (Arking 1964; Koffler et al, 1973), simplest, and still most frequently used technique for objectively extracting cloud information from digital satellite images is the threshold technique. Here a visible brightness or infrared temperature threshold is set such that if a pixel is brighter or colder than the threshold, the pixel is assumed to be cloud covered. The fractional area covered with cloud is simply the ratio of the number of cloudy pixels to the total number of pixels. The cloud height can be determined by comparing the infrared temperature of the pixel with a sounding. The simplicity of this technique makes it attractive. Thresholding method is one of the means of filtering clouds from data to be used in Sea Surface Temperature (SST) retrievals (McClain et al. 1985). In practice, two complications arise: (1) the problem of clouds smaller than the satellite scan spot and (2) the problem of how to set the threshold.

2. **Histogram Techniques** – These techniques serve as alternates to threshold techniques, which histogram of the pixels in an area will show clusters of pixels that represent cloud or surface types. The histograms have as many dimensions as the number of channels of data, which attempt to locate clusters of pixels as opposed to drawing vertical or horizontal lines. There are several methods for identifying clusters. The dynamic cluster method (Desbois et al. 1982; cf. Simmer et al. 1982; Phulpin et al. 1983) randomly chooses a maximum number of points in the histogram (roughly 15) to serve as cluster centers. Each point in the histogram is assigned to the closest cluster. Since randomly selected clusters are likely to be misplaced, the next step results in a correction. The center of gravity and standard deviations are calculated for each cluster. The center of gravity serves as the new cluster center. Points are again assigned to the clusters, this time using the standard deviations as distance measures. Any cluster that receives too few points is eliminated and its points are distributed among the remaining clusters. The correction step is repeated until the cluster centers and their standard deviations do not change. After cluster centers are identified in the histogram, they must be classified. The clusters can be identified manually or by setting thresholds or by comparing with previous classifications. Once surface clusters have been identified, cloud amount can be determined by dividing the number of pixels that belong to cloudy clusters by the total number of pixels. Cloud temperature and brightness can be estimated from the average temperature and brightness of each cloudy cluster.

3. **Pattern Recognition Techniques** – For some applications, it is important to know the type of cloud or surface. Pattern recognition techniques attempt to classify arrays of pixels in a manner analogous to the way a person might perform such a classification. Consider the following simple hypothetical example. Suppose that an analyst, armed with a visible satellite image of an oceanic area, were charged with the task of determining whether an observer on a ship would report clear, partly cloudy or cloudy skies. The analyst would examine a small area of the image centered on the location of the ship. So, to be consistent,

the size of the area should be approximately the same as that which would be seen by the surface observer. In such way, if the area in the image were uniform and bright, the analyst would classify it as cloudy. Then, if it were uniform and dark, the analyst would classify it as clear. Finally, if a portion of it were bright and a portion dark, the analyst would classify it as partly cloudy.

A simple pattern recognition algorithm might simulate this process as follows. Suppose that the mean and standard deviation of the pixels in the area were calculated. Uniform scenes, either clear or cloudy, would have a low standard deviation. Thus, partly cloudy scenes must have a high standard deviation. Cloudy scenes would have high mean brightness, whereas clear (ocean) scenes would have low mean brightness. Partly cloudy scenes would have intermediate values of mean brightness. The algorithm could first check the standard deviation. If it were sufficiently high, the scene would be classified as partly cloudy. For low standard deviation scenes, the mean brightness would discriminate between clear and cloudy cases. Note that in pattern recognition measures of the spatial variability of pixels in a scene, such as standard deviation, are referred to as texture parameters. However, spectral parameter is the term sometimes used to describe nontexture parameters such as mean or extreme radiances or differences of these quantities between channels.

4. **Multispectral Techniques** – In this category we group cloud retrieval techniques that rely on radiance measurements at two or more wavelengths and use simple models to make retrievals. Thus, radiative transfer techniques are similar, but they use the results of more complex radiative transfer calculations to make retrievals.

The best-known multispectral technique is the bispectral technique of Reynolds and Vonder Haar (1977), in which visible and 11-ms^{-1} infrared radiance observations are used to retrieve cloud amount and cloud-top temperature. Each scan spot is assumed to be partly covered with a single cloud layer. The radiance observed in the visible channel is modeled as solar radiation reflected from the cloud tops and from the surface:

$$L_{VIS} = N\gamma_{cld}E_{sun} + (1 - N)\,\gamma_{clr}E_{sun} \qquad (5.40)$$

where N is the cloud amount; γ_{cld} and γ_{clr} are the bidirectional reflectances of the clouds and the surface, respectively, which are assumed known. Then, value E_{sun} is the solar irradiance (integrated over the radiometer spectral response), which is assumed to be known and to be the same at the surface and at cloud top. The radiance observed in the infrared channel is similarly modeled as being emitted by the surface (through the holes in the cloud), emitted by the cloud, and transmitted through the cloud after being emitted by the surface with the following equation:

$$L_{IR} = (1 - N)\,L_{clr} + N\varepsilon L_{cld} + N(1 - \varepsilon)\,L_{clr} \qquad (5.41)$$

where ε is the cloud emittance (assumed known), L_{cld} is the radiance the cloud would have if it were black, and L_{clr} is the radiance of the clear area (also assumed known).

5. **Spatial Coherence** – Spatial coherence is a very precise method for determining cloud amount and cloud-top temperature done by Coakley and Bretherton (1982). The technique starts with several assumptions that preclude it from being applicable in all situations but that result in precise estimates in those cases to which it does apply. Furthermore, in part the precision results from the fact that the technique is capable of detecting cases to which it is not applicable. In fact, clouds are assumed to exist in layers with uniform cloud top and emittance (the original technique only dealt with a single layer of cloud) and to exist over a uniform background. Some portions of the image must be completely clear and portions completely cloudy. The technique is well suited to analyzing marine stratocumulus, but is applicable in other situations as well. It is not applicable, for example, near fronts or when cirrus overlies the scene. The spatial coherence technique rests on the idea that portions of the image that are either completely clear or completely cloudy will exhibit little spatial variability, whereas partly cloudy areas will exhibit large spatial variability. Thus, infrared data are analyzed over a moderately large sector (on the order of 250 km^2). The image is subdivided into blocks of at least 2 x 2 pixels.

6. **Radiative Transfer Techniques** – All cloud retrieval techniques can be broken into two steps: (1) cloud detection, and (2) parameter retrieval. Thus, radiative transfer techniques come into play in the second step, after a cloud is determined to exist, however radiative transfer calculations are used to determine its properties. The advantage of using radiative transfer calculations is that they allow the retrieval of parameters, such as cloud optical depth and microphysical properties, that is not retrievable with other methods.

Arking and Childs (1985) have used this technique with AVHRR data to retrieve visible optical depth, cloud-top temperature, cloud amount and an indication of the cloud phase and drop size. They note that given the cloud amount, visible radiances are affected primarily by the cloud optical depth, 1 1-μm radiances are affected primarily by the cloud-top temperature, and that 3.7-μm radiances (during daylight) are affected primarily by the cloud microphysical parameters (chiefly drop size and drop phase). They chose six sets (models) of microphysical conditions to approximate the range of clouds expected. All of the cloud drops were assumed to be spherical. Three models had ice spheres, and three had liquid water spheres. The ice models had modal drop radii of 4, 8, and 32 μm; the water models had modal drop radii of 4, 8 and 16 μm. Radiative transfer calculations determined the reflectance, transmittance, and absorptance of plane-parallel cloud layers as tabulated functions of optical depth, zenith angle, and azimuth angle at the three wavelengths.

7. **Geometric Techniques** – A quite different approach to cloud height retrieval involves making geometric measurements with satellite images. A technique that is restricted to low Sun angles uses observations of cloud shadows on the Earth's surface (Fujita 1963; Smith and Reynolds 1976). Knowing the positions of the Sun and the satellite, one can use the distance from a cloud to its shadow to calculate the height of the edge of the cloud. A more useful technique is stereoscopy or simply stereo, which involves simultaneously observing a cloud

with two satellites (Minzner 1976; Hasler 1981/1983). Just as two eyes allow depth perception, views by two satellites allow determination of the height of clouds. Suppose, for example, that GOES-East and GOES-West view the same cloud. The magnitude of the displacement between the locations of the cloud in the two images can be related through spherical trigonometry to the height of the cloud. Plate 6 shows a perspective view of a thunderstorm cloud top with height contours derived from stereoscopic analysis. Hasler (1981) estimates that with GOES data, stereo heights are accurate to within ± 0.5 km. Stereo images can also be formed from one GEO image and one low Earth orbiter image.

5.4.3 Clouds from Microwave Radiometry

In the MW portion of the spectrum, the liquid water content of clouds can be measured. This quantity is not measurable in other portions of the spectrum. Thus, both water vapor and cloud droplets weakly absorb microwave radiation and that scattering of microwave radiation is negligible for non-precipitating drops. For centimeter wavelengths, water vapor absorption peaks at 22.235 GHz, so absorption by water droplets increases with frequency. That is, the absorption coefficient for each is proportional to the amount of water (liquid or vapor) per unit volume. Moreover, over the oceans surface, which provides a uniform, cold background due to the low emittance of water, atmospheric water vapor and liquid water increase the brightness temperature measured by a satellite-borne radiometer. By exploiting two frequencies, one near 22.235 GHz and another in the atmospheric window near 31.000 GHz, column-integrated water vapor and liquid water content (both in units of mass per unit area) can be retrieved as follows. So, applying the integrated form of Schwarzchild's equation (5.11), the brightness temperature measured by a microwave radiometer is:

$$T_B(\lambda, \mu) = \varepsilon_0(\mu)\, T_0 \tau_0^{1/\mu} + [1 - \varepsilon_0(\mu)]\, T_{space} \tau_0^{2/\mu} + \int_{\tau_0}^{1} TW_\lambda(\tau, \mu)\, d\tau \quad (5.42)$$

where ε_0 is the surface emittance (strongly dependent on polarization and μ, and weakly dependent on T_0), T_0 is the surface temperature, and τ_0 is the vertical transmittance from the surface to the satellite.

Since the transmittance τ_0 is close to one, an adequate approximation is $T = T_0$. Ignoring the small second term and removing T_0 from the integral is giving as follows:

$$T_B \approx T_0 \left\{ 1 - [1 - \varepsilon_0(\mu)]\tau_0^{2/\mu} \right\} \quad (5.43)$$

Information on surface properties is contained in T_0 and ε_0 and atmospheric information is contained in value τ_0. If values U_L and U_V are the column-integrated liquid water content and water vapor content (U_V is also called the precipitable water), while values β_L and β_V are the mass absorption coefficients for cloud droplets and water vapor, respectively, then it is producing the following equation:

$$\tau_o^{2/\mu} \approx \exp\left[-2/\mu\left(\beta_L U_L + \beta_V U_V + \delta o_2\right)\right] \approx (T_o - T_B)/\left[1 - \varepsilon_o(\mu)\right] T_o \quad (5.44)$$

where δo_2 is the small oxygen (O_2) optical depth. Taking the logarithm is giving as follows:

$$\beta_L U_L + \beta_V U_V = -\mu/2 \, \ln \, (T_o - T_B)/\left[1 - \varepsilon_o(\mu) \, T_o\right] - \delta o_2 \quad (5.45)$$

The U_L and U_V coefficients can be found by using a more accurate radiative model or (for U_V, at least) by comparing with observations. However, it must be noted that independent measurements of liquid water with which to compare satellite estimates are essentially nonexistent; thus the accuracy of the satellite estimates is not known.

The Soviet satellite Kosmos 243 carried the first MW radiometer capable of making such measurements (Gurvich and Demin 1970). The US satellites Nimbus-5/6/7, Seasat, and two DMSP satellites have carried instruments from which liquid water and precipitable water could be retrieved. The Indian satellites Bhaskara 1/2 also carried MW sensors.

5.4.4   Stratospheric Aerosols

Aerosol measurements from satellites rely chiefly on the scattering of sunlight, although a few infrared techniques have been explored. Because somewhat different techniques are involved, here will be discussed stratospheric aerosols and tropospheric aerosols separately.

Limb scanning techniques (McCormick and Lovill 1983), with their large optical path, are well suited to the measurement of stratospheric aerosols, which may have number densities of only $1-10 \text{ cm}^{-3}$. Tropospheric aerosols are perhaps 1000 times more abundant. The most frequently used technique is solar occultation, in which the transmittance of solar radiation is measured during spacecraft sunrise or sunset. A great advantage of solar occultation is that it is nearly self-calibrating; unattenuated solar radiation is measured before each sunset or after each sunrise. Only the ratio of the attenuated to the unattenuated radiation (the transmittance) is used in the retrieval process. Variations in solar output and in instrument gain are eliminated.

In past, McCormick (1979) and McCormick (1983) review solar occultation experiments. The first such experiment was performed aboard Apollo-Soyuz in 1975. The Stratospheric Aerosol Measurement (SAM) experiment consisted of a hand-held, namely, one-channel Sun photometer with which astronauts made measurements during two sunrises and two sunsets, which measurements were centered at 0.83 μm. Thus, the provided results were sufficiently promising to prompt the development of two complementary instruments to fly on unmanned satellites.

The Stratospheric Aerosol Measurement II (SAM II) experiment was launched aboard the satellite Nimbus-7 satellite in October 1978. This instrument also was a one-channel Sun photometer, but its spectral range was centered at 55°, where absorption due to atmospheric gases is negligible. Then, the Stratospheric Aerosol and Gas Experiment I (SAGE I) was launched aboard a dedicated Applications Explorer Mission satellite (AEM 2) in February 1979. However, its orbit had an inclination of 55°, which caused its orbital plane to process with respect to the Sun. It made sunrise and sunset measurements between 79°N and 79°S (depending on season), and it sampled all longitudes in about a month. Thus the SAGE I coverage complemented that of SAM II. About 15 sunrises and 15 sunsets were observed per day for the first 4 months of operation until a problem with the power supply limited observations to sunsets only. Observations were made until November 1981. SAGE I made measurements in four bands centered.

The next instrument SAGE II was launched in October 1984 on the Earth Radiation Budget Satellite (ERBS). Like AEM 2, ERBS is in a nonsunsynchronous orbit. SAGE II is similar to SAGE I except that it has sever channels centered at 1.02, 0.936, 0.600, 0.525, 0.452, 0.448, and 0.385 μm. The 0.936-μm channel provides the ability to retrieve stratospheric water vapor. The data collected on SAGE I and the following instrument SAGE II began taking measurements in October 1984. Both were critical to the discovery of the Earth's ozone holes and the creation of 1987 Montreal Protocol, which banned ozone-depleting substances, such as Chloro-Flouro-Carbob (CFC).

The SAGE III mission is an important instrument designed by NASA in 2005 to fulfill the primary scientific objective of obtaining high quality, global measurements of key components of atmospheric composition and their long-term variability. Thus, it is a nearly exact replica of SAGE III Meteor-3M, sent into orbit in 2001, The primary focus of SAGE III on the Russian International Space Station (ISS) will be to study aerosols, clouds, water vapor, pressure and temperature, nitrogen dioxide, nitrogen trioxide, and chlorine dioxide.

5.4.5 Tropospheric Aerosols

Studies of tropospheric aerosols also rely on observations of scattered radiation. Instead of the dark or uniform background of space, thus, these techniques must deal with the Earth's surface as background. Often, observations are made over the ocean. Two basic types of aerosols have been studied using satellite data: marine aerosols (haze) and Saharan dust. Griggs (1975) compared surface-based measurements of marine aerosol optical depth (using a Volz sun photometer) with Landsat observations at 0.55, 0.65, and 0.75 μm and found that optical depth could be retrieved within ±10%. Griggs (1979) found similar results with GOES-1 and

NOAA-5 Scanning Radiometer data. Fett and Isaacs (1979) and Isaacs (1980) studied the effects of marine aerosols on DMSP visible imagery. Hindman et al. (1984) and Durkee et al. (1986) have used AVHRR data and Nimbus-7 Coastal Zone Color Scanner data to study marine aerosols off the California coast. In fact, these last three studies used the marine aerosol models of Shettle and Fenn (1979). Durkee et al. (1991) have used AVHRR channels 1 and 2 to retrieve aerosol optical depth. The combination of two channels lessens the uncertainty caused by assuming a single scatter Albedo and a scattering phase function.

Satellite data were first used to study Saharan dust by Fraser (1976). He used a radiative transfer model to convert Landsat radiances to vertically integrated aerosol mass. Carlson and Wendling (1977) and Carlson (1979) were able to map dust optical depth over the Atlantic using NOAA 3 VHRR data. Norton et al. (1980) studied Saharan dust using GOES data. They were able to estimate the dust optical depth to ± 0.1 accuracy for optical depths up to 0.4. Recently Deuze et al. (1988) studied Saharan aerosols over the Mediterranean Sea using Meteosat and AVHRR data. In January 1989, National Environmental Satellite, Data, and Information Service (NESDIS) began producing weekly maps of aerosol optical depth based on AVHRR channel 1 data (Rao et al. 1988). The maps have a 100-km grid and include all the world's oceans. The retrieval scheme relies on a lookup table calculated with a radiative transfer model. Besides, the model includes Rayleigh scattering and ozone absorption as well as aerosol scattering. The table is calculated as a function of solar zenith angle, satellite zenith angle, and the relative azimuth angle. Given these three parameters, aerosol optical depth can be interpolated on the basis of the observed radiance. To avoid Sun glint, retrievals are made only in the antisolar half of each scan line. Moreover, aerosol measurements over land areas are also possible if one knows the surface reflectance. Fräser et al. (1984) have used GOES visible data to estimate summertime aerosol optical depth over the eastern US part. They then related the optical depth to the column-integrated mass density of sulfate particles. Measurements outside the visible to near infrared portion of the spectrum also appear to be useful.

The new generation of GOES-R successfully launched aboard an Atlas V541 rocket on 19 November 2016, which became GOES-16, The Advanced Baseline Imager (ABI) onboard GOES-R can view Earth in 16 different spectral bands. It imager has a huge advance over the current system, which provides three times more differentiated spectral information, four times better spatial resolution and more than five times faster coverage of the same area. Forecasters can use the higher resolution images to track the development of storms in early stages. Six of the ABI bands are similar to the GOES imager. Starting with the shorter wavelengths, these bands are useful for the detecting daytime clouds, fog, insolation and winds, fog and low clouds at night, fire and hot spots, volcanic eruption and ash, daytime snow and ice, monitoring water vapor in the mid-troposphere and so on.

5.5 Precipitation Measuring

Precipitation is defined as the liquid or solid products of the condensation of water vapour falling from clouds or deposited from air onto the ground. Participation includes rain, hail, snow, dew, rime, hoar frost and fog precipitation. The total amount of precipitation, which reaches the ground in a stated period is expressed in terms of the vertical depth of water (or water equivalent in the case of solid forms) to which it would cover a horizontal projection of the Earth's surface. Snowfall is also expressed by the depth of fresh, newly fallen snow covering an even horizontal surface.

It is important to conclude that latent heat release from tropical convections is the driving force of the global atmospheric circulation. Observing the magnitude, spatial distribution and temporal variability of clouds and precipitation in the tropics has long since been one of the primary goals of atmospheric research. In fact, since vast of the tropical region is covered by ocean, surface-based rain gauge measurements of rainfall can merely capture a small fraction of the whole picture of the tropical precipitation. In Fig. 5.7 is shown data from rain gauge stations, satellites and sounding observations have been merged to estimate monthly rainfall on an 2.5° global grid from 1979 to 2010. Thus, the careful combination of satellite-based rainfall estimates provides the most complete analysis of rainfall available to date over the global oceans, and adds necessary spatial detail to the rainfall analyses over land. In addition to the combination of these data sets, estimates of the uncertainties in the rainfall analysis are provided as a part of the GPCP products. In August 2012 GPCP v2.2 started to use upgraded emission and scattering algorithms, the GPCC precipitation gauge analysis, and inclusion of the DMSP F17 SSM/I satellite instrument. The monthly GPCP has become the standard against

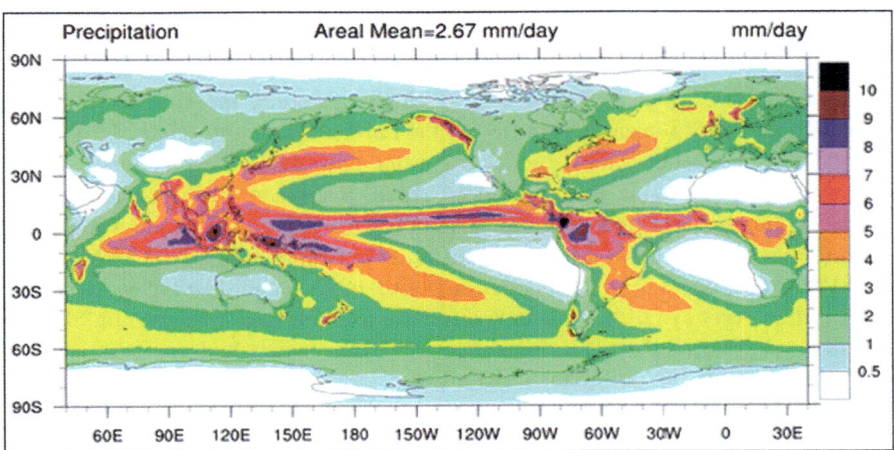

Fig. 5.7 Global Precipitation Climatology Project (GPCP) (Courtesy of Manual: by NCL)

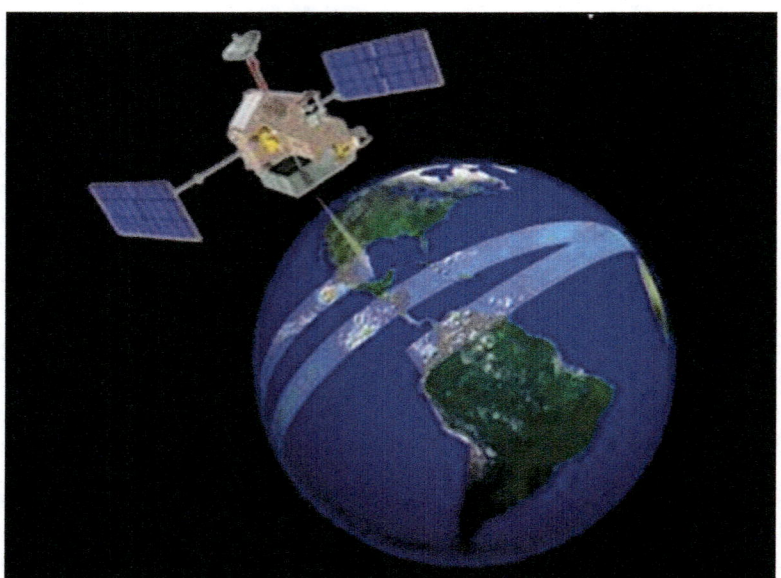

Fig. 5.8 Global Precipitation Climatology Project (Courtesy of Manual: by UCAR)

which climate model output is evaluated, which is available in netCDF (NCAR Command Language) or binary format.

The rainfall measurement in the tropics is Tropical Rainfall Measuring Mission (TRMM), which will naturally have to rely on satellite observations. In 1997 onboard TRMM satellite have been installed sensors for tropical rainfall measurements, such as precipitation radar and a multi-frequency microwave radiometer (Kummerow et al. 2000). In the following subsections, many measured results are primarily derived from these two sensors.

The satellite TRMM Microwave Imager (TMI) is a passive microwave sensor designed to provide quantitative rainfall information over a wide swatch under the TRMM satellite, which is shown in Fig. 5.8. By carefully measuring the minute amounts of microwave energy emitted by the Earth and its atmosphere, TMI is able to quantify the water vapor, the cloud water and the rainfall intensity in the atmosphere. Namely, it is a relatively small instrument that consumes little power, combined with the wide swath and the quantitative information regarding rainfall make TMI the "workhorse" of the rain-measuring package on TRMM. The TRMM project was a joint space mission between NASA and the Japan Aerospace Exploration Agency (JAXA), designed to monitor and study tropical rainfall. The satellite was launched on 27 November 1997 and deactivated on 9 April 2015, carrying onboard the following instruments: Precipitation Radar (PR), TRMM Microwave Imager (TMI), Visible and Infrared Scanner (VIRS), Clouds and the Earth's Radiant Energy Sensor (CERES) and Lightning Imaging Sensor (LIS). Many results are primarily derived from listed sensors such as: Rainfall

Climatology, Horizontal and Vertical Variability of Rain Field, Diurnal Variations and Latent Heating, (see more details in Satellite Meteorology by author G. Liu of the Department of Meteorology, Florida State University, Florida, USA).

The new Global Precipitation Measurement (GPM) is a joint mission between JAXA and NASA as well as other international space agencies to make frequent (every 2–3 h) observations of Earth's precipitation. Thus, it is part of NASA's Earth Systematic Missions program and works with a satellite constellation to provide full global coverage for global precipitation maps to assist researchers in improving the forecasting of extreme events, studying global climate and adding to current capabilities for using such satellite data to benefit society. Of course, GPM builds on the notable successes of the above introduced TRMM system. The project is managed by NASA consists of a GPM Core Observatory satellite assisted by a constellation of spacecraft from other agencies and missions, which measures the two and three-dimensional structure of Earth's precipitation patterns and provides a new calibration standard for the rest of the satellite constellation. The GPM Core Observatory was assembled, tested and launched from Tanegashima Space Centre, Japan, on 28 February 2014. Many space agencies in the US, Japan, India and France, together with Eumetsat operate the remaining satellites in the constellation for agency-specific goals, but also cooperatively provide data for GPM.

5.5.1 *Passive Visible and Infrared Techniques*

Classification of remote sensing techniques is to measure precipitation is on passive remote sensing, such as (1) Visible and infrared techniques and (2) Microwave techniques, and Active remote sensing, such as: Radar, e.g., Tropical Rainfall Measuring Mission (TRMM) radar. Visible (VIS) and infrared (IR) remote sensing techniques to measure precipitation can be used for different meteorological observations. Clouds are not transparent in the IR and visible, such as rain drops can not be sensed directly, thus the approach is to relate independent measurements of rainfall to the properties of a cloud measured by IR and visible remote sensing (called indirect measurements of precipitation).

Many meteorologists worldwide investigated a novel method of fusing both VIS and IR images with the major objective of obtaining higher-resolution IR images, which main characteristics are started in sector 5.1.3. Most existing image fusion methods focus only on visual performance and many fail to consider the thermal physical properties of the IR images, leading to spectral distortion in the fused image. The GEO meteorological satellites often collect data for both IS and VIS satellite channels. Therefore, the IR data of GEO meteorological satellites are of great importance in research and practical applications. They reflect the distribution of temperatures on the Earth's surface, and are used widely in weather forecasting, numerical weather prediction, and climate modeling. However, the infrared spatial resolution is relatively low. By contrast, the VIS data have considerably higher resolution, but do not reflect the thermal dynamics of the Earth and atmosphere. The fusion of IR and VIS data into one coherent image provides a method of improving the infrared spatial resolution.

For instance, the GOES visible wavelength channel produces images, which can be thought of as black-and-white photographs of the earth and clouds from outer space. During the daylight hours, it is the most widely used channel because it has the highest resolution of the five imaging channels, and because it approximates what we see with the human eye. In fact, the primary utility of the VIS satellite channel imagery is in the daytime monitoring of thunderstorms and tropical cyclones, which applications is presented by VIS Channel No 1, with Central Wavelength of 0.65 μm and with Sample sub-point (E/W × N/S) of 0.57 × 1.00 km. In fact, some other interesting features that can be distinguished with the GOES visible imagery can include images such as: Wet Ground, Blowing Dust, Snow and Ice, Fog and Volcanic Ash. These images can be found on Internet via Google and may be individually clicked upon with the mouse in order to bring up a larger image that shows the different features of GOES VIS images.

The IR Channel 2 is different from the other imaging channels, in that it responds to both emitted terrestrial radiation and reflected solar radiation, which Central Wavelength of 3.9 μm and with Sample sub-point (E/W × N/S) of 2.30 × 4.00 km. Thus, since the emissivity of water droplets at 3.9 μm is less than that for longer wavelengths, so it is often easier to identify fog and stratiform cloudiness in the Channel No 2 imagery, and to discriminate between water and ice clouds. Many times fog can be identified on Channel 2 imagery as cooler regions, though confusion can occur between stratus or fog, and cold ground. Thus, combining this imagery with other channels resolves most of these problems including such as: Coastal Fog and Areas of FL Wildfires. The 3.9 μm channel is also very sensitive to sub-pixel hot spots. Therefore, in cloud-free areas it can be used alone, or in combination with other channels, to identify fires when the fires are of large enough size or of great enough intensity.

5.5.2 Passive Microwave Techniques

Microwave remote sensing technique has always been recognized as a powerful tool for meteorological and oceanographic applications, because of its ability to measure water vapour and liquid water even in the presence of most clouds. However, it did not come into popular use due the poor ground resolution of microwave images and because of the fact that land surface emissivity in the microwave region was high and variable. Moreover, microwave sensors have to be flown on low earth orbiting satellites and it has not been possible so far to place them on geostationary platforms because of the weak microwave signal strength. The principles of satellite-borne microwave radiometry have been reviewed by Pandey (1995).

The Seasat, Nimbus-5, Nimbus-7 and India's Bhaskara-II satellites, were the earliest to carry passive scanning microwave radiometers and clearly demonstrated the potential of microwave measurements in the retrieval of atmospheric and oceanographic parameters. In addition, Nimbus-5 satellite had the Electrically Scanning Microwave Radiometer (ESMR) operating at a frequency of 19.35 G Hz and N imbus-7 carried the Scanning Multichannel Microwave Radiometer (SMMR). Furthermore, the DMSP constellation is a long-term US Air Force effort in space designed to monitor the meteorological, oceanographic and solar-geophysical environment of the earth. In December 1972, DMSP data was declassified and made available to the civilian and scientific community. The Special Sensor Microwave Imager (SSM/I) was first flown on the DMSP satellites in June 1987. The US maintains an operational constellation of two DMSP satellites, each in a 101 minute, sun-synchronous near-polar orbit (PEO) at an altitude of 830 km above the surface of the earth. The SSM/I satellite instrument is a 7-channel, 4-frequency, linearly-polarized, passive microwave radiometric system which measures atmospheric, ocean and terrain microwave brightness temperatures at 19.35, 22.235, 37.0 and 85.5 GHz. The footprint sizes vary from 13×15 km at 85 GHz to 43×69 km at 19 GHz.

The SSM/I satellite data are used to derive a variety of geophysical parameters such as follows: ocean surface wind speed, ice cover, cloud liquid water, integrated water vapor, precipitation over water, soil moisture, land surface temperature, snow cover and sea surface temperature. Most of the retrieval algorithms in vogue are in the form of statistical correlations between the brightness temperatures of various satellite channels or differences between channels, with these parameters.

The TMI imaging sensors onboard the TRMM satellite is a passive sensor operating at 5 microwave frequencies of 10.65, 19.35, 22.235, 37.0 and 85.5 GHz. These frequencies are very similar to those of SSM/I, but TMI has an additional 10.7 GHz channel designed to provide a more linear response for the high rainfall rates in the tropics. All TMI channels have horizontal and vertical polarizations, except the 21 GHz channels, which have only vertical polarization. There is 53° conical scanning and the footprints vary in size from 5 km at 85 GHz to 45 km at 10 GHz. TMI has an improved ground resolution resulting from the lower altitude of TRMM.

The Indian weather satellite Bhaskara-II was launched in 1981 with a Satellite Microwave Radiometer (SAMIR) with onboard three channels at 19.35, 22.235 and 31.40 GHz using vertical polarization. However, the Indian Remote Sensing Satellite IRS-P4 (also known as Oceansat-1) launched on 26 May 1999, carried a Multichannel Scanning Microwave Radiometer (MSMR) operating at 6.6, 10, 18 and 21 GHz frequencies in both H and V polarizations.

5.5.3 Active Ground and Satellite Radar Technique

Weather radar, also called Weather Surveillance Radar (WSR) and Doppler weather radar, is a type of radar used to locate precipitation, calculate its motion and estimate its type as rain, snow, hail etc. Modern weather radars are mostly pulse-Doppler radars capable of detecting the motion of rain droplets in addition to the intensity of the precipitation. Both types of data can be analyzed to determine the structure of storms and their potential to cause severe weather and tropical storms.

During WW2 operators discovered that weather was causing echoes on their screen and masking potential enemy targets. Techniques were developed to filter them, but scientists began to study the phenomenon. Soon after the war, surplus radars were used to detect precipitation. Since then, weather radar has evolved and is now used by national weather services, research departments in universities and in television newscasts. Raw images are routinely used and specialized software can take radar data to make short-term forecast of future positions and intensities of rain, snow, hail and other weather phenomena. Radar output is even incorporated into numerical weather prediction models to improve analyses, weather bulletins and forecasts.

The precipitation radar was the first spaceborne surveillance equipment designed to provide three-dimensional maps of storm structure. In fact, these measurements yield invaluable data on the intensity and distribution of the rain, on the rain type, on the storm depth and on the height at which the snow melts into rain. Thus, the estimates of the heat released into the atmosphere at different heights based on these measurements can be used to improve models of the global atmospheric circulation.

In this sector on the estimation of observed precipitation here will be reviewed the science of precipitation measurement using various platforms, such as ground and satellite radars, which are illustrated in Fig. 5.9. Radar is a remote sensing Quantitative Precipitation Estimation (QPE) tool with excellent spatial and temporal resolution. Coverage of ground radar may be inconsistent from place to place and from storm to storm, so satellite radars is another remote sensing QPE tool with much coarser resolution than ground radar.

Accurate estimates of both the spatial and temporal distribution of observed precipitation are important for input to sea, river and flash flood models, for reliable forecasting of high impact hydrometeorological events, and for analysis and forecasts of drought and water supply. Thus, topics in this module include the meaning of QPE, as measurement tools for estimating observed precipitation and using climatology in precipitation estimates.

The surveillance precipitation radar has a horizontal resolution at the ground of about 3.1 miles (5 km) and a swat width of 154 miles (247 km). One of its most important features is its ability to provide vertical profiles of the rain and snow from the surface up to a height of about 12 miles (20 km). This radar is able to detect fairly light rain rates down to about 0.027 in. (0.7 mm) per hour. At intense rain rates, where the attenuation effects can be strong, new methods of data processing have been developed that help correct for this effect. The precipitation radar is able to separate out rain echoes for vertical sample sizes of about 820 feet (250 m) when looking straight down. It carries out all these measurements while using only 224 watts of electric power, the power of just a few household light bulbs.

The space precipitation radar was built by the Japanese JAXA as part of its contribution to the joint US/Japan TRMM observation system. The next

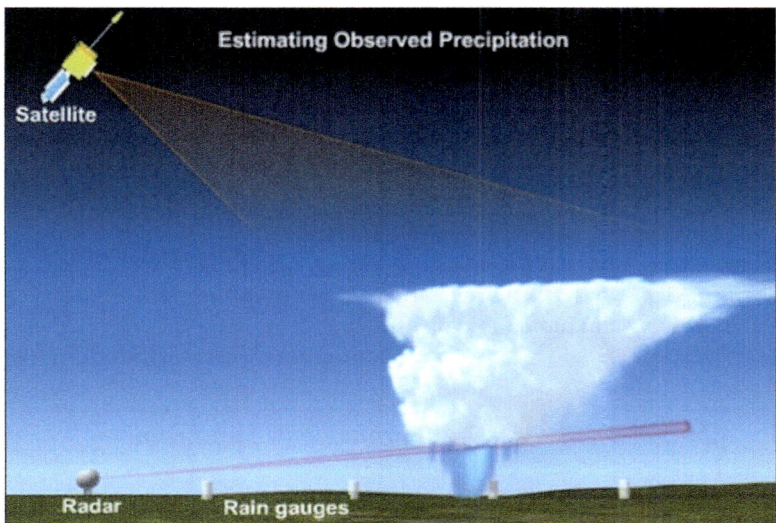

Fig. 5.9 Ground/Satellite Weather Radars (Courtesy of Manual: by Comet)

observation satellite is CloudSat, which uses radar to measure the altitude and properties of clouds and flies in formation in the "A Train" with several other satellites, such as Aqua, Aura, Calipso and French Parasol.

Among the three primary instruments on TRMM, the most innovative is the Precipitation Radar. Other instruments similar to the TMI and VIRS have operated in space before, but to date there has not been any radar in space for the purpose of measuring rainfall.

5.5.4 Severe Thunderstorms and Lighting

There are three basic conditions for the development of thunderstorms: moisture, instability and a lifting mechanism. Air is considered unstable if it continues to rise when given an upward push. An unstable air mass is characterized by warm moist air near the surface and cold dry air aloft. As a rising parcel of air-cools, some of the water vapour will condense, forming a cumulonimbus cloud that is called the thunderstorm. For a thunderstorm to develop, what is needed is an initial trigger or a mechanism that will give a start to the process of upward motion. The temperature of the lowest layers of the atmosphere increases rapidly in the afternoon or evening because of land heating and the warmest air tends to rise. Lifting is also provided by fronts, particularly cold fronts and dry lines.

Severe thunderstorms and potential tornadic thunderstorms can be identified on satellite imagery by several interesting signatures. Severe thunderstorms quickly extend to the top of the troposphere or beyond and are identifiable on infrared images by very cold cloud tops and by rapid areal expansion of the anvil region. By comparing the temperature of the cloud top to the tropopause temperature will be determined whether the top has reached or penetrated the tropopause. Although not a storm, lightning is an indicator of thunderstorms. A special low-light nighttime visible sensor on board the DMSP satellite can detect lightning, even though its designers never envisioned that it would do so. The visible sensor on the Operational Linescan System (OLS) instrument was designed to be sensitive enough to detect clouds illuminated by moonlight. Because of its sensitivity, the sensor is momentarily saturated when a bright lightning flash occurs as it scans across a thunderstorm. This produces a bright streak along a scan line, which lasts until the sensor falls below saturation a few pixels later.

5.5.5   Advanced Global Distribution of Precipitation

Global satellite-based rainfall products are currently based on microwave-only, calibrated infrared (IR), and microwave plus IR observations from various satellite missions using a variety of merging techniques (e.g., Sorooshian et al. 2000; Kuligowski 2002; Kidd et al. 2003; Turk and Miller 2005; Huffman et al. 2007; Kubota et al. 2007; Joyce et al. 2011).

The range of available new products reflects significant differences in the techniques of measurement accuracy, sampling frequency and merging methodology. While IR sensors on GEO satellites can provide precipitation estimates (inferred from cloud-top radiances) at high temporal resolutions (up to 15-min intervals on some platforms), microwave sensors remain the instrument of choice for measuring precipitation since the radiative signatures are more directly linked to the precipitating particles. Further advances in development of affordable global precipitation product require more accurate and more frequent microwave measurements within a unified observational framework.

However, in Fig. 5.10 is illustrated schematic diagram of the scanning patterns and swaths of the Dual-frequency Precipitation Radar (DPR) and GPM Microwave Imager (GMI), both instruments onboard GEO satellite, known as Global Precipitation Measurement (GPM) Core Observatory. At this point, from this platform, the DPR instrument scans cross-track in relatively narrow swaths at Ku band (13.6 GHz) and Ka band (35.5 GHz). Building upon the success of TRMM launched by NASA of the US and JAXA of Japan in 1997, NASA and JAXA successfully deployed the GPM Core Observatory on February 28, 2014.

More exactly, the dual-frequency radar reflectivity observations are nearly beam-matched over the 125 km Ka-band swath, with a horizontal resolution of approximately 5 km, and a vertical resolution of 250 m in standard observing mode. The Ku band radar scans over a wider, 245 km swath. The GPM Microwave Imager (GMI) scans conically over an 885 km wide swath at frequencies of 10.65, 18.7, 23.8, 36.5, 89.0, 165.5, 183.31 $\pm$ 7, and 183.31 $\pm$ 3 GHz. Measured brightness temperatures are in two polarizations (vertical and horizontal) at all but the

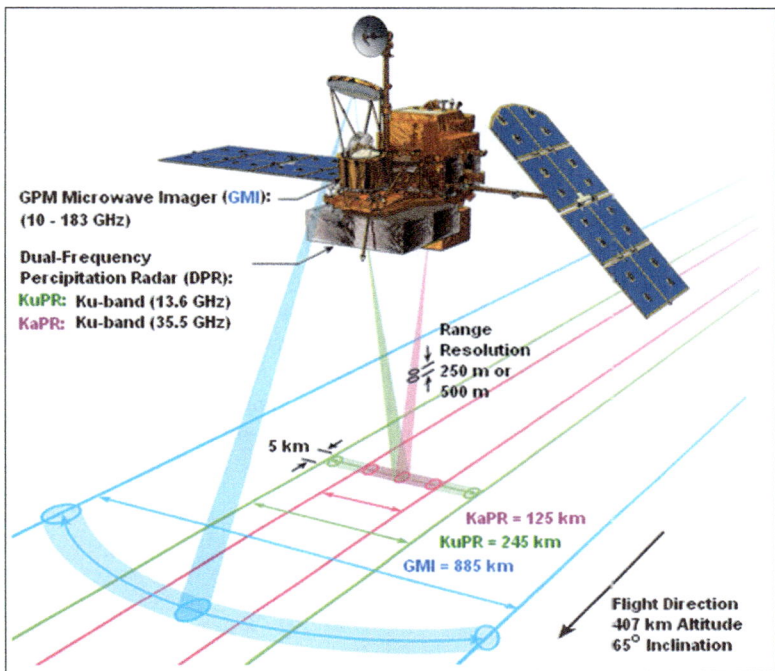

Fig. 5.10 Configuration of the GPM Core Observatory (Courtesy of Manual: by Hou)

23.8 GHz and 183.3 GHz channels, which provide only vertical polarization measurements. The GMI observations are diffraction limited, with the lowest-resolution footprints (approx. 26 km) at 10.7 GHz and the highest-resolution footprints (approx. 6 km) at the 89.0 GHz and higher frequency channels.

Thus, the GPM core observatory carries the first spaceborne dual-frequency phased array precipitation radar, the DPR, operating at Ku and Ka bands (13 and 35 GHz, respectively) and a conical-scanning multichannel (10–183 GHz) microwave imager, the GMI. This sensor package is an extension of the TRMM instruments (Kummerow et al. 1998), which focused primarily on heavy to moderate rain over tropical and subtropical oceans. In such a way, the GPM sensors will extend the measurement range attained by TRMM to include light-intensity precipitation (i.e., <0.5 mm h^{-1}) and falling snow, which accounts for a significant fraction of precipitation occurrence in the middle and high latitudes (Mugnai et al. 2007; Kulie and Bennartz 2009). In particular, the DPR and GMI measurements will refine retrieval algorithms through the construction of a unique observational database by quantifying the microphysical properties of precipitating particles. In the other words, this database will set a new standard for spaceborne precipitation measurements through its use as common reference over a variety of environmental conditions to unify measurements from the many microwave radiometers before, during and beyond the lifetime of the GPM Core Observatory.

5.6   Earth Radiation Budget

The Earth's Radiation Budget is a concept used for understanding how much energy the Earth gets from the Sun and how much energy the Earth-system radiates back in total to outer space as invisible light. Absorption of solar radiation and emission of terrestrial radiation drive the general circulation of the atmosphere and are largely responsible for the Earth's weather and climate. In this chapter we discuss satellite measurements of the constituent radiative quantities, which comprise the radiation budget of Earth. The very first successful meteorological measurements from space were of the Earth's radiation budget. They were made by a Suomi radiometer from the Explorer 7 satellite launched 13 October 1959. The long history of radiation budget measurements is reviewed by House et al. (1986). This section will concentrate on recent satellite experiments and here will be provided discussion regarding measurements of the solar constant and then to proceed with radiation budget at the top of the atmosphere and at the Earth's surface.

5.6.1   Solar Constant

The solar energy reaching the Earth is traditionally quantified as the solar constant, which in the precise manner is the annual average solar irradiance received outside the Earth's atmosphere (at the top of the atmosphere) on a surface normal to the incident radiation and at the Earth's mean distance from the Sun.

The actual solar irradiance varies by $\pm 3.4\%$ from the solar constant during the year due to the eccentricity of the Earth's orbit about the sun, but it varies as a result of other causes also. Ground-based measurements of the solar constant date back to 1913, but they have a problem is that they must be corrected for the effects of the atmosphere. On the other hand, spacebased measurements require no such correction, which date from 1975.

Both the Nimbus 6 and 7 satellites carried an Earth Radiation Budget (ERB) experiment (Smith et al. 1975; Jacobowitz et al. 1978). In fact, part of the instrument was designed to monitor the sun's output. Once each orbit, a 10-channel radiometer viewed the Sun for approximately 3 min. The 10 channels were designed to measure the solar irradiance in several spectral intervals. On the Nimbus 7 ERB, satellite channel 10 was replaced with channel 10C, an electrically self-calibrating cavity radiometer, which measures the total solar irradiance in the spectral range from less than 0.2 μm to greater than 50 μm (Hickey et al. 1988a). The Nimbus7 10C data began on 16 November 1978 and are still being collected as of this writing.

The Earth Radiation Budget Satellite (ERBS) was an NASA scientific research spacecraft launched on October 5, 1984, to study the ERB and stratospheric aerosol and gases. It was carried into LEO and the spacecraft was expected to have a 2-year operation life, but ultimately, the mission provided scientific data about the Earth's ozone layer for more than two decades.

The Solar Maximum Mission (SMM) satellite carried an active cavity radiometer, Active Cavity Radiometer Irradiance Monitor (ACRIM); Willson (1979, 1984), Willson and Hudson (1988). Launched on 16 February 1980, SMM made solar irradiance measurements until December 1989. The Earth Radiation Budget Experiment has flown on the Earth Radiation Budget Satellite (launched 5 October 1984) as well as the NOAA 9 and 10 satellites (launched 12 December 1984 and 17 September 1986, respectively).

Continued observations of the Sun are important both to resolve the differences between the various instruments and to determine the cause and effects of the variable solar output. Solar constant measurements continue to be made on Nimbus 7, the ERBS Satellite and NOAA 9 and 10. In addition, solar constant measurements are being made on the Upper Atmosphere Research Satellite (UARS). In fact, the UARS was NASA-operated orbital observatory whose mission was to study the Earth's atmosphere, particularly the protective ozone layer, was deployed in LEO on 15 September 1991.

5.6.2   Top of the Atmosphere Radiation Budget

The goal of radiation budget studies is to measure the incoming and outgoing radiation as a function of time and space. Incoming radiation is irradiance and is given the symbol E (t, Θ, Ψ, h), where t is time, Θ is latitude, Ψ is longitude and h is height above the Earth's surface. Thus, outgoing radiation is radiant exitance and is given the symbol M (t, Θ, Ψ, h). Usually we are not interested in the wavelength dependence of these quantities except for the division between shortwave and longwave components. Shortwave radiation, usually defined, as radiation with wavelengths less than 5 μm is primarily reflected solar radiation. Longwave radiation (> 5 μm) is primarily emitted terrestrial radiation.

Satellite methods to measure radiation budget quantities depend on the altitude in which one is interested. The two most common altitudes are the Top of the Atmosphere (TOA, often defined for radiation budget purposes as 30 km) and the surface. These two altitudes will be discussed separately.

At the top of the atmosphere, the quantities we want to measure are the shortwave radiant exitance (M_{SW}), and the longwave radiant exitance (M_{LW}), which is called the Outgoing Longwave Radiation (OLR). With these quantities can calculate the Albedo such as:

$$A (t, \Theta, \Psi) = M_{SW}(t, \Theta, \Psi)/(E_{sun}\mu_{sun}) \qquad (5.46)$$

The OLR data is a measure of the amount of energy emitted to space by Earth's surface, oceans and atmosphere, which values are observed from the AVHRR instrument aboard the NOAA polar orbiting spacecraft. Data are centered across equatorial areas from 160°E to 160°W longitude. The absorbed solar radiation from above equation is as follows:

$$E_{abs}(t, \Theta, \Psi) = (1 - A)\ E_{sun}\mu_{sun} \tag{5.47}$$

The net radiation can be expressed by the following equation:

$$E_{net}(t, \Theta, \Psi) = (1 - A)\ E_{sun}\mu_{sun} - M_{LW} = E_{sun}\mu_{sun} - M_{SW} - M_{LW} \tag{5.48}$$

In these equations, value $\mu_{sun} = \mu_{sun}\ (t, \Theta, \Psi)$ is the cosine of the solar zenith angle, and $E_{sun} = E_{sun}\ (t)$ is the solar constant S_{sun} adjusted for distance from the sun, such as:

$$E_{sun} = S_{sun}[d'_{sun}/d_{sun}(t)]^2 \tag{5.49}$$

where d_{sun} is the Earth-Sun distance at time t and d'_{sun} is the mean Earth-Sun distance by 1 Astronomical Unit (AU).

Instruments used for radiation budget studies can be divided into two spectral categories: Broadband sensors attempt to measure spectrally integrated radiation budget quantities and Narrowband sensors measure only narrow portions of the spectrum; extrapolation to other portions of the spectrum is necessary for radiation budget calculations. Otherwise, radiation budget instruments may also be divided into two categories Fields-of-View (FOV) such as: Wide-Field-of-View (WFOV) sensors and Narrow-Field-of-View (NFOV) sensors. In fact, WFOV sensors category measure radiation from horizon to horizon, and they are often called flat-plate sensors because they typically consist of a flat element that measures the irradiance received at the satellite.

Current PEO satellites are a very good platform for measuring the radiation budget. In such a way, a satellite-mounted instrument makes two observations daily, separated by about 12 hours, of every point on Earth. These are sufficient to calculate the monthly average radiation budget, which is usually the time scale of interest. There are some problems: (1). The measurements are made at the satellite height, not at the top of the atmosphere (the inverse problem); (2). The twice-daily observations may not adequately sample the diurnal variation, particularly of shortwave radiation. A diurnal model may be necessary (the diurnal problem); (3). Radiation budget quantities, which integration over wavelength, shortwave or longwave are necessary. Satellite sensors, however, do not measure exactly these quantities (the spectral correction problem or unfaltering problem); and (4). To measure the radiant exitance from a point, one must observe the point from all possible directions. Satellites, however, observe a single point from a single direction (the angular dependence problem). In summary, to retrieve accurate radiation budget quantities, we must know the radiance in all directions, at all times, in all wavelengths and at all heights.

5.6.3   *Surface Radiation Budget*

For some meteorological applications, particularly agricultural applications, the radiation budget at the Earth's surface is required. The observations of the surface radiation budget are not as direct as observations of the radiation budget at the top of the atmosphere, so the corrections for the atmosphere must be made. Since the Earth surface radiation budget is dominated by clouds, however, satellite data are well suited for this purpose, and they can provide worldwide estimates. Thus, the surface radiation budget may be divided into five quantities: downward shortwave radiation (called insolation), upward shortwave (reflected) radiation, downward longwave radiation, upward longwave radiation, and net radiation.

1. **Downward Shortwave Radiation (DSR)** – The DSR product is an estimate of the total amount of shortwave radiation (both direct and diffuse) that reaches the Earth's surface. The product algorithm uses spectral channels in both the visible and the infrared in addition to data regarding Albedo and atmospheric composition to compute the DSR at the Earth's surface. The DSR product has many applications both in the general and applied sciences. As one of the components of the surface energy budget, it is needed in climate studies. Used together with cloud and aerosol properties it provides estimates of cloud and aerosol effects (forcing). It is also used in surface energy budget models, land surface assimilation models such as used at NOAA National Center for Environmental Prediction (NCEP), NASA Land Data Assimilation System (LDAS) and ocean assimilation models either as an input (providing observationally-based forcing term), or as an independent data source to evaluate model performance. Thus, DSR data are also employed in estimating heat flux components over the coastal ocean to drive ocean circulation models. In agriculture, DSR is used as input in crop modeling. But in hydrology, however, it is used in watershed and run-off analysis, which is important for determining flood risks and dam monitoring. The solar energy industry also needs estimates of DSR value for both real-time and short-term forecasts for building energy usage modeling and optimization. Furthermore, since high irradiance values result in surface drying, DSR is also used in monitoring fire risk. Due to interest in the use of solar energy, downward shortwave radiation at the surface has been studied for some years. Following Fritz et al. (1964) and Hanson et al. (1967), the solar radiation incident at the top of the atmosphere suffers one of three fates: it may be reflected to space, it may be absorbed by the atmosphere, or it may be absorbed at the ground.

2. **Reflected Shortwave Radiation (RSR)** – This radiation product measures the total amount of shortwave radiation that exits the Earth through the top of the atmosphere. The algorithm will use several spectral channels in both the visible and infrared spectrum to measure the RSR. In fact, information from this product will provide an integral piece of the Earth's radiation budget, aiding in climate modeling and prediction. Namely, RSR is the product of the surface Albedo and the downward solar radiation. THUS, Knowledge of the surface Albedo is the key element in estimating reflected solar radiation. The basic approach to

estimating surface Albedo has been through the minimum Albedo technique (e.g., Raschke and Preuss 1975). The region of interest is monitored over a period of time and the minimum Albedo during that time is noted. Since few locations are likely to be cloud-covered for an entire month, the minimum Albedo is likely to represent the clear-sky planetary Albedo. Corrections for atmospheric absorption and scattering may then be applied to retrieve the surface Albedo itself (Preuss and Geleyn 1980). Surface Albedo is also produced as a byproduct of solar insolation calculations (Gautier et al. 1980). Dedieu et al. (1987) have mapped surface Albedos over Europe using Meteosat data.

3. **Downward Longwave Radiation (DLR)** – This type of radiation at the surface depends primarily on the temperature and moisture structure of the atmosphere and on the character of the clouds. Satellite soundings, therefore, are suitable to estimate downward longwave radiation. Darnell et al. (1983, 1986) have used operationally retrieved TOVS soundings to estimate downward longwave radiation at the surface. However, the NESDIS operational retrievals were processed to yield temperature and moisture information at 50 hPa vertical resolution, then a radiative transfer model was applied to calculate the clear-sky downward radiation. Cloud amount and top height information in the TOVS retrievals were used to calculate the effects of clouds. The clouds were assumed to be 50 hPa thick, thus the cloud base pressure was assumed to be 50 hPa greater than the cloud top pressure. The radiative transfer model was then run to calculate the downward longwave radiation under a plane parallel cloud. Finally, the clear-sky and overcast results were averaged using the cloud amount as a weight. Darnell et al. (1983) compared essentially instantaneous satellite estimates with ground-based pyrgeometer data, and in 1986 improved their data handling and corrected the pyrgeometer data for shortwave heating of the dome. Frouin et al. (1988) also used TOVS soundings but they experimented with different ways to estimate the effects of clouds. One method used GOES visible data to estimate cloud amount and cloud optical depth and used GOES infrared data to estimate cloud top height. The optical depth was converted into a cloud depth and finally a cloud base. Other methods used GOES cloud fractions but simpler treatment for cloud base estimation. In May 1998, the NOAA deployed advanced TOVS (ATOVS) onboard NOAA-15 PEO satellite. In April 1999, however, the ATOVS sounding products were implemented into the NESDIS operation, replacing NOAA-12.

4. **Upward Longwave Radiation (ULR)** – Little longwave radiation is reflected by the surface, so nearly all upward longwave radiation is emitted by the surface. The problem of estimating upward longwave radiation, therefore, boils down to estimating the surface temperature. The ULR product is a measure of the total OLR at the top of the atmosphere, and provides important information regarding the Earth's outgoing energy and overall energy budget at the TOA. It is one of three radiation budget parameters that determine the Earth radiation budget at the TOA. Besides, the other two parameters are the incoming solar radiation and the reflected solar radiation. The OLR is estimated directly from several ABI infrared radiances for each ABI pixel, regardless of sky condition. This product,

together with the incoming total solar radiation and the reflected solar radiation define the Earth's radiation budget at the TOA. Understanding of the Earth's radiation budget is important for climate monitoring purposes.

5.7 Measurements and Monitoring of Other Earth Observation Parameters

Satellite observations are providing our societies with effective ways of monitoring our planet and helping to improve the exploitation and management of Earth's resources. Thus, it is addressing important global environmental, social and economic challenges, and is also providing innovative services to improve quality of life and economical efficiency.

In this sector will be introduced measurements and monitoring of other Earth observation parameters such as follows: Hydrology Analyzes and Measuring, Sea Waves and Ocean Dynamic, Sea Surface Temperature, Ecosystem and Sea Pollutions, Cryosphere Detection and Measuring, Monitoring Agricultural and Forestry Landscape, Mapping Global Land Cover and Mapping Desertification.

5.7.1 Hydrology Analyzes and Measuring

Observation, measuring and modeling of terrestrial water cycle is important for sustainable management of water resources as well as understanding the impact of climate change. Satellites provide an important role in measurements of various dimensions of water cycle components in spatio-temporal domain. Satellite active and passive instruments operating in optical, thermal and microwave wavelength region help in retrieval of various hydro meteorological parameters like: rainfall, soil moisture, evapo-transpiration, groundwater, water level and water quality etc. Earth Observation systems are being used in assessment of water-spread area, elevation of water surface and temporal changes. Seasonal monitoring of reservoir spread, conjunctive water uses and water management practices associated to transplanting operations helps in irrigation scheduling. Satellite observations are regularly used to generate snow cover map which help in snow melt runoff estimation. Satellite data has been used to study the changes in the extent of glaciers inventory and their monitoring in terms of whether they are retreating or being stable over the time is another important contribution of space technology towards understanding the climate change signals.

Satellite altimetry is a technique for measuring height using SAR satellites, namely the time taken by a radar pulse to travel from the satellite antenna to the surface and back to the satellite receiver. Combined with precise satellite location data, altimetry measurements yield sea-surface heights. Satellite altimetry is one of the recent techniques, which is used for monitoring lake and reservoir volume and water level, river levels and discharge and floods modeling. Temporary or permanent flooded areas can be observed and integrated into hydrological models. With the combination of in-site measurements in parallel with satellite altimeter data for modeling can help in water resource management and measure the effects of hydrological cycle on water resources due to climate variability.

Hydrological remote sensing is being carried out using measurements from various satellite systems such as: SARAL-Altika Mission (Inland Water level) of Indian Space Research Organization (ISRO) and CNES (Space Agency of France), Indian RISAT-1 SAR Mission (surface water spread and soil moisture) and Indian Oceansat-2 and SCATSAT-1 Mission (soil wetness and inland water level). Other products that improve the accuracy of satellite data based research in hydrology include Cosmo-SkyMed from the Italian Space Agency, Radarsat-2 from the Canadian Space Agency, and Sentinel-1 from ESA (Schumann et al. 2015). Others are Global Change Observation Mission-Water (GCOM-W) from the Japan Space Agency (JAXA), Global Precipitation Measurement (GPM) from JAXA/USA, and Soil Moisture Active Passive (SMAP) from the USA.

5.7.2 Sea Waves and Ocean Dynamic Measuring

Disturbances of sea waves are a constant presence in the world's ocean waters. In fact, the vast majority of ocean waves, however, are generated by wind. Because waves travel all across the globe, transmitting vast amounts of energy, understanding their motions and characteristics is essential. The forces generated by waves are the main factor impacting the geometry of beaches, the transport of sand and other sediments in the near shore region, and the stresses and strains on coastal structures. When waves are very large, they can also pose a significant threat to commercial shipping, recreational boaters and the beach going public. Thus for ensuring sound coastal planning and public safety, sea wave measurement and analysis is of great importance.

Three factors determine how large wind-generated waves can become. The first factor is wind speed, and the second factor is wind duration or the length of time the wind blows. The final factor is the fetch, the distance over which the wind blows without a change in direction. The faster the wind, the longer it blows, and the larger the fetch, the bigger the waves that will result. When waves are being generated by strong winds in a storm, the sea surface generally looks very chaotic, with lots of short, steep waves of varying heights. In general, seas are short-crested and irregular, and their surface appears much more disturbed than for swells. Swells, on the other hand, have smooth, well-defined crests and relatively long periods. Swell is more uniform and regular than seas because wave energy becomes more organized as it travel longs distances. Longer period waves move faster than short period waves, and reach distant sites first. In addition, wave energy is dissipated as waves travel (from friction, turbulence, etc.), and short-period wave components lose their energy more readily than long-period components. As a consequence of these processes, swells form longer, smoother, more uniform waves than seas.

Ocean dynamics define and describe the motion of water within the oceans. In fact, ocean temperature and motion fields can be separated into three distinct layers: mixed (surface) layer, upper ocean and deep ocean. Looking out at the water, an ocean wave in deep water may appear to be a massive moving object, namely a wall of water traveling across the sea surface. But in fact the water is not moving along with the wave. The surface of the water and anything floating atop it, like a boat or buoy simply bobs up and down, moving in a circular, rise-and-fall pattern. In a wave, it is the disturbance and its associated energy that travel from place to place, not the ocean water. Thus, an ocean wave is therefore a flow of energy, traveling from its source to its eventual break-up. This break up may occur out in the middle of the ocean, or near the coast in the surfzone.

In order to understand the motion and behavior of waves, it helps to consider simple waves: waves that can be described in simple mathematical terms. Sinusoidal or monochromatic waves are examples of simple waves, since their surface profile can be described by a single sine or cosine function. Simple waves like these are readily measured and analyzed, since all of their basic characteristics remain constant.

Although ocean waves have been measured by satellites since the first radar altimeter was tested on NASA's Skylab space station in 1973, there has not been access to consolidated datasets on ocean waves for scientists and engineers needing this information for modeling and forecasting until today. The satellite altimeters used to measure surface geostrophic currents also measure wave-height. The altimeter technique works as follows. Radio pulse from a satellite altimeter reflect first from the wave crests, later from the wave troughs. The reflection stretches the altimeter pulse in time and the stretching is measured and used to calculate wave-height. Altimeters were flown on Seasat in 1978, Geosat from 1985–1988, ERS-1/2 from 1991, Topex/Poseidon from 1992, and Jason from 2001. Thus, altimeter data have been used to produce monthly mean maps of wave-heights and the variability of wave energy density in time and space. The next step, just begun, is to use altimeter observation with wave forecasting programs to increase the accuracy of wave forecasts.

Forecasting the state of the sea is important for shipping, offshore and coastal engineering, management of coastal zones and tourism. For instance, this is set to become easier through ESA's GlobWave project, which offers a one-stop service for satellite data on ocean waves. Safety and financial losses associated with rough seas are a real concern. Ocean waves slow down the passage of ships, endanger marine industries such as oil and gas extraction, damage aquaculture and offshore wind farms and erode coastlines.

5.7.3   Sea Surface Temperature Measuring

Sea Surface Temperature (SST) is a critical quantity in the study of both the ocean and the atmosphere as it is directly related to and often dictates the exchanges of heat, momentum and gases between the ocean and the atmosphere. The technique of satellite remote sensing can provide thermal information in a short time over a wide area of soil and sea surface. In general, temperature measurement by satellite remote sensing is based on the principle that any object emits electro-magnetic energy corresponding to the temperature, wavelength and emissivity. In fact, as one of easiest ocean variables to observe, SST has a very long history in the study of the ocean and its interchange with the atmosphere. One of the most common operational applications of SST satellite observations is as a boundary layer input and an assimilation data set for atmospheric circulation and forecast models.

The computation of SST from infrared satellite observation data started in the mid-1970's using the primary instrument called the Scanning Radiometer (SR) onboard NOAA's PEO weather satellites. On the same spacecraft the VHRR instrument, however, had a 1 km spatial resolution, which was much better than the 8 km spatial resolution of the SR. Efforts to fully automate the computation of SST from the satellite radiances produced some very artificial looking SST fields. During this period the only in site SST data were from ships and these were initially used for comparison with the satellite SST measurements.

Satellite infrared SST measurements have resulted in major advancement in oceanography, meteorology and climatology. However, the infrared SST retrievals have two significant limitations: (1) Retrievals cannot be done when clouds (which cover roughly half the Earth) are present, and (2) Atmospheric aerosols from volcanoes and significant fires can cause a spurious cooling in the SST retrieval. Furthermore, the cloud detection algorithms are not totally reliable, with some clouds going undetected.

It has long been recognized that microwave radiometry offers a solution to the cloud and aerosol problem. At frequencies below about 12 GHz, the surface radiance is proportional to SST and microwaves penetrate clouds with little attenuation, giving a clear view of the sea surface under all weather conditions except rain. In addition, at these frequencies, atmospheric aerosols have no effect, making it possible to produce a very reliable SST time series for climate studies. The first satellite microwave radiometers operating at these low frequencies were flown on SeaSat and Nimbus-7, launched in 1978.

Already introduced satellite TRMM Microwave Imager (TMI) in sector 5.5, is the first in a series of MW radiometers that is using to measure SST under nearly all weather conditions launched in 1997. In the next 2 years, two Advanced Microwave Scanning Radiometers (AMSR) were launched on US and Japanese spacecraft. The AMSR will have an additional 6.9-GHz channel that will enhance SST satellite retrieval. Later in the decade, the Conical Microwave Imager Sounder (CMIS) become a primary sensor flying on the National Polar Orbiting Environmental (NPOES) satellite system launched in 2011.

5.7.4 Sea Pollution and Ecosystem Monitoring

The development of satellite remote sensing technique allows the detection and monitoring of oil spills at sea caused by tankers and on/offshore oil operations becoming increasingly important due to the constant threat posed to marine wildlife and the ecosystem. In this sense, most man-made oil pollution comes from land-based activity, but public attention and subsequent regulation has tended to focus most sharply on seagoing oil tankers.

Today, the availability of data from Earth observation satellites offers the possibility to complement and optimize surveillance strategies, allowing a more cost-effective approach to oil-pollution monitoring. In Fig. 5.11 (Left) is shown example of oil spill detecting service by TerraSAR-X, a radar Earth observation satellite, launched on 15 June 2007, which is a joint venture being carried out under a public-private-partnership between the German Aerospace Centre (DLR) and EDAS Astrium.

The SAR constellation seems to be one of the most effective instruments onboard satellites for the detection of slicks since slicks damp strongly short waves measured by SAR and oil spills appear as a dark patch on the SAR image. SAR observations do not depend on weather (clouds and sunshine), which permits the showing of illegal discharges that most frequently appear during night. SAR can also survey storms areas, where accident risks are increased. The most suitable SAR radar configuration for oil pollution study is C-band radar frequency with VV polarization, with a 20 to 45° incident angle, which are carried onboard satellites such ERS-2 Europe (1995), Envisat Europe (2002) and Radarsat-2 (C) Canada (2007).

5.7.5 Cryosphere Detection and Monitoring

The cryosphere is the frozen water region of the Earth, and it may be found in different categories, including snow cover, freshwater ice, sea ice, permafrost and continental ice masses such as glaciers and ice sheets floating at sea. The cryosphere lowers the Earth's surface temperature by reflecting a large amount of sunlight, stores fresh water for millions of people, and provides habitat for many plants and animals.

While icebergs, glaciers and ice sheets are formed on land, sea ice is just frozen ocean water. The amount of sea ice in the oceans increases during winter and decreases during summer while some sea ice exists all year in some regions. During some part of the year, approximately 15% of the oceans are covered by sea ice. In Fig. 5.11 (Right) is shown example of sea ice monitoring image done by TerraSAR sensors for enhanced ecological and safety solutions at sea.

Sea ice observation from coastal stations and ships has a history of more than 100 years. Regular sea ice charting, however, using aircraft and satellites has

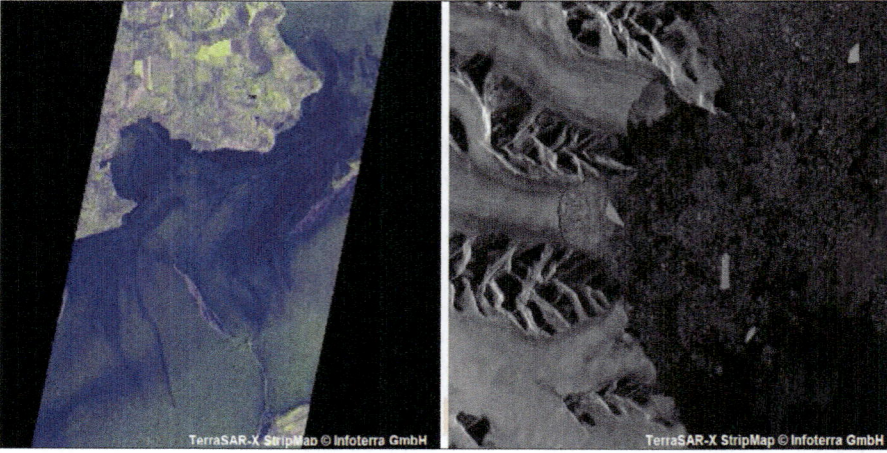

Fig. 5.11 Oil Spill Detecting and Sea Ice Monitoring Service (Courtesy of Manual: by Astrium)

developed mostly since WW2. Aircraft survey was the main observation method until the 1980s, but use of satellite data has developed over the last three decades and is now the most important method.

The first satellite sensors providing views of the large-scale structure and motion of sea ice utilized visible and infrared channels, such as those onboard the early Nimbus, Tiros and ERTS (later renamed as Landsat). By the late 1960s, it was apparent that the sequential synoptic observations needed for sea ice and climate studies could not be acquired by visible sensors, which are limited to cloud-free and well-illuminated conditions. Sea ice exists in regions that are dark for several months and are frequently cloudy in the remaining months of the year (Gloersen et al. 1992).

Therefore, it has been necessary to develop observation methods using microwaves (MW) that are able to penetrate clouds and are not dependent on light conditions. The first passive MW remote sensing systems for satellites were launched on the Russian Cosmos 243 and Cosmos 384 in 1968 and 1970, respectively. In the US, passive MW technology was first used in satellite remote sensing of sea ice during the late 1960s and early 1970s, when a prototype of the ESMR was flown on Nimbus-5 over the Polar Regions (Campbell 1973). The first atlas of Antarctic sea ice based on passive microwave data was produced by Zwally et al. (1983). The period since 1970 has been one of great advancement in remote sensing of sea ice. After the ESMR period 1973–1976, a more advanced satellite instrument, the SMMR was operated on Nimbus-7 for 9 years, from 1978 to 1987. A similar instrument, the SSM/I followed after the SMMR and has provided continuous measurements for more than 25 years. This series of similar spaceborne instruments provided the longest and first regular time series of global sea ice data, allowing studies of variability and trends of the ice area and extent in both hemispheres (Cavalieri et al. 1997; Johannessen et al. 1999). Passive MW observations have fairly coarse resolution (typically 30 km) and are more suitable for global monitoring than for regional observations.

New passive microwave systems, such as AMSR-E, provide improved resolution of ice concentration charts, typically 6–10 km. Active microwave systems, such as real-aperture Side-Looking Radars (SLR) and SAR, were developed during the 1970s and 1980s for aircraft surveillance and used in ice monitoring to provide detailed maps of the ice conditions, especially in areas of heavy ship traffic. Satellite SLR systems were used extensively in Russian ice monitoring during the 1980s and 1990s (Johannessen et al. 2000; Alexandrov et al. 2000). Spaceborne SAR with high spatial resolution and independence of cloud cover and light conditions, making it possible to observe sea ice with much better accuracy than visible and passive microwave methods. In 1978 Seasat was the first satellite that provided high-resolution SAR images of sea ice, but it only operated for about 3 months. The European Remote Sensing (ERS) program, which started in 1991, provided in 10 years a major global satellite SAR remote sensing of sea ice. Since 1996 the Canadian Radarsat-1 has delivered wide SAR images over large parts of the Northern Hemisphere sea ice. Arctic sea ice deformation fields and linear kinematics features have been derived from regular ScanSAR images and used to estimate ice area and volume production (i.e., Kwok 6 Sea Ice Monitoring by Remote Sensing and Cunningham 2002). Since 2003 the European Envisat has delivered wide-swath ASAR data, and ice observation is one of the main applications (Flett 2004; Sandven et al. 2004; Johannessen et al. 2005). Besides, other satellite microwave systems such as scatterometer and radar altimeter data have also shown promising results for observation of ice parameters.

5.7.6 Agricultural and Forestry Landscape Monitoring

Satellite imagery provides an efficient means to retrieve daily information on the status and extent of forest resources and changes thereof. The large area coverage and high spatial resolution of newly launched optical and radar satellite systems offer new opportunities to remotely estimate and access land cover information on a wall-to-wall basis. At the same time the high temporal resolution and capacity to acquire data on a weekly basis improves the capability to detect most recent land cover changes such as deforestation and forest degradation events for local, regional and global forest monitoring systems. The concept of a forest landscape as global assessment of forest alteration, as it is used here, is a mosaic of land cover types that are naturally interspersed, which landscape is dominated by forests but may include extensive naturally treeless areas (for example, small lakes, wetlands, rivers, and rocky outcroppings) as these components coexist naturally within the broader forest ecosystem.

The forest landscape zone boundary was defined using a global tree canopy cover dataset as part of the Vegetation Continuous Fields (VCF) MODIS 500 m product, (Hansen et al. 2003). The forest landscape zone was assessed in two steps. First, a preliminary forest fragmentation analysis was carried out for countries for which Geographic Information System (GIS) datasets on a scale 1:500,000 or finer for transportation infrastructure and settlements were available. Buffer zones were assigned to roads, pipelines, power lines and settlements that were subsequently eliminated from the area of study. The second step was to use high spatial resolution Landsat TM (global coverage for an average date of 1990) and ETM+ (global coverage for an average date of 2000) imagery to systematically assess all remaining candidate forest areas for human-caused alteration and to delineate developed and fragmented areas.

The Landsat agricultural and forestry environmental images were obtained free of charge from the GeoCover Landsat Orthorectified image collection (Tucker et al. 2004). The image analysis was conducted through expert-based visual interpretation, using GIS data overlays with additional thematic and topographic map layers. In fact, the purpose was to detect evidence of significant human-caused alteration and fragmentation. Patches with such evidence were eliminated from the area of study and remaining areas, if large enough, were classified as monitoring data. The utilization of dense time series of multitemporal and multi sensorial satellite imagery produced by the ESA Sentinel and other SAR satellite missions data is well suited for can address challenges caused by phonological changes of forest canopies between the seasons and data availability restrictions due to clouds.

Furthermore, satellite-gathered Earth-observation data offer a significant improvement in any country ability to reduce expenditure by providing a new and cost effective tool for agricultural monitoring and management as well. With the advent of the ERS missions in particular, SAR observation data for landscape monitoring are now available over footprint of PEO satellite. The application of satellite remote sensing techniques for accurate crop identification and crop-area estimates requires multi-temporal series of data, as each single image has to be acquired within a key time-window for optimal target discrimination. SAR data also complement optical data available from other operational observation systems.

The ERS-1 satellite has onboard an array of earth-observation instruments that gathered information about the Earth landscape, ice and atmosphere using a variety of measurement principles. These instruments include: Microwave Radiometer (MWR), Synthetic Aperture Radar (SAR), Radar Altimeter (RA) and Along-Track Scanning Radiometer (ATSR).

5.7.7   Global Land Cover Mapping

Global land cover mapping using environmental observation satellite data is an important variable for many studies involving the Earth surface, such as climate, food security, soil erosion, hydrology, atmospheric quality, conservation biology, plant functioning and so on.

Land cover not only changes with human caused land use changes, but also changes with nature. Satellite remote sensing has been widely recognized as the most economic and feasible approach to derive land-cover information and mapping over large areas (Cihlar 2000; Gong 2012). Since 2000, enormous progress has been achieved in data processing at the global scale, particularly for data acquired on board of the US satellites such as Terra, Aqua and the Landsat series. Reflectance data corrected to the top-of-the-atmosphere level are available in image composites for every 16 days at 30 m, and daily for 250 m and 500 m resolution with average geometric accuracies better than 1 pixel (e.g., Roy et al. 2008, 2010). Previous global land cover products derived using time series optical satellite data at coarse spatial resolution (300 m–1 km) did not provide sufficient thematic detail or change information for global change studies and for resource management (Giri et al. 2013).

5.7.8 Desertification Monitoring and Mapping

Desertification is the degradation of land in progressive environments of desolate areas, in which productive land becomes desert. Thus, global satellite monitoring and mapping using satellites has begun to provide information about desertification. For instance, satellite images of the same locations taken over a period of time show changes in the land. Remote sensing can be used to monitor the progress of desertification.

The synoptic view provided by the satellites, in combination with the frequent provision of spectral data, makes satellite remote sensing a very useful monitoring tool for monitoring and mapping of desertification. As desertification is a phenomenon that affects large areas on the Earth, high spatial resolution is not as important, in comparison with the need for frequent coverage of large land areas. As a result, medium spatial resolution instruments on satellites such as Terra (NASA) or ENVISAT (ESA) can provide data of sufficient quality.

The satellite Landsat TM (Thematic Mapper) has provided a sequence of images for the years in 1998, 2001, 2004, 2007 and 2010, which they were submitted to the algorithms developed to validate them.

There is no specific operational programme for the monitoring and mapping of landscape desertification, however, there are systematized as follows: (1) Observation programmes mainly based on data from low-resolution meteorological satellites, particularly NOAA with onboard AVHRR and Meteosat, combined with ground data on rainfall, agricultural yield, population location, markets and commodity prices; and (2) Long-term monitoring programmes for the production of detailed geographic maps and databases on the state and changing pattern of land resources. In fact, these programmes mainly use high-resolution satellite data (Landsat, SPOT) and increasingly, the GIS data, particularly for the studies on resource degradation and potential.

The evaluation and mapping of these various parameters should be based on a judicious concurrent utilization of:(a) Remote-sensing data; (b) Data from measuring stations and ground observation and posts; and (c) Maps, statistics, databases and summary reports on the various desertification parameters, available at local, national or regional level.

Chapter 6
Satellite Meteorological Instruments

In this chapter are introduced satellite meteorological instruments for installation onboard Polar Earth Orbit (PEO) and Geostationary Earth Orbit (GEO) satellites. In addition, it will include other satellite instruments, radiation and broadband instruments. Thus, research and technology innovation on design, implementing and exploiting new satellite meteorological instruments that will provide unprecedented details of the weather at different levels of the atmosphere is making good progress.

6.1 Introduction to PEO Satellite Meteorological System

Initially, two countries maintained systems of Polar Earth Orbit (PEO) satellites: the former Soviet Union and the United States. The Soviet satellites of Meteor program and the US PEO satellites began with Television Infrared Observation Satellite Program (TIROS-1), launched on 1 April 1960 and 10 satellites were launched in the series ending in 1965. The first operational series of meteorological satellites was the Environmental Science Services Administration (ESSA) series; 9 satellites were launched between 1966 and 1969. The second series of operational satellites were 6 satellites of Improved TIROS Operational System (ITOS) and National Oceanic and Atmospheric Administration (NOAA) 1–5 were launched between 1970 and 1976. The current polar-orbiting series began with TIROS-N, which was launched 13 October 1978. The NOAA 8, 9, 10, 11,13, and 14 satellites are modified versions of TIROS-1Í and are called Advanced TIROS-1.

The satellite bus for the TIROS-1Í meteorological series is built by RCA Corporation (now Martin Marietta Astro Space), which duties are many. It contains radio transmitters, which send data collected by the instruments to Earth stations, and radio receivers, which acquire operational commands from the ground. The satellite bus contains computers that process the data from the meteorological instruments and the tape recorders on which data are recorded for later transmission to Earth. It must supply electrical power to itself and the instruments, control the environment (particularly temperature) for the instruments and electronics, maintain proper orientation, and be capable of adjusting its orbit. These are difficult tasks in the harsh environment of space.

The Polar Operational Environmental Satellite (POES) system is project of NOAA, which is responsible for the construction, integration, launch of NOAA series satellites and as well as operational control of the spacecraft, taken normally 21 days after launch. The NOAA satellites carry several scientific instruments and two for Search and Rescue (SAR). The last satellite of POES Project NOAA-19 was launched in February 2009. The satellite data provides economic and environmental benefits on a continuous, reliable basis. The benefits that directly enhance the quality of human life and protection of Earth's environment include. At this point, over 50% of the US public utilizes 3–5 day weather forecasts for planning recreational and business activities and many agencies in US utilize NOAA meteorological satellite data products to manage resources, plan civic and industrial expansion, schedule services and monitor population growth.

Countless lives and properties have been saved by monitoring severe storm movement and forecasting national disasters. From monitoring ozone levels and animal migration patterns to forecasting and detecting forest fires, the NOAA series is a vital tool of environmental research and protection. Thus, global data collected about the earth is used to monitor the environment and trend changes over time SAR instruments carried on POES satellites contributed to saving over 28,000 lives. In fact, the SAR solution of Cospas-Sarsat system detects signals from downed oceangoing ships, land vehicle and aircraft giving a precise estimate of their location to aid in rescue operations. In addition, the Space Environment Monitor (SEM) measures energetic particles, such as protons, electrons and alpha particles for solar, ionospheric and other studies. Finally, the Data Collection System (DCS) relays meteorological and other data transmitted from ground-based instruments.

Another important PEO meteorological system MetOp (Meteorological Operational) is a series of three polar orbiting satellites developed by the European Space Agency (ESA) and operated by the European Organization for the Exploitation of Meteorological Satellites (EUMETSAT). The satellites form the space component of the overall Eumetsat Polar System (EPS), which in turn is the European half of the Eumetsat/NOAA Initial Joint Polar System (IJPS). The satellites carry a payload comprising 11 scientific instruments and two, which also support SAR services. In order to provide data continuity between MetOp and NOAA POES, several common instruments are carried on both fleets of satellites as more cost effective, efficient and convenient solutions.

The EPS (EUMETSAT Polar System) program is a part of the Global Operational Satellite Observation System (GOSOS), which is under the auspices of the World Meteorological Organization (WMO). It consists of operational and research satellites in both polar and geostationary orbits that provide a wealth of information to the global user community for operational meteorology and climate monitoring. The MetOp Eumetsat PEO satellite is part of EPS constellation launched on 19 October 2006, as Europe's first PEO satellite used for operational meteorology. In the future, subsequent satellites in the series are provisionally planned for launch at 5-year intervals to match the 5-year design life and provide continuity of PEO satellite meteorological services.

However, in case of extended mission lifetime, a period of dual operations is expected as well as an extension to the overall programme. Finally, in 2021 EPS will start to launch a second generation of MetOp satellites will be deployed, called MetOp-SG. This satellite program will be constellation in a series of six meteorological satellites developed by ESA and Eumetsat to be deployed from year 2021.

As stated above, the first Soviet Union's operational meteorological satellite Meteor-1 of PEO program was launched on 26 March 1969 on a Vostok rocket and originally placed in orbit at an altitude of 650 km. Besides, two solar panels were automatically oriented toward the sun. Meteor-1-1 was the first of a series of 25 launches of similar spacecraft (model designation Meteor M 11F614) from 1969 to 1977. The Meteor satellites were designed to monitor atmospheric and sea-surface temperatures, humidity, radiation, sea ice conditions, snow-cover and clouds. Some of the processed data and TV pictures from the satellite were distributed to meteorological centers on the world.

The next generation Meteor-2-21/Fizeau is the twenty-first and last in the Meteor-2 series of USSR meteorological satellites, which were launched 22 times from 1975 to 1993. The Meteor-3 series was launched 7 times between 1984 and 1994 after a development program that began in 1972. The Meteor-3-6/PRARE satellite is the sixth in the Russian Meteor-3 series of meteorological PEO satellites launched in 1994.

The Meteor-M1 PEO satellite is a Russian weather satellite designed to monitor the Earth's climate with six 6 onboard instruments, which will give to meteorologists a comprehensive look at the planet's weather systems, helping forecasters create more accurate climate outlooks. A suite of imagers and sounders will take pictures of cloud formations and detect sea surface temperatures, air temperatures and moisture. The spacecraft also carries radar designed for surveillance and detecting ice in the Polar Regions to aid navigation of ships. This satellite was replaced by the Meteor 3-M1 observatory as the first and only of the Meteor-3 M series Russian PEO weather satellites, which was launched on 10 December 2001 from the Baikonur Cosmodrome in Kazakhstan.

Polar orbiting weather satellites (PEO) circle the Earth at a typical altitude of 850 km (530 miles) in a North to South or vice versa path, passing over the poles in their continuous flight. Polar satellites are in Sun-synchronous orbit, which means they are able to observe any place on Earth and will view every location twice each day with the same general lighting conditions due to the near-constant local solar time. In such a way, PEO weather satellites offer a much better resolution than their geostationary counterparts due their closeness to the Earth.

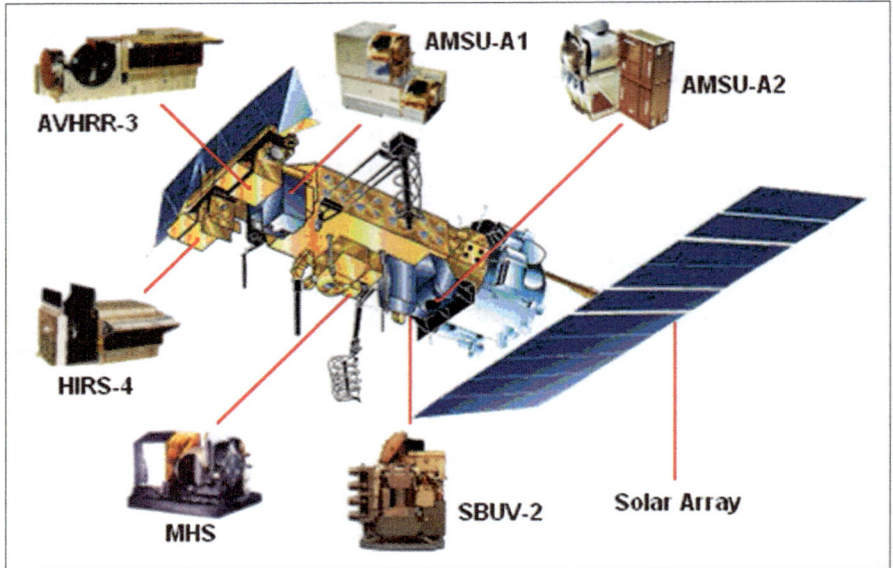

Fig. 6.1 Instruments onboard POES satellites (Courtesy of Manual: by NASA)

6.2 Meteorological Instruments Onboard POES Satellites

The primary mission of the POES constellation is to provide daily global observations of weather patterns and environmental measurements of the Earth's atmosphere, its surface and cloud cover and the proton and electron flux at satellite altitude, including to establish long-term data sets for climate monitoring and change predictions. Since the beginning of the POES program, environmental data and products acquired by its satellites have been provided to users around the globe via special weather instruments installed onboard POES platform, which is shown with 5 selected instruments in Fig. 6.1.

The United States has the NOAA series of POES meteorological satellites presently in orbit are NOAA-17, 18 and 19 as primary spacecraft, NOAA-15 and 16 as secondary spacecraft, NOAA-14 in standby and NOAA-12.

These weather satellites, with updated instruments, operate in a morning and afternoon orbit, respectively. Thus, to support the Polar (POES) mission of NOAA satellite program, these satellites carry the following satellite instruments:

1. **Advanced Very High Resolution Radiometer (AVHRR/3)** – This instrument provides daily and nighttime estimates of-land and sea surface temperature, while radiance imaging can be use to show cloud, ice, snow and vegetation cover.

2. **High Resolution Infrared Radiation Sounder (HIRS/3)** – This instrument estimates the atmosphere's vertical temperature and humidity profiles and the

pressure of Earth's surface. In addition, it estimates sea surface temperatures, total atmospheric ozone in the stratospheric layer, precipitable water, cloud height and surface Albedo.

3. **Advanced Microwave Sounding Unit (AMSU-A and B)** – This instrument is using for daily measurements of the scene radiance and estimates the global atmospheric temperature and moisture profiles from the surface to the upper stratosphere.

4. **Solar Backscatter Ultraviolet Radiometer (SBUV/2)** – This instrument is providing measurements of solar irradiance and Earth radiance (backscattered solar energy) in the near ultraviolet spectrum.

5. **Space Environment Monitor (SEM-2)** – This instrument is providing measurements of Earth's radiation belts and charged particle fluxes at satellite altitude. Besides, it provides awareness of solar and terrestrial anomalies, including warnings of solar wind occurrences that may damage or impair satellite functionality.

6. **Search and Rescue Satellite Repeater and Processor (SARR and SARP)** – This is a part of the international Cospas-Sarsat PEO satellite system designed to detect and locate maritime Emergency Position Indicating Radio Beacons (EPIRB), land mobile Personal Locator Beacons (PLB) and aeronautical Emergency Locator Transmitters (ELT).

7. **Solid State Recorder (SSR) and Digital Tape Recorder (DTR)** – Those instruments have a special purpose to complete recording and provide data storage systems that store selected sensor data during each orbit for subsequent playback

6.2.1 Advanced Very High Resolution Radiometer

The Advanced Very High Resolution Radiometer (AVHRR) is a radiation-detection imager or a scanning radiometer that can be used for remotely determining cloud cover and the surface temperature. The term surface can mean the surface of the Earth, the upper surfaces of clouds or the surface of a body of water. This radiometer uses 6 detectors that collect different bands of radiation wavelengths and it provides support to Infrared Atmospheric Sounding Interferometer (IASI), High-Resolution Infrared Radiation Sounder (HIRS), Advanced Microwave Sounding Unit (AMSU) and Microwave Humidity Sounder (MHS).

The first AVHRR was a 4-channel radiometer, first carried on TIROS-N satellite launched in October 1978). This was subsequently improved to a 5-channel instrument (AVHRR/2) that was initially carried on NOAA-7 launched in June 1981. The latest instrument version is AVHRR/3, with 6 channels, first carried on NOAA-15 launched in May 1998 is shown in Fig. 6.1. In fact, the AVHRR instrument is a satellite scanning radiometer, which means that it makes calibrated measurements of upwelling radiation from small areas, such as scan spots or pixels that are scanned across the subsatellite track. Images or pictures are constructed by displaying successive scan lines on photographic film or on a computer display. Thus, the operation of the AVHRR is representative of many scanning radiometers on PEO low Earth satellites.

The AVHRR is built by ITT Aerospace/Communication Division in Fort Wayne, Indiana. The AVHRR, like its predecessor, the Very High Resolution Radiometer (VHRR), which flew onboard the previous generation of NOAA satellites (the ITOS series), consisted of a rotating scan mirror, a telescope, internal optics, detectors and electronics. The scan mirror is elliptical with a major axis of 29.46 cm and a minor axis of 20.96 cm. It is aligned at a 45° angle to the axis of the telescope, which is nominally parallel to the satellite velocity vector. As the mirror rotates, it scans the field of view of the telescope across the Earth perpendicular to the satellite ground track. The AVHRR scan mirror rotates at a rate of 360 revolutions per minute (rpm). Since the subsatellite point of the TIROS-N series satellites moves at the rate of about 392 km min^{-1}, the distance between successive scan lines at the subsatellite point is about 1.1 km. However, the distance between successive scan lines is one parameter that describes the spatial resolution of a satellite instrument.

Two versions of the AVHRR have been flown. As already stated, the AVHRR/1 has 4 channels, while the AVHRR/2 has 5 channels. Channels 1 (0.6 μm) and 2 (1.1 μm) on both instruments use silicon detectors that are 2.54 mm on each side. In front of the detector is a 0.6-mm-square field stop, which blocks radiation except for that coming from a rectangular scans spot on Earth. Channel 3 (3.7 μm) uses an indium antimonide (InSb) detector that is 0.173 mm square. Channel 4 on the AVHRR/1 (11 μm) and channels 4 (11 μm) and 5 (12 μm) on the AVHRR/2 use mercury cadmium telluride (HgCdTe) detectors that are also 0.173 mm square. To lessen the amount of thermal noise, the detectors for channels 3, 4, and 5 are cooled to 105 K by exposing them to space (2.7-K equivalent blackbody temperature) through the side of the instrument housing. NOAA K will have a third version of the AVHRR (AVHRR/3) on board. In fact, AVHRR/3 will add a sixth channel, which is called channel 3A, sensitive to radiation between 1.58 and 1.64 μm, while Channel 3A will operate during daylight, and channel 3 will operate at night.

The US NOAA-19 POES meteorological satellite designated as NOAA-N′ was launched on 6 February 2009 with five satellite instruments. In Fig. 6.2 is illustrated 6-channel imaging radiometer AVHRR/3 instrument that detects energy in the visible and infrared (IR) portions of the electromagnetic spectrum. It measures reflected solar (visible and near-IR) energy and radiated thermal energy from land, sea, clouds, and the intervening atmosphere. The AVHRR/3 has an Instantaneous Field of View (IFOV) of 1.3 milliradians providing a nominal spatial resolution of 1.1 km (0.69 mi) at nadir. A continuously rotating elliptical scan mirror provides the cross-track scan, scanning the Earth from ±55.4¡ from nadir. The mirror scans at six revolutions per second to provide continuous coverage. This instrument provides spectral and gain improvements to the solar visible channels that provide low light energy detection. Its Channel 3A, at 1.6 microns, provides snow, ice and cloud discrimination. In fact, Channel 3A will be time-shared with the 3.7-micron channel, designated 3B, to provide five channels of continuous data.

Fig. 6.2 AVHRR (Courtesy of Manual: by NOAA)

In addition, NOAA-N′ carries a suite of instruments that provides data for weather and climate predictions. Like its predecessors, this satellite provides global images of clouds and surface features and vertical profiles of atmospheric temperature and humidity for use in numerical weather and ocean forecast models, as well as data on ozone distribution in the upper part of the atmosphere, and near-Earth space environments and information important for the marine, aviation, power generation, agriculture and other communities. Thus, an external Sun shield and an internal baffle have been added to reduce sunlight impingement into the instrument's optical cavity and detectors. For NOAA spacecraft, and later missions, the hysteris scan mirror has been replaced with a 360-RPM Brushless DC motor, which is more reliable and operates with significantly reduced jitter.

The AVHRR satellite instrument is a radiation-detection imager that can be used for remotely determining cloud cover and the surface temperature. Note that the term surface can mean the surface of the Earth, the upper surfaces of clouds, or the surface of a body of water. This scanning radiometer uses 6 detectors that collect different bands of radiation wavelengths as shown below. The AVHRR/3 instrument weighs approximately 72 pounds, measures 11.5 in. × 14.4 in. × 31.4 in., and consumes

The AVHRR/3 instrument weighs approximately 72 lb., measures 11.5 in. × 14.4 in. × 31.4 in., and consumes 28.5 watts power. Measuring the same view, this array of diverse wavelengths, after processing, permits multi spectral analysis for more precisely defining hydrologic, oceanographic and meteorological parameters. In fact, comparison of data from two channels is often used to observe features or measure various environmental parameters. However, the three channels operating entirely within the infrared band are used to detect the heat radiation from and hence, the temperature of land, water, sea surfaces and the clouds above them, which AVHRR/3 channel characteristics are presented in Table 6.1.

Therefore, the NOAA-N′ primary onboard satellite instruments are AVHRR/3, HIRS/4 and AMSU-A, which all designed for a 3-year mission. The SEM/2 instrument is fitted to the satellite and is composed of TED and MEPED detectors.

Table 6.1 AVHRR/3 channel characteristics

Channel number	Resolution at Nadir in km	Wavelength in μm	Typical applications
1 (Near IR)	1.09	0.580–0.68	Daytime clouds and land surface mapping
2 (Near IR)	1.09	0.725–1.00	Daytime land-water boundaries
3A (Near IR)	1.09	1.580–1.64	Snow and ice detection
3B (IR)	1.09	3.550–3.93	Nighttime clouds mapping and sea surface temperature
4 (IR)	1.09	10.300–11.30	Nighttime clouds mapping and sea surface temperature
5 (IR)	1.09	11.500–12.50	Sea surface temperature

The Solar Backscatter Ultraviolet Spectral Radiometer (SBUV/2) was designed for a 2-year mission, and the MHS instrument was designed for a 5-year mission.

The Channel 1 detectors of AVHRR/3 are sensitive to visible light, and thus dependent entirely on sunlight reflected off the Earth. Thus, illumination levels need to be quite high to obtain usable visible light images. Land/sea contrast is generally poor, particularly at higher latitudes. Channel 2 is reflected infrared (IR) energy. This channel is usually assumed to be the "visible" channel on Automatic Picture Transmission (APT) mode. Land/sea boundaries are very clear and cloud detail is also very good. Channel 2 is the most used daytime channel for APT images. Channel 4 is the long wave infrared channel and is effective both day and night. It is the channel offering good land/sea and cloud contrast during the night and is the channel used for nighttime APT imagery. Channel 5 has very similar characteristics to Channel 4. Channel 3B can image the Earth by both reflected infrared and emitted infrared energy.

When energy falls on the AVHRR detectors, it generates a proportional electric current, which is amplified and converted to digital information via an analog-to-digital converter. This digital information is what comprises the actual weather satellite imagery. The image is composed of 2048 picture elements (pixels) per line; the number of lines received at a station varies based on the length of time the spacecraft is above the horizon on that particular orbit. Thus, each pixel transmitted has a resolution of 1.08 km. at the satellite nadir point (point on Earth immediately below the satellite sensor). However, as the image moves away from the nadir point, the pixels become progressively distorted, and resolution decreases to approximately 5 km.

This digital data from the AVHRR/3 is processed to produce separate data streams that are transmitted by the satellite to the ground stations. These data transmissions are:

1. **High Resolution Picture Transmission (HRPT)** – Real time 1.1 km resolution images provided onboard satellite instruments containing all five spectral channels and telemetry data transmitted as high-speed digital data.

Table 6.2 Main characteristics of AVHRR/3 instrument

Parameter	Performance
Telescope	8 in. diameter afocal Cassegrain
Scan motor	360 rpm hysterisis – synchronous
Scan mirror	8.25 in. × 11.6 in. Elliptical ribbed beryllium
Cooler	Two stage radiant cooler controlled "105 K
Data output	10 bit parallel words
Video sample rate	40 kHz simultaneous sample of all channels
Output data rate	200 k word/sec max
Line sync pulse out	100 μs "6 pps
Input clock	0.9984 MHz
Overall dimensions	31.33 in. × 14.35 in. × 11.5 in.
Weight	73 lb.
Line to line scan jitter	+/− 17 μs
Scan sync drift/24 h	<3.0 μs

2. **Global Area Coverage (GAC)** – Recorded 4 km digital images that are produced over all regions of the Earth will be then are transmitted, on command, to NOAA command and control ground stations.
3. **Local Area Coverage (LAC)** – Full resolutions HRPT data that are recorded over selected regions of the Earth and then are transmitted, on command, to NOAA command and control ground stations.
4. **Automatic Picture Transmission (APT)** – Continuous real time analog transmissions of 2 channels of processed, reduced resolution AVHRR data.

The main characteristics of AVHRR/3 instrument are presented in Table 6.2:

A NOAA-18 AVHRR Infrared Window (10.8 μm) image with surface air temperatures and corresponding station identifications illustrated in Fig. 6.3 shows the signature of cold air (violet colors) settling into river valleys and other low-elevation terrain areas across the cloud-free interior of Alaska at 1916 UTC (10:16 am local time) on 18 January 2017. Thus, there was a layer of clouds (warmer cyan colors) over much of the North Slope of Alaska, which were acting to limit strong surface radiational cooling, with resulting surface air temperatures only as cold as the –20 °F. This AVHRR image was about 1 h before the low temperature at Fairbanks International Airport (PAFA) dropped to −51 °F (−46 °C), the first low of −50 °F or colder at that location since 31 December 1999 (−53 °F). While these were certainly cold temperatures, in general most were several degrees warmer than the daily record lows for 18 January 2017.

6.2.2   High Resolution Infrared Radiation Sounder

The High Resolution Infrared Radiation Sounder 2 (HIRS/2) is derived from the HIRS/1, which flew on the Nimbus-6 satellite. The HIRS/2 satellite instrument is

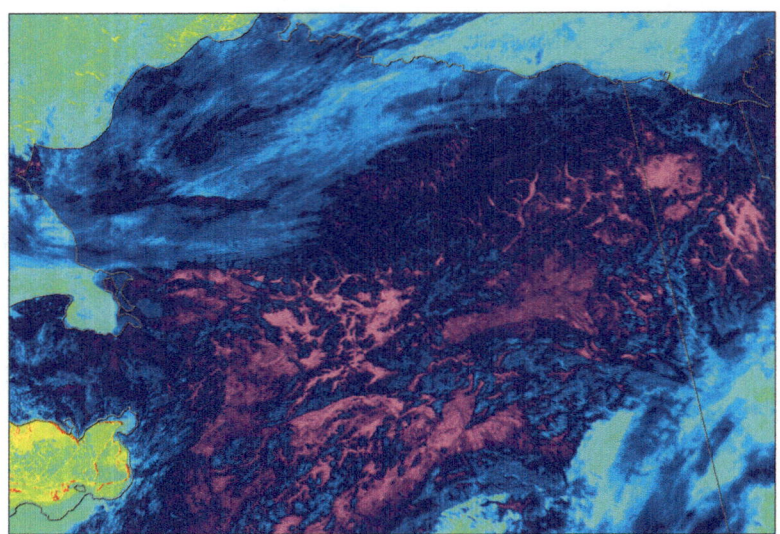

Fig. 6.3 Satellite NOAA-18 AVHRR IR image (Courtesy of Manual: by NOAA)

built by ITT Aerospace/Communication Division and operates much like the
AVHRR. The primary differences between the HIRS/2 and the AVHRR are
(1) the HIRS/2 has many more channels (20) than does the AVHRR (4 or 5), and
(2) the HIRS/2 has much coarser resolution (42 km) than the AVHRR (1.1 km).
These differences are due to the different requirements of the two instruments. The
AVHRR is designed to make images, in which the horizontal structure of the
atmosphere is most important, whereas the HIRS/2 is used for soundings, in
which the vertical structure of the atmosphere is most important.

Firstly HIRS was carried on Nimbus-6, which launched in 1975. Subsequent
integrations include TIROS-N satellite launched in October 1978, NOAA-7
launched in June 1981 and NOAA-15 launched in May 1998. The HIRS/2 instrument utilizes two carbon dioxide bands for temperature sounding: 7 channels are
located in the 15-μm band and 6 channels are located in the 4.3-μm band.

However, older sounding instruments, such as the Vertical Temperature Profile
Radiometer (VTPR), which flew on the NOAA satellites prior to the TIROS-N
series, had only 15-μm temperature sounding channels. The 4.3-μm channels were
added to improve sensitivity (change in radiance for a given change in atmospheric
temperature) at relatively warm temperatures. Besides, moisture is sensed with
3 channels in the 6.3-μm band of water vapor. The 9.7-μm channel is designed to
sense ozone an 3 channels are in atmospheric windows: the 11.11 and 3.76-μm
channels help determine the surface (skin) temperature, whereas the 0.69-μm
channel is used to detect clouds.

In Fig. 6.4 is shown image realized by NOAA meteorological satellite as a high
quality Climate Data Record (CDR) of upper tropospheric water vapor along with
HIRS brightness temperatures. In fact, this HIRS channel 12 measures upper

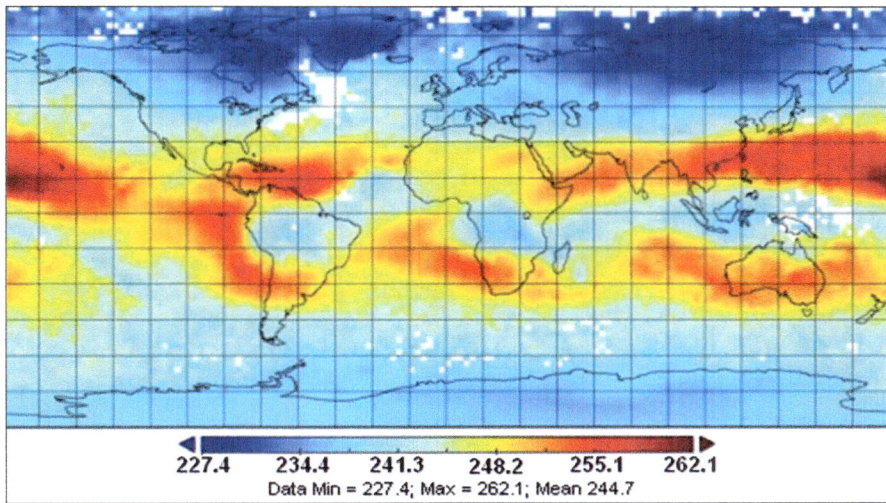

Fig. 6.4 Satellite HIRS Ch12 brightness temperature (Courtesy of Manual: by NOAA)

tropospheric humidity, while HIRS level-1b data in conjunction with cloud clearing, limb-correction and inter-satellite calibration provide necessary inputs to produce the Brightness Temperature CDR. Thus, the development of an CDR, including computer algorithms, data sets and documentation is typically a painstaking process involving NOAA and multiple scientists.

The scan mirror in the AVHRR rotates continuously, while the HIRS/2 scan mirror moves in steps. Between each step, the mirror moves 1.8°. The instrument then stares at a location for approximately 65 ms while all 20 channels are sampled. During the next 35 ms, the scan mirror steps to the next position. The total time between scan spots is 100 ms. Thus, 56 scan spots between nadir angles of ±49.5° are sampled during each scan line. The total time for a scan line, including the time for the mirror to rotate back to the first position is 6.4 s. The instantaneous field of view of the radiometer is 1.25°.

Only 3 detectors are used on the HIRS/2. The visible channel (20) is sampled by a silicon detector at ambient temperature. Channels 1–12 are called the longwave channels and are sampled by a single HgCdTe detector cooled to 105 K. The shortwave channels (13–19) are sampled by a single InSb detector also cooled to 105 K. Sampling of multiple channels with a single detector is accomplished with a rotating filter wheel. The angular length of each filter and the rotation rate of the filter wheel determine the integration time for each channel and are chosen to provide an acceptable signal to noise ratio.

Unlike the AVHRR, the HIRS/2 is not calibrated on each scan line. Instead, every 256 s the instrument goes into calibrate mode. First it looks at space, then at the internal hot and cold calibration sources. Each of these is viewed for a time equivalent to one full scan line.

Table 6.3 Main characteristics of HIRS/3 instrument

Channel	Channel frequency	(cm^{-1}) micron	Channel	Channel frequency	(cm^{-1}) micron
1	669	14.96	11	1365	7.33
2	680	14.71	12	1533	6.52
3	690	14.49	13	2188	4.57
4	703	14.22	14	2210	4.52
5	716	3.97	15	2235	4.47
6	733	13.64	16	2245	4.45
7	749	13.35	17	2420	4.13
8	900	11.11	18	2515	4.00
9	1030	9.71	19	2660	3.76
10	802	12.47	20[*]	14,500	0.690

The HIRS/3 instrument is a distinct stepping, line-scan instrument designed to measure scene radiance in 20 spectral bands to permit the calculation of the vertical temperature profile from Earth's surface to about 40 km. Some of the other objectives of HIRS/3 are the measurement of temperature profiles, moisture content, cloud height and surface Albedo.

This instrument measures scene radiance in the IR spectrum. Data from the instrument is used, in conjunction with the AMSU instruments, to calculate the atmosphere's vertical temperature profile from the Earth's surface to about 40 km (24.9 mi) altitude. In fact, the data generated by HIRS/3 is also used to determine ocean surface temperatures, total atmospheric ozone levels, precipitable water, cloud height and coverage and surface radiance, which are presented in Table 6.3.

The NOAA-19 satellite is the last of the NOAA POES series, which carries a suite of instruments that provides data for weather and climate predictions. As stated above, the NOAA-N' primary instruments includes, shown in Fig. 6.2, the AVHRR/3, HIRS/4 and AMSU-A, were all designed for a 3-year mission, while SBUV/2 was designed for a 2-year mission and MHS was designed for a 5-year mission, which positions.

The newer differences with the HIRS/3 and HIRS/4 is that the radiant cooler operates at 95 K, IFOV size is now reduced to 10 km, the inclusion of a fifth Internal Warm Target (IWT) temperature sensor, and finally tertiary telescope temperature sensor. The latest HIRS/4 instruments is illustrated in Fig. 6.5

The HIRS/4 satellite atmospheric sounding instrument provides multi-spectral data from 1 visible channel (0.69 micron), 7 shortwave channels (3.7–4.6 microns), and 12 longwave channels (6.7–15 microns) using a single telescope and a rotating filter wheel containing 20 individual spectral filters. The IFOV for each channel is approximately 7.0° that, from a spacecraft altitude of 870 km (470 mi), encompasses a circular area of 10 km (6.2 mi) in diameter at nadir on Earth. This is an improvement in resolution over the 20-km (12.4 mi) HIRS/3 instrument that was flown on NOAA-KLM, where KLM means NOAA satellites beyond NOAA-14. An elliptical scan mirror provides a cross-track scan of 56 steps of 1.8° each. The

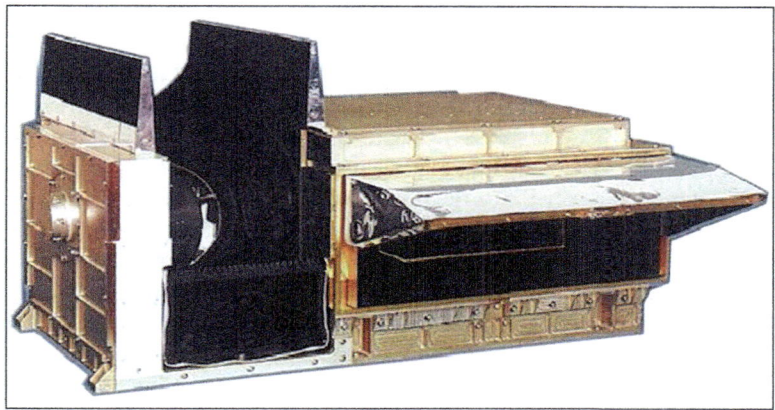

Fig. 6.5 HIRS/4 (Courtesy of Manual: by NOAA)

mirror steps rapidly, then holds at each position while the optical radiation passing through 20 spectral filters is sampled. Each Earth scan takes 6.4 s and covers ±49.5° from nadir. IR calibration of the HIRS/4 is provided by views of space and the internal warm target, each viewed once per 38 Earth scans.

6.2.3   Advanced Microwave Sounding Unit

The second sounding instrument on the NOAA satellites is the Microwave Sounding Unit (MSU) or its new generation known as Advanced MSU (AMSU). Its primary purpose is to make temperature soundings in the presence of clouds. The instrument is built by the Jet Propulsion Laboratory (JPL) of the California Institute of Technology. Its predecessor was the Scanning Microwave Spectrometer (SCAMS), which flew on the Nimbus-6 satellite.

External the MSU instrument is similar to the HIRS/2 in that it has rotating scan mirrors, which step perpendicular to the satellite track. In all there are 11 steps each of 9.47°. The total scan line, therefore covers ±47.35° from nadir. The time for each step is 1.84 s, and the total time for each scan line is 26.6 s. The MSU field of view is 7.5°, so its resolution is much coarser than that of the HIRS/2. This is a general property of microwave instruments. The Rayleigh criterion for the resolving power of a lens is that two points can just be resolved (distinguished as two points rather than one) when their angular separation λ is given by the following equation:

$$\theta = 1.22\lambda/D \qquad (6.1)$$

where λ is the wavelength of radiation and D is the diameter of the lens. The same principle applies to a microwave-receiving antenna. Consider an infrared instrument ($\lambda \approx 10$ μm) and a microwave instrument ($\lambda \approx 1$ cm) with comparably sized optics. Besides, other factors being equal, the size of a scan spot for the microwave instrument will be approximately 1000 times that of the infrared instrument, due to the ratio of their wavelengths. In fact, that the MSU resolution is not 1000 times worse than the HIRS/2 resolution indicates that the optics of the two instruments are not comparable. However, it should be kept in mind that the Rayleigh criterion severely limits the resolution of all microwave instruments. Only by using very large antenna can the resolution of microwave radiometers be made comparable to the resolutions of visible or infrared radiometers.

Although externally the HIRS/2 and the MSU instruments are similar, internally they are quite different. In the MSU, radiation from each of the two scan mirrors enters a feedhorn where it encounters a transducer that separates the beam into two beams (4 channels total). Thus, each of these beams travels via a waveguide to a device called a Dicke switch, which switches between viewing radiation from the feedhorn or from an internal microwave source of known temperature. In the MSU this switching occurs at the rate of 1 kHz.

The radiation is then mixed with an internally generated signal at the frequency of the channel and processed by a super-heterodyne (radio) receiver. The output of the receiver is proportional to the difference between the brightness temperature of the scene being viewed and the temperature of the internal radiation source. Thus, calibration of the MSU is accomplished by viewing a blackbody, attached to the instrument housing, and space once each scans. Since no scan lines are devoted exclusively to calibration, the MSU has no breaks in its scan pattern.

The new generation of MSU instrument is AMSU is a 20-channel microwave radiometer. Its primary mission objective is to obtain global temperature and humidity profiles. AMSU instrument is comprised of three separate units: AMSU-A1 (channels 3–15) illustrated in Fig. 6.6 (Left), AMSU-A2 (channels 1 and 2) illustrated in Fig. 6.6 (Right) and AMSU-B (channels 16–20). Channels 3–14 use the 50–60 GHz oxygen band to provide data for vertical temperature profiles up to 50 km. The "window" channels (1, 2, 15, and 16) provide data to enhance temperature sounding by correcting for surface emissivity, atmospheric liquid water and total precipitable water. However, channels 18–20 use the 183.3 GHz water vapor absorption line to provide data for the humidity profile. The first flight of AMSU took place on the NOAA-15 satellite with a launch in 1998.

The AMSU-A instrument always consisting of AMSU-A1 and AMSU-A2 was built by the US Aerojet, a subsidiary of GenCorp of Sacramento, CA for NASA/GSFC. The AMSU-A replaces the MSU and SSU (Stratospheric Sounding Unit) instruments flown on previous missions. It was first flown on NOAA-15 satellite launched in May 1998, which de facto is a cross-track, line-scanning meteorological instrument (whiskbroom type) designed to measure scene radiances, i.e. brightness temperatures. Its atmospheric temperature profile measurements from the surface are up to 50 km in 15 channels. Temperature resolution is 0.25–1.2 K. It is configured by the antenna/drive/calibration that is integrating a conical corrugated horn-fed shrouded reflector, multiplexer, closed-loop antenna

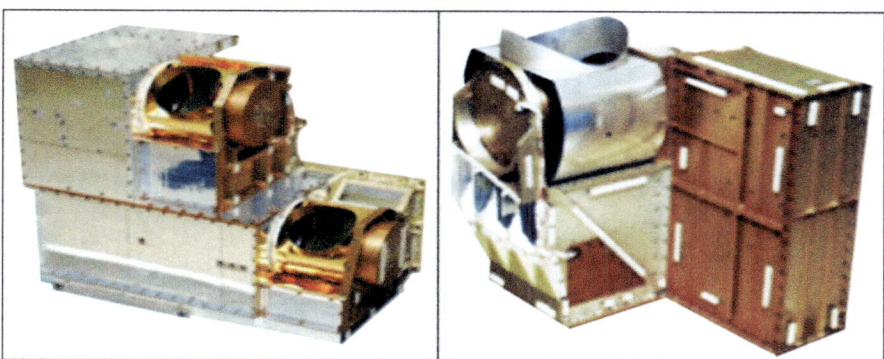

Fig. 6.6 AMSU-A1and AMSU-A2 (Courtesy of Manual: by NOAA)

scan drive assembly and closed path calibration assembly. Then AMSU-A consists receiver, signal processor and Structural/Thermal subsystem. The shrouded reflector is rotated once every scan line (8 s) for: each of 30 Earth viewing scene observations, a view of the cosmic background (about 2.73 K) and a view of a warm calibration load (about 300 K).

During the rotation cycle, the shroud prevents solar reflections from interacting with the warm load and also ensures maximum coupling of the source radiation to the antenna feed. A complete end-to-end in-flight calibration is achieved in a through-the-antenna method, which provides maximum in-flight calibration accuracy. Thus, the accuracy of the warm calibration load brightness temperature is $> \pm 0.2$ K. The closed loop antenna scan drive provides beam-pointing accuracy within $\pm 0.2°$.

The AMSU-A utilizes an 8-s scan period with a step and settle scan across the Earth scene, while AMSU-B utilizes an 8/3-s scan period with a constant speed scan across the Earth scene. In Table 6.4 are presented main characteristics of AMSU-A instrument.

The AMSU-A1 measures scene radiance in the microwave spectrum. The data from this instrument is deployed in conjunction with the HIRS instrument to calculate the global atmospheric temperature and humidity profiles from the Earth's surface to the upper stratosphere, approximately a 2-millibar pressure altitude (48 km or 29.8 mi). The data is used to provide precipitation and surface measurements including snow cover, sea ice concentration and soil moisture. This instrument has two 8 cm diameter satellite antennas (reflectors without momentum compensation), each with a $3.3°$ nominal IFOV at the half power points (FWHM) providing a resolution of about 50 km at nadir. Thus, each antenna provides a cross-track scan of $\pm 48.33°$ from nadir with a total of 30 contiguous Earth views (stepped scan positions) per scan line. The total scan period is 8 s. The footprint (resolution) at nadir is 50 km. The swath width is approximately 2100 km. Channels 11 through 14 contain 4 pass bands each.

Table 6.4 Main
characteristics of AMSU-A
instrument

Parameter	Performance
IFOV	3.3°
Sampling interval	3.3°
Spatial resolution	48 km at nadir
Scan period	8 s
Data rate	3.2 kbps
Power	90 W
Mass	104 kg

The AMSU-A2 is a cross-track scanning total power radiometer. It is divided into two physically separate modules, each of which operates and interfaces with the spacecraft independently. Module A-1 contains 13 channels and Module A-2 contains 2 channels. The instrument has an IFOV of 3.3° at the half-power points providing a nominal spatial resolution at nadir of 48 km (29.8 mi). The antenna provides a cross-track step scan, scanning ±48.3° from nadir with a total of 30 Earth fields-of view per scan line, and the instrument completes one scan every 8 s. AMSU-A2 has a single 17 cm diameter antenna (reflector with momentum compensation) with a 3.3° nominal IFOV. All other instrument/observation parameters are the same as for AMSU-A1.

Three AMSU-B flight instruments onboard NOAA-K, L and M satellites are being built by the UK EADS-Astrium Ltd. It obtains humidity profiles in 5 channels spanning the height range from the surface to about 42 km. It covers 16–20 channels, the highest channels 18, 19 and 20 span the strongly opaque water vapor absorption line at 183 GHz and provide data on the atmosphere's humidity level. It is a scanning electro-optical sounder based on a rotating mirror platform. It consists of a scanning 30 cm parabolic reflector antenna, which is rotated once every 8/3 s and focuses incoming radiation into a quasi-optic system and separates the frequencies of interest into three separate feed horns of the receiver assembly. The receiver subsystem provides further demultiplexing of the 183 GHz signal in order to selectively acquire three defined double-sided bands around the 183 GHz signal. Thus, the antenna provides a cross-track scan of ±48.95° (FOV) from nadir with a total of 90 Earth views per scan line. The instrument's IFOV is 1.1° and the separation between the center of one Earth view to the next is 1.1°. The total scan period is 8/3 s.

6.2.4 Solar Backscatter Ultraviolet Radiometer

The successor of Solar Backscatter Ultraviolet Radiometer (SBUV/1), which flew on the Nimbus-7 satellite, is a nadir pointing non-spatial SBUV/2 spectrally scanning ultraviolet radiometer carried in two modules, which is integrated with two modules such as the sensor module of optical elements/detectors and the electronics module. The overall radiometric resolution is approximately

Fig. 6.7   SBUV/2 and SEM/2 (Courtesy of Manual: by NOAA)

1 nanometer (nm). Besides, two optical radiometers form the heart of the instrument: a monochrometer and a "Cloud Cover Radiometer" (CCR). The monochrometer measures the Earth radiance directly and selectively the Sun when a diffuser is deployed. The CCR module measures the 379-nm wavelength and is coaligned to the monochrometer. The output of the CCR instrument is representing the amount of cloud cover in a scene and is used to remove cloud effects in the monochrometer data.

The SBUV/2 instrument, shown in Fig. 6.7 (Left), is a is a series of operational remote sensors installed onboard the NOAA PEOS weather satellites in Sun-synchronous orbits, which have been providing global measurements of stratospheric total ozone, as well as ozone profiles, since March 1985. The SBUV/2 instruments were developed from the SBUV experiment flown on theNimbus-7 spacecraft, which improved on the design of the original BUV instrument on Nimbus-4. These are nadir viewing radiometric instruments operating at mid to near UV wavelengths. The SBUV/2 data sets overlap with data from SBUV and Total Ozone Mapping Spectrometer (TOMS) instruments on the Nimbus-7 spacecraft. These extensive data sets (January 1979 to the present) measure the density and vertical distribution of ozone in the Earth's atmosphere from 6 to 30 miles.

The SBUV/2 measures solar irradiance and Earth radiance (backscattered solar energy) in the near ultraviolet spectrum (160–400 nm). Thus, the following atmospheric properties are measured from this data:

– The global ozone concentration in the stratosphere to an absolute accuracy of 1%;
– The vertical distribution of atmospheric ozone to an absolute accuracy of 5%; and
– The long-term solar spectral irradiance from 160 to 400 nm Photochemical processes and the influence of "trace" constituents on the ozone layer.

The Ball Aerospace-built SBUV/2 helped to discover the ozone hole over Antarctica in 1987, and continues to monitor this phenomenon. Atmospheric ozone absorbs the Sun's ultraviolet rays, which are believed to cause gene mutations, skin cancer, and cataracts in humans. Ultraviolet rays may also damage crops and aquatic ecosystems. The first SBUV/2 instrument was launched on NOAA-9 in December 1984 and the last instrument in this series was launched in February 2009 aboard the NOAA-19 spacecraft.

6.2.5 Space Environment Monitor

The Space Environment Monitor (SEM/2), shown in Fig. 6.7 (Right) is a spectrometer and provides measurements to determine the intensity of the Earth's radiation (Van Allen) belts and the flux of charged particles at the satellite altitude. It provides knowledge of solar terrestrial phenomena and also provides warnings of solar wind occurrences that may impair long-range communications, high-altitude operations, damage to satellite circuits and solar panels, or cause changes in drag and magnetic torque on satellites. Otherwise, the SEM/2 satellite instrument consists of two separate sensor units including a common Data Processing Unit (DPU), such as follows:

1. Total Energy Detector (TED) sensor unit with its performance requirements determines heat energy input into upper atmosphere from absorption of electrons, protons and positive ions. Its energy levels are for electrons from 0.05 to 20 keV and for protons is 0.05 to 20 keV, while its filed of view is two at $15°$ full angle with $-x$, $-x + 30°$; and
2. Medium Energy Proton and Electron Detector (MEPED) sensor unit has performance requirements the same as TED sensor unit. Its energy levels are for electrons from 30 to 7000 keV and for protons from 30 to 6900 keV, above >16, >35, >70, >140 MeV. Its fields of view are at $15°$ full angle with $-x$, $-x + 90°$, $15°$ full angle with $-x$, $-x + 90$ degrees and $120°$ full angle with $-x$.

The DPU device serves as the interface between the sensors and the spacecraft. It combines outputs into a 2-s, 40-word format and also provide command, calibrate and timing interfaces. In addition, the TED unit senses and quantifies the intensity in the sequentially selected energy bands. The particles of interest have energies ranging from 0.05 to 20 keV. The MEPED senses protons, electrons, and ions with energies from 30 keV to levels exceeding 6.9 MeV.

6.2.6 Search and Rescue Satellite Repeater and Processor

The US/Canada/France Sarsat and the Russian Cospas systems as a part of the international Search and Rescue (SAR) system known as Cospas-Sarsat network

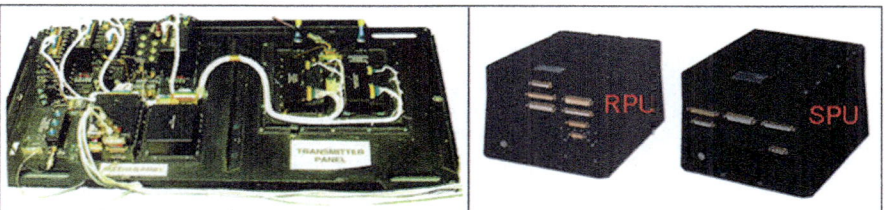

Fig. 6.8 Sarsat SARR and SARP (RPU/SPU) devices (Courtesy of Manual: by NOAA)

are designed to detect and locate ships via Emergency Position Indicating Radio Beacons (EPIRB), persons and land vehicles via Personal Locator Beacons (PLB) and aircraft via Emergency Locator Transmitters (ELT) operating at 406.05 MHz.

The Sarsat NOAA spacecraft carries two instruments to detect these emergency beacons: the Search and Rescue Repeater (SARR) provided by Canada, shown in Fig. 6.8 (Left), and the Search and Rescue Processor (SARP-2) provided by France, which Receiver Power Unit (RPU) and SARP Signal Processing Unit (SPU) are illustrated in Fig. 6.8 (Right). Otherwise, similar instruments are carried by the Russian Cospas PEO satellites.

The SARR transponds the signals of 406.05 MHz emergency beacons, which are detected on the ground only when the satellite is in view of a ground station known as a Local User Terminal (LUT). The SARP detects the signal from 406.05-MHz beacons but stores the information for subsequent downlink to a LUT, so global detection of emergency beacons is provided. This processor consists of a Receiver Power Unit (RPU) and Signal Processing Unit (SPA). The fishing fleet is required to carry 406.05-MHz emergency beacons. Thus, the 406.05-MHz beacons are also carried on most large international ships, some aircraft and pleasure vessels, as well as on terrestrial carriers. Otherwise, as of 01 February 2009, the Cospas-Sarsat system will no longer detect 121.5 or 243 MHz distress signals.

6.2.7   Digital Tape Recorder and Solid State Recorder

The Digital Tape Recorders (DTR) and Solid State Recorder (SSR) are complete recording and data storage systems that store selected sensor data during each orbit for subsequent playback. Thus, the recorders are part of the Command and Data Handling subsystem of the spacecraft that downloads to the Command and Data Acquisition Stations.

The DTR, illustrated in Fig. 6.9 (Left), consists of an electronics unit containing the data conditioning circuitry, command and telemetry subsystems, power supply, timing references, and spacecraft interfaces. Two pressurized transport units each contain a coaxial reel-to-reel tape transport with associated gearless and motorless negator-spring tape tensioning system, bearing assemblies, motor/capstan, record/

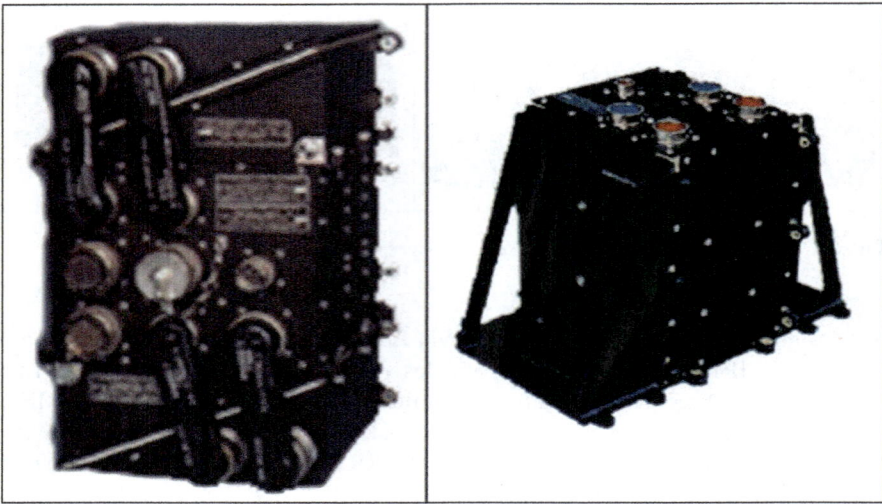

Fig. 6.9 DTR and SSR units (Courtesy of Manual: by NOAA)

playback and erase heads, and record/playback electronics. Each DTR has a nominal storage capacity of 1 GBit and selectable playback rates up to 2.66 Mb/s.

The SSR, illustrated in Fig. 6.9 (Right), performs identical functions to the DTR using solid state Dynamic Random Access Memory (DRAM) devices instead of magnetic tape and its associated electromechanical elements. It provides an increased storage capacity of 2.8 GBit and superior bit error rate performance with its custom Error Detection and Correction (EDAC) circuitry.

The NOAA-M contains a hybrid of four DTR units and one SSR unit, which are installed onboard NOAA-M POES spacecraft dedicated to provide multiple data streams, storage capacity and hardware redundancy.

However, the NOAA-N and N′ POES onboard satellite contain only solid-state recorders such that the three SSR units are configured as five recording devices, while each of them with two memory groups.

6.3   Meteorological Instruments Onboard European PEO Satellites

The Meteorological Operational satellite programme (MetOp) is a European undertaking providing weather data services that are used to monitor climate and improve weather forecasts. The MetOp program's series of three satellites has been jointly established by ESA and the Eumetsat, forming the space segment of Eumetsat's Polar System (EPS). The programme also represents the European contribution to a

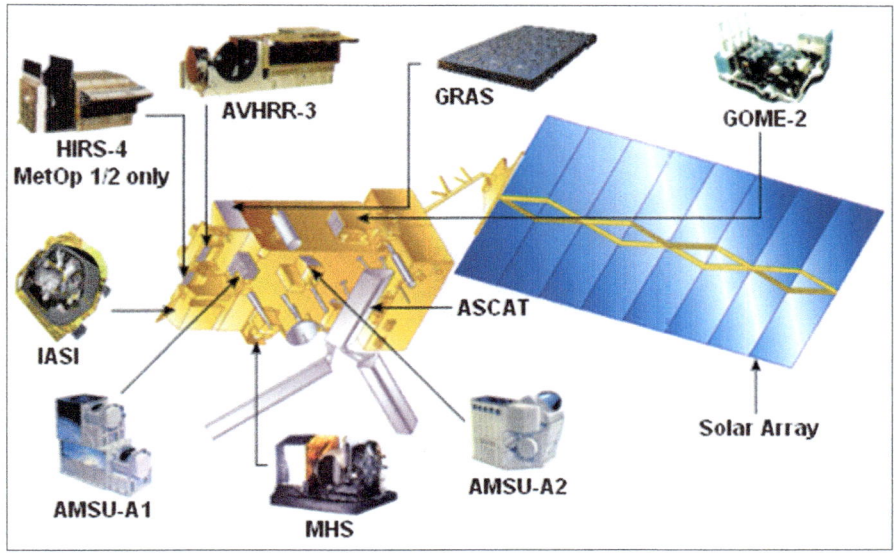

Fig. 6.10   Instruments onboard MetOp satellites (Courtesy of Manual: by ESA)

new cooperative venture with the US NOAA, which for the last 40 years has been delivering meteorological data from polar orbit, free of charge, to users worldwide.

With an array of sophisticated onboard instrumentation, illustrated in Fig. 6.10, MetOp meteorological satellites can provide data of unprecedented accuracy and resolution on a host of different variables such as temperature and humidity, ocean surface wind speed and direction and concentrations of ozone and other trace gases, thus marking a major advance in global weather forecasting and climate monitoring capabilities.

The current MetOp series of Eumetsat has three identical PEO satellites. Launched in 2006, MetOp-A is Europe's first PEO mission dedicated to operational meteorology and will be followed by MetOp-B in 2012 and MetOp-C in 2016. These first three MetOp satellites guarantee the continuous delivery of high-quality data for medium and long-term weather forecasting and for climate monitoring until at least 2020. The MetOp-A program was originally planned as a much larger satellite concept, called POEM (Polar-Orbit Earth-Observation Mission), a successor mission series to ERS-1/2 on the Columbus Polar Platform (PPF design). However, this idea was abandoned at the ESA Ministerial Council in Granada, Spain, in 1992. Instead, Envisat and MetOp were born. Full approval of the EPS (Eumetsat Polar System) program was granted in September 1998.

The MetOp-SG (Second Generation) program is being implemented in collaboration with Eumetsat, responsible for the overall mission, funds the recurrent satellites, develops the ground segment, procures the launch and Launch and Early Operations (LEOP) services and performs the satellites operations, while

Fig. 6.11 MHS (Courtesy of Manual: by ESA)

ESA will develop the prototype MetOp-SG satellites including associated instruments and procures, on behalf of Eumetsat.

6.3.1   Microwave Humidity Sounder

The Microwave Humidity Sounder (MHS) is a part of NOAA POES and MetOp weather satellite programs dedicated to measure profiles of atmospheric humidity as well as cloud liquid water content and to provide qualitative estimates of the precipitation rate. Twelve years ago, on 20 May 2005, the first MHS instrument was launched into orbit onboard the NOAA-18 satellite, which is depicted in Fig. 6.11. Four MHS instruments are currently in orbit, such as onboard satellites MetOp-A, MetOp-B, NOAA-18 and NOAA-19. All MHS instruments are in good health, while the only exception being one channel on NOAA-19. A fifth instrument will be launched on board MetOp-C in 2018.

The MHS weather instrument is a passive microwave radiometer onboard PEO satellites used to retrieve vertical profiles of atmospheric water vapour and to provide operational data from five-microwave channels radiometer, with channels from 89 to 190 GHz. Thus, it is very similar in design to the AMSU-B instrument, but some channel frequencies have been altered. Besides, it is used to study profiles of atmospheric water vapor and provide improved input data to the cloud-clearing algorithms in IR/MW sounder suites. These are key inputs for numerical weather prediction models, which are used operationally for weather forecasting worldwide.

The MHS instrument is total power, microwave radiometer designed to scan through the atmosphere to measure the apparent upwelling microwave radiation from the Earth at specific frequency bands. Since humidity in the atmosphere ice, cloud cover, rain and snow attenuate microwave radiation emitted from the surface of the Earth, it is possible, from the observations made by MHS, to derive a detailed picture of atmospheric humidity with the different channels relating to different altitudes. The MHS unit works in conjunction with four of the US instruments provided by the NOAA, namely the AMSU-A1, AMSU-A2, AVHRR and HIRS. Along with these instruments, MHS is already in operation on the NOAA-18 satellite, which was launched in May 2005, and it also forms part of the payload on NOAA-N′ launched in 2008. This instrument represents a significant enhancement in performance over the AMSU-B currently flying on the earlier NOAA-15/16/17 satellites.

Therefore, the MHS instrument will provide improved data for weather prediction models with a resulting improvement in weather forecasting. Finally, it also will help to ensure the continuous and improved availability of operational meteorological observations from polar orbit whilst providing Europe with an enhanced capability for the routine observation of the Earth from space, and in particular, to further increase Europe's capability for long term climate monitoring.

As stated before, the MHS instrument is a five-channel microwave operating in the 89 to 190 GHz region, but self-calibrating, cross-track scanning and full-power radiometer. Its channels H1 at 89.0 GHz and H2 (157 GHz are window channels that detect water vapour in the very lowest layers of the atmosphere and also observe the Earth's surface. In fact, channel H1 provides information on surface temperature and emissivity in conjunction with AMSU-A data and detects low altitude cloud and precipitation. Channels H5 (190.3 GHz), H4 (183.3 $\pm$ 3.0 GHz) and H3 (183.3 $\pm$ 1.0 GHz) measure water vapour at increasing heights in the atmosphere.

The MHS instrument scans the surface of the Earth three times every 8 s, taking 90 pixels across the Earth view each scan. The five channels are co-registered, with each pixel being separated by 1.111° in angle. At nadir, the instrument footprint corresponds to a circle of diameter approximately 16 km. The full swath of the instrument is approximately 1920 km. The instrument views a hot on-board calibration target and cold space each scan to provide a two-point calibration. Using data from these calibration views, the Earth view pixels can be converted into calibrated radiances or brightness temperatures, which data mapped by AMSU-B/MHS sensors is shown in Fig. 6.12. This dataset provides chart advanced scientific researchers a high quality Climate Data Record (CDR) of global brightness temperatures. In fact, the data period of record begins in 1998 with monthly updates. The data resolution is roughly at 16 km over the entire globe, with 90 observations per scan. Channels utilized are 89, 150, 157, 183 $\pm$ 1, 183 $\pm$ 3, 183 $\pm$ 7, 190 GHz.

The MHS data produced by MetOp satellite from is used in Numerical Weather Prediction models to improve the accuracy of future weather forecasts. Besides, the MHS data is also used to generate specific products, such as cloud liquid water content and total precipitable water in the atmosphere, as well as rain rates. Otherwise, the MHS satellite sensor has been designed and developed by Airbus Defense Space (formerly EADS Astrium), but under contract to Eumetsat. Therefore, the MHS instrument is part of the ATOVS (Advanced TIROS Operational Sounder) package, and is follow-on to the AMSU-B provided by the MetOffice flowing on the MetOp and NOAA-K/L/M satellites.

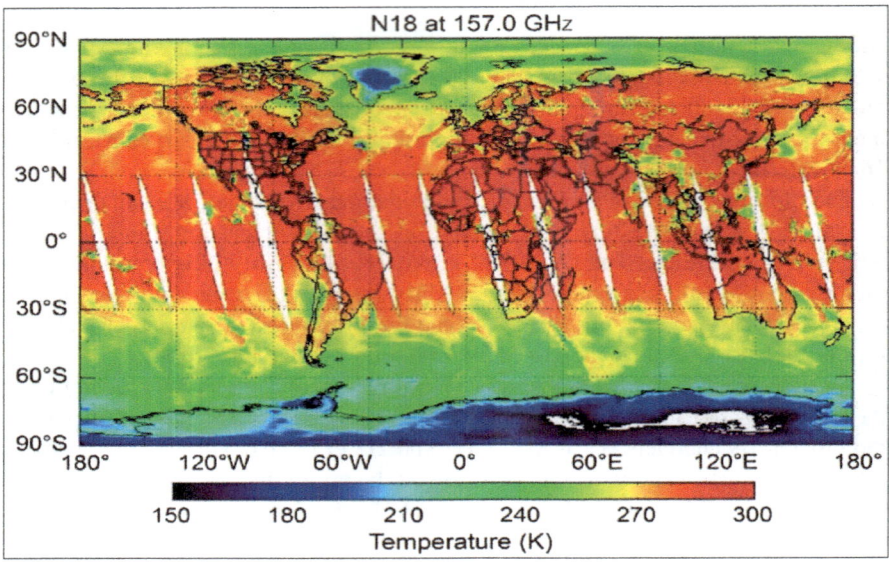

Fig. 6.12 AMSU-B/MHS mapping (Courtesy of Manual: by NOAA/ESA)

6.3.2   Infrared Atmospheric Sounding Interferometer

The Infrared Atmospheric Sounding Interferometer (IASI) is 1 of 11 instruments onboard the MetOp-A satellite that was launched on 19 October 2006. This instrument measures infrared energy emitted by the Earth-atmosphere system in thousands of spectral channels, which is illustrated in Fig. 6.13. It is the largest and heaviest instrument on the MetOp satellite and is provided by the French Space Agency CNES (Centre National d'Etudes Spatiales). CNES also provides the operational Level 1 processing software for the Core Ground Segment (CGS) and has a technical expertise centre known as IASI TEC in Toulouse that analyzes instrument performance.

The IASI satellite instrument is an infrared Fourier transform spectrometer associated with an imaging system, which uses nadir-viewing geometry passively to measure the radiance from Earth. The radiation emitted from both Earth and the atmosphere is affected by the emission, absorption and scattering of the atmospheric molecules along its optical path. In fact, it is the radiation resulting from these interactions that IASI instrument measures as the atmospheric spectrum, containing thousands of emission/absorption features. Besides, this instrument is based on a Michelson interferometer, which decomposes the emitted spectrum, after which an inverse Fourier transform and radiometric calibration are applied. The calibrated data are subsequently transmitted to the ground segment and its essential components are as follows: A scanning mirror; An afocal telescope to transfer the image onto the scan mirror; The Michelson interferometer, including a

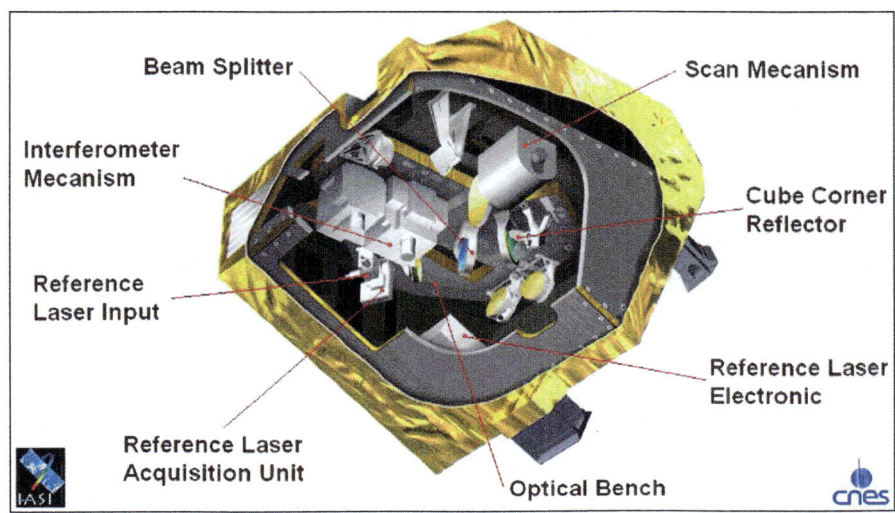

Fig. 6.13 IASI instrument (Courtesy of Manual: by ESA/CNES)

beam-splitter and corner cube mirrors that are designed to create the optical path difference necessary for the specified spectral resolution; The cold box, which the recombined beams are directed into, contains refractive optics that divides the spectral range into three bands; and The Digital Processing Subsystem that carries out the inverse Fourier transform and calibration on the interferograms before being transmitted to the ground.

The IASI instrument covers the spectral range 645–2760 cm^{-1} with a spectral resolution of 0.5 cm^{-1} (apodized) giving a total of 8461 channels in each IASI spectrum. This results in a large quantity of data that can cause problems for both retrievals and data assimilation. In such a way, choosing an optimal subset of the channels is an established method to reduce the amount of retrieval data.

The spectral range is divided into three the following bands: Band 1, from 645 to 1210 cm^{-1}, is primarily for atmospheric temperature and ozone sounding; Band 2, from 1210 to 2000 cm^{-1}, is primarily for water vapour sounding and retrieving N_2O and CH_4 column amounts; and Band 3, from 2000 to 2760 cm^{-1}, is for temperature sounding and the retrieval of N_2O and CO column amounts.

The IASI instrument achieves global coverage and provides measurements twice a day at each location, at a local time of 09:30, from a low altitude, sun synchronous orbit. Thus, this instrument scans perpendicularly to the motion of the satellite between angles $-47.85''$ and $+47.85''$, with respect to the view nadir. Within each scan there are 30 steps at which measurements are taken, as well as views to the calibration targets, an internal hot black body and cold deep space. Each view consists of a 2×2 matrix containing independent circular pixels with diameters of 12 km, as seen in Fig. 6.14, to increase the probability of obtaining cloud-free

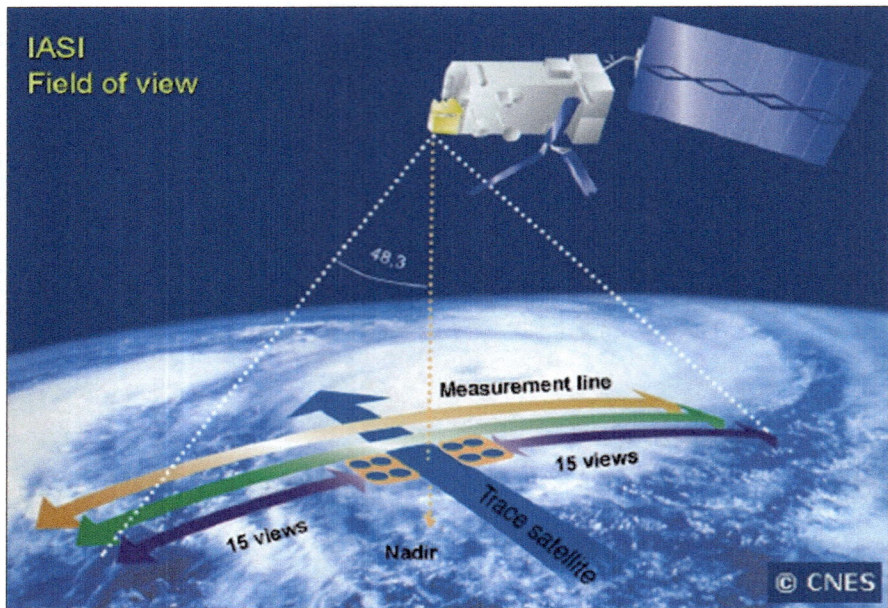

Fig. 6.14 IASI instrument field of view (Courtesy of Manual: by ESA/CNES)

views. The IASI instrument scan pattern provides measurement locations compatible with other instruments onboard the MetOp satellite.

The important ability of IASI is to detect and accurately measure the levels and circulation patterns of atmospheric gases that are known to influence the climate will then significantly contribute to climate change monitoring. In fact, IASI will also deliver data on land-surface emissivity and sea-surface temperature (in cloud-free conditions), such as:

- The temperature of the troposphere and lower stratosphere is measured by IASI sensor under cloud-free conditions with a vertical resolution of 1 km in the lower troposphere, a horizontal resolution of 25 km, and an accuracy of 1 Kelvin;
- The humidity of the troposphere is measured under cloud-free conditions, with a vertical resolution of 1–2 km in the lower troposphere and a horizontal resolution of 25 km, with an accuracy of 10%;
- Tropospheric pollution arising from industrial activity and biomass burning;
- Troposphere and stratosphere exchange with special events such as volcanic eruptions, solar proton events and related regional and global phenomena
- The total amount of ozone in stratosphere under cloud-free conditions is measured with a horizontal resolution of 25 km and an accuracy of 5%, and total column-integrated content of CO, CH_4 and N_2O with an accuracy of 10% and a horizontal resolution of 100 km. The focus being the behavior of the "ozone hole" and mid-latitude ozone as the halogen loading of the stratosphere reaches its maximum; and

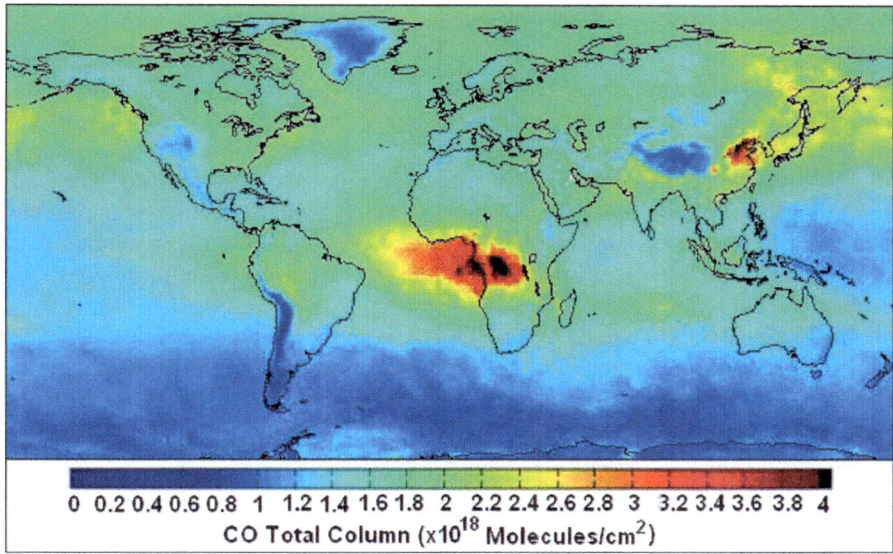

CO Total Column (x10^{18} Molecules/cm^2)

Fig. 6.15   IASI CO distribution (Courtesy of Manual: by LATMOS)

– IASI also measures the fractional cloud cover and cloud top temperature.

In Fig. 6.15 are illustrated distributions of shown IASI CO instrument for August 2011. High CO concentrations are represented in yellow and red. In such a way, the high CO levels above central Africa are due to biomass burning. The CO above US and China is related to industrial processes. The data of IASI sensor is used to improve pollution forecast, which are assimilated daily in the European Centre for Medium-Range Weather Forecast (ECMWF) by model to provide operational CO pollution forecast.

Typically, IASI CO data are available for ECMWF assimilation less than three and a half hours after observation, allowing the model to integrate new data every 30 min. Carbon monoxide contributes to climate change through its effect on ozone and methane chemistry, and is currently regulated by air quality standards. Carbon monoxide is an important ozone precursor and an important tracer for air pollution, with important connections to climate, and it is produced by incomplete combustion of fossil and bio-fuels caused predominantly by industrial processes and biomass burning.

6.3.3 Global Ozone Monitoring Experiemnt-2

The Global Ozone Monitoring Experiemnt-2 (GOME-2) instrument is a spectrometer that provides the capability to monitor high layer atmospheric ozone, nitrogen dioxide, sulphur dioxide and other trace gases, in near real-time.

The first GOME meteorological instrument for global ozone monitoring, including water vapor measurements was launched on 21 April 1995 onboard the second European Remote Sensing Satellite (ERS-2). In such a way, this instrument was able to measure a range of atmospheric trace constituents, with an emphasis on global ozone distribution.

On 30 January 1998, the ESA Earth Observation Programme Board gave its final go-ahead for the MetOp satellite meteorological program. The instruments on the MetOp weather satellites will produce high-resolution images of the Earth's surface, vertical temperature and humidity profiles, and temperatures of the land and ocean surface on a global basis. In addition, there will be instruments for monitoring trace gases and wind flow over the oceans. This instrument payload will be of significant value to meteorologists and other scientists, particularly to those studying the global climate. Given the need for global-scale routine monitoring of the abundance and distribution of stratospheric ozone and associated trace gas species, a proposal was put forward for the inclusion of GOME-2 instrument on the MetOp weather satellites.

The GOME-2 sensor is one of the new-generation European instruments carried onboard MetOp-A launched in October 2006, which is illustrated in Fig. 6.16. It continues the long-term monitoring of atmospheric trace gases started by first generation GOME sensor flew onboard ERS-2 satellite and similar Sciamachy sensor onboard Envisat satellite. The GOME-2 maps concentrations of atmospheric ozone, nitrogen dioxide, sulphur dioxide and further trace gases. Furthermore cloud properties and intensities of ultraviolet radiation are retrieved. These measured data are crucial for monitoring the atmospheric composition and the detection of pollutants.

Thus, the GOME-2 instruments will ensure the operational and accurate monitoring of atmospheric ozone and trace gases for next 15 years, contributing greatly to improving weather forecasting, atmospheric chemistry research, and climate monitoring. Observations of levels of other gases such as nitrogen dioxide and sulphur dioxide are used for the monitoring and forecasting of air quality while recorded levels of ozone are used for warnings and Ultra Violet (UV) forecasting.

The GOME-2 instrument field of view of each step may be varied in size from 5 km × 40 km to 80 km × 40 km. The mode with the largest footprint, 24 steps with a total coverage of 1920 km × 40 km, provides daily near global coverage at the equator. Based on the successful work with the GOME data processors, the German Aerospace Centre (DLR) plays a major role in the design, implementation and operation of the GOME-2 ground segment for total column products.

The DLR centre is a partner in the Satellite Application Facility on Ozone and Atmospheric Chemistry Monitoring (O3M-SAF), which is part of the EPS weather

Fig. 6.16 GOME-2
(Courtesy of Manual: by
ESA)

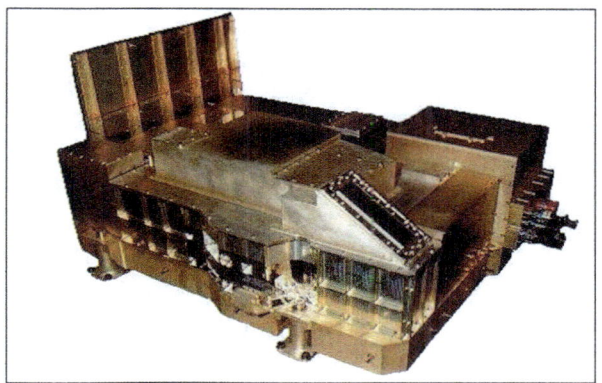

ground segment. At this point, this scientific Centre is responsible in this project for the generation of total column amounts of the various trace gases, water vapor and cloud properties, which may be retrieved from GOME-2 level 1b products.

The GOMME-2 instrument is mounted to the flight direction side of the MetOp-A satellite. A mirror of instrument scans a swath towards the surface of the Earth and directs incoming light into a telescope. The light is reflected by a number of mirrors before being directed towards a quartz prism, which separates the light into four main beams. The four beams are split further by holographic gratings to resolve the light to 0.5 nm wavelengths or better. In such a way, each scan back and forth across the satellite track takes just 6 s with a scan-width of 1920 km. Global coverage can be achieved within 1 day. The advanced GOME-2 observes four times smaller ground pixels (80 km × 40 km) than previous GOME on ERS-2 and has better polarization and the calibration capabilities.

The GOME Data Processor (GDP) operational algorithm is the baseline algorithm for the trace gas column retrievals from GOME-2/MetOp. The GDP 4.2 is a classical DOAS-AMF fitting algorithm for the generation of total column amounts of ozone, NO_2, BrO, SO_2, H_2O, HCHO and OClO [Van Roozendael et al., 2006]. In Fig. 6.17 is shown H_2O total column distribution in March 2010 as measured by GOME-2 onboard MetOp-A satellite. In fact, the global water vapour patterns over land and ocean are clearly visible with moist Intertropical Convergence Zone near the equatorial regions and dry Polar Regions.

Atmospheric water vapour (H_2O) is the most important natural (as opposed to man-made) greenhouse gas, accounting for about two-thirds of the natural greenhouse effect. Despite this importance, its role in climate and its reaction to climate change are still difficult to assess. Many details of the hydrological cycle are poorly understood, such as the process of cloud formation and the transport and release of latent heat contained in the water vapour. In contrast to other important greenhouse gases like carbon dioxide (CO_2) and methane, water vapour has a much higher temporal and spatial variability. Global monitoring of H_2O by MetOp-A is therefore a key to understanding its impact on climate.

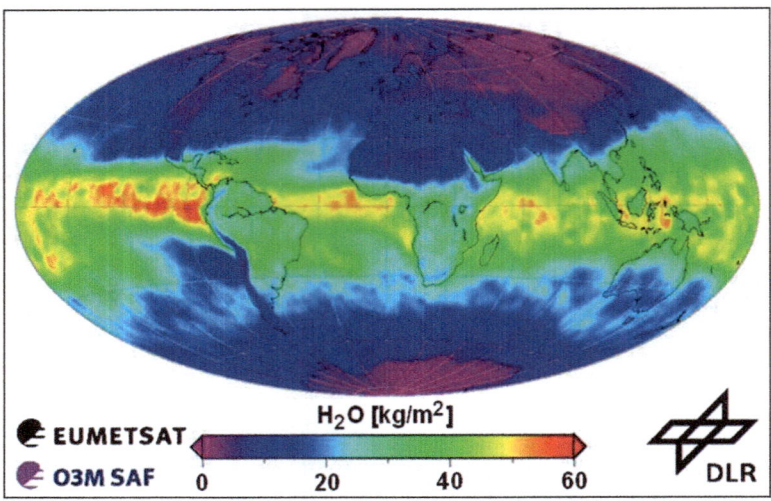

Fig. 6.17 GOME-2 global water vapour patterns (Courtesy of Manual: by DLR)

Formaldehyde (HCHO) is one of the most abundant hydrocarbons in the atmosphere and is an important indicator of so-called Non-Methane Volatile Organic Compound (NMVOC) emissions and photochemical activity. As such, it is an indicator of the presence of volatile organic compounds in the atmosphere, which in turn play an important role in the formation of toxic ozone close to the surface and also have an important influence on climate through the formation of large aerosol particles. MetOp-A measurements of HCHO can be used to constrain NMVOC emissions in current chemical transport models used in the forecasting and analysis of pollution events and also in modeling climate change.

Otherwise, GOME-2 level 1 products are generated at Eumetsat. The GOME-2 level 2 total columns products of ozone, minor trace gases and cloud properties are generated at DLR in the framework of the Satellite Application Facility on Ozone and Atmospheric Chemistry Monitoring (O3M SAF). Near-real-time products are disseminated via EumetCast and Internet. GOME-2 Level 2 off-line products can be order on-line via EOWEB or Unified Meteorological ARchive Facility (UMARF). The GOME-2 level 3 (composites) and level 4 assimilated products are generated at DLR in the framework of WDC.

6.3.4 Global Navigation Satellite System Receiver for Atmospheric Sounding

The service of Global Navigation Satellite System (GNSS) is employed for MetOp special instrument known as GNSS Receiver for Atmospheric Sounding (GRAS), which is shown in Fig. 6.18 (Left), while its antenna is illustrated in Fig. 6.18

Fig. 6.18 GRAS instrument and antenna (Courtesy of Manual: by ESA)

(Right). This satellite instrument provides information on vertical profiles of temperature and humidity, which are used for climate monitoring, and in particular, it generates sounding data for numerical weather prediction models operates as an atmospheric sounder.

The GRAS instrument uses radio occultation to measure vertical profiles of atmospheric temperature and humidity by tracking signals received from a constellation of operational GNSS satellites, while they are setting or rising behind the Earth's atmosphere. In fact, as of December 2016 only the USA NAVSTAR GPS and the Russian GLONASS are fully operational GNSS-1 network. The second-generation GNSS-2 systems are Chinese BeiDou (Compass) and European Galileo, which both will be fully operational by 2020.

Radio occultation is based on the fact that when radio waves pass through the atmosphere, either during a rise event or during a set event as seen by the receiver, they are refracted along the atmospheric path. The degree of refraction depends on gradients of air density, which in turn depend on temperature and water vapor. Thus, measurement of the refracted angle contains information about these atmospheric variables. As the measurements are made tangentially to the atmosphere, the profiles will be provided with a resolution within a few hundred meters to 1.5 km, while horizontal coverage of each profile is in the order of a few hundred kilometers. With the nominal number of GNSS satellites, GRAS will provide 500 precise atmospheric profiles every day nearly equally distributed over the Earth's surface. The GRAS sensor can track up to eight satellites for navigation purposes, two additional satellites for rise and two others for set occultation measurements. Then, GRAS has onboard GNSS satellite prediction for optimizing the navigation and occultation measurements. Besides, GRAS level 1b products contain information on a per occultation basis, along with auxiliary information.

The GRAS instrument is a satellite GNSS receiver. It was developed by RUAG Space AB (formerly SAAB Ericsson Space) in Sweden, under contract to ESA/Eumetsat and is being used on the series of MetOp satellites. As stated above, GRAS provides a minimum of 500 atmospheric profiles per day through the process of radio occultation via GNSS satellite, yielding information of the temperature and humidity of the Earth atmosphere. In addition, GRAS provides navigation solutions of the MetOp satellite position along its orbit.

Therefore, the GRAS instrument receives high quality radio signals from GPS navigation satellites, occulting the Earth atmospheric limb, via a tangential path through the Earth's atmosphere. As such, the GRAS sensor is not just an instrument, but requires a full system to provide products. Moreover, it needs a GPS constellation of satellites, and a precise orbit determination is required for the MetOp satellite, and at this point GRAS tracks the phase of GPS signals at MetOp over an occultation interval. An occultation occurs whenever a GPS satellite raises or sets on the Earth limbs, as seen from the Low Earth orbit (LEO) satellite. The GPS signal is refracted and slowed as it traverses the Earth's atmospheric limb. This causes a phase delay that relates to characteristics of the Earth's atmosphere. GRAS then compares the measured phase with the phase that would be expected in the absence of an atmosphere in order to derive bending angles.

The Doppler shift in the received GNSS signals can be processed to obtain vertical profiles (at least from 5 to 30 km) of atmospheric parameters, such as temperature and pressure, with a high degree of accuracy. In the stratosphere and upper troposphere, where water vapour density is low, refraction is dominated by the vertical temperature gradients, and an accurate temperature profile can be retrieved. In the lower troposphere, the water vapour effects are dominant and the combined temperature/water vapour profile can be retrieved. Measurement data from GRAS will be combined with data received in GPS ground based receivers, providing precise orbit determination, to retrieve the final atmospheric profiles.

In Fig. 6.19 is illustrated the GNSS (GPS/GLONASS) type of radio occultation, which relies on radio transmissions from GNSS satellites. This is a relatively new technique, first applied in 1995, for performing atmospheric measurements, used as a water forecasting tool, and could also be harnessed in monitoring climate. The technique involves an LEO satellite, such as MetOp, receiving a signal from an GNSS satellite. The signal has to pass through the atmosphere and gets refracted along the way. The magnitude of the refraction depends on the temperature and water vapor concentration in the atmosphere. GNSS Radio occultation amounts to an almost instantaneous depiction of the atmospheric state.

The relative position between the GNSS satellite and the LEO satellite changes over time, allowing for a vertical scanning of successive layers of the atmosphere. In particular, Radio Occultation observations from GPS (GPSRO) satellites observations can also be conducted from aircraft or on high mountaintops. The nominal GPS and GLONASS constellation has 24 satellites distributed in six or three Medium Earth Orbit (MEO) orbital planes around the globe at an altitude of 21,150 or 19,130 km, respectively.

The radio occultation measurements are achieved by the precise tracking of GNSS signals while transecting the atmosphere at very low elevation angles. Thus, these signals slightly refracted pass through the atmosphere, which refraction angle depending on temperature and the amount of water vapour in the atmosphere. In fact, this is accomplished through high gain antennas pointed in the velocity and anti-velocity direction of the satellite. The signal bending due to atmospheric

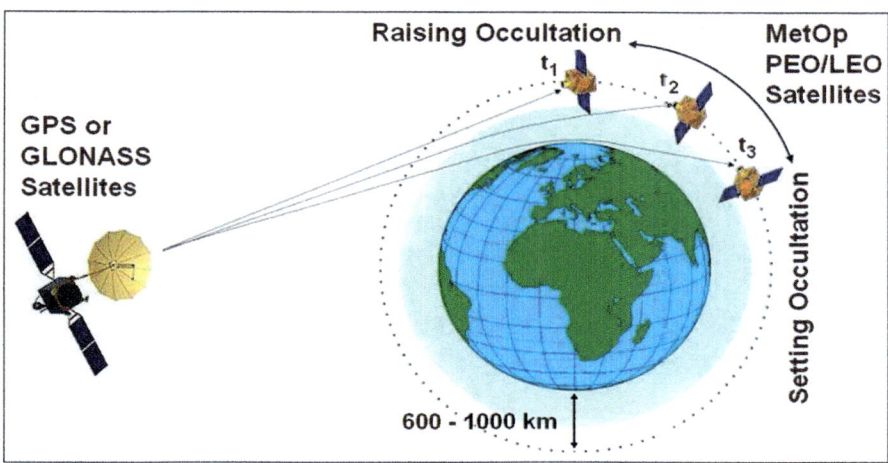

Fig. 6.19 GNSS radio occultation (Courtesy of Manual: by ESA)

refraction can then be derived. These measurements contain information that can be used in numerical weather models to retrieve atmospheric parameters such as temperature and humidity.

As stated above, the RUAG Space developed the GRAS instrument, which is in operation on the MetOp series of satellites since 2006 and provides 700 daily atmosphere soundings for numerical weather prediction. It is characterized by its accurate measurements and its high availability. In addition, the RUAG Space has also been selected for the development of the next generation of RO instrument for the MetOp Second Generation (MetOp-SG) programme. The GRAS-2 for MetOp-SG has the capability to track signals from several GNSS constellations: Galileo, GPS, GLONASS, and Compass. It uses advanced signal processing technique to allow continuous measurement coverage to the Earth's surface. Besides, it will also provide ionospheric electron density measurements for space weather applications. The first flight unit will be provided 2019.

In Fig. 6.20 is shown geographical coverage of the GRAS occultations geometry within a day period for 1 November 2006. On that day 660 occultations were recorded, 338 setting ones and 322 rising ones. The coverage is globally homogenous, which demonstrate that GRAS will provide precious profiles over sparsely covered regions such as oceans and Polar Regions. Otherwise, the required number of occultation per day is 500, which is met with substantial margins. All performances for GRAS as a simple GNSS receiver and as an atmospheric sounder are well within specifications.

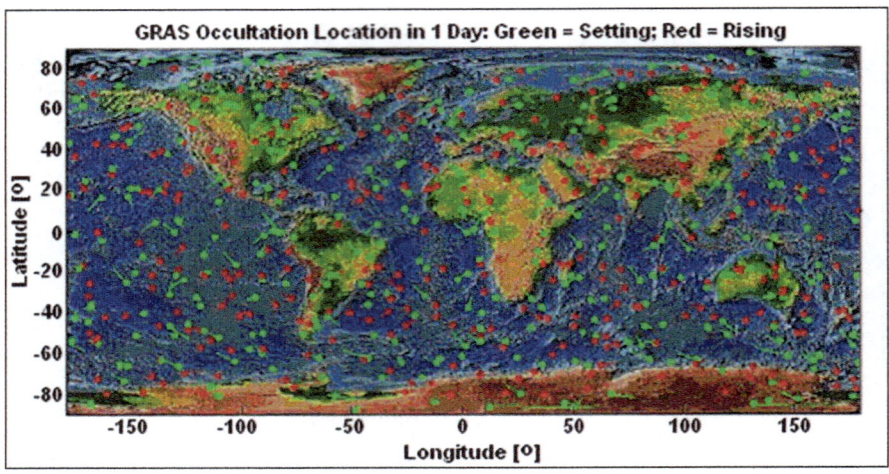

Fig. 6.20 Coverage of the GRAS occultations (Courtesy of Manual: by ESA)

6.3.5   Advanced Scatterometer

The Advanced Scatterometer (ASCAT) is de facto a radar instrument monitors near surface wind speed and direction over the global oceans and will follow the development of storms and hurricanes, typhoons and cyclones, which is illustrated in Fig. 6.21. In fact, this is an advanced scatterometer that will fly as part of the payload of the MetOp satellites, which in turn form part of the Eumetsat PEO system is developed by Dornier Satellitensysteme under the leadership of Matra Marconi Space, the satellite Prime Contractor for ESA. From its polar orbit, ASCAT will measure sea surface winds in two 500 km wide swaths and will achieve global coverage in a period of just 5 days.

Wind scatterometers already flown onboard ESA's ERS-1 and ERS-2 PEO remote sensing satellites have demonstrated the value of such instruments for the global determination of sea-surface wind vectors. Wind scatterometers are used to infer data on wind speed and direction from radar measurements of the sea surface. These successful instruments as a part of the Active Microwave Instrument (AMI) were conceived about 20 years ago. During the intervening period, there has been a considerable evolution in the capabilities of spaceborne hardware. The ASCAT instrument is the successor to these lactometers, which are flown as part of the payloads of Metop-1, 2 and 3 PEO satellites. It represents advances in increased coverage, reduced data rate and power-efficient low-mass technology.

The ASCAT data has also proved to be very useful in a variety of studies, including polar ice and tropical vegetation. Thus, water plays a unique role at microwave frequencies at which scatterometers are operated. It is the only naturally abundant medium with a high dielectric constant and provides increasing the fraction of liquid water contained in soil, snow and vegetation increases the

Fig. 6.21 ASCAT
(Courtesy of Manual: by
ESA)

dielectric properties of these media, thereby significantly altering their scattering and absorption behavior.

The backscattering coefficient, measured with scatterometers, is dependent on the dielectric properties of the soil surface layer, surface roughness and vegetation. Besides, ASCAT is providing useful data for ice and land applications, such as sea ice extent, permafrost boundary, desertification and scn on. Because the scatterometer radar signal can penetrate the surface, ASCAT can also observe subsurface/subcanopy climate-related features. With the rapid global coverage, day or night and all-weather operation, ASCAT offers a unique tool for long-term climate studies.

The ASCAT instrument is real aperture radar, operating at 5.255 GHz (C-band) and using vertically polarized antennas. It transmits a long pulse with Linear Frequency Modulation ("chirp"). Thereafter, ground echoes are received by the instrument and, after de-chirping, the backscattered signal is spectrally analyzed and detected. Thus, in the power spectrum, frequency can be mapped into slant range, provided the chirp rate and the Doppler effect frequency are known. The processing is, in effect, a pulse compression, which provides range resolution. From around 837 km altitude, the instrument transmits well-characterized pulses of microwave energy towards the sea surface. In that manner, winds over the sea cause small-scale (centimetric) disturbances of the sea surface, which modify its radar backscattering characteristics in a particular way.

These backscattering properties are well known and are dependent on both the wind speed over the oceans and the direction of the wind, with respect to the point from which the sea surface is observed. Onboard ASCAT, two sets of three antennas measure the resultant electromagnetic backscatter from the wind-roughened ocean surface, yielding two swaths 550 km wide with a gap of 670 km between them, on each side of the satellite ground track, which scenario is shown in Fig. 6.22. The three antennas on each side are oriented to broadside and ±45° of

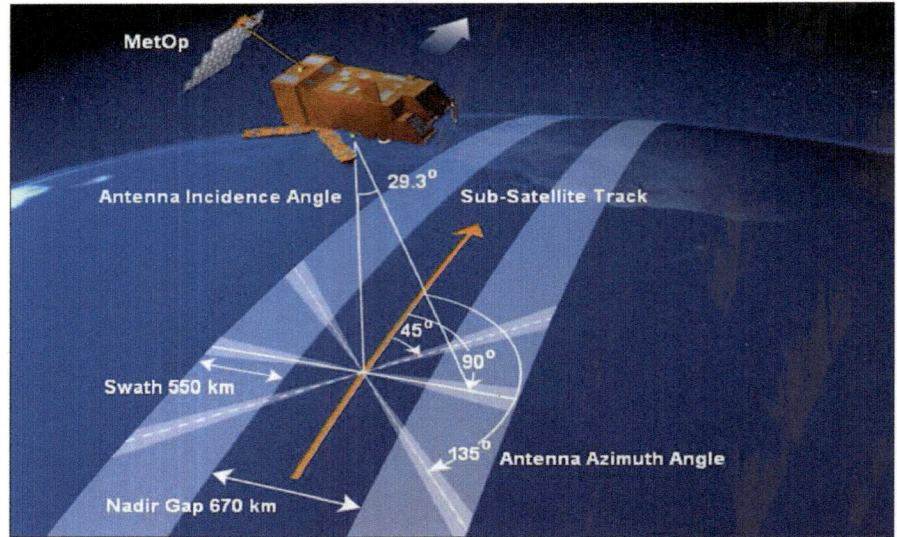

Fig. 6.22 Instrument ASCAT onboard MetOp satellite (Courtesy of Manual: by ESA)

broadside, and so, make sequential observations of the backscattering coefficient of each point of interest from three directions. These three antennas correspond to three different angles (azimuth angles). For instance, for the set of antennas facing to the right, the azimuth angles are 45° (fore beam), 90° (mid beam) and 135° (aft beam). This results, as the METOP satellite moves forward at 7 km/s, in three backscatter measurements for the same area, the so-called sigma-naught triplets. Among the rest, the three directions are needed to resolve the wind direction ambiguity.

The ASCAT ocean surface winds are a 10 m neutral stability wind. These products are processed by NOAA/NESDIS (National Environmental Satellite, Data and Information Service) utilizing measurements from ASCAT aboard the Eumetsat MetOp satellite. The current Geophysical Model Function (GMF) being used is CMOD5.5, where the GMF relates the normalized radar cross-section to the ocean surface wind speed and direction.

The useful ASCAT observations are as follows: Ocean surface wind vectors of 50, 25 and 12.5 km resolution with speed and direction, Storm data, Global ambiguity and Ice data. In Europe, the Nederland KNMI institute is responsible for the ASCAT wind products, and additional data and information can be found at the Eumetsat Ocean and Sea Ice Satellite Application Facility (OSI SAF). In Fig. 6.23 (Left) is depicted ASCAT wind product, which is showing directions of cyclone wind of 25 km resolution over the Atlantic Ocean, close to Ireland and Scotland. In Fig. 6.23 (Right) is depicted the same ASCAT wind product of 12.5 km resolution. Here are presented ASCAT storm and ice data as follows:

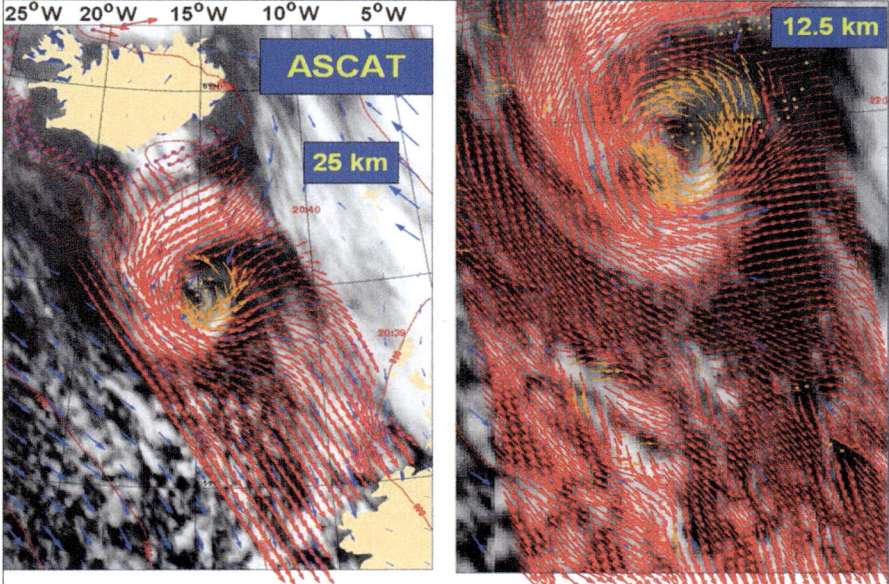

Fig. 6.23 ASCAT onboard MetOp satellite (Courtesy of Manual: by ESA/KNMI)

1. **ASCAT Storm Data** – The storm data and centered imagery are generated based on storm centres around the world. These files are updated whenever new information about active storms become available, and the imagery is updated shortly after receipt of these files. The ambiguity image plots all the ambiguities resulting from the wind retrieval process, with 4 being the maximum number of ambiguities. The NRCS (Normalized Radar Cross-Section or Sigma-0) images shown here are gridded into an approximately 3 km grid in latitude and longitude, currently showing only vertical polarization from the forward look. In the NRCS images, brighter means higher backscattered power levels, which in ocean scenes generally implies increased small-scale (~2-5 cm) roughness on the surface. The hypothesis is that the NRCS images will prove useful in certain situations in locating features, such as storm centers on the ocean surface, especially in cases where the wind vector retrievals are inconclusive. There may be information in these images, which will require experience in fully exploiting. Any feedback on these images would be appreciated. Details about these and other enhanced resolution scatterometer image products can be found at the NASA Scatterometer Climate Record Pathfinder Project (NSCRPP).

2. **ASCAT Ice Data** – The daily ice image products are generated from the ASCAT Level 1B sigma-0 data, which are operationally obtained from Eumetsat. The ASCAT ice images are produced using software that was developed as part of the NSCRPP The data product contains the most recent images

covering the Antarctic, the Arctic, the Ross Ice Shelf, South Georgia Island and the Waddell Sea.

6.4 Introduction to GEO Satellite Meteorological System

The meteorological satellites that are placed in Geostationary Earth Orbit (GEO) are called geostationary meteorological or weather satellites. The GEO meteorological satellites are in the orbit around the Earth at a height of 35,880 km. The orbiting speed of the GEO weather satellite is equal to the speed of Earth rotation, namely it is spinning of the Earth on its own axis. Hence, when we look the Earth from the GEO satellite, Earth looks like it is at rest position. Thus, weather satellites easily capture the images of the Earth with their sensors.

Meteorological GEO satellites are operated globally by the US NOAA/NASA with GOES satellites, by the European Eumetsat – Meteosat satellites, by the Russian Roskosmos and RosHydroMet – Electro satellites, by the Chinese CMA/NRSCC – FY-4 satellites and by the Japanese JMA – GMS/Himawari satellites, which global coverage map is shown in Fig. 6.24. Here are not presented GEO satellites developed by Indian ISRO – Insat and Kalpana satellites and by South Korean KMA – COMS satellites.

The GEO satellite orbit is in the Earth's equatorial plane at a height of 38,500 km. At this height, the speed of the satellite is the same as the Earth's rotation, so the satellite appears to be stationary over a certain point on the equator. In fact, this orbit allows the satellite to continually observe the same area: 42% of the earth's surface. To get global coverage you need a network of 5–6 satellites. These satellites, however, do not see the poles at all.

The Applications Technology Satellite (ATS-1) was launched on 7 December 7 1966, as the first of six spacecrafts used to test the feasibility of placing a satellite into GEO. It was originally intended to be a communications satellite, but de facto also provided a platform for meteorological and navigation equipment. Otherwise, for all these years, measurements from GEO satellite sensors have faced severe limitations because of the low radiances reaching up to the height of the satellites and the comparatively low signal-to-ratio ratio of the instruments. Recent advances in sensor technology have, however, helped to overcome these limitations and new sensors of various types are under active consideration for being placed at geostationary altitudes.

For weather analysis and forecasting GEO satellites have an advantage over PEO satellites, because images may be made frequently rather than once or twice per day. Thus the motion and rate of change of weather systems may be observed.

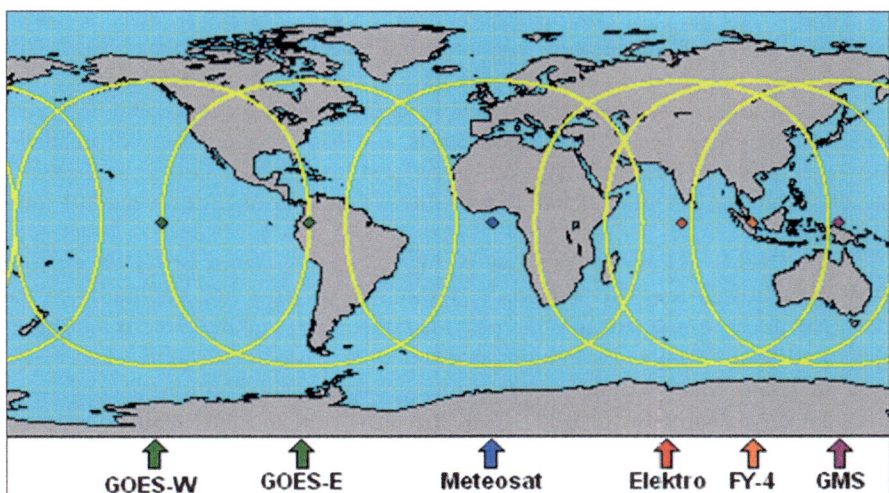

Fig. 6.24 GEO weather satellite coverage map (Courtesy of Manual: by NCAR)

6.5 Meteorological Instruments Onboard GOES Satellites

The Geostationary Operational Environmental Satellite (GOES) series of meteorological satellites is owned and operated by the National Oceanic and Atmospheric
Administration (NOAA). The objective of the GOES system is to maintain a
continuous data stream from two satellites to support the requirements of the US
National Weather Service. Currently, GOES satellites provide half-hourly observations of the Earth and its environment, which are continuously transmitted to
ground terminals and processed for rebroadcast to primary weather services and
research communities around the world. The GOES satellites operate from two
primary locations: GOES East is located at 75° W and GOES West is located at
135°W over the Pacific Ocean. NOAA also maintains an on-orbit spare GOES
satellite in the event of an anomaly or failure of GOES East or GOES West.

The series of GOES satellites began with the Synchronous Meteorological
Satellite (SMS) launched on 17 May 1974 as SMS-1. In this case, like TIROS-N,
both SMS-1 and SMS-2 satellites were considered experimental satellites and were
initially operated by NASA. The GOES were commissioned and are operated by
NOAA as well, while SMS 1/2 and GOES 1, 2, and 3 were designed and built by
Ford Aerospace based on NASA's ATS 1–3, while GOES 4, 5, 6, and 7 were
designed and built by Hughes Aircraft.

The early GOES 1–3 or A-C satellites were spin-stabilized, viewing Earth only
about 10% of the time and provided data in only two dimensions (three if you
consider time), which coverage is depicted in Fig. 6.24. There was no indication of
cloud thickness, moisture content, temperature variation with altitude or any data in
the vertical dimension.

However, on the 1980s the capability of GOES 4–7 or D-H was added to obtain vertical profiles of temperature and moisture throughout the atmosphere. This added dimension gave forecasters a more accurate picture of the intensity and extent of storms, allowed them to monitor rapidly changing events, and to predict fog, frost and freeze, dust storms, flash floods, and even the likelihood of tornadoes. However, as in the 70s, the imager and sounder still shared the same optics system, which meant the instruments had to take turns. Also, the satellites were still spin-stabilized.

The GOES-8 spacecraft launched in 1994, including series of GOES 8-12 or I-M, brought real improvement in the resolution, quantity and continuity of the data. Advances in two technologies provided: three-axis stabilization of the space-craft and separate optics for imaging and sounding. Three-axis stabilization meant that the imager and sounder could work simultaneously. Forecasters had more accurate data and better pinpoint locations of storms and potentially dangerous weather events such as lightning and tornadoes. Thus, the satellites could tempo-rarily suspend their routine scans of the hemisphere to concentrate on a small area of quickly evolving events to improve short-term weather forecasts.

The GOES 13-15 or N-P satellites further improved the imager and sounder resolution with the Image Navigation and Registration subsystem, which uses geographic landmarks and star locations to better pinpoint the coordinates of intense storms. Optics of detectors was improved and because of better batteries and more available power, imaging is continuous.

Each GOES satellite carries two major instruments: an imager and a sounder. Imager data, which consist of measurements of five different wavelengths of visible and infrared (IR) energy, are acquired on a three-hourly basis via Forecast Systems Laboratory of NPAA. Of the five wavelengths sampled by the imager, the visible, water vapor IR, and thermal IR wavelengths are used to monitor clouds, atmo-spheric water vapor, and (in cloud-free areas) sea surface temperatures.

6.5.1 GOES 4-7 (D-H) Instruments

The GOES/SMS first-generation GEO satellites have three principal subsystems: the Space Environment Monitor (SEM), the Telemetry, Tracking, and Command (TTC) subsystem and a meteorological instrument. The SEM is quite different from the SEM onboard the NOAA satellites. It includes a magnetometer, a solar X-ray sensor and an energetic particle monitor. Thus, the instrument package is designed to study solar activity and the Earth's magnetic field. The TTC has several functions including transmission to Earth of data collected by the satellite instruments, satellite command, relay of data from Earth-based Data Collection Platforms (DCP) and relay of weather facsimile (WEFAX) charts.

Prior to GOES-4 the meteorological instrument on GOES was the Visible and Infrared Spin Scan Radiometer (VISSR). Onboard GOES 4-7 satellite, illustrated in Fig. 6.25, the primary instrument has been the VISSR Atmospheric Sounder (VAS),

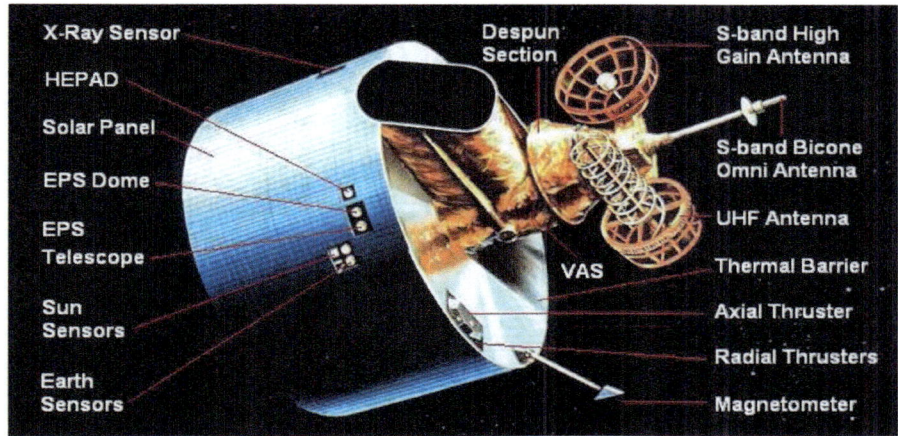

Fig. 6.25 GOES 4–7 weather satellite (Courtesy of Manual: by NOAA)

which has sounding capabilities in addition to the VISSR capabilities, and it was designed and built by Santa Barbara Research Center (subsidiary of Hughes Aircraft. It has spun and de-spun portions, which includes the antennas for communication with Earth. The VAS resides in the spun portion of the satellite, and its optical axis coincides with the principal axis of the satellite.

The spin axis is maintained closely parallel to the Earth's spin axis. Otherwise, the satellite rotates at the rate of 100 rpm (10.4720 rad s^{-1}, which provides East-West optical scanning, while North-South optical scanning is provided by a scan mirror that moves in 192-/xrad steps. At GEO altitude, the Earth subtends only 17.4°. Therefore, the mirror need only scan ±10° from the subsatellite point to view the entire hemisphere. A complete hemispherical scan is made in 1821 rotations of the satellite with one mirror step after each rotation. At 100 rpm, a hemispherical image, called a full-disk image, is completed in 18.21 min. Usually full-disk images are terminated short of 1821 scans thus eliminating a portion of the Southern Hemisphere. At the end of each image, the mirror rapidly retraces to its northernmost position to start a new image.

The normal GOES 4-7 satellite operating schedule calls for an image to be initiated each 30 min. GOES-East images start on the hour and 30 min past the hour; GOES-West images start at 15 and 45 min past the hour. During special rapid-scan periods, sectors of limited North-south extent (but full East-West extent) are scanned in less time. Scanning periods of 15, 7.5, 5, and 3 min have been used.

In Fig. 6.26 is illustrated VAS instrument onboard GOES 4-7 with all main components.

After being reflected by the VAS scan mirror, radiation enters a 40.64-cm diameter, f/7.2, Ritchey-Chretien telescope. At the principal focal plane, the radiation is intercepted by an array of eight optical fibers connected to eight photomultiplier tubes sensitive to radiation in the wavelength range

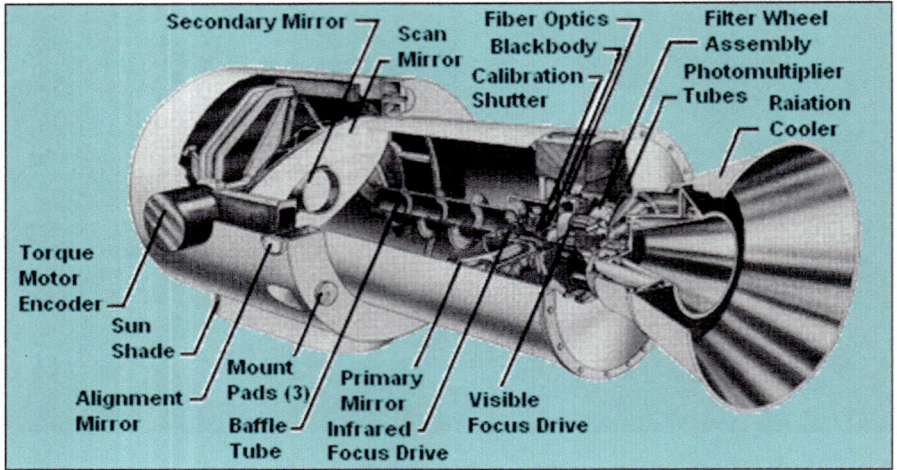

Fig. 6.26 VAS instrument onboard GOES 4–7 (Courtesy of Manual: by NOAA)

0.54–0.70 μm. In fact, this array of optical fibers is arranged in a North-South line, and each fiber has a field of view of 25 μrad (E-W) by 24 μrad (N-S).

Thus, with each rotation of the satellite, eight lines of visible data are swept out. The 25 × 24-μrad field of view results in a ground resolution of about 0.9 km at the subsatellite point. The visible channels are sampled at the rate of 500 kHz, which, at a satellite rotation rate of 10.47 rad s^{-1}, means that visible scan spots are separated by 21 μrad or 0.75 km East-West. Visible scan lines are separated by 24 μrad or 0.86 km North-South. In such a way, visible scan lines are contiguous, but adjacent scan spots overlap slightly.

The visible data of VAS instrument are digitized using 6 bits, but the output count is not proportional to radiance; rather, it is approximately proportional to the square root of the radiance. However, this choice was made because the noise from the photomultiplier tubes is proportional to the square root of the radiance. By making the output proportional to the square root of radiance, the noise level, expressed in counts, is nearly independent of the count value. The visible channels are calibrated before launch, and calibration curves are available from NESDIS.

On each rotation of the GOES 4–7 satellite, 15,288 samples are taken for each of the eight visible channels. Since a complete full-disk image is made in 1821 rotations, it contains 14,568 lines (8 × 1821). A single full-disk visible image, therefore, consists of 15,288 × 14,568 × 6 = 1,336,293,504 bits (159.3 megabytes), not counting the documentation and navigation information, which is normally part of the image file. Archiving all of these data is a challenge.

Since there are eight visible channels, each digitized to 6 bits, and each sampled at the rate of 500 kHz, the visible data are taken at the rate of 24 Mb/s s^{-1} (megabits per second). The addition of the infrared data boosts the total data rate to 28 Mb/s s^{-1}. This rate only occurs, however, while the instrument is viewing the Earth. Ninety-five percent of the time the instrument views space and collects no data. To decrease the data rate so that users may receive it more easily, the following arrangement is employed.

The high-data-rate, raw VAS data are transmitted in real time to the Command and Data Acquisition (CDA) station located at Wallops Island, Virginia. The data are transmitted back to the satellite at a data rate of approximately 2.1 Mb/s s^{-1}, which means that the data that were collected only while the instrument viewed Earth, take one full rotation of the satellite to transmit. This data, called stretched data, are retransmitted by the GOES satellite to anyone with the proper receiving equipment. Real-time ground processing, in addition to stretching the data, allows some useful modifications of the data stream:

- Offsets are added to the visible data to correct for differences in the eight-photomultiplier tubes. (This reduces the "striping" that can sometimes be seen);
- Raw output voltages from the infrared sensors are converted into calibrated, 10-bit counts that are proportional to radiance;
- The data are reordered so that adjacent fields of view are adjacent in the data stream (see below). Satellite orbit and attitude information is embedded in the data stream; and.
- A grid is included in the data stream for those who print images.

Thus, the exact format of the retransmitted VAS data stream is detailed in the Operational VAS Mode AAA Format (NESDIS, 1987). In many satellite instruments a beamsplitter is used to separate the visible from the infrared radiation. A beamsplitter allows the visible and infrared detectors to view the same point on Earth at the same time, but it necessarily causes some loss of signal. In the VAS, no beamsplitter is used. The infrared field of view is displaced slightly from the visible field of view. This displacement is corrected in real time by reordering the data during ground processing.

The VAS has three operating modes:

1. **VISSR Mode** – The VISSR is a backup mode, in which onboard computer, called the VAS Processor, is turned off, and the instrument operates much like the VISSR on the first generation of GOES satellites. Normally the filter wheel is set to channel 8 (11 μm, which although can be set to any position), the two small detectors are active, and full-resolution visible and 3 × 7-km IR data are collected.
2. **MSI Mode** – This is the normal VAS imaging mode. Here, that full-resolution visible data are collected along with two, three or four channels of infrared data. In addition, two images may be collected at 3 × 7-km resolution, one image may be collected at 3 × 7-km resolution and two images at 3 × 14-km resolution, or four images may be collected at 3 × 14-km resolution. In this mode, the filter wheel is under control of the VAS Processor, and the scan mirror steps 192 μrad on each revolution of the satellite.
3. **Dwell Sounding Mode** – This mode is used to make soundings, while no visible data are collected. In this mode, the filter wheel and the scan mirror are under

control of the VAS Processor. All 12 positions of the filter wheel are sampled sequentially. Normally, the small detectors are used for channels 3–5, 7, and 8, and the large detectors are used for the remaining channels. The scan mirror does not step with each rotation of the satellite. Thus, to reduce any noise, the instrument dwells on the same scan line for several programmable rotations. In such a way, the number of rotations of the satellite at each mirror position for each filter wheel position is called the spin budget. Finally, the noise is further reduced by averaging the radiance over an area. Because of the multiple spins on each scan line, only a limited latitude band can be dwell sounded during the 30-min GOES duty cycle.

Since 1978, the digital GOES data has been archived by specially adapted video tape recorders contracted by NESDIS. In addition, a number of hard copy prints and negatives are archived by NESDIS. The GOES data from these archives may be requested from the Satellite Data Services Division of NESDIS.

6.5.2   GOES 8-12 (l-M) Instruments

The five satellites that constitute the second generation of GOES I-M satellite series, such as GOES I, J, K, L and M or GOES 8, 9, 10, 11 and 12 after launch, are being built by the prime contractor, Space Systems/Loral, which sample of satellite with major elements is depicted in Fig. 6.27. The GOES-8 satellite was launched on 13 April 1994. Menzel and Purdom in 1994 described the GOES I-M satellites and instruments.

The most significant change from previous GOES satellites was that GOES I-M series are a three-axis stabilized satellite designed after Insat. Three-axis stabilization means that the instruments always point toward Earth, whereas the previous GOES spin causing their instruments to sense the Earth only about 5% of the time. The improved duty cycle of the GOES I-M instruments allow better radiometric accuracy and higher spatial resolutions than is possible on GOES 4–7. Thus, other improvements include better pointing accuracy, such as within 4 km at the subsatellite point, and the routine capability to track stars for navigational purposes.

The earlier stated VAS combines imaging and sounding in one instrument, which means that imaging has to be abandoned while soundings are being made. The Imager and Sounder instruments onboard GOES I-M are separate instruments capable of independent and simultaneous operation. This increases the rate at which soundings are available.

6.5.2.1   GOES 8-12 (l-M) Imager Instrument

The GOES I-M Imager is built by the ITT Aerospace/Communication Division and is similar to the AVHRR, which like AVHRR has 5 channels: 0.65, 3.9, 6.75, 10.7

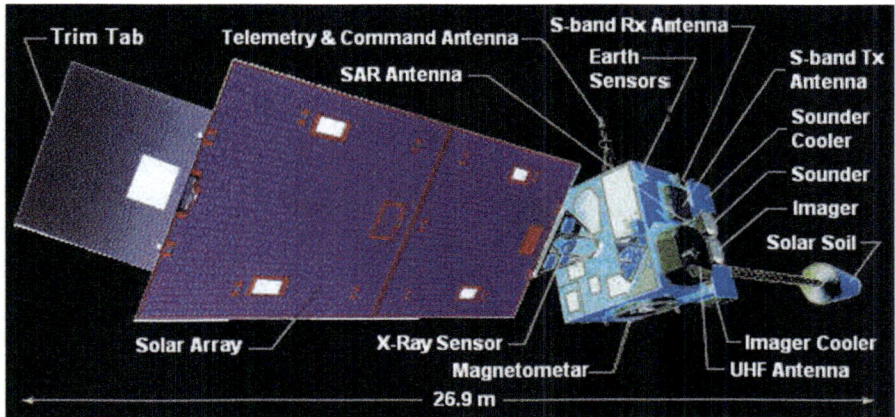

Fig. 6.27   GOES 8–12 weather satellites (Courtesy of Manual: by NOAA)

and 12.0 µm, however, a water vapor channel of 6.75 µm has been substituted with the 0.9-µm AVHRR channel. Thus, all five channels are scanned during every Imager frame.

The Imager scans West-to-East on one scan line followed by an east-to-west scan on the next line. Visible data are swept out eight lines at a time with eight silicon photodiode detectors. The 3.9, 10.7 and 12.0-µm channels are swept out two scan lines at a time, and the 6.75-µm channel is swept out one line at a time. At the end of each scan line, the mirror steps down (South) 224 µrad (8 km) to begin a new scan.

The ground resolution of some of the channels is improved over that possible with VAS instrument. The 3.9-µm channel has approximately four times the resolution, and the 10.7 and 12.0-µm channels have about twice the resolution, whereas the 0.65 and 6.75-µm channels have about the same resolution

The radiometric accuracy of some channels also increase and digitization for the visible channel increases from 6 bits to 10 bits and also the output is linear in radiance instead of proportional to the square root of radiance. This should improve the low-light detection capabilities of the visible channel.

In addition, the scan patterns for the GOES I-M satellite Imager, shown with all containing components in Fig. 6.28, are much more flexible than those previously available. For example, it is possible to suspend a full-disk scan to perform a rapid scan of a small area and then to resume the full-disk scan, while rapid scanning must be halted during full disk scanning. Even though three-axis stabilization makes it unnecessary to stretch the data as is done with the VAS instrument, the raw data are still transmitted to Wallops Island, where they are corrected, registered, tagged with ground coordinates and merged with sounder data before being transmitted back to the satellite. The GOES satellite then retransmits the processed data to users.

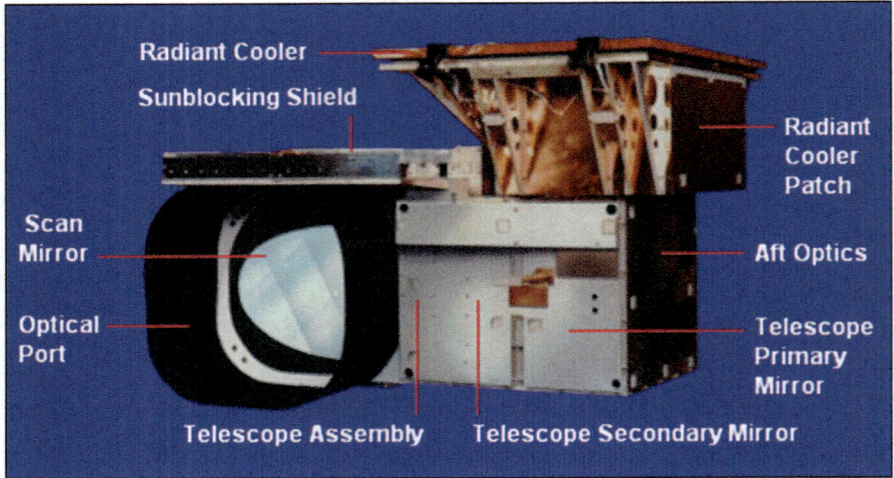

Fig. 6.28 GOES 8–12 imager (Courtesy of Manual: by NOAA)

6.5.2.2 GOES 8-12 (l-M) Sounder Instrument

The GOES I -M Sounder is also built by ITT Aerospace/Communication Division. It looks much like the Imager because both instruments have nearly identical telescopes and it has 19 channels including a visible channel for cloud detection.

Like the HIRS/2 instrument, a rotating filter wheel provides channel selection. Four scan spots, on four successive scan lines, are simultaneously sampled during each rotation of the filter wheel (100 ms). The scan mirror steps 280 μrad (10 km at nadir) in the East-West direction with each rotation of the filter wheel. At the end of a scan line, the scan mirror drops 1120 μrad (40 km at nadir) in the north-south direction to begin the next scan. The scans are alternately East-West and West-East directions. The Sounder has roughly the same radiometric accuracy as the VAS instrument, but the Sounder does not have to dwell on a point as VAS does to reduce noise. This improves flexibility in scanning, and also, the digitization increases from 10 bits to 13 bits per channel.

A significant improvement in the GOES I-M satellite Sounder is that it has more channels (19 versus 12) to retrieve atmospheric parameters, which is depicted with all containing components in Fig. 6.29. In particular, it has more surface-sensing channels than VAS (6 versus 3) and more channels in the less-water-vapor-sensitive shortwave IR (<6 μm) channels (6 versus 3). These improvements should allow the GOES I-M Sounder to better separate surface, moisture, and temperature effects (Hayden, 1989). The extensive changes in the way the GOES I-M instruments work requires a new data format called GVAR (GOES Variable, Komajda and McKenzie, 1987). This change requires the modification of the ground equipment for all users who receive the retransmitted data.

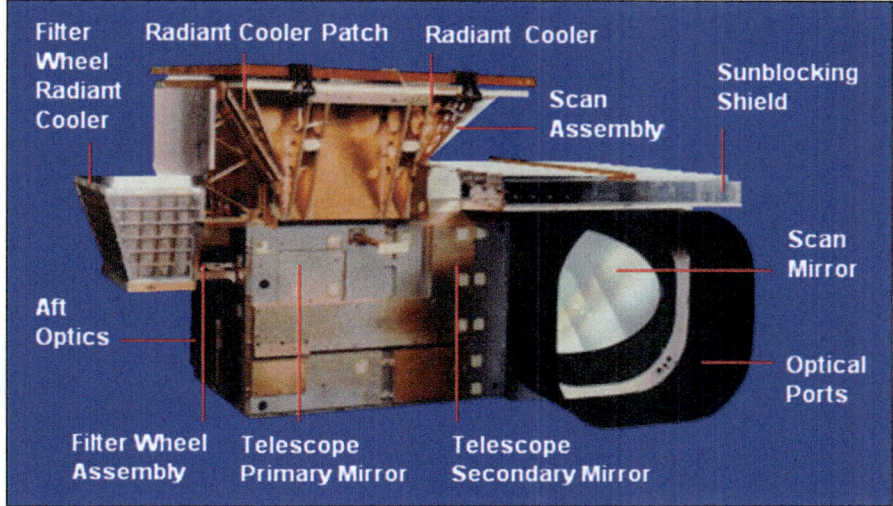

Fig. 6.29 GOES 8–12 Sounder (Courtesy of Manual: by NOAA)

An interesting point is that the motion of the mirrors causes perturbations in the satellite attitude. The independent motion of the Sounder mirror would cause the Imager scan lines to be crooked (and vice versa). However, to prevent this, the GOES I -M satellite has a Mirror Motion Compensation system that automatically anticipates attitude changes of the satellite (caused by the Sounder scan mirror, for example) and makes mirror adjustments (in the Imager, for example) to compensate for them (Savides and Reseck, 1989).

6.5.3   GOES 13-15 (N-P) Instruments

The three GOES 13, 14 and15 or N, O and P are representing third generation of weather satellites constitutes of five satellites that constitute the second generation. The GOES-P is the third and last spacecraft in the GOES N-P series launched in March 2010.The GOES-N/O/P satellites are built by Boeing Satellite Systems, Inc. (formerly Hughes Space and Communications), are based on the Boeing 601 space-craft and are the latest in the series of three-axis body stabilized geosynchronous weather and environmental satellites.

The new satellites will more accurately locate severe storms and other weather phenomena, resulting in precise warnings to the public. The spacecraft enable the primary Imager and Sounder sensors to constantly face the Earth and thus frequently image clouds, monitor the Earth's surface and ocean surface temperatures and sound the Earth's atmosphere for its vertical temperature and water vapor distribution. Atmospheric phenomena can be tracked, ensuring real-time coverage

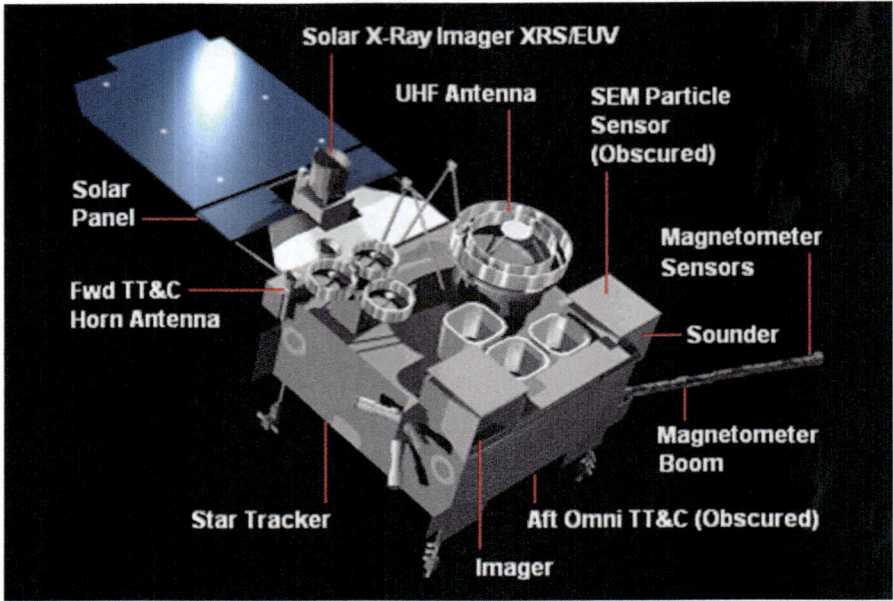

Fig. 6.30 GOES 13–15 weather satellites (Courtesy of Manual: by NOAA)

of events such as severe local storms, tropical hurricanes and cyclones, events that directly affect public safety, property, economic and development.

The GOES-13/14/15 meteorological satellite subsystems, such as Telemetry, Tracking and Command (TT&C), communications payload, mechanical, electrical, power, propulsion, attitude control and image navigation and registration have been designed to support the requirements of the on-board instruments and the data products and services that are described in this brochure. In Fig. 6.30 is illustrated the GOES spacecraft in its on-orbit configuration, a summary of the spacecraft description and components.

Same as the second generation of GOES 8-12, the third generation of GOES 13-15 satellites also carries an Imager, a Sounder and a collection of other space environment monitoring instruments. Both the Imager and the Sounder have a flexible scan control mechanism that allows the instruments to scan small areas as well as all of North and South America and global scenes (called full-disk images). In fact, small areas scan selection permits rapid and continuous viewing of local areas for monitoring of regional phenomena and accurate wind determination. The scan control mechanism also allows continuous observations of severe storms and changing, short-lived weather phenomena. Thus, commands from the NOAA Satellite Operations Control Center select the position and size of the area for observation.

The resolution or clarity of each image depends on the size of the picture elements (pixels) in the image that is being acquired. The size of each pixel, in turn, depends on which band is being used. Otherwise, different bands take images with different-size pixels. An image with pixels that are 4 km (2.5 miles) on each side provides twice the resolution of an image whose pixels are 8 km (5 miles) on each side. All Sounder channels generate data that have circular pixels with 10-km (6.2-mile) diameters.

1. **Imager Instrument** – This GOES 13-15 instrument shown in Fig. 6.28 is an imaging radiometer uses data obtained from its five channels to continuously produce images of the Earth's surface, oceans, severe storm development, cloud cover, cloud temperature and height, surface temperature and water vapor. In addition, this instrument also is allowing users to identify fog at night, distinguish between water and ice clouds during daytime hours, detect hot spots (such as volcanoes and forest fires), locate a hurricane eye and acquires measurements of ground and sea surface temperatures.

    The imager on GOES-P, as on GOES-O before it, has improved resolution, allowing more accurate observation of clouds, vertical winds and volcanic ash. It simultaneously senses emitted thermal and reflected solar energy from selected areas of the Earth. It uses a scan mirror system to alternately sweep East-West and West-East perpendicular to North-South path. The rate of scanning allows the instrument to gather data in its five spectral channels while stepping north-to-south from a 1864 × 1864-mile area (3000 × 3000 km) in 3 min and from an area of 621 × 621 miles (1000 × 1000 km) in just 41 s.

2. **Sounder Instrument** – The Sounder instrument illustrated in Fig. 6.29 gathers atmospheric data over an area from 60° North to 60° South latitude. This information provides meteorologists with a detailed description of conditions in the atmosphere at any time and allows to deduce atmospheric temperature and moisture profiles, surface and cloud-top temperatures, including ozone distributions by mathematical analysis and by adding to data from the Imager. Sounder data is also used in computer models that produce mid-and long-range weather forecasts. The Sounder instrument operates independently of and simultaneously with the Imager, using a similar type of flexible scan mechanism system. The Sounder scans the full Earth and can be commanded to scan and provide soundings of local regions of interest. It can provide imagery over the entire area that is visible to the Sounder, sector imagery, and scans of local regions. Sounder data got from the chosen scan area is fed into powerful computer programs that develop advanced numerical weather prediction models for use by the NOAA national weather service and weather forecasters.

The Space Environment Monitor (SEM) consists of three instrument groups as follows:

1. **Energetic Particle Sensor (EPS)** – An EPS instrument package, developed by the US Assurance Technology Corporation (ATC), (Formerly GE Panametrics Corp.) in Carlisle, MA measures the energetic particles at geosynchronous orbit, including protons, electrons and alpha particles. The Energetic Proton, Electron and Alpha Detector (EPEAD) shown in Fig. 6.31 (a), High Energy Proton and

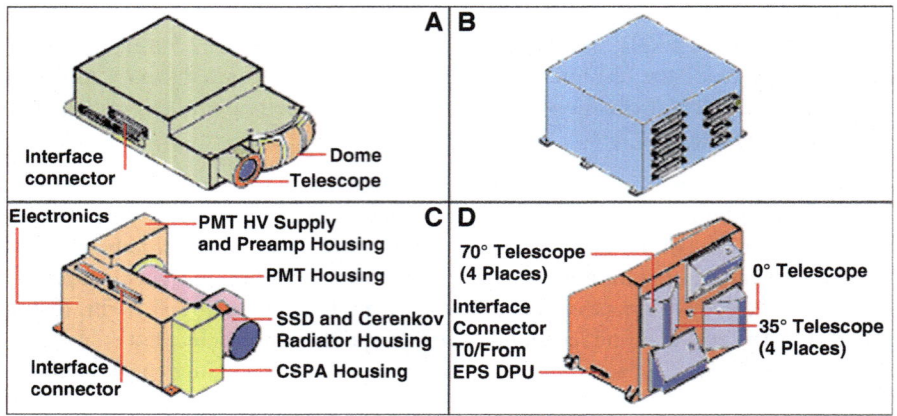

Fig. 6.31 Space environment monitor (Courtesy of Manual: by NOAA)

Alpha Detector (HEPAD) shown in Fig. 6.1 (c) and Magnetosphere Proton Detector (MAGPD), shown in Fig. 6.31 (d), three EPS elements, detect protons over the energy range of 80 keV to greater than 700 MeV, in 16 channels. The EPEAD and HEPAD detect alpha particles over the energy range 3.8 MeV to greater than 3400 MeV, in eight channels.

The EPEAD and Magnetospheric Electron Detector (MAGED) sensors depicted in Fig. 6.31 (d) are additional two elements of the EPS, detect electrons over the energy range of 30,000 electron volts (30 keV) to greater than four million electron volts (4 MeV), in eight channels. The Data Processing Unit (DPU) for processing all data is illustrated in Fig. 6.31 (b). The radiation in the environment consists of particles trapped within the Earth's magnetosphere as well as particles arriving directly from the Sun and cosmic rays that have been accelerated deep in space. The sensors accurately measure the number of particles over a broad energy range and are the basis for operational alerts and warnings of hazardous conditions. Thus, energetic particles pose a risk to satellites and to astronauts and they can disrupt navigation and communications systems used on the ground and in aircraft. The continuous long-term monitoring of the environment provided by these sensors forms the basis for engineering guidelines for satellite design, for analyzing satellite failure and anomalous behavior, for assessing the risk of human exposure to radiation, and for research leading to improved models of the radiation environment. The sensors on the EPS have been expanded on GOES-N, O and P to provide coverage over an extended energy range and with improved directional accuracy.

2. **Magnetometer Sensors** – Each of the GOES-N, O and P satellites have two identical magnetometers, provided by the Science Applications International Corporation (SAIC), Inc. in Columbia, Maryland, US. They can operate independently and simultaneously to measure the magnitude and direction of the Earth's geomagnetic field, detect variations in the magnetic field near the

spacecraft, provide alerts of solar wind shocks or sudden impulses that impact the magnetosphere and assess the level of geomagnetic activity. Thus, all magnetometer data is archived for the scientific community and other interested users. One magnetometer sensor is mounted on the end of the boom 27.9 ft (8.5 m) from the GOES spacecraft. The other is positioned on the same boom 2.6 ft (0.8 m) a little closer to the spacecraft. Otherwise, the second magnetometer sensor serves as a backup in case the first magnetometer sensor fails and provides for better calibration of the magnetometer data channel.

3. **Solar Instruments** – These instruments can be: (a) Solar X-ray Sensor (XRS) dedicated to observe and measure solar x-ray emissions in two bands. In real time, it measures the intensity and duration of solar flares in order to provide warnings of potential disruption of radio communications; (b) Solar Extreme Ultraviolet Sensor (EUVS) is designed to monitor solar extreme ultra-violet emissions to provide a measure of the solar impact on satellite orbit drag and radio communications. GOES-P has the same 5-channel configuration that flew on GOES-N/13; and Solar X-ray Imager (SXI) uses a telescope assembly to observe the Sun's x-ray emissions and to provide early detection and location of solar disturbances. These observations allow space weather forecasters to monitor solar features and activities such as solar flares, loops, coronal holes and coronal mass ejections-clouds of charged particles shooting toward Earth from the Sun.

The GOES-P also carries Data Collection System (DCS), which uses the GOES spacecraft to relay data from remotely located in-situ sites at or near the Earth's surface. The NOAA Low-Rate Information Transmission (LRIT) System is making the GOES data widely accessible by users with low-cost receivers. The GOES is a vital part of the Emergency Managers' Weather Information Network (EMWIN) system, transmitting a live stream of weather and other critical emergency information. Finally, GOES-P is also part of the SAR Cospas-Sarsat service to relay emergency signals transmitted from aircraft, marine vessels or individual emergency locator transmitters to the ground Mission Control Centres.

In this section will be presented only two samples done by GOES 13–15 Satellites:

In Fig. 6.32 is depicted by GOES-13 images of reduced three-panel display over the Gulf of Mexico shows a visible image at sunset and 3.9 and 10.7 images 2 h later in the middle and right panels, respectively. Notice that the 3.9 image is much crisper in appearance than the 10.7 image. This is in part due to diffraction, which is less by a factor of 3 at 3.9 μm and in part because the 3.9 channel is a cleaner window. Also, notice how the cumulus region appears cooler at 10.7 μm. This is because many of the 4-km fields of view are only partially filled with clouds.

In Fig. 6.33 is illustrated shown GOES-11 6.7 μm (Left) vs. GOES-13 6.5 μm (Right) Water Vapor (WV) images at 1215 to 2332 UTC on 15 April 2009. This first comparison demonstrates the improved resolution with GOES-12 to15. The water vapor boundaries associated with two upper-level jet streaks are enhanced

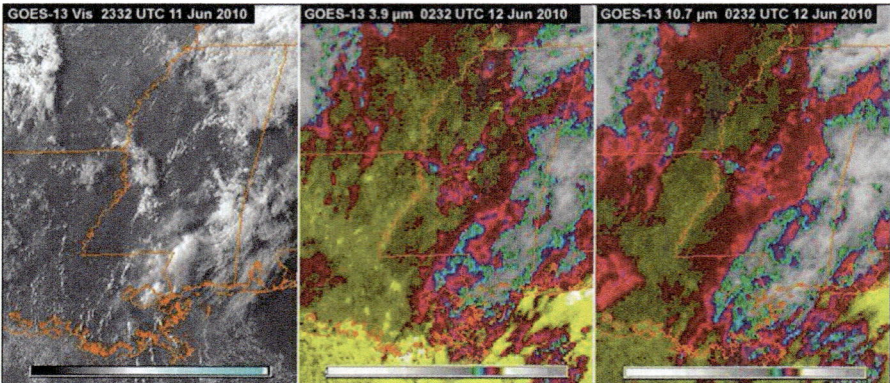

Fig. 6.32 Images of GOES-13 satellite (Courtesy of Manual: by NOAA)

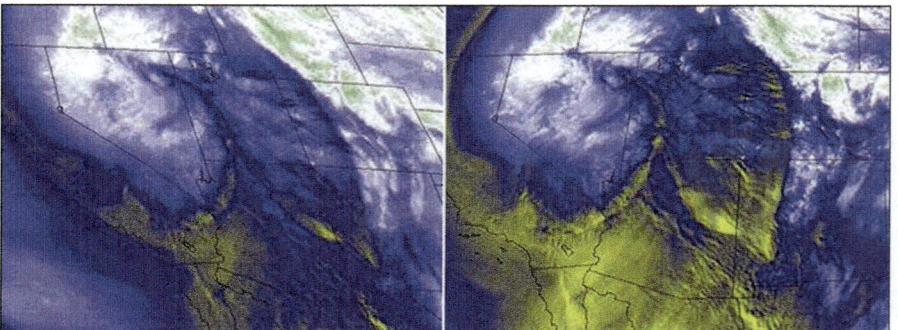

Fig. 6.33 GOES-11 (*Left*) vs. GOES-13 (*Right*) WV images (Courtesy of Manual: by NOAA)

and appear smoother than with GOES-11. The higher resolution GOES-13 imagery makes it more evident that both jet streaks are gradually weakening, as seen by the brightening/warming over the course of the animation. Higher-resolution water vapor imagery can also aid in the analysis of smaller scale features and associated mesoscale forcing mechanisms, such as the interactions between mesoscale jets, terrain, and wintertime precipitation.

Animation of the GOES-13 imagery on the right more clearly depicts development of banded clouds and precipitation to the right (East and North) of the second jet streak axis from NE Nevada south into Northeastern Arizona and Northwestern New Mexico. At the same time, strong Westerly and Southwesterly flow around the base of the upper-level trough is interacting with mountainous terrain to produce mountain wave signatures that extend across much of the southwestern US. Depiction of the finer scale structure and horizontal extent of the mountain waves is much improved with the higher resolution GOES-13 imagery when compared to GOES-11.

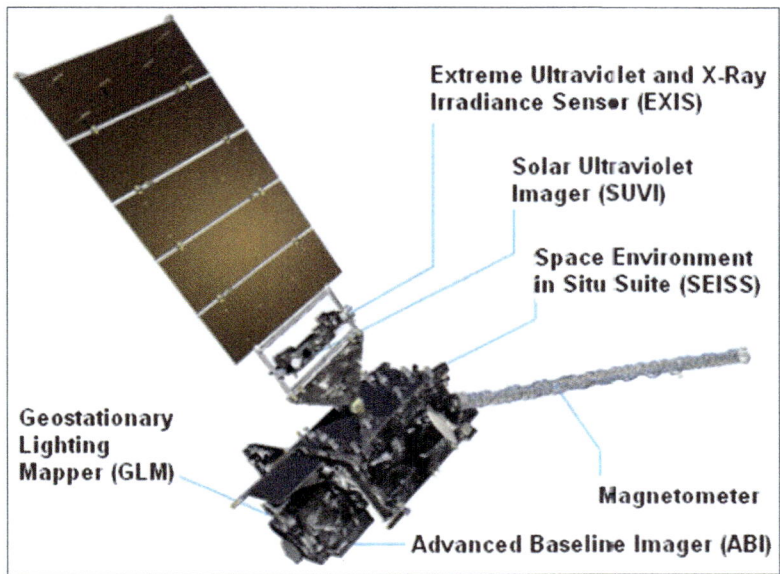

Extreme Ultraviolet and X-Ray
Irradiance Sensor (EXIS)

Solar Ultraviolet
Imager (SUVI)

Space Environment
in Situ Suite (SEISS)

Geostationary
Lighting
Mapper (GLM)

Magnetometer

Advanced Baseline Imager (ABI)

Fig. 6.34   GOES R-S-T-U satellite series (Courtesy of Manual: by NOAA)

6.5.4   GOES 16-19 (R-U) Instruments

The series of the US GOES 16, 17, 18 and 19 or R, S, T and U are the fourth generation of GEO meteorological satellites. The GOES-16 launched on 19 November 2016 is currently undergoing on orbit testing and commissioning, which with instruments is shown in Fig. 6.34. The next GOES-S is scheduled for launch in March 2018, GOES-T for launch in 2019 and GOES-U according with schedule has to be launched sometimes in 2024.

The GOES-16 satellite, previously known as GOES-R new generation project, is a weather satellite, which upon completion of testing will form part of the GOES system operated by the US NOAA. It is expected to provide atmospheric and surface measurements of the Earth's Western Hemisphere for weather forecasting, severe storm tracking, space weather monitoring and meteorological research.

The GOES-16 satellite is a follow-on to the current third generation of GOES 13–15 (N-P) system that is used by NOAA centre for meteorological monitoring and weather forecasting operations as well as by researchers for understanding interactions between land, ocean, atmosphere and climate. The GOES-R series program is a collaborative effort between NOAA and NASA to develop, deploy and operate the meteorological satellites. They are managed from Goddard Space Flight Centre in Greenbelt, Maryland.

The US NOAA normally operates two meteorological satellites in GEO over the equator. Each satellite views almost a third of the Earth's surface: one monitors North and South America and most of the Atlantic Ocean, the other North America

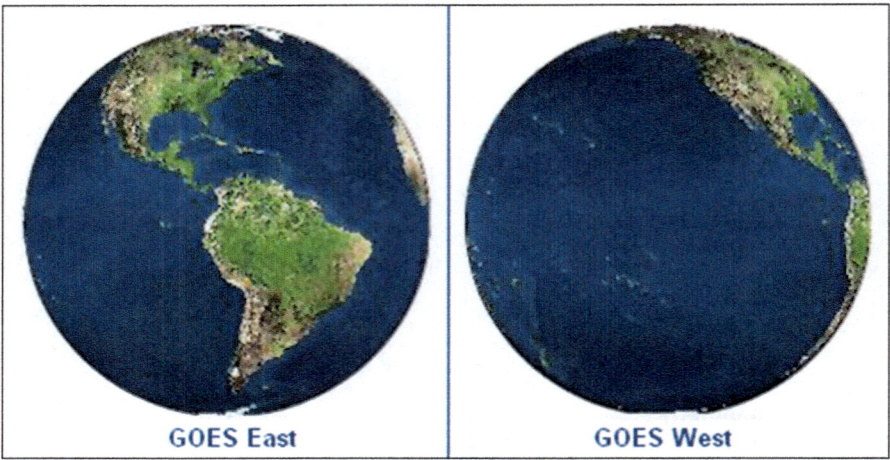

Fig. 6.35 GOES east and GOES west coverages (Courtesy of Manual: by NOAA)

and the Pacific Ocean basin. In fact, GOES-13 (or GOES-East) is positioned at 75°
W longitude and the equator, shown in Fig. 6.35 (Left), while GOES-15 (or GOES-
West) is positioned at 135° W longitude and the equator, shown in Fig. 6.35 (Right).
The two operate together to produce a full-face picture of the Earth, day and night.
Coverage extends approximately from 20° W longitude to 165° E longitude. This
figure shows the coverage provided by each satellite.

The series of GOES-R and GOES S, T and U will extend the availability of the
operational GOES satellite system through 2036. In November 2017, GOES-16 will
be positioned as GOES-East at 75° West longitude, has several improvements over
the old GOES system. Its advanced instruments and data processing provides:

1. Three times more special information known as the special resolution of a
   spectrograph, or more generally, of a frequency spectrum, is a measure of its
   ability to resolve features in the electromagnetic spectrum. It is denoted by $\Delta\lambda$,
   and is closely related to the resolving power of the spectrograph, defined as
   follows:

$$R = \lambda/\Delta\lambda \qquad (6.2)$$

where value $\Delta\lambda$ is the smallest difference in wavelengths that can be distinguished
at a wavelength of λ. For example, the Space Telescope Imaging Spectrograph
(STIS) can distinguish features 0.17 nm (nanometre) apart at a wavelength of
1000 nm, giving it a resolution of 0.17 nm and a resolving power of about 5900.
In such a way, an example of a high-resolution spectrograph is the Cryogenic High-
Resolution IR Echelle Spectrograph (CRIRES) installed at ESO's Very Large
Telescope, which has a spectral resolving power of up to 100,000.

2. Four times greater spatial resolution or angular resolution is a value case of optics and imaging systems for an angular resolution in graph drawings. In fact, this resolution is value describes the ability of any image-forming device or instrument, such as an optical or radiotelescope, a microscope, a camera or an eye, to distinguish small details of an object, thereby making it a major determinant of image resolution.
3. Five times faster coverage then three previous generations of GOES weather satellites.
4. Real-time mapping of total lighting activity that occurs during thunderstorm discharges.
5. Increased large thunderstorms activities and tornado or twisters warnings lead-time.
6. Improved hurricane (storm) track and intensity forecasts of its speed and direction.
7. Improved monitoring of solar x-ray flux (photos) used to track solar activity and flares.
8. Improved monitoring of solar flares during Sun eruptions and unusually large coronal mass ejections of plasma and magnetic fields from the solar corona.
9. Improved geomagnetic storm forecasting, which is a special temporary disturbance of the Earth's magnetosphere caused by a solar wind shock wave or cloud of magnetic field.

The GOES-16 spacecraft A2100 bus is 3-axis stabilized and designed for about 10 years of on-orbit operation preceded by up to 5 years of on-orbit storage. In addition, it will provide near-continuous observations as well as vibration isolation for the Earth-pointed optical bench and high-speed spacecraft-to-instrument interfaces designed to maximize different data collection. Thus, the cumulative time that GOES-16 science data collection (including imaging) is interrupted by momentum management, which for station keeping and yaw flip maneuvers will be under 120 min/year. This represents a nearly two order of magnitude improvement compared to older GOES satellites.

The GOES-16 satellite instrument suite includes three types of instruments: Earth Sensing, Solar Imaging and Space Environment Measuring.

6.5.4.1 Earth Facing Instruments (EFI)

The EFI solution provides two special instruments pointed toward Earth, such as:

1. **Advanced Baseline Imager (ABI)** – The ABI sensor is designed and built by Exelis Geospatial Systems (now Harris Space & Intelligence Systems), is the primary instrument on GOES-16 for imaging Earth's weather, climate and environment, which is shown in Fig. 6.40 (Below), in Fig. 6.36, artist picture (Left) and product (Right). It will be able to view the Earth across 16 special bands, including two visible channels, four near-infrared channels and ten infrared channels. It will provide three times more spectral information, up to

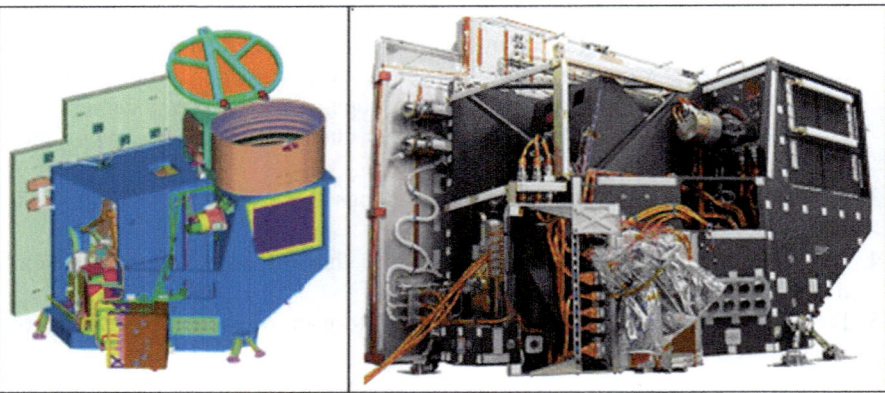

Fig. 6.36 GOES ABI instrument (Courtesy of Manual: by NOAA)

four times the spatial resolution (depending on the band), and more than five times faster coverage. Forecasters will be able to use the higher resolution images to track the development of storms in their early stages. Otherwise, instruments nearly identical to ABI have been delivered to Japan for use on Himawari-8 and 9.

The ABI satellite sensor is the next 4th-generation multispectral imager, 2-axis scanning radiometric imager, intended to begin a new era in an environmental remote sensing with greatly improved capabilities and features, more spectral bands, faster imaging cycles, and higher spatial resolution than the current imager generation of GOES-N to P. In fact, this instrument is the 16-channel or bands imager of multispectral data, with 2 bands in VIS (0.47 μm and 0.64 μm), 4 near IR and 10 bands in IR (0.86 μm to 13.3 μm), and with 3.2X greater spectral coverage. In a more determined sense, the ABI spatial resolution is band-dependent, the IGFOV (Instantaneous Geometric Field of View) ranges from 0.5 km at nadir for broadband visible, 1.0 km for Short-Wave Infrared (SWIR), and 2.0 km for Mid-Wave Infrared (MWIR) and Thermal Infrared (TIR) data.

The ABI instrument improves on the current GOES imager, GOES 8-12 and 13-15, with 4 times higher spatial resolution, 5 times faster imaging, increased spectral coverage, due to 3.2 times more channels and provides more accurate measurements for observing subtle features. This sensor is designed to provide better products for forecasting, severe weather warning, numerical weather prediction and climate and environmental monitoring.

Imagery at visible, near infrared, and infrared wavelengths, along with associated products will be produced for most of the Western Hemisphere every 15 min compared to the current three-hour interval. In the contiguous US, the viewing interval will shrink from the current 15 min to every 5 min. And for a moveable smaller-scale area during severe weather and other environmental emergencies,

Fig. 6.37 Comparison of GOES-16 and GOES-13 images (Courtesy of Manual: by NOAA)

imagery and a subset of products will be available every 30 s. As the 2-h ABI simulation shows here, all three imaging modes will be producing product sets at the same time.

In Fig. 6.37 (Left) is depicted the composite color full-disk visible image of the Western Hemisphere, which was captured from NOAA GOES-16 satellite on 15 January 2017 and was created using several of the 16 spectral channels available on the satellite's ABI. This image shows North and South America and the surrounding oceans, which in comparison with the second image depicted in Fig. 6.37 (Right) shows the new image alongside one taken by GOES-13 at the same time. Note that cloud and continental features are slightly offset because of the different positions of the satellites, because GOES-16 is located at 89.5° Wt longitude, while GOES-13 observes from 75° W. The last image shows a close-up of North America that day, as a vast winter storm spread across the continent

The overall objectives of ABI instrument of GOES-16 are to provide high-resolution imagery and radiometric information of the Earth's surface, the atmosphere and the cloud cover or measurement of the emitted and solar reflected radiance simultaneously in all spectral channels. More exactly, data availability, radiometric quality, simultaneous data collection, coverage rates, scan flexibility and minimizing data loss due to the Sun, are prime requirements of the ABI system.

In Table 6.5 presents a key performance parameter comparison of 3rd generation of GOES 8-12 and 4th generation of GOES-16 imagers.

The ABI instrument can provide a full-disk image of Earth every 15 min and an image of the continental US every 5 min. The ABI instrument also allows the GOES team to target areas of severe weather, wildfires, volcanic eruptions and other weather and environmental phenomena as often as every 30 s. As stated, the ABI sensor observes the planet in 16 spectral bands, compared to 5 bands observed by existing NOAA GOES meteorological satellites. In addition, the spatial resolution of ABI instrument is 4 times better and the scanning rate 5 times faster than

Table 6.5 Key performance parameter comparison of 3rd and 4th generation imagers

Requirement	3rd generation GOES-13/14/15 Imager	ABI of GOES-R
No of spectral bands	5	16
Data rate	2.6 Mbit/s	75 Mbit/s
Spatial resolution:		
0.64 µm (VIS)	~ 1 km	0.5 km
Other VNIR bands <2 µm	N/A	1.0 km
Bands >2 µm	4 km	2.0 km
Time for full disk scan	26 min	15 or 5 min
Absolute INR (image navigation and registration)	54 µrad	21 µrad (EW), 21 µrad (NS)
Registration between images (15 min)	36 µrad	16 µrad (0.5, 1.0 km)
		21 µrad (2.0 km)
Cross-channel image co-registration	50 µrad (VIS to IR)	6.3 µrad (0.5, 1.0 to 2 km)
	28 µrad (IR to IR)	5.2 µrad (0.5, 1.0 to 1 km)
VIS (reflective bands) calibration	No	Yes

previous GOES satellites. Thus, those improvements should help meteorologists better distinguish between clouds, water vapor, smoke ice and volcanic ash in the atmosphere. The ABI camera was built by the same team that made the Advanced Himawari Imager (AHI) onboard Japan's Himawari-8 satellite.

2. **Geostationary Lightning Mapper (GLM)** – The GLM sensor will provide continuous lightning measurements over a large portion of the Western Hemisphere, mapping total lightning (cloud-to-cloud and cloud-to-ground) flash rates and trends, which is shown in Fig. 6.40 (Below). Studies show that a sudden increase in to**tal lightning activity** or flash rate correlates with impending tornadoes and severe storms. The GOES-R satellite series of GLM sensor takes continuous day and night measurements of the frequent intra-cloud lightning that accompanies many severe storms and will do so even when the high-level cirrus clouds atop mature thunderstorms may obscure the underlying convection from the imager. The ABI and GLM will offer complementary and reinforcing sources of information on the intensity and development of potentially severe storms.

The GOES-16 GLM satellite sensor, illustrated with its main components in Fig. 6.38, will take important role in prediction of storms, gales and tornado warnings. In Fig. 6.39 is shown combined field-of-view of the GLM sensors from GOES-E and GOES-W, which include GOES-R as well. The GLM sensor consists of a telescopic Charge-Coupled Device (CCD) camera sensitive to 777.4 nm lights. It has a spatial resolution of 8 km (5.0 mi) (at nadir) to 14 km (8.7 mi) (at edge of field of view) and captures 500 frames per second. The CCD's pixel pitch varies

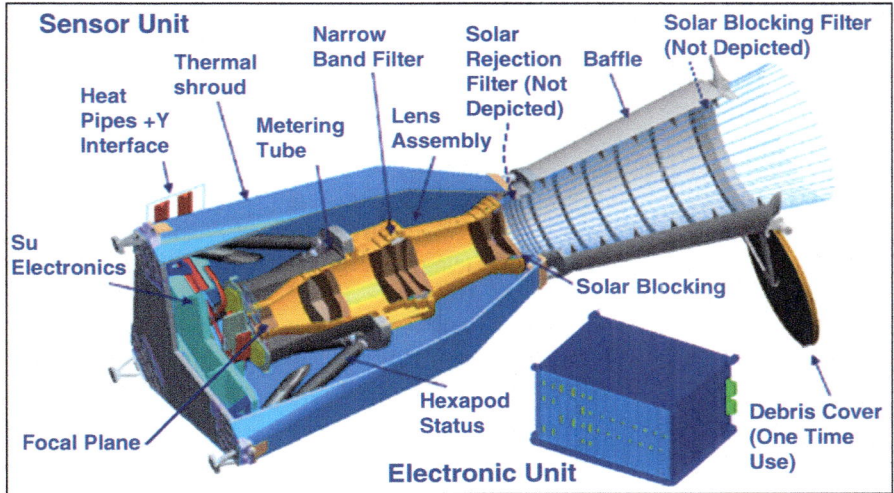

Fig. 6.38 GLM sensor with main comparison (Courtesy of Manual: by NOAA)

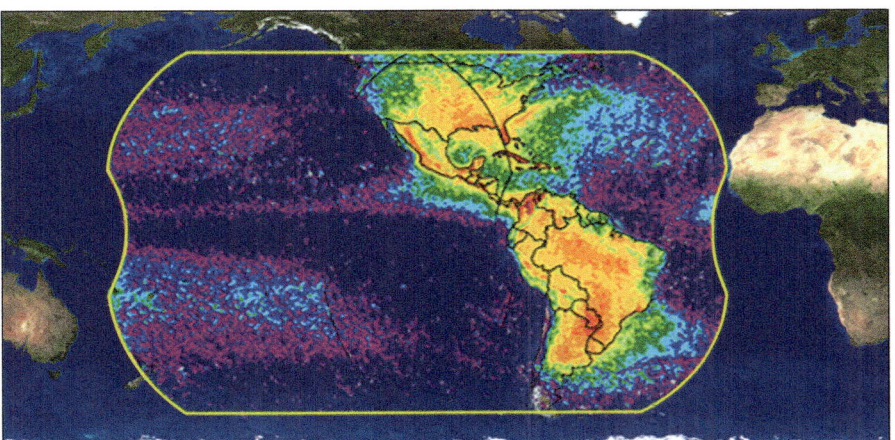

Fig. 6.39 GLM annual lighting coverage map (Courtesy of Manual: by NOAA)

across its area. The GLM is a single-channel, near-infrared optical transient detector that can detect the momentary changes in an optical scene, indicating the presence of lightning. It detects and maps total lightning activity continuously over the Americas and adjacent ocean regions with near-uniform spatial resolution of approximately 10 km. It will collect information such as the frequency, location and extent of lightning discharges to identify intensifying storms. It will also provide improvement in tornado warning lead-time and false alarm rate reduction. In addition, it is anticipated that GLM data will have applications to aviation weather services, climatological studies and severe thunderstorm forecasts and warnings. This sensor will provide information to identify growing, active and potentially destructive thunderstorms over land as well as ocean areas.

6.5.4.2 Sun Facing Instruments (SFI)

In Fig. 6.40 (Below) are shown locations of ABI and GLM instruments as Earth Facing Instruments (EFI). Two Sun-Facing Instruments (SFI) are mounted to the arm holding the solar panel: the Solar Ultraviolet Imager (SUVI) and the Extreme Ultra Violet (EUV)/X-Ray Irradiance Sensors (EXIS), which both are depicted in Fig. 6.40 (Right). Besides, two Space Environment Instruments (SEI) sensors, such as Space Environment In-Situ Suite (SEISS) and Magnetometer (MAG) are illustrated in Fig. 6.40 (Left).

The SFI instruments are located on the Nadir Pointing Platform (NPP) looking to the Earth, while EXIS sensors are located on the Sun-Pointing Platform (SPP) of the satellite, which is located on the solar array yoke. The SPP provides a stable foundation and will track the daily and seasonal movement of the Sun, which is critical to the success of SUVI. SUVI monitors the Sun in the extreme ultraviolet wavelength range and will capture full-disk solar images around-the-clock. Thus, it will be able to see more of the environment around the Sun than earlier NOAA geostationary satellites. The solar corona is so hot that it is best observed with X-ray and extreme-ultraviolet (EUV) cameras. Various elements emit light at specific

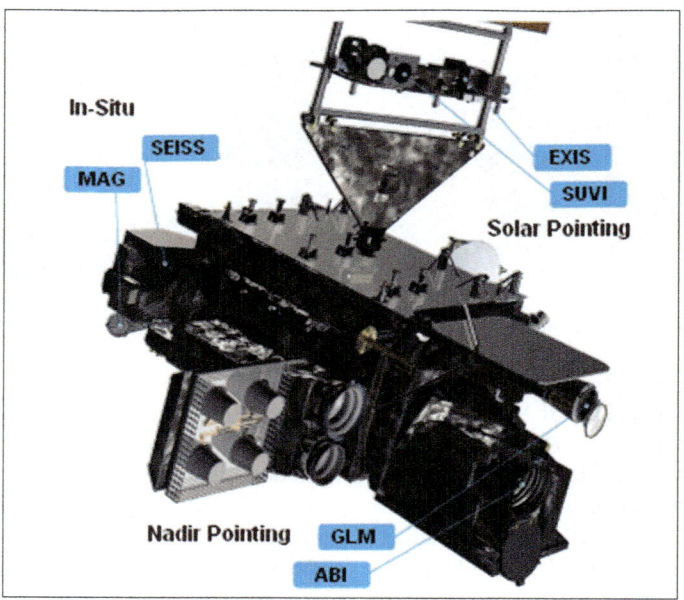

Fig. 6.40 GOES-R instruments (Courtesy of Manual: by NOAA)

EUV and X-ray wavelengths depending on their temperature, so by observing in several different wavelengths, a picture of the complete temperature structure of the corona can be made. The GOES-16 SUVI observes the Sun in six EUV channels.

1. **Solar Ultraviolet Imager (SUVI)** – This is a telescope that observes the Sun in the EUV wavelength range. It observes and characterizes complex, active regions of the Sun, solar flares and the eruptions of solar filaments that may give rise to coronal mass ejections. So, depending on the size and the trajectory, solar eruptions may effect Earth environment, referred to as space weather, include the disruption of power utilities, communication and navigation systems and possible damage to orbiting satellites and the ISS. Observations of flares and solar eruptions will provide an early warning of possible impacts to the Earth environment and enable better forecasting of potentially disruptive events.

   The GOES-R Series SUVI instrument will monitor the Sun in the EUV wavelength range, which is depicted in Fig. 6.40 (Right) and 6.41 (Left). By observing the Sun, it will be able to compile full disk solar images around the clock. The Solar EUV Imagery products will provide space weather scientists with images of the Sun in several different EUV spectral bands. Its high-resolution images will reveal details about the structure of active regions, filaments and solar prominences. Also of interest to space and solar weather, scientists are the boundaries of coronal holes and how the entire surface of the Sun behaves during solar flares. Therefore, Higher-level products made from these imagery products by the NOAA Space Weather Prediction Center along with other organizations will provide early warning of potential radiation hazards, such as SEP events, flares, geomagnetic storms and radio blackouts. Currently on orbit are two instruments, which produce imagery similar to what SUVI will be capable of. The Solar Dynamics Observatory-Atmospheric Imaging Assembly (SDO-AIA) instrument has a higher resolution, but the Solar and Heliospheric Observatory-Extreme UV Imaging Telescope (SOHO-EIT) instrument has a plate scale nearly identical to SUVI. Between these two instruments, it can be possible fairly to reproduce the L1b products, which SUVI will provide, which colors may vary.

2. **Extreme Ultra Violet/X-Ray Irradiance Sensors (EXIS)** – This instrument detects solar soft X-ray irradiance and solar extreme ultraviolet spectral irradiance in the 5–127 nm ranges. The EXIS satellite sensor contains two full disk instruments, the Extreme Ultra Violet (EUV) and the X-Ray Sensor (XRS) or X-Ray Irradiance Sensor (XIRS), which is shown in Fig. 6.40 (Right) and 6.41 (Right). The EXIS instrument has been designed and developed at Laboratory for Atmospheric and Space Physics (LASP) at the University of Colorado, Boulder, CO (PI: Frank Eparvier). The NOAA scientific agency requires the real-time monitoring of the solar irradiance variability that controls the variability of the terrestrial upper atmosphere, namely ionosphere and thermosphere. This device monitors solar variations that directly affect satellite drag/tracking and ionospheric changes, which impact communications and navigation

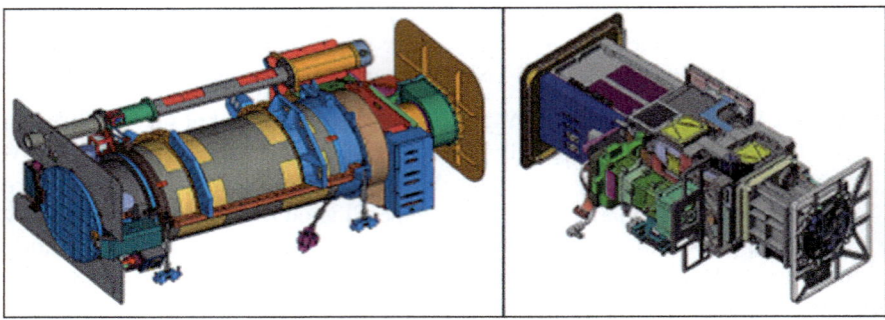

Fig. 6.41 SUVI and EXIS instruments (Courtesy of Manual: by NOAA)

operations. In fact, pre-GOES-R EUV instrument is a full disk detector measuring EUV flux in the 5–127 nm as compared to the 10–126 nm range for GOES-N range. Namely, third generation of GOES EUV instrument provides transmission grating spectrographs covering five broad band-passes, while EUV instrument onboard GOES-R provides three reflection grating spectrographs measuring specific solar emission lines from which full spectrum is reconstructed with a model. On the other hand, the XRS instrument also monitors solar flares that can disrupt space communications and degrade navigational accuracy, affecting satellites, astronauts, high latitude airline passengers and power grid performance. The GOES-R XRS measures the solar soft x-ray irradiance in two band-passes at 0.05–0.4 nm and 0.1–0.8 nm. All these particulars about EUV and XRS are critical to understanding the outer layers of the Earth's atmosphere.

6.5.4.3   Space Environment Instruments (SEI)

Two Space Environment Instruments (SEI) in-situ satellite sensors will monitor the space environment: the Space Environment In-Situ Suite SEISS) and a magnetometer.

1. **Space Environment In-Situ Suite (SEISS)** – This instrument consists of an array of sensors that will monitor the proton, electron and heavy ion fluxes at geosynchronous orbit, which is illustrated in Fig. 6.40 (Left) and its five sensors are depicted in Fig. 6.42. In more determined sense, the data provided by those sensors will be used for assessing radiation hazards to astronauts and satellites. In addition to hazard assessment, the data can be used to warn of high flux events, mitigating damage to radio communication systems. This instrument suite consists of the Energetic Heavy Ion Sensor (EHIS), the Solar and Galactic Proton Sensor (SGPS), the Data Processing Unit Magnetospheric Particle Sensor-High and Low (MPS-HI and MPS-LO). Data will drive the solar

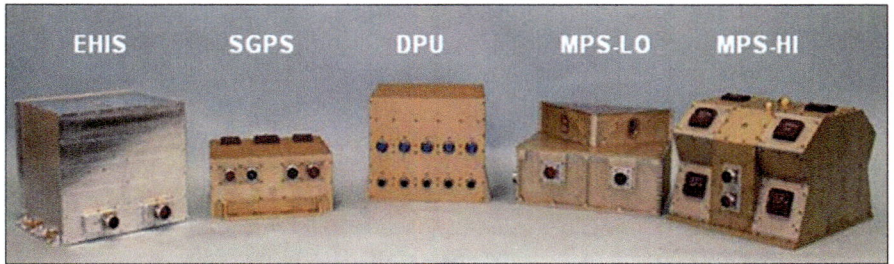

Fig. 6.42 SEISS instruments (Courtesy of Manual: by NOAA)

radiation storm portion of NOAA space weather scales and other alerts and warnings and will improve solar energetic forecasts. In August 2006, NASA in coordination with NOAA awarded a contract to ATC (Assurance Technology Corporation) of Carlisle, MA to design and develop the first SEISS package. The SEISS instrument package onboard GEOS satellites monitors the near-Earth particle and electromagnetic environment in real-time. It monitors geomagnetically trapped electrons and protons, electrons, protons and heavy ions of direct solar origin and galactic background particles. Data from SEISS will drive solar radiation storm portion of NOAA space weather scales and other warnings and will improve energetic particle forecasts.

(a) Energetic Heavy Ion Sensor (EHIS) was designed and developed at New Hampshire University (NHU). The objective of EHIS is to measure the proton, electron and alpha particle fluxes at GEO. This includes particles trapped within Earth's magnetosphere and particles arriving directly from the sun and cosmic rays, which have been accelerated by electromagnetic fields in space. The information will be used to help scientists protect astronauts and high altitude aircraft from high levels of harmful ionizing radiation. This unit incorporates a unique system design called Angle Detecting Inclined Sensor (ADIS).

(b) Solar and Galactic Proton Sensor (SGPS) measures the solar and galactic protons found in the Earth's magnetosphere. The data provided by SGPS will assist the Space Weather Prediction Center's Solar Radiation Storm Warnings. These particular measurements are crucial to the health of astronauts on space missions, though passengers on certain airline routes may experience increased radiation exposure as well. In addition, these protons can cause blackouts of radio communication near the Earth's poles and can disrupt commercial air transportation flying polar routes. The warning system allows airlines to reroute planes that would normally fly over Earth's poles.

(c) Data Processing Unit (DPU) is special instrument suite also included in SEISS.

Magnetospheric Particle Sensor (MPS) is a three-axis vector magnetometer to measure the magnitude and direction of the Earth's ambient magnetic

field in 3 orthogonal directions in an Earth referenced coordinate system. It
will provide a map of the space environment that controls charged particle
dynamics in the outer region of the magnetosphere.

(d) Data Processing Unit Magnetospheric Particle Sensor-Low (MPS-LO) sensor measures electron and proton flux over an energy range of 30 ev to
30 kev. MPS-LO will be able to tell scientists the amount of charging by low
energy electrons that the GOES-R spacecraft is undergoing. Thus, spacecraft
charging can cause Electrostatic Discharge (ESD) and arcing between two
differently charged parts of the spacecraft. This unit discharge arc can cause
serious and permanent damage to the hardware on board a spacecraft, which
affects operation, navigation and interferes with measurements being taken.

(e) Data Processing Unit Magnetospheric Particle Sensor-High (MPS) sensor
will monitor medium and high-energy protons and electrons which can
shorten the life of a satellite. High-energy electrons are extremely damaging
to spacecraft because they can penetrate and pass through objects, which can
cause dielectric breakdowns and result in discharge damage inside of
equipment.

2. **Magnetometer (MAG)** – This sensor measures the space environment magnetic
field that controls charged particle dynamics in the outer region of the magnetosphere, which is illustrated in Fig. 6.40 (Left) and 6.43. These particles can be
dangerous to spacecraft and human space flight. Thus, the geometric field
measurements will provide alerts and warnings to satellite operators and
power utilities. This data will also be used in research, being among the most
widely used GOES spacecraft data by the national and international research
community. Besides, it will be part of NOAA space weather operations, providing information on the general level of geomagnetic activity and permitting
detection of sudden magnetic storms and measurements will be used to validate
large-scale space environment models that are used in operations. Its requirements are similar to the tri-axial fluxgates that have previously flown and to
measure three components of the geomagnetic field with a resolution of 0.016 nT
and response frequency of 2.5 Hz. The MAG device is provided by Lockheed
Martin, Newton, PA and is boom mounted on GOES-R.

6.5.4.4   Unique Payload Services (UPS)

In December 2008, NASA, in coordination with NOAA, selected Lockheed Martin
Space Systems Company of Denver to build the GOES-R series spacecraft. The
contractor has to design and develop GOES-R series of spacecraft in its Newtown
PA, Sunnyvale CA, and Denver CO facilities and provide pre-launch, launch and
post-launch support. The basic contract is for two satellites with options for two
additional satellites. In Fig. 6.44 are illustrated all main components with payloads
and antenna onboard GOES-R one from the series of next fourth negations.

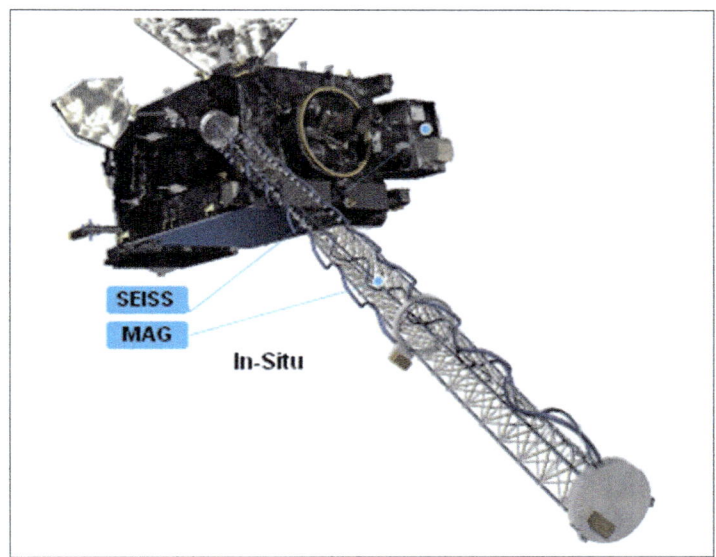

Fig. 6.43 MAG instruments (Courtesy of Manual: by NOAA)

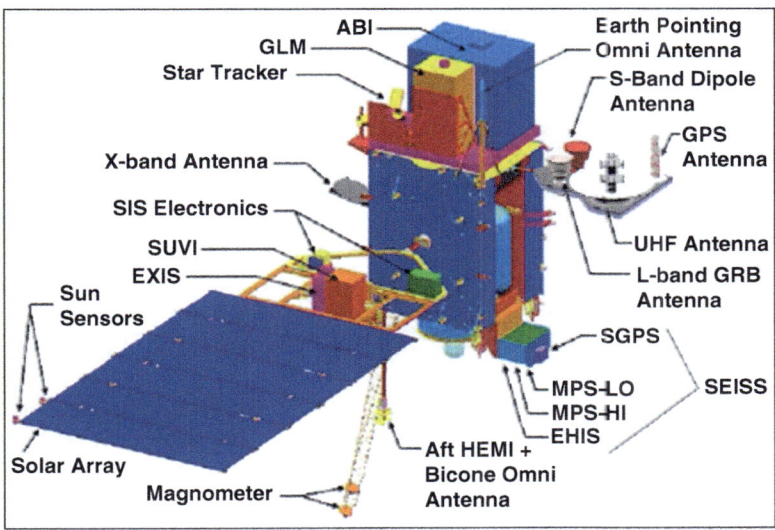

Fig. 6.44 GOES-R payloads (Courtesy of Manual: by NOAA)

The core of GOES-R or 16 spacecraft is the Unique Payload Services (UPS) suite includes of transponder payloads with adequate antenna that provide commu-nications relay services in addition to primary mission data. The UPS suite consists of the Data Collection System (DCS) for collecting and relaying environmental data from remotely located platforms; the GOES Re–Broadcast (GRB) data service for transmitting real-time data products to users; the special system of Low Rate Information Transmission service (LRIT) or the High Rate Information Transmission (HRIT) for providing data from a variety of satellite and other ground environmental data to different ground users; the Emergency Managers Weather Information Network (EMWIN) system containing data specifically to meet the needs of emergency managers; and the Search and Rescue Satellite Aided Tracking (SARSAT) system to detect, locate and help to assist mariners, aviators and others in distress.

1. **Data Collection System (DCS)** – The DCS equipment is a relay satellite system used to collect information from a large number of Earth-based platforms that transmit in-situ environmental or mobile sensor data on predefined frequencies and schedules, in response to thresholds in sensed conditions or in response to requests. Thus, enhancements to the DCS program during the GOES-R era include expansion in the number of user-platform channels from 266 to 433. Thus, the DCS is an environmental data collection relay system that adds the benefits of providing global coverage and platform location. The GOES-R series data will be used in real time for critical life and property forecasting and warning applications primarily by the US National Weather Service (NWS) and globally as well, so these users will be able to monitor the rapid development and interaction of severe storms.

2. **GOES Re–Broadcast (GRB) Data Service** – The GRB transmission system is the primary space relay of Level 1b products and will replace the GOES VARiable (GVAR) service. It will provide full resolution, calibrated, navigated, near-real-time direct broadcast data. The data distributed via GRB service will be the full set of Level 1b products from all instruments aboard the GOES-R and next series of spacecraft. This concept for GRB system is based on analysis that a dual-pole circularly-polarized L-band link of 12 MHz bandwidth can support up to a 31 Mb/s data rate, enough to include all ABI channels in a lossless compressed format as well as data from GLM, SUVI, EXIS, SEISS and MAG.

   The LRIT and HRIT systems are determined to define a standard for dissemination of data, preferably from GEO spacecraft towards LRIT/HRIT user ground stations.

3. **LRIT/HRIT Systems** – The main approach of LRIT/HRIT is to disseminate rasterized image data mapped to the surface of the Earth, preferably those generated by or deducted from satellite remote sensing data. Additionally, LRIT/HRIT system shall provide means to forward other types of graphical information, alphanumeric data or binary data. Thus, LRIT and HRIT data is an important source of information for nowcasting, numerical weather prediction, climate monitoring and research. In particular, the digital LRIT service evolves from the current analog WEFAX system. It provides a wide dissemination of all GOES satellite imagery and other data at the relatively low information rate of 128 Kb/s, and has a requirement to upgrade the user information rate to

256 Kb/s. The HRIT system is intended for use on high rate communication links, mainly at 0.256 Mb/s through 10 Mb/s.

4. **Emergency Managers Weather Information Network (EMWIN)** – This network is a direct meteorological service that provides users with weather forecasts, warnings, graphics and other information directly from the US NWS in near-real time. The GOES EMWIN relay service is one of a suite of methods to transmit these products to end-users. However, the HRIT service provides broadcast of low-resolution GOES satellite imagery data and selected products to remotely located user HRIT terminals. A service provided though a transponder onboard the GOES satellite. EMWIN is a suite of data access methods that make available a live stream of weather and other critical information to Local Emergency Managers and the Federal Emergency Management Agency (FEMA). In addition to the primary data streams, the GOES-R series antennas will support a set of Unique Payload Services. The HRIT/EMWIN information is uplinked in S-band and downlinked in L-band at ground infrastructures of Wallops Command Data Acquisition Station (WCDAS) and Remote Backup Facility (RBU). The HRIT/EMWIN broadcasting provides low-resolution GOES imagery and products, along with emergency weather forecasts and warnings generated by the NWS. In parallel, the GOES-R series system will support the collection of in-situ environmental sensor data from DCS platforms and will transponds commands to DCS platforms using the GOES-R antennas at WCDAS. Interfaces between the Antenna System and the HRIT/EMWIN and DCS systems will mirror those in place at WCDAS today, but with new and upgraded capabilities to support more DCS terminals and higher data rate signals for HRIT/EMWIN.

5. **Search and Rescue Satellite Aided Tracking (SARSAT) System** – This is an integral part of the international SAR satellite program Cospas-Sarsat, which detects and locates ships, land vehicles or persons and aircraft in distress almost anywhere in the world at any time and under almost any conditions. The SARSAT transponder on the GOES-R satellite will provide the capability to immediately detect distress signals from emergency beacons and relay them to ground stations, called Local User Terminals (LUT). These signals are routed to the Mission Control Center (MCC) and SAR forces to locate distress.

6.6   Other GEO Satellite Instruments

Several other GEO satellite meteorological platforms exist that are designed primarily for weather imaging and atmospheric research. These include the Meteosat satellites operated by the European Community (EU), the GOMS satellites operated by Russia, the Feng-Yun satellites operated by China, the Indian INSAT satellites and Japanese GMS satellites.

6.6.1   European Meteosat Instruments

In the two decades between 1977 and 1997, the EU Eumetsat had launched seven GEO meteorological satellites of the Meteosat series, which coverage is depicted in Fig. 6.24. The prime Meteosat spacecraft is usually kept positioned at 0° longitude on the equator. In the spring of 1998, Meteosat-5 was moved to 63°E to cover the Indian Ocean and support the Indoex program. The Meteosat provides a spinning radiometer, which has 3 spectral bands, VIS, WV and TIR, with a wavelength range of 0.45–1.0, 5.7–7.1 and 10.5–12.5, respectively. The VIS channel resolution is 2.5 km and for the other two channels is 5 km.

The Meteosat Second Generation (MSG) satellite (Schmetz et al. 2002) carries a 12-channel spinning radiometer called SEVIRI, with a 1 km resolution for the VIS band and 3 km resolution for the JR bands. The MSG satellite is capable of providing full-disc images every 15 min. The 12-channel configuration of SEVIRI includes a new channel in the ozone absorption band, a triple TJR window and three WV channels to serve various new applications. The MSG-1 satellite was successfully launched on 28 August 2002. When MSG-1 became operational in 2004, it was renamed Meteosat-8. The MSG-2 satellite was launched on 21 December 2005.

The Meteosat Third Generation (MTG) programme aims at developing the next generation of the European geostationary Meteosat satellite system for numerical weather prediction and nowcasting. The MTG satellite system will be operated by the EU Eumetsat and will provide Europe's national meteorological services as well as international users and science communities with improved imaging and advanced infrared sounding capabilities for both meteorological and climate applications. The MTG programme is established as a common undertaking between Eumetsat (responsible for the development of the ground segment the launch services, the procurement of recurring satellites and) and ESA, responsible for the development of the space segment. Eumetsat and ESA designed 6 satellites of MTG, and this sector will be introduced 2 main types, such as MTG-I illustrated in Fig. 6.45 (Left) and MTG-S or Sentinel 4 depicted in Fig. 6.45 (Right).

The objective of the MTG system is to provide continuous high-resolution observations and geophysical parameters of the Earth system derived from direct measurements of the radiation, which it emits and reflects using satellite-based sensors from a geostationary orbit. Thanks to advances in space technology, MTG will also provide an enhanced service compared to the current MSG system, contributing significant improvements to the existing service with an improved imagery mission and introducing new sounding and lightning missions from a geostationary orbit.

(1) The MTG space segment consists of six satellites carrying two different payload suites. The system is composed of four MTG-I satellites dedicated to the imaging applications and two MTG-S satellites dedicated to sounding applications. The MTG-I satellite embarks the FCI, LI, SAR and DCS payloads. The MTG-S satellite embarks the IRS and UV/VIS/NIR (UVN or Sentinel 4) payloads. Both MTG-I and MTG-S are based on a common three-axis stabilized platform on the

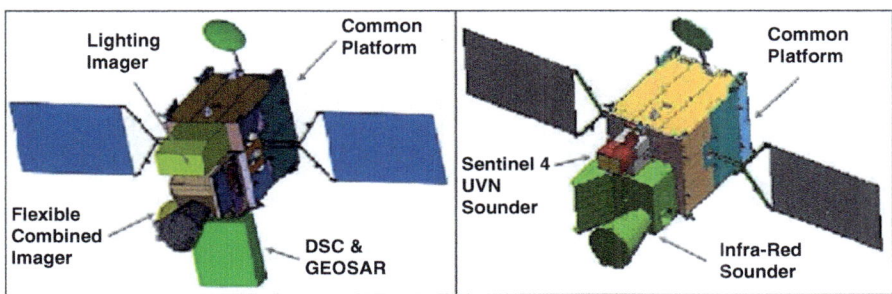

Fig. 6.45 Eumetsat MTG-I/S satellites (Courtesy of Manual: by ESA)

basis of OHB's SGEO platform. The MTG space segment design and development program entered the implementation phase in November 2010.

This generation of Meteosat satellites will be based on three-axis stabilized platforms. The operational configuration of MTG will be a system of two imaging satellites (MTG-I) and one sounding satellite (MTG-S). The launch of the first MTG-I satellite shown in Fig. 6.46 (Left) is planned in 2019 and the launch of the first MTG-S satellite shown in Fig. 6.46 (Right) is scheduled for 2021. The MTG payload will provide the following service:

All in all, the MTG system will be supporting the following missions, services and associated payloads:

1. **Flexible Combined Imager (FCI)** – This instrument onboard MTG-I will continue the very successful operation of the Spinning Enhanced Visible and Infrared Imager (SEVIRI) onboard MSG satellites, which is depicted in Fig. 6.46 (Left). The satellite's three axes stabilized platform will be capable of providing additional channels with better spatial, temporal and radiometric resolution, compared to the current MSG satellites. Requirements for the FCI have been formulated by regional and global Numerical Weather Prediction (NWP) and Nowcasting communities. These requirements are reflected in the design, which allows for Full Disk Scan (FDS), with a basic repeat cycle of 10 min, and a European Regional-Rapid-Scan (RRS), which covers one-quarter of the full disk with a repeat cycle of 2.5 min. Thus, the FCI instrument takes measurements in 16 channels, of which eight are placed in the solar spectral domain between 0.4 and 2.1 μm, delivering data with a 1 km spatial resolution. The additional eight channels are in the thermal spectral domain between 3.8 and 13.3 μm delivering data with a 2 km spatial resolution. In the RRS mode there will be two additional channels in the solar domain, with a spatial resolution of 0.5 km, and two in the thermal domain, with a spatial resolution of 1 km.

   The FCI instrument will outperform the SEVERI observations on cloud, aerosol, moisture and fire detection by adding new channels and by significant improving temporal, spatial and radiometric resolution of the data. In such a way, the FCI instrument onboard MTG-I satellites will continue to play very

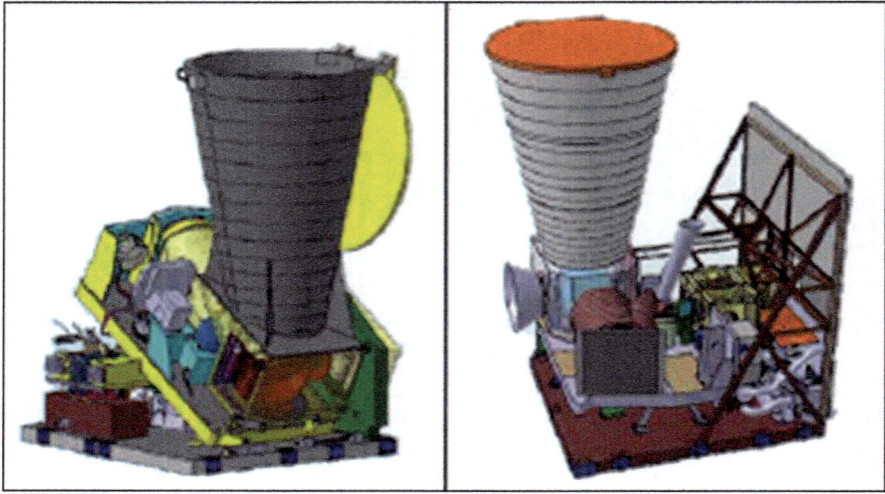

Fig. 6.46 FCI and IRS (Courtesy of Manual: by ESA)

important role in imaging radiometry from the GEO satellite platform in the decades to come.

2. **Infrared Sounding (IRS)** – The IRS sensor onboard MTG-S meteorological satellites will be able to provide unprecedented information and data on horizontally, vertically and temporally, namely 4-dimensional, resolved water vapour and temperature structures of the atmosphere, which is illustrated in Fig. 6.46 (Right). Thus, retrieving highly resolved vertical structures of humidity (~2 km resolution with 10% accuracy) and temperature (~1 km with $0.5° - 1.5°$ accuracy) by remote sensing techniques does require measurements within the water vapour and CO_2 absorption bands with extremely high spectral resolution and accuracy. The IRS is based on an imaging Fourier-interferometer with a hyperspectral resolution of 0.625 cm^{-1} wave number, taking measurements in two bands, the Long-Wave InfraRed (LWIR) and the Mid-Wave InfraRed (MWIR), with a spatial resolution of 4 km. The IRS will deliver over the Full Disk in the LWIR (700–1210 cm^{-1} or 14.3–8.3 μm) 800 spectral channels and in the MWIR (1600–2175 cm^{-1} or 6.25–4.6 μm) 920 channels with a basic repeat cycle of 60 min. The IRS includes the ozone band within LWIR and the carbon monoxide band within MWIR. This will allow measurement within the free troposphere, leading to information on enhanced levels of pollution in the boundary layer below. In fact by providing operational measurements of carbon monoxide and ozone, IRS will also make a significant contribution to the space segment of the Copernicus initiative.

3. **Lightning Imagery (LI)** – This satellite sensor is detecting continuously almost the full Earth disc, lightning discharges occurring in clouds or between cloud and ground with a resolution around 10 km, which with main components is shown in Fig. 6.47 (Left), and its Field of View (FOV) with Optical Channels (OC) is

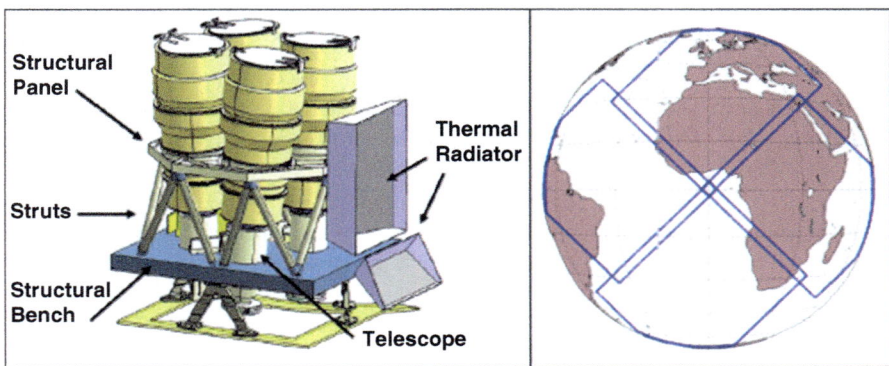

Fig. 6.47 LI optical head and FOV of optical channels (Courtesy of Manual: by ESA)

illustrated in Fig. 6.47 (Right). The overall objective of the LI system is to deliver on a continuous basis information on total lightning over the full disk (high timeliness, data quality homogeneity in time and space), allowing to extend "locally developed" algorithms for NWC severe weather and storm warning to be applied over wider areas like Europe or the full Earth Disk. The LI has no heritage in Europe, while two USA LEO missions (LIS and OTD) have already flown and one, the Global Lightning Mapper (GLM) is installed onboard of GOES-R satellite. The LI instrument is designed and determined to provide the following service:

- The LI sensor measurements of total lightning, Intra Cloud (IC) and Cloud Ground (CG), are complementing the global measurements of CG lightning provided by ground based systems to improve the quality of information essential for air traffic routing and safety.
- The IC and CG data and other information allows assessing the impact of climate change on thunderstorm activity by monitoring and long-term analyzing lightning characteristics in cooperation with the two NOAA GLM on GOES-R and GOES-S satellites. A major part of the globe is covered by a long-term committed GEO lightning (IC + GC) observing system.
- Providing GEO lighting IC + CG information on a global scale will be a prerequisite for studying and monitoring the physical and chemical processes in the atmosphere regarding Nitrogen Oxides (NC and NO_2), referred together as NOx, which is playing a key role in the ozone conversion process and acid rain generation.
- Error characterized IC + CG information can be assimilated to improve very short range forecasts of severe convective events or used to verify/validate other satellite data based NWC algorithms to forecast time and location of initiation of lightning.

The GEO LI system is conceived for the detection of lightning in Earth's atmosphere from a geostationary orbit, where the FOV of Earth's full disk extents to an angle of 17.5°. The basic LI observation concept is to cover this

FOV with four lenses and four detectors, each one covering a square FOV of 8.7° × 8.7°, corresponding to 12.3° diagonal FOV, in order to reduce the incidence angle on the interferential filter placed on the pupil of each subsystem. An important advantage in splitting the interference filter in four parts is related to the possibility to reduce its dimensions and to ease its manufacturing.

The Thales Alenia Space (TAS) Company as the MTG prime contractor is responsible for the procurement of the LI instrument developed and manufactured by Selex Galileo of Campi Bisenzio, Italy, a Finmeccanica company, with the following overview:

- The LI instrument is composed of one LI Optical Head (LOH) and one electronics unit, the LI Main Electronics (LME), which components are illustrated in Fig. 6.47 (Left). The LOH consists of four identical Optical Channels (OC), each one including a protective cover on the baffle aperture to prevent baffle and optics contamination during launch and prelaunch activities.
- A baffle for stray light suppression and thermal load minimization, while a special Solar Rejection Filter (SRF) is designed to minimize both the background level and the thermal load inside the OC.
- A special Narrow Band Filter (NBF) is using to reduce the bandwidth in the range of the lightning spectral pulse and a kind of optical system with F# 1.73, 110 mm entrance pupil diameter, which is determined by radiometry required to achieve the Instrument Average Detection Probability IADP performance and 190 mm effective focal length, and which is determined by the targeted Ground Sampling Distance (GSD) of 4.5 km at Sub Satellite Point (SSP) and the size of detector pixels.
- A Complementary Metal Oxide Semiconductor (CMOS) image sensor with detection of with 1000 × 1170 pixels, 24-μm pitch, 1000 frame/s and a processing electronics device is implementing the detection functions.

Therefore, each OC images a different portion of the visible Earth surface with the four line of sights tilted 4.75° from the SSP toward North, South, West and East in order to achieve the required coverage, which is illustrated in Fig. 6.47 (Right). The LME performs the overall payload functions, the interface to the platform, the configuration of the processing electronics, the data flow regulation and finally compacts and packetizes the scientific data, which consists of the Power Units and the Single Board Computer.

4. **Ultraviolet, Visible and Near-Infrared (UVN)** – This instrument is providing sounding mission covering Europe every hour taking measurements in three spectral bands (UV: 290–400 nm; VIS: 400–500 nm, NIR: 755–775 nm) with a resolution of around 10 km. The UVN mission will be implemented with the GME-S Sentinel-4/UVN payload. Thus, the UVN sounder on MTG-S is the GEO component of the joint Copernicus Sentinel-4 (GEO) and Sentinel-5 (LEO) spacecraft concept for climate protocol monitoring and air quality applications expected to deliver data products on ozone, nitrogen dioxide, sulphur dioxide,

formaldehyde, aerosol optical depth, and aerosol scattering height. Furthermore, a separate file, Copernicus: Sentinel-4 (or simply Sentinel-4), is provided on the eoPortal, which describes only the UVN Sounder, a hosted payload on the MTG-S spacecraft series.

5. **Search And Rescue (SAR) Relay Service** – This system is providing the Geostationary Search and Rescue (GEOSAR), same as GEOS Sarsat network, as a part of the global Cospas-Sarsat international system, whose aim is to provide distress alert and location information for rescue authorities for maritime, aviation and land users in distress.

6. **Data collection system (DCS) Mission** – This special service involves the collection and transmission of observations and data detected by surface, buoy, sea ship, balloon or airborne Data Collection Platforms (DCP).

6.6.2   Russian Electro Instruments

The Russian GOMS-1, also referred to as Electro-1 (and a spelling of Elektro-1), is the first experimental GEO weather satellite. Initial program planning started in the early 1970s by the VPK Military-Industrial Commission for development of a unified meteorological system to be operational by 1979 and continued through the 1980s with involvement of Roshydromet (Committee for Hydrometeorology and Environmental Monitoring), Russian Federal Space Agency (Roscosmos) and other institutions in Russia. In fact, the program's numerous delays were due to both equipment and software problems.

The Elektro-L No.1 (in Russian: Электро-L), also known as Geostationary Operational Meteorological Satellite No.2 (GOMS No.2) is a Russian GEO weather satellite, which was launched in 2011. This satellite became the first Russian geostationary weather satellite to be launched since Elektro No.1 in 1994. Thus, Elektro-L No.1 was manufactured by NPO Lavochkin, based on the Navigator satellite bus with a mass at launch of 1740 kg or 3840 lb. Designed to operate for 10 years, the satellite is positioned over the Indian Ocean at longitude 76° degrees East.

The MSU-GS imaging scanner onboard Elektro satellite is the primary instrument aboard the spacecraft, which camera passes additional tests to improve performance in infrared channels It is designed to produce visible light and infrared images of a full disc of the Earth. It can produce an image every half-hour, with the visible light images having a resolution of 1 km and the infrared images having a resolution of 4 km. The satellite also carries GGAK-E, a heliophisics payload designed to study radiation from the Sun and it will be used to relay data between Russian weather stations. This satellite will also be used to relay signals as part of the Russian Cospas subsystem as a part of international Cospas-Sarsat mission for SAR of missing ships, vehicles with persons and aircraft. It carries seven infrared channels and three visible channels and its overall mission objectives are:

- To provide on an operational basis multispectral imagery, such as hydro-meteorological data of the atmosphere, including the cloud-covered sky and of the Earth's surface within the coverage region (visible disk) of the spacecraft;
- To collect heliospheric, ionospheric, and magnetospheric data;
- To provide the needed communication services for the transmission/exchange of all data with the ground segment;
- To provide the services of data collection for the DCP (Data Collection Platforms) in the ground segment as well as to provide the services of GEOSAR COSPAS/SARSAT system;
- To acquire, in real-time, television images of the Earth's surface and cloud cover within a radius of 60° centered at the sub-satellite point in the VSB and IR regions of the spectrum;
- To measure temperature profiles of the Earth's surface (land and ocean) as well as cloud cover and to measure the radiation state and magnetic field of the space environment at the geostationary position;
- To provide the exchange of high-speed digital meteorological data retransmissions via the satellite and to transmit via digital radio channels TV, images, temperature, radiation and magnetometric information to the independent receiving stations via the main and regional data receiving and processing centers; and
- To provide an exchange of high-speed digital data (retransmission via GOMS) among all regional centers within Russian Roshydromet.

The Elektro-L satellites are designed to operate for 10 years each, which main components and instruments are depicted in Fig. 6.48. They have the following main instruments:

1. **Multispectral Scanner-Geostationary (MSU-GS)** – This is an 10-channel radiometer instrument being designed and developed by the Federal State Unitary Enterprise/Russian Scientific Institute of Space Device Engineering (FSUE/RISDE), Moscow, Russia (Note: the Russian abbreviation of FSUE/RISDE is RNIIKP). The objectives of these instruments are to obtain solar reflected imagery and brightness temperature measurements from Top of Atmosphere (TOA) and of the ocean and land Earth's surface. In addition, the tropospheric moisture content is determined. An image of a subregion of the disk of size "yy" km x "zz" km can be obtained in parallel. The instrument data rate is 2.56–15.36 Mb/s. The MSU-GS instrument is a multi-zonal scanner to take imagery in 3 visible and 7 infrared bands. In full accordance with specifications of Roshydromet, MSU-GS parameters will be close to those of the SEVIRI (Spinning Enhanced Visible and Infrared Imager) instrument now operating onboard the MeteoSat-8/MSG-1 spacecraft of Eumetsat.

In Fig. 6.49 (Left) is illustrated MSU-GS instrument with main components, such as Blind, Visible and Near Infrared (VNIR), IR Module and Radiative Cooler. This satellite instrument is capable of producing images of the Earth's whole hemisphere in both VIS and IR frequencies, providing data for climate

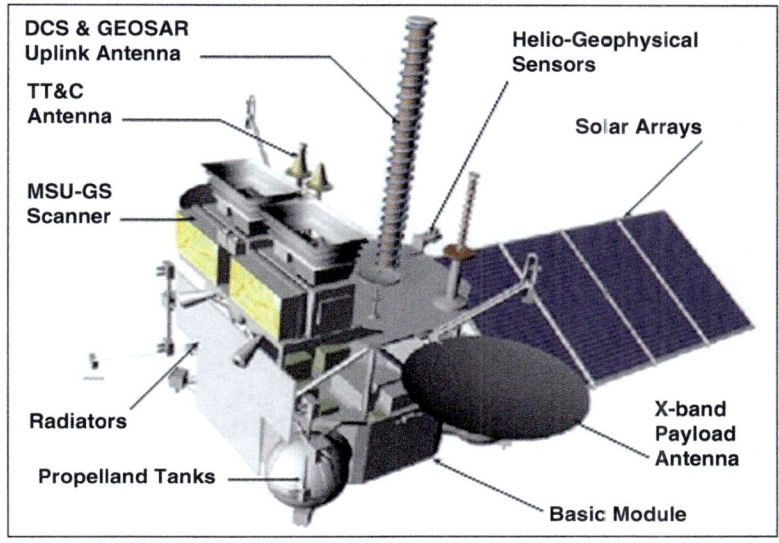

Fig. 6.48 Elektro-L sensors (Courtesy of Manual: by Roshydromet)

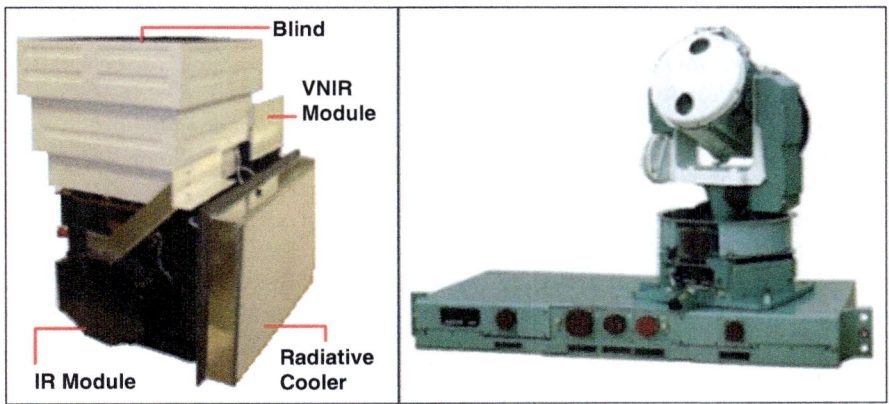

Fig. 6.49 MSU-GS and ISP-2M (Courtesy of Manual: by Roshydromet)

change and ocean monitoring in addition to their primary weather forecasting role.

In Fig. 6.50 (Left) is illustrated full-disk image of Earth lit from Russia's Elektro-L spacecraft during the equinox on 22 September 2013, and in Fig. 6.50 (Right) is shown another full-disk image of Earth from Russia's Elektro-L spacecraft.

2. **Heliogeophysical Instrument Complex on Electro-L (GGAK-E)** – The objective of this instrument suite is to obtain particles, such as particle count of

Fig. 6.50 Full-disk image of Earth (Courtesy of Manual: by Roshydromet)

protons, electrons and alpha particles, to measure the Sun's X-ray radiation, the solar constant, and to measure the magnetic field components as: Monitoring and forecasting of solar activity, Monitoring and forecasting radiation and magnetic field in the near-Earth space, and Diagnostics and monitoring of the natural and modified magnetosphere, ionosphere and upper atmosphere. The GGAK-E instrument components are: (1) SKIF-6: Corpuscular radiation spectrometer; (2) SKL-E: Solar cosmic rays spectrometer; (3) GALS-E: Detector of galactic cosmic rays; (4) ISP-2 M: Solar constant sensor, which is depicted in Fig. 6.49 (Right); (5) VUSS-E: Solar UV radiation sensor; (6) DIR-E: Solar X-ray radiation flux sensor; and (7) FM-E: Magnetometer instrument.

3. **Onboard Data Sampling System (ODSS)** – The ODSS suite is a data collection system provided by Roshydromet, Moscow.

4. **Geostationary Search and Rescue (GEOSAR)** – The GEOSAR satellite system is part of Cospas-Sarsat for emergency and distress alert via special satellite beacons at 406 MHz. These distress signals sent seaman, personal or airman beacon transmitters can be received by the Cospas transponder onboard Elektro-L satellite and relayed on 1544.5 MHz to the ground GEOLUT stations. From there distress data are sent via Mission Control centres to the relevant authorities for SAR of mobiles or persons in distress.

6.6.3   Chinese Feng-Yun Instruments

Feng-Yun satellites operated by China Meteorological Administration (CMA) is the first GEO meteorological satellite of China launched on June 10, 1997. The

Feng-Yun 4 (FY-4) GEO meteorological satellites is representing China's second generation mission, which was launched on 10 December 2016 with intention to build the high-altitude component of the country's space-based weather now- and forecasting system through the 2030s.

This satellite carry onboard the following weather instruments: (1) Advanced Geostationary Radiation Imager (AGRI) as multi-purpose imagery and wind derivation by tracking clouds and water vapour features; (2) Advanced Geostationary Radiation Imager (AGRI) serving as multi-purpose imagery and wind derivation by tracking clouds and water vapour features with 14 channels. balanced VIS, NIR, SWIR, MWIR and TIR; (3) Data Collection Service (DCS) from in-situ Data Collection Platforms (FCP); (4) Geostationary Interferometric Infrared Sounder (GIIRS) Temperature/humidity sounding and wind profile derivation by tracking water vapour features; (5) Lightning Mapping Imager (LMI) with an CCD camera operating at 777.4 nm (O_2) to count thunderstorm flashes and measure their intensity; (6) Space Environment Monitoring Instrument Package (SEMIP) – Fields for space weather and monitoring magnetic fields; and Space Environment Monitoring Instrument Package (SEMIP/SEM) – Space for space environment monitoring instrument package – space environment Monitor.

6.6.4   Indian INSAT Instruments

The Indian INSAT GEO Satellites of Indian Meteorological Department are located above the equator at approximately 90o East longitude, providing meteorological images of the Indian Ocean and central Asia.

6.6.5   Japanese GMS Instruments

The Japanese Geostationary Meteorological Satellites (GMS) Himawari-9 identical to the previous satellite Himawari-8, was launched on 2 November 2016 and will be put stand-by in orbit until 2022, then succeeds the observation from operated by Japan Meteorological Agency (JMA). The Himawari-9 meteorological satellite is located above the equator at 140.7° East longitude, providing good coverage of the western Pacific. The GMS satellite is similar to the U.S. GCES-West and GOES-East satellites, except that it does not carry a sounding instrument. In such a way, it will be possible to use GMS imager data in much the same way is using GOES imager data.

Chapter 7
Antenna Systems and Propagation

In the beginning of radio development, radio communication systems were conceived for the transmission and receiving of telegraphy and telephony signals via antenna systems. The consideration of antenna transmission is inevitable where their propagation characteristics are much affected by different and changeable local environments during movement of mobile and differ greatly from those observed in fixed satellite systems. To create antenna hardware for mobile and fixed systems, engineers have to consider all related factors in order to realize full mechanical and transmission potentials.

This chapter describes antenna characteristics, requirements and basic relations of antenna systems for mobile and fixed applications, such as low-gain omnidirectional antennas, three principal divisions of medium-gain directional antennas and high-gain directional aperture antennas and finally, are presented the major practical type of antenna used in meteorological satellite communications for transmitting and receiving Ground Earth Stations (GES).

In addition will be introduced propagation and interference characteristics very important for providing quality and reliability of satellite propagation channels. To design an effective satellite communications model it is necessary to consider the quantum of all propagation characteristics, such as signal lost in normal environment, path depolarization causes, transionospheric contribution, propagation effects important for mobile systems, including reflection from the Earth's surface, fading due to sea and land reflection, signal blockage and to the different local environmental interferences for all mobile and handheld applications. At any rate, the local propagation characteristics on the determinate geographical position have very specific statistical proprieties and results for ground transmitting and receiving satellite stations and infrastructures, which are related to the Carrier-to-Noise-density ratio (C/N_o).

© Springer International Publishing AG 2018   441
S.D. Ilčev, *Global Satellite Meteorological Observation (GSMO) Theory*,
DOI 10.1007/978-3-319-67119-2_7

7.1 Evolution of Antenna Systems for Radio Communications

The oldest existing antennas was used by Heinrich Hertz in 1888 at a very short distance in his first experiments for proving the existence of electromagnetic (EM) waves, were neither physically nor functionally separated from high-frequency generators, and up to the present day resonant circuits are taken as models for illustrating certain antenna characteristics.

The Russian professor of physics Popov designed his first world's ever radio receiver in 1895 with antenna in the shape of wire mounted on a balloon in the air and transmitter with a lightning conductor as an antenna, including a metal filings coherer and a detector element with telegraph relay and a bell. Namely, the antenna was clearly separated and regarded an independent unit in a radio system as transmitting and receiving stations were set up. Soon later, Marconi started commercially to deploy radio and antenna equipment and to establish his own company for the production of radio and antenna equipment.

7.1.1 Overview of Antennas for Radio and Satellite Communications

The satellite systems introduced new complexities into the design of ground-based antennas. The direct line-of-sight between antenna and satellite requires the antenna to see from horizon to overhead (zenith – 90°) in elevation and 360° in azimuth angle, with total hemispherical coverage. This is fulfilled in the case of transceiver antenna through the use of tracking rotatable high-gain antennas often installed in the premises of GES.

Antennas act as converters between conducted waves and electromagnetic waves propagating freely in space, shown in Fig. 7.1. Their name is borrowed from zoology, in which the Latin word antennae are used to describe the long, thin feelers possessed by many insects. In Fig. 7.2 it can be seen that the antenna is an important element in any radio system because it acts like a link of a chain. So the overall performance is significantly influenced by the performance of transmit and receive antennas.

At first glance, modern antennas may still look very similar to the ancient model developed by Hertz and Prof. Popov. However, they are nowadays optimized at great expense for their intended application. Communications antenna technology primarily strives to transform one wave type into another with as little loss as possible.

This requirement is less important in the case of test antennas, which are intended to provide a precise measurement of the field strength at the installation site to a connected test receiver; instead, their physical properties need to be known with high accuracy. The explanation of the physical parameters by which the

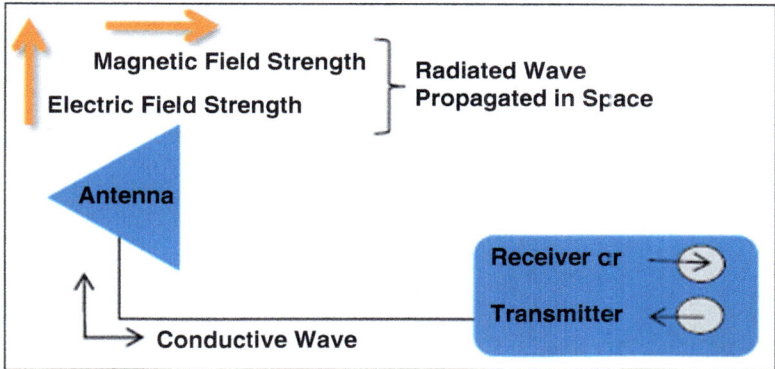

Fig. 7.1 Basic antenna functionality (Courtesy of Manual: by Rohde & Schwarz)

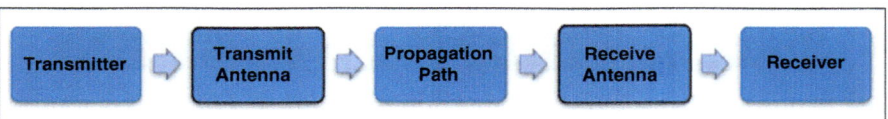

Fig. 7.2 Block diagram of radio link (Courtesy of Manual: by Rohde & Schwarz)

behavior of each antenna can be both described and evaluated is probably of wider general use; however the following chapters can describe only a few of the many forms of antenna that are in use today.

7.1.2 Satellite Antennas Geometry

Communication via satellites is always done using different types of antenna and in particular is employed a directional parabolic reflector. Since it is directional, the antenna must be accurately pointed toward the satellite and then fixed in place. On the other hand, some ground and mobile antennas have automatical tracking and pointing mechanism, which always is providing antenna in the focus of satellite. Tracking mechanism of mobile antenna is most sophisticated, because has to follow the heading of moving ship or aircraft.

All Geostationary Earth Orbit (GEO) satellites are positioned over equator and at the same altitude in the region of space known as the Clarke belt. Before attempting to align an antenna with GEO satellite, it is useful to imagine what the Clarke belt would look like if observer could see in the sky from different locations on the Earth.

Antenna using Low Earth Orbiting (LEO) satellites presents a different sort of challenge. While a geostationary satellite appears to "hang" in a single spot in the

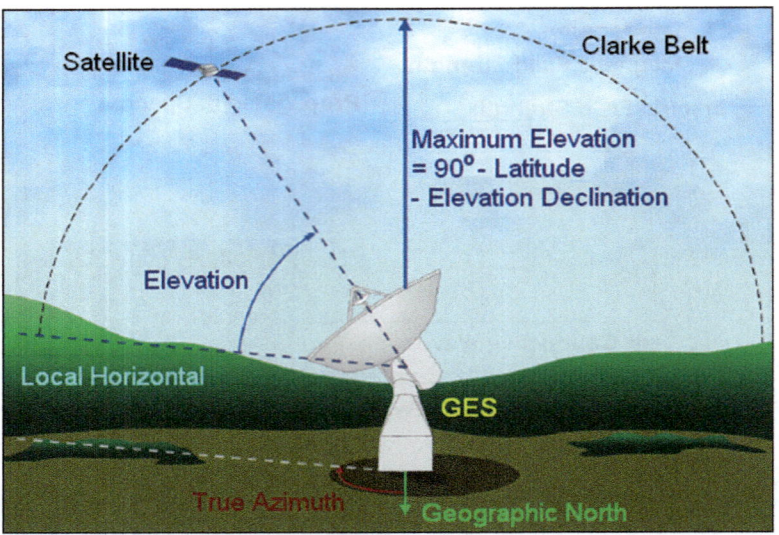

Fig. 7.3 Clarke Belt and Antenna Look Angles in the Northern Hemisphere (Courtesy of Manual: by Rohde & Schwarz)

sky all the time, LEO satellites "rise" and "set" again and again relative to a single location on the ground.

In most cases, it is impractical to install an antenna that tracks the movements of these spacecraft, and in this point omnidirectional antennas are normally used. Power levels for these types of installations must be carefully calculated so that a sufficient level of energy reaches the satellite regardless of its position in the sky at any given moment.

The Clarke belt forms an imaginary arc in the sky, which imagine in Fig. 7.3 is showing Azimuth and Elevation angles of antennas in Local Horizontal plane in the geographical North. If an observer is at the equator, the belt passes directly over his overhead. If an observer is in the Northern Hemisphere, the belt stretches from the horizon a little South of due east, rising to its maximum height directly towards the South, and then reaching the horizon a few degrees South of due West. If an observer is in the Southern hemisphere, the maximum height of the arc appears directly North of his location.

The true Azimuths of all GEO satellites than are visible from a location in the Northern Hemisphere range from a little more than 90° to a little less than 270°. For locations in the Southern Hemisphere, the true Azimuths are greater than 270° or less than 90°.

The Elevation of GEO satellite can range from a little more than 0° in local horizon to a maximum Elevation that depends on the latitude of the Earth station antenna. For antenna at the equator, the maximum possible elevation is equal to 90° minus the latitude (φ) minus the apparent declination (d).

7.1.3 Antennas Requirements and Technical Characteristics

This section describes important general requirements for fixed and mobile antenna solutions used in any satellite transmissions, such as follows:

1. **Mechanical Characteristics** – The radio and satellite communication antennas have to satisfy significant requirements of mechanical characteristics in relation to the construction strength and easy installation. Easy installation and appropriate physical shape are very important requirements in addition to compactness and lightweight.

   In the case of fixed antennas, the installation requirements are not as severe compared to that of mobile antennas because even in big ships there is no a comfortable space to install an antenna set.

2. **Electrical Characteristics** – Mechanical design of antenna is perfect because of some functional or electrical characteristics; however, designers of antenna have to keep in mind that the compact design of antenna has two major disadvantages in electrical characteristics, such as low-gain and wide beam coverage. The gain is closely related to the beamwidth and a Low-Gain Antenna (LGA) should have a wide beamwidth. As the gain of antenna is theoretically determined by its physical dimensions, reducing the size of antenna means decreasing its gain. Because of low-gain and limited electric power supply, it is very difficult for antennas to have enough receiving capability of Gain-to-noise-Temperature (G/T) and transmission Equivalent Isotropically Radiated Power (EIRP). These disadvantages of GES antenna can be compensated by a satellite that has a large antenna and High Power Amplifier (HPA) with enough electrical power. A powerful satellite with high G/T and EIRP performance should permit the fabrication of compact and lightweight antennas. The next disadvantage is that a wide beam antenna is likely to transmit undesired signals to and receive them from an undesired direction, which will cause interference in and from other systems.

7.2 Basic Relations of Antennas

The basic relations of antenna systems are very important parameters to easily understand the mode of antenna functions in two-way (duplex) satellite transmission systems, such as GES transceiving antennas. Moreover, these characteristics of antenna systems are needed for link budget calculations and for good satellite up and downlink design, which can provide reliable and acceptable quality satellite communications. At this point, this implies that the signal transmitted via the Transmitted (Tx) antenna must reach the Receiving (Rx) antenna of other communication antennas at a carrier level sufficiently above the unwanted signals generated by various unavoidable sources of noise and interference.

7.2.1 Frequency and Bandwidth in Meteorological Satellite Communications

In almost all present and forthcoming satellite communication systems using GEO and LEO or Polar Earth Orbit (PEO) satellites, the L-band 1.6/1.5 GHz is used for a link between the satellite and GES terminals. The required frequency bandwidth in L-band in mobile satellite systems is about 8% to cover transmission and receiving channels. In using a narrow-band satellite antenna, such as an omnidirectional patch antenna, some efforts have to be made to widen the bandwidth. Thus, the S and L-band are allocated in WARC-92 for the Big LEO Iridium and Globalstar systems, which require frequency bandwidths of about 5%.

In general, Radio Frequency (RF) is any of the Electromagnetic Wave (EW) frequencies that lie in the range extending from around 3 kHz to 300 GHz, which include those frequencies used for communications or radar signals. The RF usually refers to electrical rather than mechanical oscillations, while mechanical RF systems do exist.

In particular, Meteorological Satellite (MetSat) systems are using the following RF spectar:

– 137–138 MHz in Satellite to Earth (S-E) direction – Dissemination of low rate data from Non-GEO (PEO or LEO) MetSat to user stations, GEO or User Earth Station (UES);
– 401–403 MHz in Earth to Satellite (E-S) direction – Data uplink from Data Collection Platform (DCP) to GEO MetSat, which frequency band is divided into regional and international channels;
– **1670 – 1710 MHz (S-E)** – The band is divided into several parts and is used for downlink of raw data, DCP data, and dissemination;
– **2025–2110 MHz (E-S)** – Telecommand and ranging signals from main Earth and uplink of processes data for dissemination;
– **2200–2290 MHz (S-E)** – Telemetry and downlink of raw data to main station;
– **7450–7550 MHz (S-E)** – Downlink of medium rate raw data from GEO MetSat to main Earth station (not used on current generation MetSat systems);
– **7750–7850 MHz (S-E)** – Downlink of raw data from Non-GEO (PEO or LEO) MetSat to main Earth station, but also dissemination;
– **8025–8400 MHz (S-E)** – Downlink of sensor data from GEO and Non-GEO MetSat to main Earth station;
– **18.1 – 18.3 GHz (S-E)** – Downlink of high rate raw data to main Earth station; and
– **25.5 – 27 GHz (S-E)** – Downlink of high rate raw data to main Earth station (currently no plans for next generation MetSat).

The bandwidth of an antenna is defined as the range of usable frequencies within which the performance of the antenna with respect to some characteristics conforms to a specified standard. The parameter most commonly taken into account here is the impedance match, where. Voltage Standing Wave Ratio (VSWR) < 1.5, – but other parameters like gain or side lobe suppression may serve as a bandwidth criteria here, too.

Radio and satellite antennas are usually designed with a specific frequency, and their performance characteristics, e.g., gain and radiation pattern, are generally specified with that specific frequency. These characteristics change, generally deteriorate, as the operating frequency deviates from the specified frequency. The frequency range within which the performance characteristics maintain certain level with respect to the specified performance characteristics, e.g., 90%, is referred to as the antenna bandwidth.

Many different types of antennas, e.g., wire or aperture antenna have different bandwidth characteristics. In general, wire antennas have narrower bandwidths than aperture antennas. For instance, the bandwidth of a simple linear wire antenna is about 8–16% of its designed frequency. Therefore, much effort had been spent to develop broadband wire antennas by changing the physical shapes of the antennas.

For broadband antennas, the bandwidth is usually expressed as the ratio of the upper-to-lower frequencies of acceptable operation. For example, a 10:1 bandwidth indicates that the upper frequency is 10 times greater than the lower. A ratio of 2:1 is called an octave and a ratio of 10:1 is a decade, which broadband bandwidth is expressed by relation:

$$BW_b = f_H/f_L \qquad\qquad (7.1)$$

where f_H is the highest usable frequency and f_L is the lowest usable frequency. An antenna is said to be broadband when BW is equal or greater than 2.

For narrowband antennas, the bandwidth is expressed as a percentage of the frequency difference (upper minus lower) over the center frequency of the bandwidth. For example, a 5% bandwidth indicates that the frequency difference of acceptable operation is 5% of the center frequency of the bandwidth.

There exists also a different definition of antenna bandwidth, which values is valid only for narrowband antennas:

$$BW_n(in\%) = (f_H - f_L/f_C) \cdot 100 \qquad\qquad (7.2)$$

where f_C is the center frequency. Values here can range from 0 to 200%, thus in practice this definition is only used up to about 100%.

7.2.2   Gain and Directivity

The required antenna gain is determined by a link budget, which can be calculated by taking into consideration the required channel quality and the satellite capability. The channels are expressed as C/N_o and depend on the G/T and EIRP values of the satellite and GES.

Table 7.1 Classification of L-Band satellite antenna systems

Type of antenna	Gain class	Typical gain [dBi]	Typical G/T [dBK]	Typical antenna (dimension)	Typical satellite communication services
Omnidirectional	Low	0–4	−27 to −23	Quadrifilar Drooping-dipole patch	LSD (messages)
Semidirectional (only in azimuth)	Medium	4–8	−23 to −18	Array (2–4 elements) Helical, patch SBF (0.4 m Φ)	Voice/HSD
		8–16	−18 to −10	Phased array (20 elements)	
Directional	High	17–20	− 8 to −6	Dish (0.8 m Φ)	Voice/HSD
		20–24	−4	Dish (1 m Φ)	

In the case of present GEO satellite systems a medium gain between 8 to 15 dBi is required for voice and High Speed Data (HSD) channels of 24 Kb/s. On the other hand, in the case of High Gain Antenna (HGA) minimum 24 dBi is required, due to the difference in satellite capabilities. Thus, Low Gain Antenna (LGA) of about 0–4 dBi used in omnidirectional antenna systems to provide Low Speed Data (LSD) of only about 600–1200 b/s.

There are no exact definitions to differentiate between characteristics of Low, Medium and High-gain antenna systems, except by the gain quantum, shape of the antenna and type of service. For instance, classification of L-Band GES antenna systems by their receiving and service capabilities is illustrated in Table 7.1.

The ideal antenna gain can be defined with an isotropic (hypothetical) antenna, which has an isotropic radiation pattern without any losses and therefore radiates power in all directions in uniform intensities. Thus, if input power (P_{in}) is put into an isotropic antenna, the power flux-density per ideal unit area (P_{id}) at distance (r) from the antenna is given by the following relation:

$$P_{id} = P_{in}/4\pi r^2 \ \left[W/m^2\right] \qquad (7.3)$$

However, if radiated power density is $P(\theta, \varphi)/r^2$ in directions (θ = angle between the considered direction and the one in which maximum power is radiated, known as boresight; and φ = phase) at distance (r) from the antenna under elevation, the gain of the antenna can be defined by the following equations:

$$G(\theta, \varphi) = P(\theta, \varphi)/r^2/P_{id} = P(\theta, \varphi)/r^2/P_{in}/4\pi r^2 = 4\pi \, P(\theta, \varphi)/P_{in}$$
$$= P(\theta, \varphi)/P_{in}/4\pi \ [dBi] \qquad (7.4)$$

The above-defined gain is called an absolute gain or directive gain, which is determined only by the directivity (radiation pattern) of the antenna without taking account of any losses in the antenna system, such as impedance mismatch loss or spillover loss. Thus, if direction is not specified and the gain is not given a function of (θ, ϕ), it is assumed to be maximum gain. There is a general relationship between absolute gain and the physical dimensions of the antenna and this is given by the equation as follows:

$$G = 4\pi/\lambda^2 \eta a \tag{7.5}$$

where η = aperture efficiency and a = physical aperture, which will denote the effective aperture of the antenna. According to the above relation it can be realized that compact antennas with small apertures must have low gain. If an antenna aperture is a dish a known diameter (d), can be written in normal and in decibel expression as follows:

$$G = (\pi d/\lambda)^2 \eta = 10 \log \eta (\pi d/\lambda)^2 \ [dBi] \tag{7.6}$$

Thus, it can be calculated that the gain in the mobile shipborne antenna with a diameter of d = 1 m operated at 1.5 GHz is about 21 dBi.

The directivity of the antenna $D(\theta, \phi)$ does not include dissipative losses and is defined as the ratio of $P(\theta, \phi)$ to the power per unit solid angle from an isotropic antenna radiation, the same total antenna radiated power (P_-). The antenna directivity can be expressed by:

$$D(\theta, \phi) = P(\theta, \phi)/P_r/4\pi \tag{7.7}$$

The definition of antenna directivity does not take the efficiency of an antenna into account because ($P_r/4\pi$) is related to the actual power launched into space. The ratio of $G(\theta, \phi)$ to $D(\theta, \phi)$ is termed the radiation efficiency of the antenna.

7.2.3 Radiation Pattern, Beamwidth and Sidelobes

Radiation calculation is possible in principle if the EM field can be described quantitatively at all points of the antenna surface whose boundaries are those of the apertures. For an antenna that generates a single focused beam, the principal parameter affecting the antenna radiation pattern $E(\theta, \phi)$, after the aperture size (a), is the aperture illumination distribution $E_a(r, \psi)$, which is the amplitude of the far field radiation pattern E, at the point (θ, ϕ), being essentially the Fourier transform of the illumination distribution and is given by:

$$E(\theta, \phi) = 1/\pi a^2 \int_0^{2\pi} \int_0^a E_a(r, \psi) \exp\left[-jkr \sin \theta \cos (\phi - \psi)\right] r dr d\psi \tag{7.8}$$

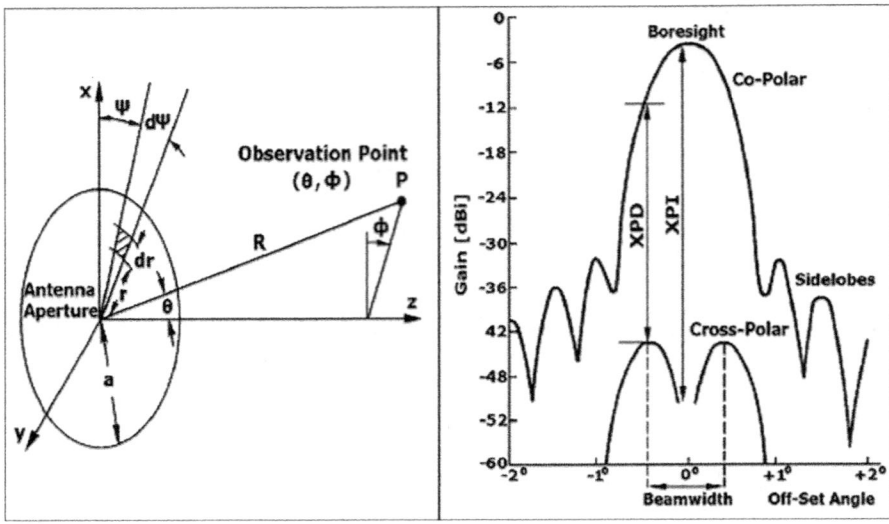

Fig. 7.4 Geometrical parameters of antenna pattern and gain characteristics (Courtesy of Book: by Evans)

An example considers a satellite communication antenna, which utilizes a circular aperture, where for circularly symmetric aperture illumination distribution this relation reduces:

$$E(\theta) = 2/a^2 \int_0^a E_a(r) \, J_0(kr \, \sin \, \theta \, \cos \, (\phi - \psi)] \, rdr \qquad (7.9)$$

where a = d/2 denotes the radius of antenna aperture; J_0 = first kind and order zero of the Bassel function and k = 2 $\pi/2$ denotes the wave number.

The other notations that denote distance and angles in coordinates are defined in the geometry illustrated in Fig. 7.4 (Left). The antenna radiation pattern is three-dimensional in nature, so it usually has to be represented from the point of view of a single-axis plot.

The characteristics of the satellite antenna radiation pattern affect interference levels directly. Any improvement in the pattern will therefore be fully reflected in the interference level and such improvement constitutes a very effective means of solving interference problems. In fact, to improve the pattern, one can either increase the antenna diameter or, with a constant diameter, use a specific technique for reducing the sidelobes. This method is therefore applicable when the satellite network is in the initial stages of development.

The antenna gain is normally calculated with reference to the boresight, i.e., the direction at which the maximum antenna gain occurs, in the case when θ, $\phi = 0°$. Gain is usually expressed in (dBi), where component (i) refers to the fact that it is relative to the isotropic gain. In this instance, the matter of moment in a dual polarization frequency re-use satellite communication system is polarization discrimination between the co-polar and cross-polar signals, especially in the antenna main beam region, as illustrated in Fig. 7.4 (Right). An important parameter that is used in an antenna's specification is the beam width evaluated by Half Power Beam Width (HPBW) $2\theta_{HP}$, where θ_{HP} is the half-power angle when radiated power becomes half the maximum level (-3 dB). The HPBW (θ_0) is given by the following equation:

$$\theta_0 = 65 \; (\lambda/d) \tag{7.10}$$

In such a way, here it is possible to realize that the half-power antenna bandwidth is inversely proportional to the operating frequency and the diameter of the antenna. For example, a 1 m receiver antenna operating in the C-band (4 GHz) has a 3 dB bandwidth of roughly 4.9°, while the same antenna operating in the Ku-band (11 GHz) has a 3 dB bandwidth of approximately 1.8°.

The antenna systems have co-polar and cross-polar gains, where the reception of unwanted, orthogonally polarized cross-polar signals will add as interference to the co-polar signal. The ability of an antenna to discriminate between a wanted polarized waveform and its unwanted orthogonal component is termed as its Cross-Polar Discrimination (XPD). When dual polarization is employed and the antenna's ability to differentiate between the wanted polarized waveform and the unwanted signal of the same polarization, introduced by the orthogonal polarized wave, it is termed as the Cross-Polar Isolation (XPI). In this context, an antenna typically would have an $XPI > 30$ dB.

The level of the antenna pattern's sidelobes is also important, as this tends to represent gain in an unwanted direction. For a transmitting gain this leads to the transmission of unwanted power, resulting in interference to other systems, or in the case of a receiving antenna, the reception of unwanted signals or noise. The sidelobe characteristic of GES is one of the main factors in determining the minimum spacing between satellites and therefore the orbit and spectrum utilization efficiency. The ITU-R S.465–5 recommendation gives a reference radiation diagram for use in coordination and interference assessment, which is defined by:

$$\begin{aligned} G &= 32 - 25 \log \phi \;\; [\text{dBi}] & &\text{for } \phi_{min} \le \phi < 48° \\ &= 10 \text{ dBi} & &\text{for } 48° \le \phi \le 180° \end{aligned} \tag{7.11}$$

where G = gain relative to an isotropic antenna; ϕ = off-axis angle referred to the main lobe axis and $\phi_{min} = 1°$ or $100 \; \lambda/d$ degrees, whichever is the greater. In this context, most of the effective power radiated by an antenna is contained in the so-called main lobes of the radiation pattern, while some residual power is radiated in the sidelobes. Sidelobes are an intrinsic property of antenna radiation and diffraction theory shows that they cannot be completely suppressed. However, sidelobes are also due partly to antenna defects which can be minimized by proper design. Conversely, due to the reciprocity theorem, the receive antenna gains and radiation patterns at the same frequency, are identical to the transmit antenna gains and radiation patterns. Unlikely, unwanted power can also be picked up by the antenna sidelobes during reception.

For large satellite antennas, with a diameter over 100 λ (wavelengths), a reference radiation pattern is recommended by the CCIR for interference to and from other satellite and terrestrial communication systems. At this point, the diameters of the vehicle antennas under discussion are, in many cases, below five wavelengths in the L-band. Further CCIR action is expected to define a reference radiation pattern for mobile satellite antennas.

7.2.4   Polarization and Axial Ratio

The antenna and the EM field received or transmitted have polarization properties. Thus, the polarization of an EM wave describes the shape and orientation of the locus of the extremities of the field vectors as a function of time. A wave may be described as linearly, circularly or elliptically polarized. Linear polarization is such that the electric E-field is oriented at a constant angle as it is propagated that can be either vertically or horizontally.

If a plane wave is propagated along the (z) axis and electric field (E) is on the (x-z) or (y-z) planes, relations for linear vertical and horizontal polarization can be written as follows:

$$E_x = E_a \cdot e^{j(\omega t - kz + \phi a)} \quad \text{and} \quad E_y = E_b \cdot e^{j(\omega t - kz + \phi b)} \qquad (7.12)$$

where E_a; ω; k and ϕ_a denote the maximum amplitude of electric field, angular frequency ($2\pi f$), wave number and initial phase, respectively, while E_b and ϕ_b are the maximum amplitude and the initial phase of the wave. Circular polarization is the superposition of two orthogonal linear polarizations, such as vertical and horizontal, with a 90° ($\pi/2$) phase difference. The tip of the resultant E-field vector may be imagined to rotate as it propagates in a helical path. There is a Left-Hand Circularly Polarized (LHCP) wave with anticlockwise rotation and a Right-Hand Circularly Polarized (RHCP) wave with clockwise rotation. An elliptically polarized wave may be regarded as the result either of two linearly or two circularly polarized waves with opposite directions. This type of polarization is the case when the amplitudes and phase difference between the two waves are not equal ($\pi/2$).

As discussed in a previous section, the signals fields can contain co- and cross-polar components. In this way, the cross-polarization of a source becomes of increasing interest to satellite antenna designers. In the case of Tx or Rx antennas with a linearly polarized field, the cross-polar component is the field at right angles to this co-polar component. Namely, if the co-polar component is vertical, then the cross-polar component is horizontal. Circular cross-polarization is that of the opposite hand to the desired principal or reference polarization. Impure circular polarization is in fact elliptical. The level of impurity is measured by the elliptical and known as the Axial Ratio (AR). The AR can be defined as the ratio of the major axis electric component to that of the minor axis by:

$$|AR| = |E_1/E_2| \ (1 \leq |AR| \leq \infty) \tag{7.13}$$

The signal for AR denotes the direction of rotation, however, an absolute value is usually used to evaluate circular polarized radiated waves and can be expressed in decibels by the following equation:

$$|AR| = 20 \cdot \log \left(|E_1/E_2|\right) \ [\text{dB}] \ \text{ for } \ (0 \leq |AR| \leq \infty) \tag{7.14}$$

Accordingly, the AR is determined by the performance of the antenna, so the AR is one of the most important parameters of circular polarized antennas. It can easily be understood that the AR depends on direction with respect to the axis of the antenna. In general, the AR is best (smallest) in the boresight direction and is progressively worse further away from the boresight. Circular polarized waves are used in order to eliminate the need for polarization tracking. RHCP has been used in the mobile transmission system. Otherwise, in the case of aperture-type antennas, such as the parabolic reflector antenna, which is commonly used as a shipborne antenna in the current Inmarsat-A and B terminal, an axial ratio of below 1. 5 dB in the boresight direction is so easy to achieve that polarization mismatch loss is almost negligible. However, in the case of phased array antennas, a degradation of the axis ratio caused by beam scanning must be taken into account.

7.2.5 Figure of Merit (G/T) and EIRP

Although gain is an essential factor in considering antennas, the figure of merit ratio of a Gain-to-Noise Temperature (G/T) is more commonly specified from the standpoint of MSS and satellite communications in general. Thus, the figure of merit for the receiving station is defined as the ratio between the gain of the antenna in the direction of the receiving signal and the receiving system noise temperature, the gain-to-noise temperature ratio (G/T) is generally given for the maximum gain derived from gain formula (7.4) as follows:

$$G_{\text{max}} = P_{\text{max}}/P_0/4\pi = 10 \log G \ [\text{dB}] \tag{7.15}$$

The G_{max} is often called the antenna gain expressed in dB, where the total radiated power in all directions can be determined by the following integration:

$$P_0 = \int_0^{2\pi} \int_0^{\pi} P(\theta, \phi) \ \sin \theta \ d\theta \ d\phi \tag{7.16}$$

The G/T value for satellite antenna is expressed in decibels per Kelvin [dB(K^{-1})] by the following relation:

$$(G/T) = 10 \log G - \log T_{SA} = 10 \log G - \log T_S \ \left[dB\left(K^{-1}\right)\right] \qquad (7.17)$$

The Earth station G/T typical values range from 35 dB(K^{-1}), for instance an GES receiving antenna with a 15–18 m diameter has some 15.5 dB(K^{-1}). The G/T is a very important parameter of an Earth station, so the methods used for its measurement and the contribution to the noise temperature are the subject of the ITU-R S.733 Recommendation.

The noise temperature measured at the terminals of antenna pointed to the sky depends upon frequency of operation, elevation angle and the antenna sidelobe structure. In more formal terms, the noise temperature will be derived from a complete solid angle integration of the noise power received from all noise sources (terrestrial and galactic) and determined for clear weather conditions by the following integral:

$$T_A = 1/4\pi \int_\Omega P(\theta, \phi) \ T(\theta, \phi) \ d\theta \ d\phi \qquad (7.18)$$

Thus, to produce a low noise antenna, its sidelobes must be minimized, especially in the direction of the Earth's surface, where T = noise temperature.

System total noise temperature of the system (T_{SR}) at an input port of receiver LNA or at the antenna output (T_A), taking account of losses caused by tracking, feed lines and a radome is defined by:

$$T_{SR} = T_R + T_a(1 - 1/a) + T_A/a \ \text{ or } \ T_{SA} = T_A + T_a(a - 1) + a \, T_A \qquad (7.19)$$

where T_R = noise temperature of the receiver (LNA) with a typical value of about 80 K to 100 K in the L-band; T_a = temperature of the environment of about 300 K; L_f = total loss of feed lines and components such as diplexer, cables and phase shifters if a phased-array antenna is used; a = attenuation expressed as a power ratio (a $\geq$ 1 or in decibels a_{dB} = 10 log a); T_A = antenna noise temperature that comes from such effects as the ionosphere and the Earth, which value of about 200 K depends on factors such as frequency and bandwidth and T_{SA} = antenna with a noise temperature. In such a manner, the noise of the antenna temperature must be kept as low as possible by proper design solution in order to obtain a high figure of merit (G/T).

With reference to the previously expressed formula ($P_{id} = P_{in}/4\pi r^2$) of transmitting antenna power density on the spherical surface if it has a transmitting gain (G_T) and where (P_{in}) is equal to the transmitted power (P_T), the power density (P_D) can be written by the following equation:

$$P_D = G_T \cdot P_T / 4\pi r^2 \ \left[W/m^2\right] \tag{7.20}$$

where $(G_T \cdot P_T)$ related values are considered to be the radiation power transmitted by an ideal omnidirectional antenna. Therefore, this term is considered as an Effective (or also Equivalent) Isotropically Radiated Power (EIRP), which can be expressed in antilogarithm and decibel expressions respectively as follows:

$$EIRP = G_T \cdot P_T \ [W] \ \ and \ \ EIRP = [G_T] + [P_T] \ [dBW] \tag{7.21}$$

The EIRP value is an important parameter in evaluating the transmitting performance of an GES terminal including an antenna. However, the EIRP amount (dBW) is defined by the sum of the antenna gain (dB) and the output power of HPA (dBW), taking account of feed losses such as feed lines, cable and a diplexer.

7.2.6   Classification of Satellite Antennas

In many respects the mobile satellite antennas currently available for satellite applications constitute the weakest links of the system. If the mobile antenna has a high gain, it has to track the satellite, following both mobiles and satellite orbital motions. Thus, sometimes this is difficult and expensive to synchronize.

Therefore, if the vehicular antenna has low gain, it does not need to perform tracking but the capacity of the communications link is limited. In general, according to the transmission direction, there are three types of satellite antennas: (1) transmitting and receiving or so-called transceiving, as a part of all types of MetSat antennas; (2) only receiving is part of the special MetSat receivers in spacecraft payloads or on the ground and (3) only transmitting is built in spacecraft payloads and ground transmitters of Data Collection Platform (DCP).

7.3   Omnidirectional Low-Gain Antennas (LGA)

As stated earlier, satellite antenna systems are classified into omnidirectional and directional. The gain of omnidirectional antennas is low and generally from 0 to 4 dB in the L-band, which does not require the capability of satellite tracking. There are three types of low-gain omnidirectional antennas used in mobile and meteorological satellite systems. They are very attractive owing to the small size, lightweight and circular polarization properties. These antennas are also used as elements of directional antennas for special configurations.

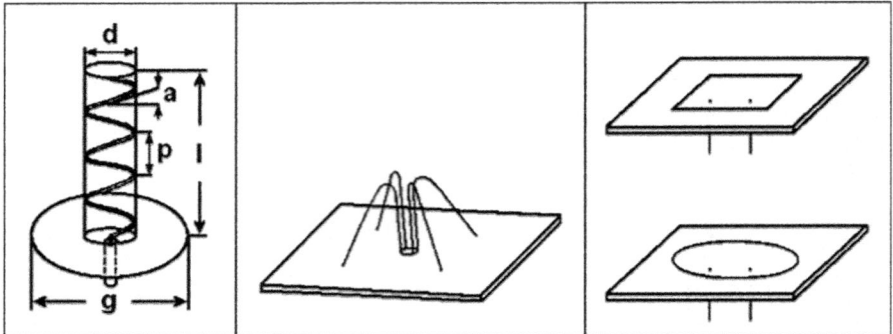

Fig. 7.5 Types of low-gain omnidirectional antennas (Courtesy of Book: by Ilcev)

7.3.1 Quadrifilar Helix Antenna (QHA)

The QHA low-gain model is composed of four identical helixes wound, equally spaced, on a cylindrical surface. The helix elements are fed with signals equal in amplitude and 0, 90, 180 and 270° in relative phase. This antenna can easily generate circular polarized waves without a balloon or a 3 dB power divider, which are required to excite a balanced fed dipole and circular polarized cross-dipoles. It can also be operated on a wide frequency bandwidth of up to 200% because it is a traveling-wave-type antenna.

The components of QHA are ground plane (g), pitch (p), pitch angle (a), length (l) and diameter (d), illustrated in Fig. 7.5 (Left). The diameter of the ground plane is usually selected to be larger than one wavelength and the number of turns is $N = l/p$. However, it is well known that the parameters for (a) are about 12–15 and the circumference of the helix (πd) is about 0.75–1.25 wavelengths. Circular polarized waves with good axial ratios can be transmitted along the (z) axis direction (axial mode).

The gain of a helical antenna depends on the number of (N) turns and typical gain and half-power beam width are about 8 dBi and 50° when $N = 12 \sim 12$ but is usually about 3 dBi.

In general, QHA has two advantages over a conventional unifilar helical antenna. The first is an increase in bandwidth, namely, it can generate axial mode circular polarized waves in the frequency range from 0.4 to 2.0 wavelengths of the helix circumference. The second is lowered frequency for axial mode operation. The principle disadvantage is an increase in the complexity of the feed system. The area of the ground plane is usually about three times the diameter of the helix.

7.3.2 *Crossed-Drooping Dipole Antenna (CDDA)*

A dipole antenna with a half-wavelength ($\lambda/2$) is the most widely used and it is also the most popular, having been used in systems such as the parabolic antenna for mobile and other satellite communications. In fact, a half-wavelength dipole is a linear antenna whose current amplitude varies one-half of a sine wave, with a maximum at the centre. As a dipole antenna radiates linearly polarized waves, two crossed-dipole antennas have been used in order to generate circular polarized waves. The two dipoles are geometrically orthogonal and equal amplitude signals are fed to them with $\pi/2$ in-phase difference. In order to optimize the radiation pattern, a set of dipole antennas is bent toward the ground, as shown in Fig. 7.5 (Middle) and for that reason it is called a drooping dipole antenna. Otherwise, the CCDA as a transceiver antenna for L-band satellite applications can be mounted inside a radome.

The CDDA is the most interesting for ground applications, where required angular coverage is narrow in elevation and is almost constant in azimuth angle. By varying the separation between the dipole elements and the ground plane, the elevation pattern can be adjusted for optimum coverage for the region of interest. The general characteristics of this antenna are: gain is 4 dBi minimums, axial ratio is 6 dB maximum and the height of the antenna is about 15 cm. This antenna has a maximum gain in the boresight direction.

7.3.3 *Microstrip Patch Antenna (MPA)*

A microstrip disc (patch) antenna is very low profile and has mechanical strength, so it is considered to be the best type for mobiles such as cars and especially in aircraft at the hybrid L to Ku-band, including UHF-band meteorological radars, which requires low air drag. In general, a circular disk antenna element has a circular metallic disc supported by a dielectric substrate material and printed on a thin dielectric substrate with a ground plane.

In order to produce a circularly polarized wave, a patch antenna is excited at two points orthogonal to each other and fed with signals equal in amplitude and 0 and $90°$ in relative phase. Thus, a higher mode patch antenna can also be designed to have a similar radiation pattern to the drooping dipole. To produce conical radiation patterns (null on axis) suitable for land mobile satellite applications, the antenna is excited at higher mode orders. In Fig. 7.5 (Right) is depicted the basic configuration for a circular patch antenna (above it is shown square patch with the same characteristics), which has two feed points to generate circular polarized waves. The resonant frequency excited by basic mode and given as:

$$f = 1.84c/2\pi a\sqrt{\epsilon_r} \qquad (7.22)$$

where (a), (c) and (ϵ_r) are the radius of circular disc, the velocity of light in free space and the relative dielectric constant of the substrate, respectively. The MPA antenna with higher order excitation is considered better because it can optimize the gain in elevation angle to the satellite in the same way as a CDDA. In fact, the area of a higher mode circular MPA is about 1.7 times larger in radius on the gain is about 6–8 dBi. The circular patch is also suitable as a satellite navigation-receiving antenna for GPS receivers.

7.4 Directional Medium-Gain Antennas (MGA)

The medium-gain directional antennas are solutions with a typical gain between 12 and 15 dBi, although some antennas can have even bigger gains. These antennas can provide voice, Fax and data (HSD) in satellite communications and transfer of meteorological data.

7.4.1 Aperture Reflector Antennas

The aperture reflector antennas are good solutions with medium-gain characteristics used in satellite communications, with three basic representatives such as SBF, modified SBF and improved SBF antennas, illustrated in Fig. 7.6 (Left), (Middle) and (Right), respectively. The main characteristics of these three antennas are shown in Table 7.2.

Moreover, due to the excellent radiation characteristics of SBF antennas, all three types of aperture antenna with half-power beam width of about 34° have been in their time proposed for shipborne antenna of Inmarsat-M. The SBF antennas consist in the stabilized platform with two gyroscopes for azimuth and elevation angles, diplexer, HPA and LNA, which are enclosed under the protective cupola of a radome. In order to stabilize the antenna, two gyro wheels rotate in opposite directions on a platform.

7.4.1.1 Short Backfire (SBF) Plane Reflector Antenna

The SBF plane reflector antenna developed experimentally by H.W. Ehrenspeck in the 1960s is well known as a highly efficient antenna of distinctly simple and compact construction. Moreover, its high directivity and low sidelobe characteristics make it a single antenna with high, even values, which is applicable to satellite communications, tracking and telemetry. Thus, an SBF antenna is very attractive

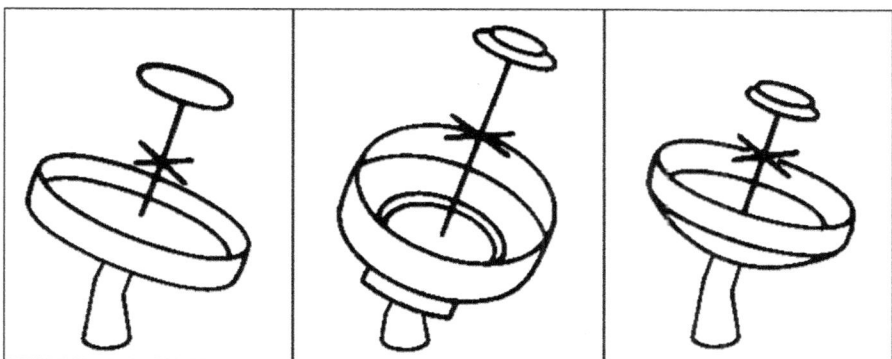

Fig. 7.6 Types of directional medium-gain aperture antennas (Courtesy of Book: by Fujimoto)

Table 7.2 Particulars of aperture types of antennas

Characteristics	SBF antenna	Modified SBF antenna	Improved SBF antenna
Effective gain	14.5 dB	15 dB	15 dB
Half-power bandwidth	34°	34°	34°
Directive gain	14.8 dB	15.5 dB	15.5 dB
First Sidelobe level	−21 dB	−22 dB	−22.5 dB
Axial ratio	−1.3 dB	−1.1 dB	−1.1 dB
Aperture efficiency:			
Effective − directive gain	65% − 75%	75% − 80%	76% − 85%
RF/VSWR bandwidth, under 1.5	3%	7%	9%
Diameter of large reflector (D$_R$):			
Bigger (D$_{R\ 1}$)	40 cm (2.05λ)	40 cm (2.05λ)	40 cm (2.05λ)
Smaller (D$_{R\ 2}$)	–	27 cm (1.38λ)	–
Diameter of small reflector (D$_r$):			
Bigger (D$_{r1}$)	9 cm (0.46λ)	–	9 cm (0.46λ)
Smaller (D$_{r2}$)	–	9.5 cm (0.48λ)	8 cm (0.41λ)
Width of a rim	4.9 (0.25 λ)	8.5 cm (0.43λ)	4.9 (0.25 λ)
Distance between (D$_R$) and (D$_r$)	9.7 cm (0.49λ)	19.5 cm (0.99λ)	12.9 cm (0.66λ)
Distance between Exciter & D$_r$.	4.9 cm (0.25λ)	–	5.7 cm (0.29λ)
Distance between (D$_{r1}$) & (D$_{r2}$)	–	–	1.8 cm (0.09λ)
Slanting angle of a (D$_R$)	0°	–	15°

for gains in the order of 13–15 dBi peak RHCP and can be mounted primarily any type of environment.

Otherwise, this type of satellite antenna consists in two circular planar reflectors of different diameter, separated generally by about one-half wavelength, forming a shallow leaky cavity resonator with a radiation beam normal to the small reflector. Namely, the antenna is fed by a dipole at around the midpoint between two reflectors and it has almost a quarter-wavelength rim on the larger reflector. It has the problem of a narrow bandwidth of about 3% because of its leaky cavity operation.

The Rx terminal G/T is -12 dBK and the EIRP of the Tx terminal is 28 dBW. The basic configuration of the SBF antenna consists in a cross-dipole element, which is required to generate a circularly polarized wave, large and small reflectors and a circular metallic rim. The antenna has the strong directivity normal to the reflector and its performance is superior to that of other types of mobile antennas with the same diameter, however, it has the problem of narrow frequency band characteristics. This antenna has many beneficial characteristics, such as efficiency and the simplicity of construction and is also considered a favorite option for a compact and high-efficiency shipboard antenna.

7.4.1.2   Modified SBF Plane Reflector Antenna

A modified SBF antenna differs from the conventional SBF antenna in that there is either an additional step on the large reflector or a change in the shape of the large reflector from a circular to a conical plate in order to improve the gain characteristics and the frequency bandwidth of the VSWR.

The dual reflector improves the input impedance characteristics covering the frequency range between transmitting and receiving sides. The conventional SBF model is a resonant-type antenna, producing input impedance characteristics that are narrow in bandwidth, so wider bandwidth is required to cover the 1.6/1.5 GHz RF range. Thus, the improvement in the antenna input impedance is greatly dependent on the size and the separation of the small reflectors. The VSWR can be reduced from 1.7 and 1.5 (at 1.54 and 1.64 GHz) to below 1.2 for each RF band.

7.4.1.3   Improved SBF Conical Reflector Antenna

The main research activities of the ETS-V satellite program have been focused on studying the reduction of fading, using compact and high-efficiency antennas with a gain of around 15 dBi, so the electrical characteristics of a simple SBF antenna have been improved by changing its main reflector from a flat disk to a conical or a step plate and by adding a second small reflector. The gain is improved by 1 dB without changing sidelobe levels.

7.4.2   Wire Antennas

The wire antenna systems are monosyllabic construction or combinations of elements, such as different shapes of wire spirals and helixes, dipoles and patches. These types of antennas have a very simple construction, with any reflector specified for medium-gain directional antennas, and with some modification, respond well to the demands of satellite systems. The antenna stabilization is obtained by a two-axis stabilized method.

7.4.2.1   Helical Wire Antennas

Since an axial mode of helical wire antenna has good circular polarization characteristics over a wide frequency range, it has been put into practical use as a single wire antenna or as an array element. With respect to the structure, this antenna can be considered a compromise between the dipole and the loop antennas and the radiation mode varies with the pitch angle and the circumference of the helix. In particular, a helix with a pitch angle of $12–15°$ and a circumference of about $1\ \lambda$, has a sharp directivity towards the axial direction of the antenna. This radiation mode is called the axial mode, which is the most important mode in helical antennas. Several studies have been carried out on the properties of the axial mode helical antenna with a finite reflector. The current induced on the helix is composed of four major waves, which are two rapidly attenuating waves and two uniform waves along the helical wire. These waves include the traveling wave and the reflected wave.

Thus, in a conventional helical antenna, the uniform traveling wave will be dominant when the antenna length is fairly large, with typical versions such as a conical helix shown in Fig. 7.7 (a) and a cylindrical helix in Fig. 7.7 (b). A conical helix is interesting for all RF bands and especially for L-band satellite communications enabling HPBW in the order of

$100°$ and circular polarization without hybrid gains of $4–7$ dBi. Cylindrical antennas can be monofilar or multifilar, also suited for satellite communications, while in a short-cut cylindrical helix antenna, the rapidly attenuating traveling wave will be dominant, especially in a two-turn ($N = 2$) helical antenna.

1. **Conical Helix Antenna** – This antenna can be regarded as a low-gain development of the cylindrical helix antenna and is suitable for wide-beam width applications with good efficiency. Thus, with suitable choices of cone angle and turn spacing, it is possible to achieve a beam width in the order of 100o. This type of antenna can also achieve an input VSWR of 1.5:1 or better than 5% frequency bandwidth merely by incorporating a simple quarter-wavelength transformer. The typical size for an L-band application is in the order of 15 cm in length and the ground plane is about 20 cm in diameter. The resultant gain is approximately 4 to 7 dBi, which is between low and medium-gain requirements.

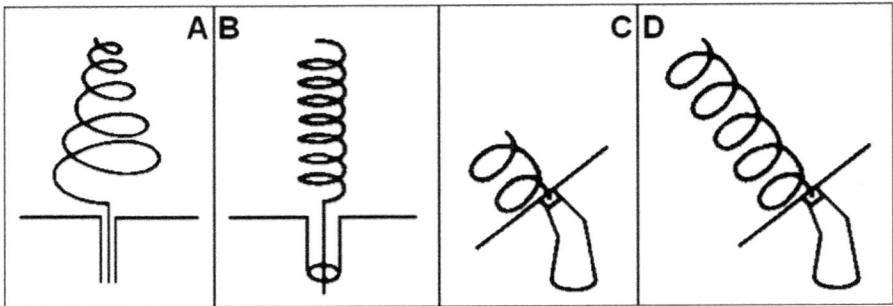

Fig. 7.7 Types of helical wire antennas (Courtesy of Book: by Fujimoto)

2. Two-Turn Cylindrical Helix Antenna has two-turns of wires, forming a simple helical antenna solution with reflector, illustrated in Fig. 7.7 (c). This model has relatively high antenna gain and excellent polarization characteristics for its size. Radiation patterns characteristically are calculated with respect to (E_θ) and (E_ϕ) planes. The gain of this antenna is 9 dBi and the axial ratio is about 1 dB, with reflector diameter (d) around 1λ. Such types of antenna have comparatively high performance in spite of their small size and compact construction.

3. Five-Turn Cylindrical Helix Antenna is shown in Fig. 7.7 (d), which main electrical characteristics are: gain is 12.5 dBi of peak RHCP for Tx and 11.5 dBi for Rx; sidelobe level has value of about -13 dB; axial ratio is 3 dB; beam width of 3 dB has angle of $-47°$; terminal G/T has -16 dBK and terminal EIRP has 29 dBW. This antenna solution is designed and developed by the European research institution ESTEC. In addition, stabilization of this antenna is obtained by gravity elevation on double-gimbaled suspension.

7.4.2.2 Inverted V-Form Cross Dipole Antenna

The inverted V-type Crossed Dipole Antenna is an advanced circularly polarized antenna with tick V-elements, shown in Fig. 7.8 (Left). The resonance of this antenna is obtained when the length is somewhat shorter than a free-space half wavelength. Thus, as the thickness is increased, the resonant length is reduced. Circular polarization can be produced by a pair of orthogonally positioned dipoles driven in quadrature phase with equal amplitudes. The crossed-dipole antenna arrangement cannot provide a good axial ratio off boresight because the radiation patterns for the straight dipole are different in both principal planes, called the H and E-planes.

This shortcoming can be improved by modifying the straight dipoles to a nonstraight version, such as the V and U-forms. The improved dipoles are called V and U-type dipoles. According to some conducted measurements, the U-type provides better electrical performance than the V-type, though the V-type is simpler in mechanical structure and is less complex. The crossed-dipole can also produce

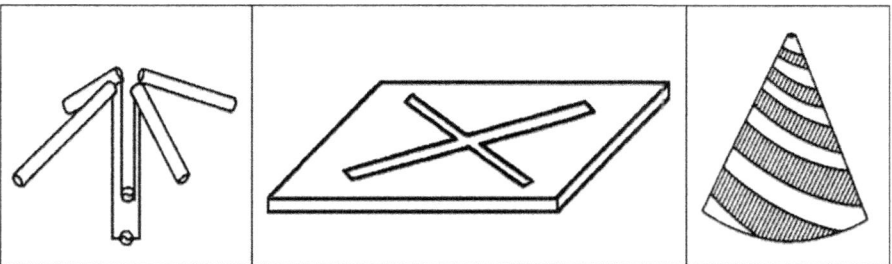

Fig. 7.8 Types of cross dipole, slot and conical wire antennas (Courtesy of Book: by Ilcev)

circular polarization without using any external circuits, such as the hybrid component. The condition to excite the circularly polarized waves can be established by a balun and the self-phasing of four radiating elements. Two of the elements are at a $0°$ phase angle and the other two are at an $180°$ phase angle.

Thus, the desired $90°$ phase difference is obtained by designing the orthogonal elements such that one is larger relative to making it inductive, while the other is smaller to make it capacitive. This type of antenna is a good model for Ku-band satellite communications.

7.4.2.3   Crossed-Slot Antenna

These antennas are useful for L-band satellite communications, which shape is illustrated in Fig. 7.8 (Middle). The slot antenna is circularly polarized and is complementary with the corresponding dipole antenna, so that the radiation pattern is the same as that for the horizontal dipole. There are only two differences: first is the property that the electric and magnetic fields are interchanged and second, is that the slot electric field component normal to the perfectly conducting sheet is discontinuous from one side of the sheet to the other because the direction of the field reverses.

In this case, the tangential component of the magnetic field is, likewise, discontinuous. This antenna can be also complementary with the corresponding crossed-dipole antenna, although the feeding method for the circular polarization is more complicated. Thus, on a model of this antenna known as a cavity-backed it needs one $90°$ hybrid to produce the circular polarization. This feed technique is effective not only to suppress undesired coupling between the cross slots but also to match the input impedance over a wider frequency band.

7.4.2.4   Conical Spiral Antenna

This type of antenna has spiral wire elements on a cone with circular polarization and is suitable for L-band mobile and GPS applications, while the bifilar version is also used in Ku-band satellite communications, as is depicted in Fig. 7.8 (Right). In comparison with a conical helix antenna, this type of antenna provides better performance and is more versatile, though the geometry is somewhat complex. Such antenna is independent of frequency and its geometry can be presented mathematically in spherical coordinates (r, θ, ɸ) as:

$$r = e^{a\phi}g(\theta) \tag{7.23}$$

where (a) and g(θ) are an arbitrary constant and angular function, respectively. Its radiation mechanism can be understood by regarding the two spirals as a transmission line. When two conductor arms are fed in antiphase at the cone apex, waves travel out from the feed point and propagate along the spirals without radiating until a resonant length has been traversed. Strong radiation occurs at that point and very little energy is reflected by the outer limits of the spiral.

Conveniently, two conductor arms can also be fed directly at the centre point or apex from a coaxial cable bonded to one of the spiral arms without any external baluns because the spiral arm can itself act as a balun. In this case, a dummy cable may be bonded to another arm to maintain the symmetrical performance. If the width of arm is decreased to a narrow constant value, the arms can be formed by the cable alone.

7.4.2.5 Planar Spiral Antennas

Cavity-backed planar spiral antennas are commonly divided into three main categories: equiangular, logarithmic and Archimedean spiral antennas. These types of antennas are well suited for flush mounting on aircraft for L to Ku-band satellite communications, synthetic aperture radars and weather forecasting. In general, this antenna has been fed by using the external balun but it can also be fed at the centre point, or apex, from a coaxial cable bonded to one of the arms, without any external baluns, like the conical spiral antenna.

1. **Equiangular Spiral Antenna** – The geometry of this antenna corresponds to the special case of the conical spiral antenna, bifilar with logarithmic period, cavity-backed and can be obtained by substituting a π/2 into θ₀ to give:

$$r_1 = r_0e^{a\phi}; r_2 = r_0e^{a(\phi-\Delta)}; r_3 = r_0e^{a(\phi\pm\pi)}; r_4 = r_0e^{a(\phi-\Delta\pm\pi)} \tag{7.24}$$

This antenna needs no external hybrid circuits to produce circular polarization and the example shown in Fig. 7.9 (Left) can radiate LHCP waves outward from the page and RHCP waves into the page when the pair of spirals is excited in antiphase at the centre. Otherwise, according to experimental measurements, the axial ratio is near unity and the HPBW is in the order of 90° over a decade bandwidth or even more. As for the input impedance the resistive part on the thickness of the antenna elements and thin elements lead to high impedance values. This implies that the

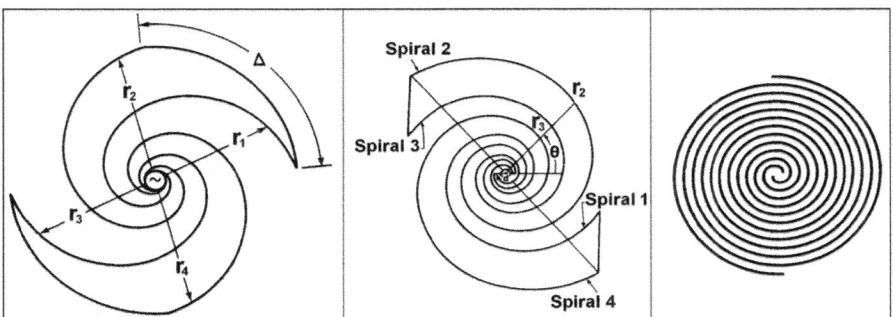

Fig. 7.9 Types of spiral wire antennas (Courtesy of Book: by Fujimoto)

impedance depends on the arm width when the structure is planar. If the angular extent (Δ) is chosen to be 90°, the geometries of the arm and the space between arms are identical, except for a rotation of 90° around an axis.

This structure is defined as self-complementary, just like the conical spiral antenna but it should be noted that the planar spiral antenna has a constant impedance of 60π [Ω] for the two arm configurations.

2. **Logarithmic Spiral Antenna** – This bifilar antenna design with logarithmic period and cavity-backed, illustrated in Fig. 7.9 (Middle), can be presented mathematically by the following equation:

$$r_1 = a^{\phi}; \quad r_2 = a^{(\phi-\Delta)}; \quad r_3 = a^{(\phi\pm\pi)}; \quad r_4 = a^{(\phi-\Delta\pm\pi)} \tag{7.25}$$

This antenna can radiate RHCP waves outward from the page and LHCP waves into the page without any external hybrid circuits, as a pair of spirals is excited with an antiphase at the centre.

3. **Archimedean Spiral Antenna** – The Archimedean spiral thin-wire bifilar cavity-backed antenna, which is illustrated in Fig. 7.9 (Right), is another geometry of the planar spiral antenna. This spiral antenna has superior bandwidth properties when fully optimized and typically consists in a pair of thin wire arms, of which the geometry can be presented by the following relation:

$$r_1 = r_0\phi; \quad r_2 = r_0(\phi - \pi) \tag{7.26}$$

This antenna also needs no external hybrid circuits to produce circular polarization and can radiate RHCP waves outward from the page and the LHCP waves into the page if the pair of thin wire arms is excited in antiphase at the centre. It is a broadband antenna and has properties similar to the standard planar spiral antenna, although it is not theoretically a frequency independent structure. When placed in a quarter-wave cavity, this antenna can achieve near-octave bandwidth, even when the cavity consists in a metal-based cylinder without any absorber.

Thus, if an absorber-loaded cylinder is employed in the cavity, a greater-than decade bandwidth may be achieved, although about half the power is dissipated into heat by the absorber. A typical Archimedean spiral antenna has an octave bandwidth for a VSWR less then 2, an axial ratio of less then 2 dB and a beam width of about 70°, while a gain of 7 to 8 dBi is achieved without an absorber. The structure has several mechanical advantages: it is compact and fairly simple to construct and the spiral arms can be easily fed, using a suitable impendence-transforming balun.

7.4.3   Array Antennas

Several different type of antenna can be arrayed in space to make a directional pattern or one with a desired radiation pattern. This type of integrated and combined antenna is called an array antenna consisting in more than two elements, such as microstrip, cross-slot, cross-dipole, helixes or other wire elements and is suitable for mobile and other satellite systems. Thus, the individual antenna elements are connected to a single receiver or transmitter by feedlines that feed the power to the elements in a specific phase relationship. Each element of an array antenna is excited by equal amplitude and phase and its radiation pattern is fixed.

7.4.3.1   Microstrip Array Antenna

The Microstrip Array Antenna (MAA) is a nine-element flat antenna disposed in three lines spaced at 94 mm, namely about a half wavelength at 1.6/1.5 GHz and whose antenna volume is about $300 \times 300 \times 10$ mm, which is shown in Fig. 7.10 (Left). As shown in this figure, the element arrangements of the MAA solutions are 3×3 rows square arrays in order to obtain similar radiation patterns in different cut planes.

The MAA beam scanning is performed by controlling four-bit variable phase shifters attached to each antenna element. This type of antenna is very applicable for the Mobile Earth Station (MES) Inmarsat-M, Inmarsat-Aero and other satellite standards.

7.4.3.2   Cross-Slot Array Antenna

The Cross-Slot Array (XSA) antenna is a 16-element solution with 97 mm spacing and their volume is about $560 \times 560 \times 20$ mm, shown in Fig. 7.10 (Middle). Evident is the element arrangement of the XSA, which is a modified 4×4 square array in order to obtain similar radiation patterns in different cut planes.

The XSA antenna beam scanning is carried into effect to control four-bit variable phase shifters associated to each antenna element. This antenna is suitable for the MES Inmarsat-M, including Inmarsat-Aero and other satellite solutions.

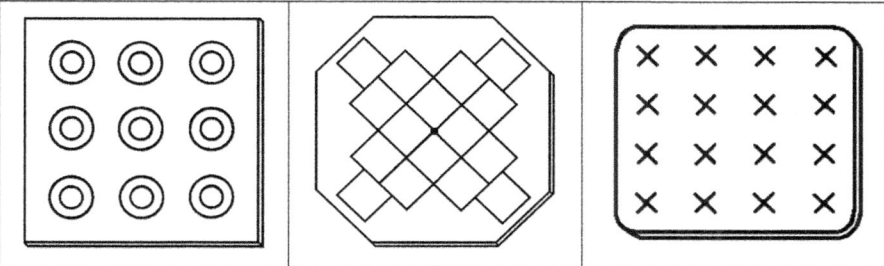

Fig. 7.10 Microstrip, cross-slot and dipole array antennas (Courtesy of Book: by Fujimoto)

7.4.3.3 Cross-Dipole Array Antenna

The cross-dipole array antenna is composed of 16 crossed-dipoles fed in phase with a peak gain of 17 dBi and with the feeding circuit behind the radiating aperture, shown in Fig. 7.10 (Right). The main electrical characteristics of this antenna are: gain is 15–17 dBi with peak RHCP transmit; axial ratio has a value of 0.7 dB; beam width of 3 dB is −34°; terminal G/T is −9.5 dBK and EIRP terminal value is 32 dBW.

Otherwise, the antenna system consists in a stabilization mechanism platform for tracking and satellite focusing, flat antenna array, diplexer, HPA and LNA, which are all protected by a plastic radome. Stabilization of the antenna is obtained by a single-wheel gyroscope, when the azimuth pointing is controlled by the output from the ship's gyrocompass. Similar to the previous two models, this satellite antenna is also suitable for maritime, land and aeronautical applications.

7.4.3.4 Four-Element Array Antennas

There have been several four-element antenna models developed, such as Yagi-Uda, Quad-Helix and four elements SBF array. Each of four-element arrays is used to cover an angular sector of 120°.

3. **Yagi-Uda Crossed-Dipole Array Antenna** – This type of array antenna has been developed for use on board ships and is protected with a radome, shown in Fig. 7.11 (Left). The feeder of this antenna is a simple formation of four in-line crossed-dipoles fixed in the middle of the reflector. This endfire array has circular polarization and the gain is between 8 and 15 dBi.
4. **Quad-Helix Array Antenna** – The quad-helix array antenna solution is composed of four identical two-turn helical wire antennas in the shape of a square and whose elements are oriented in the manner shown in Fig. 7.11 (Middle).

   According to previous studies, the effect of mutual coupling between each element of this antenna is not negligible and this mutual coupling mainly degrades the axial ratio. The axial ratio of a single helical antenna is about

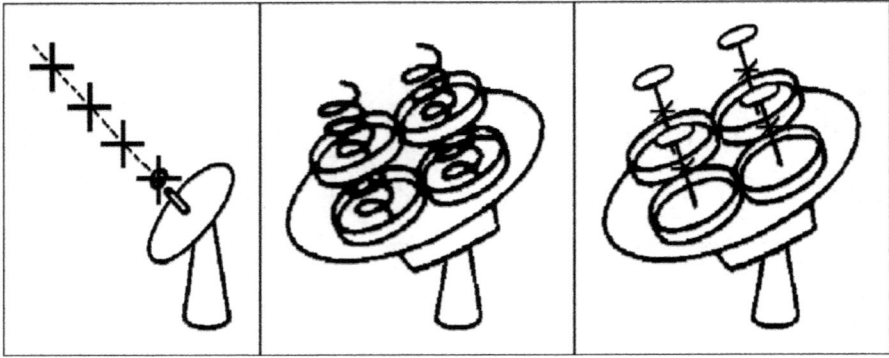

Fig. 7.11 Four-element arrays (Courtesy of Book: by Ilcev)

1 dB but this value is degraded to about 4.5 dB in the case of the array antenna with array spacing of 0.7λ. However, the best properties of antenna gain and axial ratio can be obtained at a rim height of about 0.25λ. The antenna gain is improved by 0.4 dB and the axial ratio is also improved by 3.5 dB, compared to that of the quad-helix array antenna without rims. The performance characteristics of this small antenna are essentially, gain is about 13 dB (HPBW is 38°) and aperture efficiency is about 100%. It appears that this type of helix-integrated antenna is also well suited for the shipborne Inmarsat-M standards.

5. **Four-Element SBF Array** – This antenna is developed on the basis of a conventional SBF antenna as an integrated array with four SBF elements, see Fig. 7.11 (Right). The antenna provides high aperture efficiency, circular polarization and almost high-performance gain between 18 and 20 dBi. Because of the high gain characteristics, this array is very suitable for maritime applications as a shipborne antenna.

7.4.3.5 Spiral Array Antenna

Directional antennas for mobile and satellite systems have been expected to provide voice and HSD links not only for long haul tracks but also for private cars. From that point of view, cost is an important factor to be taken into account in designing antenna systems. In the early stage, a mechanical steering antenna system was considered the best candidate for vehicles, however, it will be replaced by a phased array antenna in the near future because it has many attractive advantages, such as low profile, high-speed tracking and potential low cost.

The mechanical steering antenna with eight spiral elements and with adopted closed loop tracking method gives about 15 dBi in system gain, which is shown in Fig. 7.12 (Left). The antenna is 30 cm in radius, 35 cm in height and 1.5 kg in weight. The array consists in 2 × 4 spiral elements and it forms a fan beam with a half-power beam width of 21° in the azimuth and 39° in the elevation plane at L-band. Its peak gain is about 15 dBi, including the feeder losses, and is suitable to track the satellite for mobile satellite and other systems, because elevation angles to the satellite are not as varied as those of the azimuth angles. In effect, the antenna beam direction can be shifted in two azimuth directions, from the E or W side, by switching the pin diode phase shifters. Consequently, the difference between the received signals in both directions is used to drive the antenna system towards the satellite. The beam-shifting angle is set to approximately 4°.

7.4.3.6 Patch Array Antennas

The main feature of the future satellite communication and other systems will be portability and transportability, which means that a person or operator can directly access the satellite to establish a link using very small transportable or portable satellite transceivers with small antenna system, such as patch array antenna. However, even in the present Inmarsat L-band satellite communicational system, other satellite designers and service providers have made significant efforts to develop transportable and portable terminals with corresponding satellite antenna systems.

1. **Two-Patch Array Antenna** – The special two-patch antenna is suitable for integration in transportable and briefcase portable satellite terminals developed in the Inmarsat and ETS-V programs, which sample with two patch arrays situated on the platform is depicted in Fig. 7.12 (Middle). Therefore, this satellite antenna has two microstrip patch elements (one for Rx and another for Tx), gain is 6 dBi, EIRP is 6 dBW and G/T is −21 dBK. The reason for adopting separate Rx and Tx antennas is to eliminate a diplexer, which is too large and heavy for a compact and lightweight terminal. The antenna beam width on the lid is wide enough to point to a satellite by manual tracking. Two microstrip patch array antennas mounted on the lid of the briefcase transportable transceiver for low-speed data transmission serves the Inmarsat and ETS-V transportable and portable mobile satellite terminals.

Fig. 7.12 Spiral and Four-Patch array antennas (Courtesy of Book: by Fujimoto)

2. **Four-Patch Array Antenna** – The four-element patch array transportable antennas were developed for universal Inmarsat-C and other transportable and portable satellite terminals. On the other hand, JPL and NASA designed similar L-band types of mobile antennas, mainly for mobile satellite regional utilities in the USA. These include a mechanically steered, tilted 1 × 4 patch array and two electrically steered planar-phase array antennas. The mechanically steered four-square patch array antennas can be fixed in one line, similar to two-patch array, or can have the shape of a four-circular patch array manually steered satellite antenna, which arrangement with the platform is illustrated in Fig. 7.12 (Right). Thus, a dither-tracking four-element, circular, polarized array for land and aeronautical mobile terminals has been designed, 10.16 cm high, 50.8 cm in diameter, with 20–60° elevation coverage and with a minimum of 10 dBi gain. In fact, this antenna employs a kind of closed-loop for tracking the satellite in azimuth. The rotating antenna platform is mounted on the fixed platform that includes the motor drive and pointing system hardware.

At the end of this sector, it is necessary to conclude that all three already stated medium-gain antennas, such as Aperture Reflector Antennas, Wire Antennas and Array Antennas, feature beams that are narrow in azimuth angle, hence, they require azimuth steering to keep the beam pointed toward the desired satellite as a mobile changes its azimuth orientation. This type of medium-gain antennas provide 9–12 dBi gain, reject multipath signals outside their beam pattern and allow two communication satellites separated by 30° in a GEO coverage arc to reuse the same frequency to cover the great areas of ocean regions.

7.5   High-Gain Directional Aperture Antennas

High-gain directional aperture antennas are more powerful transmission reflectors and panels used for Inmarsat maritime and transportable applications. The typical gain of these antennas is more than 20 dBi, EIRP is a maximum of 33 dBW and G/T is about −4 dBK. There are few basic types of directional parabolic antennas: parabolic dish antenna, which uses parabolic reflector outside or inside radome, and umbrella antenna. The new antenna solution for Inmarsat transportable satellite units is known as a Quad flat panel antenna designed by the South African-based company OmniPless. However, this type of antenna can be also used in many civilian and military applications for fixed and mobile communication, including for satellite meteorological transmitting and receiving antennas.

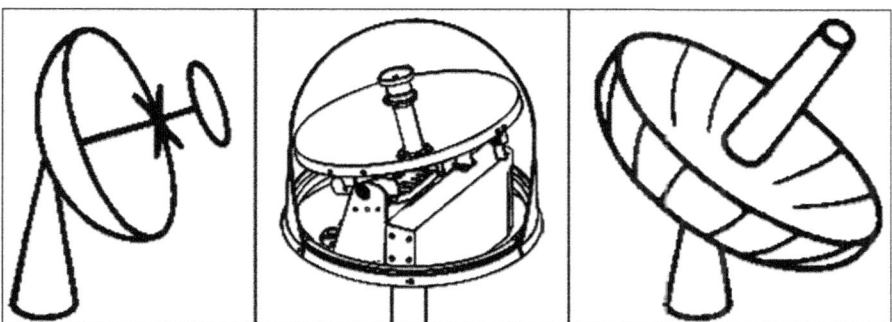

Fig. 7.13   High-gain directional aperture antennas (Courtesy of Book: by Ilcev)

7.5.1   Parabolic Dish Antenna

The first generation of parabolic dish antennas used a reflector in diameter of max 1.2 m, whereas on newer models it is likely to have reduced in size to approximately 0.7–0.8 m, which is illustrated in Fig. 7.13 (Left). Because a large proportion of the Rx signal gain and Tx EIRP is produced by the antenna, the area of the dish can only be reduced if the transmitting power from the satellite transponder is increased, when the receive preamplifier gain can be increased without an appreciable increase in noise.

The parabolic reflector is most often used for high directivity for radio signals traveling in straight lines, as do light rays. They can also be focused and reflected just as light rays can, namely, a microwave source can be placed at focal point of antenna reflector. The field leaves this antenna as a spherical wave front. As each part of the wave front reaches the reflecting surface it is phase-shifted 180°. Each part is then sent outward at an angle that results in all parts of the field traveling in parallel paths. Due to the special shape of a parabolic surface, all paths from the focus to the reflector and back into space line are the same length. When the parts of the field are reflected from the parabolic surface, they travel to the space line in the same amount of time. This antenna is a large microwave parabolic consisting in the reflector of dish shape, feeder structure, waveguide assembly, servo and drive system and protective radome.

Stable platform supports an antenna assembly, which must remain perfectly stable when MES is pitching and rolling in extremely bad weather conditions. Namely, it is essential that the stable platform hold the reflector in its A/E angular positions despite movement of the ship. The platform usually consists in a large solid bed mounted in such a way that four gyro compasses are able to sense movement and correct any errors detected, holding the platform level. In practice, it is a form of electronic gimbal.

The antenna tracking system is controlled in A/E angles by stepping motors, which in turn are electronically controlled in a simple feedback system. This electromechanical antenna enables the dish to maintain a lock on a satellite despite navigation course changes. As the MES changes course, both A and E control corrections is made automatically.

The antenna unit processor controls all necessary functions, which include satellite tracking and electronic control. The RF Electronics segment contains the Tx HPA and the Rx RF front-end LNA stage, plus all the critical bandpass signal filter stage. In modern equipment it is common practice to reduce the number of cables between antenna and main unit. Hence, this is achieved by multiplexing up/down signals or commands between antenna and main unit onto the one coaxial feeder. This type of parabolic satellite antenna systems can be used for mobile and fixed broadcasting and broadband satellite and DVB transmissions, including for satellite meteorological communications antennas between satellites and direct reedout ground stations.

7.5.2 Parabolic Dish Antenna in Radome

For mobile and especially maritime satellite communications service, antenna radomes are widely used to protect dish antennas, which are continually tracking satellites while the ship or aircraft experiences pitch, roll and yaw movements. The sample of shipborne reflector dish antenna is illustrated in Fig. 7.13 (Middle), which can be also used onboard ships for receiving meteorological data. In fact, the motorized elevation drive system of this shipborne antenna is configured to selectively adjust an elevation of the satellite dish and a motorized azimuth drive system can be configured to selectively rotate the satellite dish. A control system is connected to the elevation and the azimuth drive systems managing automated operation of the satellite antenna system. Large ships may have radomes over 3 m in diameter covering antennas for broadband, TV, voice, data, video, Internet and transmission weather data. Recent developments allow similar services from smaller installations such as up to 85 cm motorized dish used in the maritime and land applications, while aircraftborne antenna are using more aerodynamically shaped radomes for installation atop fuselage or inside tail without radome. Small ships, fishing vessels and private yachts may use radomes as small as 26 cm in diameter for voice and low-speed data. However, buses and trains can have low profile parabolic TV antenna in radome.

7.5.3 Parabolic Umbrella Antenna

The portable umbrella satellite antenna is a deployable, compact and lightweight parabolic type suitable for portable and transportable Inmarsat transceivers and other satellite systems including meteorological transmissions, which sample is illustrated in Fig. 7.13 (Right). Otherwise, this portable satellite antenna has almost all the same technical characteristics as a parabolic dish antenna.

7.5.4   Horn Antennas

Horn antennas are very popular at UHF (300 MHz–3 GHz) and higher frequencies, such as up to 140 GHz. Horn antennas often have a directional radiation pattern with a high antenna gain, which can range up to 25 dB in some cases, with 10–20 dB being typical. Horn antennas have a wide impedance bandwidth, implying that the input impedance is slowly varying over a wide frequency range.

However, the bandwidth for practical horn antennas can be on the order of 20:1 (for instance, operating from 1 GHz–20 GHz), with a 10:1 bandwidth not being uncommon. The gain of horn antennas often increases, while beamwidth decreases as the frequency of operation is increased. This is because the size of the horn aperture is always measured in wavelengths; at higher frequencies the horn antenna is "electrically larger", this is because a higher frequency has a smaller wavelength. Since the horn antenna has a fixed physical size (say a square aperture of 20 cm across, for instance), the aperture is more wavelengths across at higher frequencies. And, a recurring theme in antenna theory is that larger antennas (in terms of wavelengths in size) have higher directivities.

Horn antennas have very little loss, so the directivity of a horn is roughly equal to its gain. Horn antennas are somewhat intuitive and relatively simple to manufacture. In addition, acoustic horn antennas are also used in transmitting sound waves (for example, with a megaphone). Horn antennas are also often used to feed a dish antenna, or as a "standard gain" antenna in measurements. Horns can have different flare angles as well as different expansion curves (elliptic, hyperbolic, etc.) in the E-field and H-field directions, making possible a wide variety of different beam profiles.

Below are introduced the following main types of horn antennas:

7.5.4.1   Pyramidal Horn Antenna

Pyramidal horn is a major antenna with the horn in the shape of a four-sided pyramid, with a rectangular cross section, shown in Fig. 7.14 (Left). They are a common type of this type of antenna, used with rectangular waveguides and radiate linearly polarized radio waves.

The pyramidal horn is probably the most popular antenna in the microwave frequency ranges of 1 GHz to 18 GHz. The gain of a horn is usually very close to its directivity because the radiation efficiency is very good. The horns are often used as standards gain measurement in antenna development. Horn antennas are used as a feeder for larger antenna structure such as parabolic antennas, as standard calibration antennas to measure the gain of other antennas, and as a directive antenna for such devices as microwave radiometric, radar antenna, satellite antenna, dish antenna.

The advantages of horn antenna are moderate directivity, low standing wave ratio and broad bandwidth easy to construct and adjustment One of the first horn

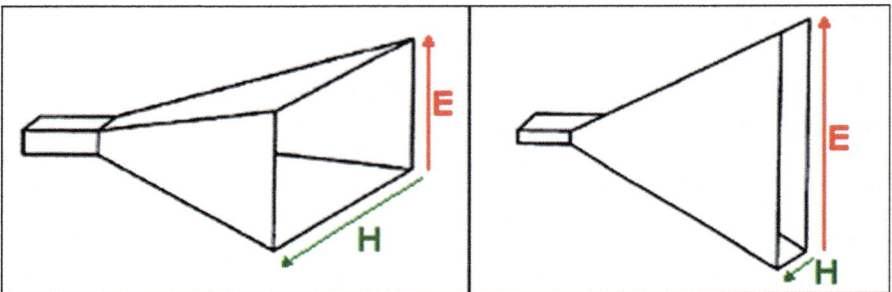

Fig. 7.14 Pyramidal and E-plane horn antennas (Courtesy of Book: by Ilcev)

antennas was constructed in 1897 by Indian radio researcher J. D. Bose in his pioneering experiments with microwaves. The development of radar in World War 2 stimulated horn research to design feed horns for radar antennas. The corrugated horn invented by Kay in 1962 has become widely used as a feed horn for microwave antennas such as satellite dishes and radio telescopes. Initially, this antenna used lower frequencies for long distance transmission, but apparently the horn was inadequate for this purpose.

7.5.4.2 E-Plane Horn Antenna

The E-plane horn antenna is a sectoral four-sided horn antenna flared in the direction of the electric or E-filed in the waveguide, which is illustrated in Fig. 7.14 (Right). This form of antenna is one that is flared in the direction of the electric or E-field in the waveguide. Thus, an E-plane sectoral feed-horn antenna is the one where the narrower dimension of the input waveguide is broadened remaining constant the other dimension. The universal radiation patterns for the E-plane sectoral feed-horn are represented with the phase error as parameter for E-plane.

For a linearly polarized antenna, this is the plane containing the electric field vector, which is sometimes called the E aperture, and the direction of maximum radiation. The electric filed or "E" plane determines the polarization or orientation of the radio wave. Thus, for a vertically polarized antenna, the E-plane usually coincides with the vertical and elevation plane. For a horizontally polarized antenna, the E-plane usually coincides with the horizontal and azimuth plane. In fact, E-plane and H-plane should be 90° apart.

7.5.4.3 H-Plane Horn Antenna

The H-plane horn antenna is also a sectoral four-sided horn antenna flared in the direction of the magnetic or H-field in the waveguide, which is depicted in Fig. 7.15

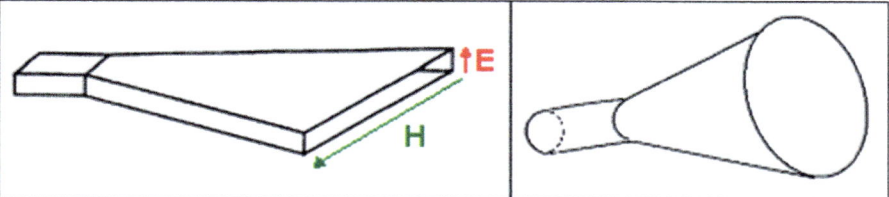

Fig. 7.15 H-plane and conical horn antennas (Courtesy of Book: by Ilcev)

(Left). In the case of the same linearly polarized antenna, this is the plane containing the magnetic field vector (sometimes called the H aperture) and the direction of maximum radiation. The magnetizing field or "H" plane lies at a right angle to the "E" plane. For a vertically polarized antenna, the H-plane usually coincides with the horizontal and azimuth plane. For a horizontally polarized antenna, the H-plane usually coincides with the vertical and elevation plane. This antenna is well suited for improving capacity and conceals connectivity without an additional ground plane. There are many solutions of this antenna in multiple gain, bandwidth, polarization and frequency, which ensures quality and flawlessness.

7.5.4.4   Conical Horn Antenna

The conical horn has a special horn in the shape of a cone with a circular cross section, which is shown in Fig. 7.15 (Right). It is used with cylindrical waveguides, and as the name indicates, the conical horn antenna has a circular cross section and end into it. This antenna is normally used with circular waveguide and is seen less frequently than the rectangular version. Pure mode conical horns do not make good feed for reflectors because the pattern symmetry is pure and the cross-polar radiation level is high. The exception is a small diameter horn where low cross polarization is possible. A conical corrugated-horn antenna design was chosen because of its properties of extremely low sidelobes, low insertion loss and excellent beam pattern symmetry. This antenna design was modified by the placement of extra grooves on the aperture of the horns to further reduce the far-sidelobes levels.

7.6   Ground Antennas for Particular Satellite Meteorological Systems

The ground satellite meteorological antennae are designed to receive signals from PEO and GEO meteorological satellites for processing weather data as Direct Readout Service (DRS) at Ground Earth Stations (GES) satellite meteorological infrastructures and also to transmit signals from special Data Collection Platform

(DCP) via GEO meteorological satellites to the different ground observation users suites.

The DRS ground infrastructures are an integral component of both the PEO and GEO satellite meteorological systems. Each of these satellite platforms can provide a high resolution and lower resolution image data product. In fact, DRS include Automatic Picture Transmission (APT), High Resolution Picture Transmission (HRPT), and Direct Sounder Broadcasts (DSB) from the POES satellites, and Low Rate Information Transmission (LRIT), and GOES Variable Format (GVAR) data from the GOES satellites. Today, the majority of the world's users of weather satellite imagery acquire them through the use of these DRS solutions. However, over 120 countries and approximately 8000 known (and an estimated several thousand more unknown) ground stations rely on these daily transmissions of meteorological data.

7.6.1 Direct Readout PEO Directional Ground Antenna Systems (GAS)

Due to the rotation of the Earth, satellites in PEO can be received in GES not more often than three to four times per day. Data received at those times thus belong to measurements made 6 h (mean value for 4 contacts being equally distributed over the day) or more before the contact. The same satellites appear over the poles appear during each orbit, about 15 times per day. If all these possible satellite contacts are used for data reception, then the measurement data is available after not more than 95 min. This is fast availability, which allows to use the data in time critical applications.

The PEO GES Terminal is situated in any given position of certain PEO satellite footprint together with adequate tracking satellite dish antenna. Depending on construction of the GES infrastructure each satellite antenna can be installed next to the GES building or on its roof. The PEO meteorological GES infrastructure may employ different antenna systems such as omnidirectional Quadrifilar Helix Antenna (QHA), directional Parabolic or Crossed Yagi antenna and so on. In this context will be introduced several types of PEO ground antenna infrastructures with their principal characteristics.

7.6.1.1 Tracking PEO Satellite L-Band GES Receiving Antenna

From meteorological, oceanography and environmental observations to disaster management, TeraScan GES of the US SeaSpace Corporation with L-band antenna system are the complete acquisition and data processing solution for every major direct-broadcast PEO satellite, which radome of L-band tracking antenna is shown in Fig. 7.16 (Left). The PEO receiver utilizes a 1.5 m antenna reflector mounted on a

Fig. 7.16 PEO GES antennas (Courtesy of Manual: by SeaSpace)

3-axis positioner and installed inside a 2.0 m radome. Radome of antenna is built by composite material, which height is 2 m, weight 41 kg and its attenuation is 0.15 dB at 1.7 GHz. Its elevation range is 0–90° and azimuth range is ±265°.

Low-loss characteristics and impedance-matched radomes of this antenna enable operation in extreme environments without sacrificing RF performance. Each antenna system is perfectly balanced in all three axes and exhibits no inherent backlash. There are no pointing and tracking errors associated with torsional stiffness since there is no wind or ice loading on the tracking system and the velocities and accelerations are kept extremely low.

Key features of this PEO antenna are: high tracking accuracies; low velocity and acceleration requirements; no keyhole losses at any elevations; 3-axis; elevation over cross-Elevation and over azimuth configuration; completely balanced system; low-loss radome enclosure for all installations; integrated tracking receiver; and Pulse Width Modulation (PWM) servo control system. Integrated Feed/LNA/ Downconverter is feed type crossed dipole, its polarization is RHCP and L-band Input Frequency 1670–1720 MHz, while RF performance antenna gives gain 25.9 dBic at 1700 MHz and G/T is 5.0 dB-K Min at 1700 MHz.

The acquisition module specifications are: operating system Linux (RHEL and CentOS); input signal power is −90 to −50 dBm; input frequency is 126–154 MHz; demodulation types are BPSK and PSK, supported telemetries are NOAA HRPT, SeaWiFS SWHRPT, FY-1D CHRPT and MetOp; and supported data encoding is NRZ, NRZ-S, NRZ-M, biphase-L. The TeraScan antenna is also available with an S-band feed and also as a shipboard system.

This type of GES ground antenna with receiver provides Weather Decision Support (WDS) for local and hemispheric images from PEO satellite meteorological data that are directly broadcast to the ground. Same as previous solutions, this WDS system also automatically receives, processes and displays meteorological and maritime forecast data tailored to the certain vessel's operating areas of interest.

7.6.1.2 Tracking PEO Satellite Multi-Band GES Receiving Antenna

The 2.4 m multi-band telemetry GES ground system is configured with a tracking antenna and receiver for the reception of real-time satellite data, which configuration is illustrated in Fig. 7.16 (Right). This receiving system includes the necessary proprietary hardware and special TeraScan software for the automated reception, data processing and visualization of the resulting products. TeraScan performs automated image navigation and geolocation for every major remote sensing satellite. Besides, a World Vector Coastline (WVC) database and the TeraNav interactive navigation tool are included with each system.

The EODS-P2400 series can currently receive and process telemetries such as NOAA HRPT, MODIS, FY-3, MetOp AHRPT, FY-1 CHRPT, NPP, EOS DB and Oceansat data, and once launched, will be able to receive and process JPSS series of satellites. This series of PEO ground satellite meteorological receivers use the following PEO antennas: EODS-P2400-X of 2.4 m diameter at X-band, EODS-P2400-LX of 2.4 m diameter at L/X-band and EODS-P2400-LSX of 2.4 m diameter at L/S/X-band.

Key features of this PEO antenna are: easy to install, operate and maintain; no keyhole losses at any elevation; operation in extreme environments and environmentally friendly, program track and auto track modes; high reliability for mission critical operations; end-to-end system solutions – turn key system; high tracking accuracies; automated capture and processing 24/7; remote access control; and low maintenance cost and power consumption.

The acquisition module specifications are: operating system Linux (RHEL and CentOS); input signal powers are -60 to -20 dBm for X-band and -90 for -50 dBm at L-band; input frequencies are 720 MHz (X-band) and 126–154 MHz (L-Band); demodulation types are BPSK, SQPSK, USQPSK and QPSK; BER performance is <0.5 dB typical; and data rates are 0.665–20.8 Mb/s (13.125 Mb/s Terra; 15.0 Mb/s Aqua; 15.0 Mb/s NPP/JPSS; and 18.7 Mb/s FY-3). Besides, clock and data outputs are ECL, TTL, RS-422 and LVDS, while its supported telemetries are NOAA HRPT, MetOp AHRPT, FY-1D CHRPT, EOS DB, Oceansat DB, DMSP RTD and NPP/JPSS.

Radome of antenna is built by composite material, which diameter is 3.2 m, height is 3.6 m, weight 364 kg and its attenuation is <0.25 dB at 8.2 GHz, and its elevation range is 0–180° and azimuth range is unlimited. Dimensions of reflector F/D Ratio if 0.375, surface tolerance is 0.030 in., effective diameter is 2.4 m and is built by solid aluminum. Its RF performance of L-band antenna gain is 30.1 dBic at 1700 MHz and G/T 12.2 dB-K at1700 MHz.

Fig. 7.17 PEO GES multiband L/S/X-band antennas (Courtesy of Manuals: by SCISYS/ Kongsberg)

7.6.1.3   Tracking PEO X/Y Satellite L/S/X-Band GES Receiving Antenna

This GES meteorological satellite tracking antenna is a parabolic antenna, designed by the German producer SCISYS Deutschland GmbH to receive satellite meteorological data and images from NOAA, Feng-Yun, OrbView and MetOp PEO satellites, which is illustrated in Fig. 7.17 (Left). Thus, the GES infrastructure is using 2.4, 2.8 and 3.0 m X/Y axis tracking satellite S, L and X-band reflector antenna system. In fact, the antenna system is operated in program track mode, and so, the pass of the satellite will be pre-calculated to reduce the cost of the tracking system by avoiding the need for expensive auto tracking systems.

The satellite receiver is prepared for the reception of PEO satellite data and it contains a QPSK Demodulator and a Viterbi Decoder. The data are sent to the station computer via a serial to parallel converter. In such a way, the software on the GES computer reads the data via a high-speed interface board and executes the Consultative Committee for Space Data Systems (CCSDS) processing.

The data are feed into the higher level processing software like IMAPP, IPOPP or CSPP. The realization of the frame synchronizer and Reed-Solomon decoder as software solution reduces the total system costs. The background processing performs all activities like pass prediction, antenna control and reception automatically, so in such a way the system operates unattended. The PEO tracking antenna consists the following components: parabolic reflector, X/Y type pedestal, DC servo motor drives, limit switches, electromechanical brake, shaft encoders, Motor Control Unit (MCU), Antenna Interface Unit (AIU) and power and control cabling. The X/Y antenna pedestal provides a program-controlled automatic tracking system allowing for movement in both, X and Y directions. DC antenna motors driven by a 4-quadrant pulse width modulation system in conjunction with digital shaft encoders allow full servo position control. The motors drive the axes through a primary gearbox and a cycloid zero backlash transmission unit. Primary and secondary limit switches at all extreme positions increase overall safety. The pedestal is designed to be mounted on a stable foundation, such as a concrete pad, fastened in place by suitable threatened rods and is connected to the control equipment by the appropriate power and control cables.

The maximum distance between the pedestal and the motor control unit is 75 m. Using special cabling set, it may be extended up to 130 m. The tracking satellite antenna can be used with unshielded parabolic reflectors ranging from 2.4 up to 3.0 m sizes. Electrical features of this antenna are: Frequency Range 1.7–2.1/ 7.7–8.3 GHz (L/S/X-band), Gain 26 dB (X-band), Beam Width 5.0° (L-band) and 1.0° (X-band). Environmental features of this antenna are: Temperature −30° to +50 °C; Operational Wind 100 km/h and Survival Wind 150 km/h.

7.6.1.4   Tracking PEO MEOS Satellite GES Receiving L/S/X-Band Antenna

The MEOS Polar GES multi-band receiving antenna is designed by Kongsberg Spacetec AS for acquisition of meteorological data from the following PEO satellites: NOAA sensors AVHRR/TOVS/ATOVS via HRPT/Eumetcast transmissions; FY-1, FY-3A or 3B sensor via CHRPT; MetOp sensors AVHRR/ATOVS via AHRPT/Eumetcast; and others. This antenna supports direct broadcast reception in L, S and X-band, including C and Ku-band reception through Eumetcast for NOAA and MetOp satellites. This antenna provides the functionality to track the satellite, receive the radio frequency and deliver data to the ingest system, which includes the following units: Antenna, Feed/downconverter, Digital receiver/bit synchronizer and Satellite tracking controller. Parabolic reflector MEOS Antenna is available from 2.4 to 4.2 m in L, S and X-band, which is illustrated in Fig. 7.17 (Right).

The basic GES and antenna package ingests data from the front-end system and provides all the necessary tools for processing and operations. Data are pre-processed and stored into a Unix file system in mission specific formats or as Level 0, 1 and map-projected products in HDF 5 format. All data is archived in a product database. Map-projected products can be viewed with the visualization software package known as MEOS VImSat. Thus, it is a fast, operational viewing tool containing functions, such as accessing archived products, zooming, printing, image enhancements, format converting and overlaying graphics.

Raw data files and higher-level products may be distributed over LAN/WAN to other users. All operations are automatic and easily configurable, including management of disk space and retrieval of processing parameter files. All status information is written to memory disk as log reports. This gives a unique capability to do diagnostics locally as well as remotely, and to generate reception quality reports. This basic package contains a Quick Look Viewer showing incoming data in real time, with possibility to show selected channels, perform image enhancement and view a previous dissemination and to display multiple missions.

7.6.1.5 Tracking PEO Satellite GES Receiving L/S-Band Antenna

The PEO GES ground satellite antenna from Dartcom provide reliable, high performance land-based and marine solutions for receiving, archiving, processing and displaying HRPT, AHRPT, Low Data Rate (LRD) data from NOAA, Defense Meteorological Support Program (DMSP), MetOp, FY-3 and Joint Polar Satellite System (JPSS) satellites. Thus, the Dartcom HRPT/AHRPT antenna system receives, archives, processes and displays data from NOAA and MetOp satellites, and optionally from FY-3 (L-Band) and DMSP (S-Band). A number of land-based antennas are available at size of frequencies 1.2 m, 1.8 m and radome-enclosed 1.5 m or for marine use a radome-enclosed 1.3 m antenna with active stabilization. In Table 7.3 are presented values of three land-based antenna systems.

Ingested data can be viewed and processed using the Dartcom iDAP/MacroPro software and outputs are also available for image processing software packages such as PCI Geomatica, ERDAS IMAGINE and ENVI/IDL, including standard interchange formats of NOAA level 1B, EPS level 0 and GeoTIFF. Direct broadcast LRIT reception from MSG, GOES, MTSAT and COMS-1 is also possible with optional extra hardware and software. The hardware itself is LRD ready, requiring only firmware and software updates to support the proposed JPSS L-band service. The system comprises main components such as: Land-based and marine antenna options available; Modular receiver rack that is containing plug-in modules for easy maintenance and upgrade; Ingest and visualization PC that is running Dartcom Polar Orbiter Ingester and Dartcom iDAP/MacroPro software. Besides, Dartcom can also provide on-site installation and training services, and supply additional PCs and other peripherals.

These tracking PEO land-based satellite antennas provide continuous tracking of satellites with no "cone of silence" (data loss at high elevations). A state-of-the-art dual-axis rotator controller with closed-loop feedback gives excellent pointing accuracy and smooth tracking. In this case, an RS-232–RS-422 link allows the ingest PC to control the rotator and provide diagnostics. The reflector is a prime focus aluminium parabolic dish finished in matt white paint (RAL 9010). An Integrated Feed/Downconverter (IFD) is mounted at the focal point in a hermetically sealed unit. Here will be introduced three types of Dartcom antennas.

1. **Parabolic 1.2 m Dish and Rotator** – The 1.2 m antenna has a 0.38 F/D ratio and 24.4 dBi gain to achieve a system G/T of better than 2.6 dBK at 1.7GHz and 5° elevation and a bit error rate of better than 1:106 from 3.5° elevation, which is depicted in Fig. 7.18 (Left). The dish is fixed to the rotator using a counterweighted aluminium frame assembly. The rotator is a compact unit with a

Table 7.3 Land-based antenna specifications

	1.2 m antenna	1.8 m antenna	1.5 m antenna and radome
Bit error rate	$1{:}10^6$ from 3.5° elevation	$1{:}10^6$ from 1° elevation	$1{:}10^6$ from 2° elevation
Azimuth range	0° to 359.9° (minimum)	0° to 359.9° (minimum)	0° to 359.9° (minimum)
Elevation range	0° to 180.0° (minimum)	0° to 180.0° (minimum)	0° to 180.0° (minimum)
Azimuth rate	10.0°/s ($\pm$10%)	12.0°/s ($\pm$10%)	48.0°/s ($\pm$10%)
Elevation rate	10.0°/s ($\pm$10%)	1.0°/s ($\pm$10%)	10.0°/s ($\pm$10%)
Mechanical tolerance	$\pm$0.15° azimuth and elevation	$\pm$0.1° azimuth and elevation	$\pm$0.15° azimuth and elevation
Tracking accuracy	$\pm$0.1°	$\pm$0.1°	$\pm$0.1°
Survival temperature	−20 °C to +60 °C	−35 °C to +63 °C	−20 °C to +60 °C
Wind speeds	85 km/h operational 145 km/h survival	100 km/h operational 210 km/h survival	185 km/h operational 185 km/h survival

Fig. 7.18 Land GES HRPT/AHRPT antennas (Courtesy of Manuals: by Dartcom)

separate housing for the power supply and controller. The dish and rotator assembly is mounted on a galvanized steel pedestal.

2. **Parabolic 1.8 m Dish and Rotator** – The 1.8 m antenna has a 0.42 F/D ratio and 27.5 dBi gain to achieve a G/T of better than 6.0 dBK at 5° elevation and a bit error rate of better than 1:106 from 1° elevation, which is depicted in Fig. 7.18 (Middle). The dish is fixed to the rotator using interface steelwork galvanized to BS729. The rotator is precision-engineered using aluminium castings and machined steel for strength. The controller is mounted inside the rotator with a separate remote housing for the power supply. The dish and rotator assembly is mounted on a durable galvanized steel pedestal.

3. **Radome-enclosed Parabolic 1.5 m Dish and Rotator** – This antenna shown in Fig. 7.18 (Right) has a 0.36 F/D ratio and 26.0 dBi gain to achieve G/T better than 4.3 dBK at 1.7 GHz at 5° elevation and a bit error rate better than 1:106 from 2° elevation. It allows future upgrade to receive X-Band data from Terra, Aqua, Suomi-NPP and FengYun-3. The dish is fixed to the rotator using a counter-weighted aluminium frame assembly. The rotator is a compact unit with a separate housing for the power supply and controller. The dish and rotator assembly is mounted on a zinc plated, powder coated steel pedestal and enclosed in a two-part (plus base) composite radome with inspection hatch on the bottom or side.

7.6.2   Direct Readout PEO Multidirectional Ground Antenna Systems (GAS)

Several different types of satellite antennas may be used for polar orbiting reception of APT imagery. One antenna is already introduced directional that requires tracking of the moving satellite and the second type is omnidirectional and less expensive giving a slightly reduced reception range.

Near the antenna, the signal is processed by a small preamplifier, which increases the strength of the desired frequencies and also filters out unwanted frequencies, the signal is then passed to the radio through a transmission line. The radio receiver used in most ground stations is a crystal controlled FM receiver with good sensitivity capable of detecting radio frequencies between 137 and 138 MHz. Since each satellite operates at slightly different frequencies, a specific crystal inside receiver is needed for each satellite that is to be accessed. Some of the more modem radios have synthesized frequency capabilities and do not require a crystal for each satellite.

The radio receives the FM signal and detects the 2400 Hz amplitude modulated sub-carrier, which carries the satellite image. Thus, if this 2400 Hz tone is inserted into an appropriate display reproduction system the satellite image can be viewed. In such a way, some stations use a stereo tape recorder to record the 2400 Hz sub-carrier at this stage of the processing so that the transmission can be played back later.

The majority of image display systems are based on personal computers that use specialized demodulator cards and software programs to ingest the audio signal from the satellite, convert it to a digital signal, display the image on the screen, and store the image to disk. Some direct reedout data users deploy the sound card provided with most PC terminals to demodulate the subcarriers. Other software features on the PC may be used to schedule data ingests, digitally enhance images and add geographic gridding or analyze the temperature data variations in the infrared imagery.

Considering the frequencies, satellite signal strength and polarization factors of the transmissions via PEO satellites a number of receiving antenna designs can accomplish adequate APT reception when used in conjunction with a good preamplifier and a properly designed radio receiver. These designs include both omnidirectional antennas and higher-gain beam antennas. At this point, with the advent of modern electronics antenna construction remains as one of the few areas that radio amateurs can build adequate hardware to yield significant savings and performance equivalent to commercially available antennas. In fact, there are numerous amateur webs sites with detailed instructions to construct antennas that are suitable for APT reception.

7.6.2.1   Omnidirectional Direct Readout PEO Satellite Antennas

In satellite communication systems is using special omnidirectional antenna as a class that radiates radio waves power uniformly in all directions in one plane, with the radiated power decreasing with elevation angle above or below the plane, dropping to zero on the antenna's axis. These antennas his oriented vertically are widely used for nondirectional antennas on the surface of the Earth because they radiate equally in all horizontal directions, while the power radiated drops off with elevation angle so little radio energy is aimed into the sky or down toward the earth and wasted. Omnidirectional antennas are widely used for radio and satellite broadcasting and receiving of signals in space and terrestrial systems including in satellite reception of meteorological data. Thus, here will be shortly introduced turnstile and quadrifilar helix antennas.

1. **Turnstile Reflector** – The turnstile or crossed-dipole antenna is one of the simplest and least expensive antennas to use for APT weather satellite imagery, which is illustrated in Fig. 7.19 (Left). It is omnidirectional, theoretically receiving the satellite signal from all directions, easy to mount and does not require the more complex tracking guidance needed by high-gain directional antennas. It is consisting of a set of two identical dipole antenna mounted at right angles to each other and fed in phase quadrature with two currents applied to the dipoles are 90° out of phase. The name reflects the notion the antenna looks like a turnstile when mounted horizontally.

   This antenna can be used in two possible modes such as: (1) In normal mode the antenna radiates horizontally polarized radio waves perpendicular to its axis; and (2) In axial mode the antenna radiates circularly polarized radiation along its axis. The disadvantage of these antennas is that they offer little and if any signal gain and this can result in a reduced area of coverage compared to the higher signal gain offered by directional antennas. Due to the design and placement of the antenna elements in a turnstile, the satellite signal is often received better in one direction than others. This can create some loss of signal strength and a "fading" of the received APT transmission. Fading often results in noise bursts or "sparklies" in the APT video image.

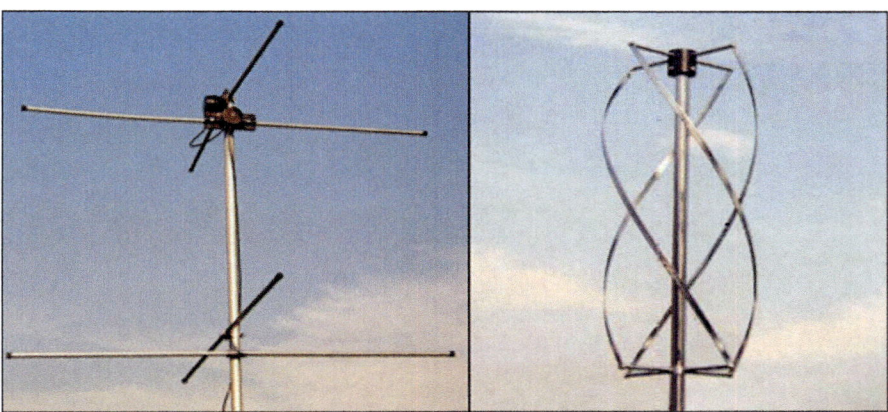

Fig. 7.19 Omnidirectional turnstile and QHA GES antennas (Courtesy of Manuals: by NOAA)

2. **Quadrifilar Helix Antenna (QHA)** – This is a special type of omnidirectional antenna that provides a much better "radiation pattern" compared to turnstile reflectors, and does not suffer from the loss of signal strength exhibited in simple turnstile antennas, which is shown in Fig. 7.19 (Right). The quadrifilar helix radiator usually consists of four 1/2-turn helices equally spaced around the circumference of a common cylinder, while the radiation pattern is omnidirectional in the plane perpendicular to its main axis. Radiation of the signal is nearly circularly polarized over the entire hemisphere irradiated, which is making it almost ideal for receiving signals from PEO weather satellites. Well-designed quadrifilar helix antennas often exhibit inherent gain from 3 dB to 5 dB. In such a way, a low physical profile combined with high performance makes the quadrifilar antenna an excellent antenna for APT reception in the home. Its small diameter also allows it to be enclosed in a waterproof housing or made from corrosion resistant materials.

3. **Double Cross Antenna (DCA)** – This omnidirectional antenna is composed by two pairs of crossed dipoles shown in Fig. 7.20 (Left). It is designed to improve the performance of the radiation pattern with respect to the turnstile antenna and its purpose of the four dipoles is to produce a radiation pattern with great RHCP reception at the zenith axis. Thus, in order to achieve this pattern, the two dipoles have to be crossed, spaced a quarter wavelength and fed in phase. These four dipoles can be mounted as shown in to produce a radiation pattern with excellent RHCP at 0° elevation in the free space radiation pattern. The DCA solution design concept for this four dipole array with hemispheric coverage with RHCP is derived from the fundamental concept that two dipoles, crossed, spaced a quarter wave. Same as previous two types of antenna, DCA is also an ideal antenna for APT reception in the home.

4. **Modified V-2 FM Antenna** – The simple omnidirectional V-2 antenna depicted in Fig. 7.20 (Right) can be constructed by modifying a commercial FM receiving antenna with a pair of crossed, folded dipole elements. These antennas can be

Fig. 7.20 Omnidirectional double cross and V-2 modified FM antenna (Courtesy of Manuals: by NOAA)

found in many stores and on web sites selling television and radio antennas. Such antennas are low cost, and are made of thin wall aluminum tubing that can be cut to modified dimensions quite easily. Thus, similar antennas designed specifically for APT can also be purchased commercially. The antenna can be modified as follows: (1) The length of the two folded dipoles are reduced by trimming the longer FM element tubing to 103 cm (40.3 in.) to provide an approximate 1/4 wavelength match for the 137.5 MHz center frequency of the APT transmission; and (2) Two reflectors at right angles to each other were made from 6.4 mm (1/4 in.) diameter aluminum tubing, cut to 113 cm (44.1 in.) length. These reflectors were mounted 43.6 cm (17 in.) below and parallel to the folded dipoles. In fact, these reflectors create a broad beam antenna that, when pointed vertically, allows a wide angle of antenna reception with no need for pointing toward the satellite as it passes over the ground station.

7.6.2.2 Directional Direct Readout PEO Satellite Antennas

Directional PEO satellite weather antennas offer much higher gain and signal-to-noise ratios compared to simple omnidirectional antennas. In fact, such directional antennas have very sharp radiation patterns, and therefore require accurate tracking of the PEO satellites to take advantage of the gain provided by the antenna system. This tracking requirement presents an additional complexity to the receiving station both in construction and operation since the satellites must be actively tracked from

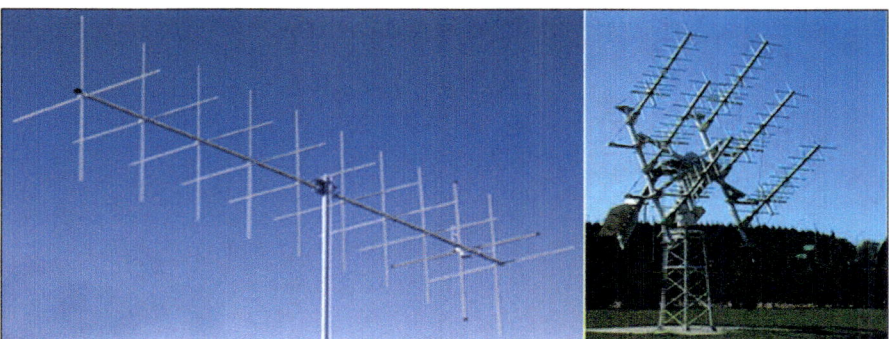

Fig. 7.21 Commercial crossed Yagi and high gain axial mode Yagi turnstile antenna – (Courtesy of Manuals: by NOAA)

horizon to horizon. Thus, directional antennas can be positioned using TV type rotors mounted in an azimuth and elevation configuration or using commercially available or home built antenna positioner controlled by a PC.

1. **Crossed Yagi Antenna** – The crossed yagi directional weather antenna has been a popular choice for many amateur constructed APT stations. In Fig. 7.21 (Left) is illustrated a commercially available crossed Yagi antenna without satellite tracking positioners, while it has to be integrated additionally. This antenna design functions well for APT reception, is relatively inexpensive and can be purchased commercially or constructed without much difficulty. The crossed yagi antenna consists of a number of elements similar to a multi-element commercial TV antenna. The major difference is that the elements are arranged at right angles to each other. This crossed element design eliminates fading of the circular polarized RF signal transmitted by the PEO satellites.

2. **Yagi Turnstile Array Antenna** – The high gain axial yagi turnstile array shown in Fig. 7.21 (Right) is used to communicate with weather satellites on 136–137 MHz at Bedu GES in Belgium. Each of the six components of the array consists of two 9-element yagi antennas mounted on the same axis at right angles and fed in quadrature to radiate a narrow beam of circularly polarized radio waves. This yagi array is directional antenna that is consisting of six or multiple parallel elements in a line, usually half-wave dipoles made of metal rods.

Fig. 7.22 Next generation GOES-R GES with antenna infrastructures (Courtesy of Manual: by NOAA)

7.6.3 Direct Readout GEO Directional Ground Antenna Systems (GAS)

The GEO GES Terminal is situated in any given position of certain GEO satellite footprint between latitudes of 75° North and 75° South together with satellite parabolic reflector dish antenna and tracking systems. Depending on construction of the GES site each satellite antenna infrastructure can be installed next to the GES building or on its roof. For instance, in Fig. 7.22 is illustrated the ground segment of newest fourth generation GOES-R satellites.

The GOES-R ground segment is an enterprise-wide backbone capable of ingesting, receiving, processing and distributing 40 times more data to the US NWS and more than 10,000 other direct users.

In January 2014, Harris manufacturer began making ground segment deliveries that included 2100 servers, 149 racks of network equipment, 317 workstations and huge storage services totaling three petabytes. Through testing and verification, Harris demonstrated the ability of the GOES-R ground segment with GEO antenna system to generate weather products at all operational sites. Harris successfully completed the final contractor testing and turned over operations of this highly reliable and secure infrastructure to NOAA on 30 April 2015 for launch preparations, and this company will continue to support the GOES-R satellite system with enhancement activities through 2019.

Additionally, the Harris team developed and deployed the 16.4 m tri-band satellite dish antennas needed to support the GOES-R satellite meteorological mission. The team also performed electronic upgrades to the 9.1 m antennas at Satellite Operations Facility of NOAA in Suitland, MD, to enable the facility to receive the GOES-R Rebroadcast Signal. The ground antenna infrastructures were installed and they are functioning as an operational, supporting current GOES satellites in operation as well. In this context will be introduced two types of GEO infrastructures with their main components and antenna systems.

7.6.3.1   Tracking GEO Satellite GES Receiving L/C/Ku-Band Antenna

The early 2016 launch of the first GOES-R satellite begins a new era in satellite weather data and information. This system will provide valuable earth imagery at four times the pixel resolution, three times the spectral data, and five times faster coverage than the current satellites. As a result, the new-generation GOES-R Rebroadcast (GRB) system will provide substantially greater nowcasting capability than the legacy GVAR service. Thus, Harris Corporation's WxConnect GRB system delivers a direct connection from the GOES-R series satellites to the GES terminal, enabling access to all collected data, at full resolution and in the shortest time feasible over all other methods of obtaining next generation of GOES meteorological data. This exclusive processing architecture and software was designed and implemented for low latency processing to support mission critical applications with fast access to newly received weather data.

Modular and scalable GES terminal of WxConnect GRB system includes options for a 3.7, 4.5 or 6.5 m satellite L-band reflector parabolic antenna and a high-power processing system providing high availability and secure reception of enhanced environmental data from the GOES-R satellite system of Harris, which GES antenna is illustrated in Fig. 7.23 (Left). The deployment services include ground antenna and processing equipment installation, both connected to the GES terminal. Optional product generation and algorithm development support are available to complete your GOES-R weather data solution. In addition, this GES terminal and ground antenna can provide high-performance DVB-S2 demodulation for robust operation in high adjacent signal environments.

The similar MEOS GES with parabolic antenna of Norway producer Kongsberg Spaced AS is a multimission, flexible and modular turnkey system for acquisition, archiving, processing, analysis and distribution of meteorological data using advanced package software. Thus, the system provides the functionality to receive the radio frequency and deliver data to the ingest system, so the front-end system of GEO GES includes: Antenna, Feed/Downconverter and Digital receiver/bit-synchronizer.

Therefore, all three GEO meteorological satellites, the Japanese MTSAT, Chinese FY-2 and European MSG are using L-band, while the MEOS MSG Ground Station supports reception relayed via Ku-band (Eutelsat 9A/10A) and C-band (AtlanticBird) satellites for DVB-S2 transmissions. The ground station also supports L-band reception directly from the MSG satellite, an option, which may become available in the future.

Depending on the customer's requirements in Table 7.4 is shown type of meteorological satellites, their antenna sizes and working frequency bands:

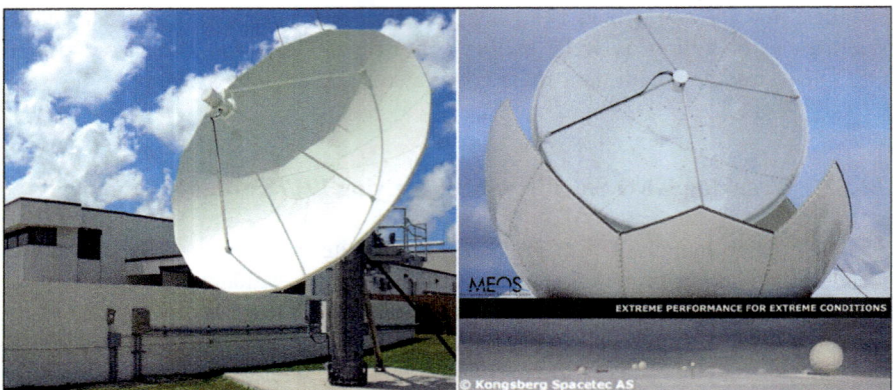

Fig. 7.23 Main GEO GES terminals and antennas (Courtesy of Manual: by Harris/Kongsberg)

Table 7.4 Type of antennas and frequency bands

Frequencies	L-band	Ku-Band (Europe)	C-band (Africa/America)
Antenna sizes in m	3.8	1 to >	2.4 to >
MTSAT	+		
FY-2	+		
MSG	*	+	+

7.6.3.2   Tracking GEO Satellite GES Receiving L/S/X-Band Antenna

The MEOS GEO GES antenna system of producer Kongsberg Spaced AS comes with dish sizes from 3.0, 3.8, 4.3 to 5.0 m is depicted in Fig. 7.23 (Right). This antenna is situated in plastic radome to be protected from influence of very harsh environment. This GES gives sufficient margin for data reception from direct readout and remote sensing satellites. Thus, designed for optimal maintainability and reliability, the MEOS satellite antenna utilizes the most modern industrial components available today. In Table 7.5 are shown different sizes of optional GES antenna, frequency bands and other characteristics:

Key features of this antenna are as follows: Pedestal connected to indoor unit by optical fiber; TCP-IP on fiber cable between pedestal and indoor unit Cable length is about 3 km; Single or dual band configuration available; Designed for L, S, and X-band missions; Dish size is 3.0–5.0; Employed X/Y geometry for lower stress to drive chain with pedestal for elimination of overhead keyhole; Extensive monitoring and control capabilities; In-field diagnostics and alignment tools; Remote and local monitoring and control available; Real time and historic status available; Java based graphical user interface; MEOS Connect ready for integration in a ground station network; Stable antenna control unit running Linux with 24/7 operations without operator intervention; Resumes operation automatically after a power

Table 7.5 Size of antennas and frequency rands

Standard reflector sizes				
	3.0 m	3.3 m	4.3 m	5.0 m
L band G/T $(1700)^2$	10 dB/K	12 sB/K	13 dB/K	14 dB/K
S band G/T $(2400)^2$	13 dB/K	15 dB/K	16 dB/K	17 dB/K
X band G/T $(8200)^2$	25 dB/K	27 dB/K	28 dB/K	29 dB/K
Pointing error	$0.09°$ rms[3]			
Pointing resolution	$0.005°$ on both axis			
Velocity	6 deg./s			
Wind speed operational	40 m/s	27 m/s	Radome recommended	
Wind speed survival	56 m/s	56 m/s	Radome recommended	
Travel				
X-axis: Mechanical ±90 deg., tracking ±87 deg				
Y-axis: Mechanical ±90 deg., tracking ±87 deg				

break; Self test and remote diagnostics; Robust computers and servo units; and Low failure probability.

7.6.3.3   Tracking GEO Satellite GES Receiving LRIT/HRIT L-Band Antenna

The LRIT/HRIT L-band tracking antenna system is reliable and high performance solution for receiving, archiving, processing and displaying LRIT and HRIT data from GEO weather Eumetcast, HimawariCast, MSG, GOES, COMS-1 and Electro satellites. The GEO parabolic 1.2 m/1.8 m/2.4 m L-Band dish antenna with LNB is illustrated in Fig. 7.24 (Left), which dish of 1.2 and 1.8 m is built of solid powder-coated aluminium and 2.4 m dish is built of glass-fiber reinforced precision compression moulded polyester reflector with three segments. The antenna dish of 3.0 and 3.7 m build of eight segments also supports GVAR system, which is illustrated in Fig. 7.24 (Right). Both antennas are supplied with galvanized steel azimuth/elevation mount and pedestal, and up to 100 m of RG213 50Ω co-axial cable.

Satellite meteorological LRIT and HRIT data is an important source of information received for nowcasting, numerical weather prediction, climate monitoring and research. The latest imagery and products are available as frequently as every 15 minutes that is allowing major improvements in the forecasting of severe weather. Antenna with receiver options is available to suit different services, geographical locations and requirements. All types of transmission are supported L-band direct broadcast services. Ingested data can be viewed and processed using the Dartcom iDAP/MacroPro software. Outputs are also available for popular image processing software packages such as PCI Geomatica, ERDAS IMAGINE and ENVI/IDL, as well as standard interchange formats such as GeoTIFF.

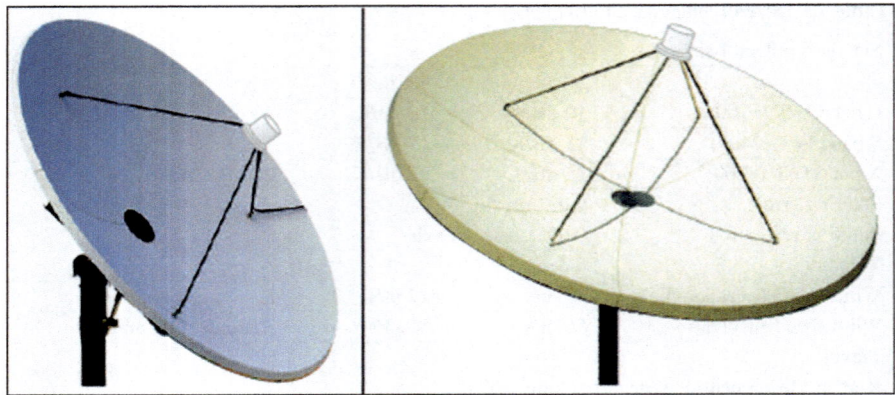

Fig. 7.24 GEO GES LRIT/HRIT L-band 1.2 m/1.8 m/2.4 m and 3.0 m/3.7 m antennas (Courtesy of Manuals: by Dartcom)

Main features of both L-band antennas are as follows: Reflector is prime focus parabolic; F/D ratio specifications are staring from 0.37 up to 0.42; Gain characteristics are between 24.0 and 34.1 dBi; Polarization is linear; and G/T at 5° elevation is between 2.2 and 13.5d B/K.

7.6.3.4 Tracking GEO Satellite GES Receiving LRIT/HRIT C-Band Antenna

The LRIT/HRIT C-band meteorological antenna is required for Eumetcast LRIT/HRIT and HimawariCast data reception. Similar to L-band antenna in Fig. 7.24 (Left) is depicted 2.4 m C-band dish antenna with LNB, while in Fig. 7.24 (Right) is depicted 3.0 m/3.7 m parabolic dish with LNB. Three-segment (2.4 m) or eight-segment (3.0 m, 3.7 m) glass-fiber reinforced precision compression moulded polyester reflector. Supplied with galvanized steel azimuth/elevation mount and pedestal, and 50 m of CT100 75Ω co-axial cable, while phase locked loop LNB in weatherproof powder-coated housing.

Main features of both L-band antennas are as follows: Reflector is prime focus parabolic; F/D ratio specifications are staring from 0.30 up to 0.37; Gain characteristics are between 37.5 to 40.9 dBi; Polarization is circular; and G/T at 5° elevation is between 17.7 and 20.7 dB/K.

The C-band is supporting reception of data for DVB receiver and software such as: TechniSat SkyStar 2 internal PCI DVB receiver card for Eumetcast; TBS 6983 internal PCIe DVB-S2 receiver card for HimawariCast; and all Supplied with drivers and data acquisition software (Tellicast for EUMETCast, KenCast FAZZT for HimawariCast).

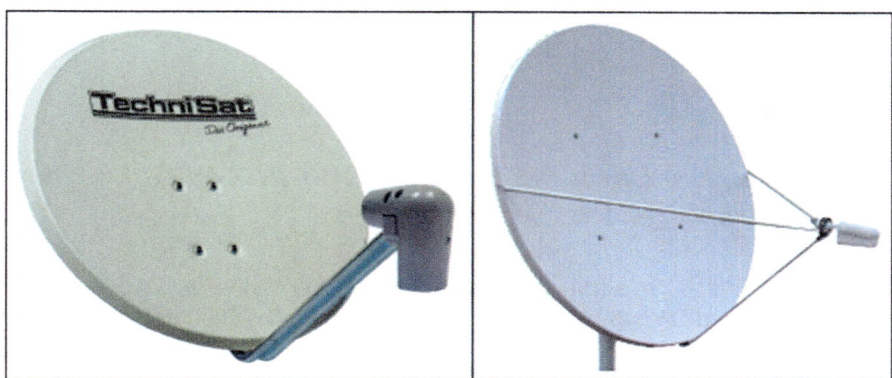

Fig. 7.25 GEO GES LRIT/HRIT Ku-band 0.85 m/1.2 m and 1.8 m DVB antennas (Courtesy of Manuals: by Dartcom)

7.6.3.5   Tracking GEO Satellite GES Receiving LRIT/HRIT Ku-Band Antenna

The LRIT/HRIT Ku-band meteorological antenna is required just for Eumetcast LRIT/HRIT both DVB and DVB-S2 data reception. In Fig. 7.25 (Left) is illustrated Ku-band 85 cm Ku-Band LRIT/HRIT antenna for Eumetcast DVB service and in Fig. 7.25 (Right) is depicted 1.8 m Ku-Band LRIT/HRIT antenna also dedicated for EUMETCast DVB service.

Thus, the service availability for DVB/DVB-S2 coverage is using 85 cm, 1.2 m or 1.8 m offset dish with LNB, which are built of high-quality powder-coated aluminium (85 cm, 1.2 m) or glass-fiber reinforced compression moulded polyester (1.8 m) reflector. In addition, 2.4–4.5 m antennas are also available for fringe area reception. Those ground satellite meteorological receiving antennas are supplied with azimuth/elevation mounting bracket (85 cm, 1.2 m), 50 m of CT100 75Ω co-axial cable and F-type connectors, while 1.8 m antenna supplied with 4″ tubular pedestal (optional non-penetrating roof mount is also available). Those antennas have integrated state-of-the-art weatherproof LNB with a noise figure of 0.3 dB.

Those ground satellite antennas are supporting DVB receiver with desk or 19″ rack mount and software Ayecka SR1 or TBS 6983 (PCIe card) DVB-S2 receiver, both recommended by Eumetsat, and they are supplied with drivers and T-Systems Business TV-IP software. Thus, the Ayecka SR1 DVB-S2 receiver specifications are as follows: RF input frequency is 950 to 2150 MHz; RF input connector is about 75Ω F-type; Symbol rates are 100 ks/s to 45 Ms./s; and channel rate up to 120 Mb/s.

Main features of both Ku-band antennas are as follows: Reflector type is 21° offset for 85 cm antenna, 21.3° offset for 1.2 m antenna and offset for 1.8 m antenna; Reflective material is Solid aluminium and powder coated for building 85 cm and 1.2 m antenna, while for 1.8 m antenna is used glass-fiber reinforced polyester; F/D ratio specifications are staring from 0.66 up to 0.67; Gain characteristics are between 382.0 and 45.5 dBi; Polarization is linear; and G/T at 5° elevation is between 17.2 and 26.0 dB/K.

7.6.4 Meteocast Direct Readout GEO DVB-RCS GES Antenna Systems

The Digital Video Broadcasting-Return Channel via Satellite (DVB-RCS) or DVB-S2/S2X and DVB-S2 broadcasting technique are the ETSI standards, which define the complete air interface specification for two-way satellite broadband Very Small Aperture Terminal (VSAT). Low cost and very effective VSAT infrastructure can provide highly dynamic and demand-assigned transmissions, such as fixed and mobile communications, meteorological data transfer etc. The existing commercial DVB-RCS/RCS2, DVB-S/S2/S2X broadcasting service provides Voice, Data and Video (VDV) transmissions via commercial GEO satellites for fixed, mobile, portable and other applications with adequate antenna systems, including satellite SCADA (M2M) system, which network configurations is depicted in Fig. 7.26.

For instance, deploying DVB-RCS transmission system can be implement direct readout and broadcasting facilities of for weather data from Automatic Weather Station (AWS) via fixed, mobile and portable VSAT antennas. In addition, satellite SCADA terminals with VSAT or other type of mobile satellites equipment with antennas (Inmarsat, Iridium, Globalstar or Orbcomm mobile satellite systems) can be also used for automatic transmissions of AWS meteorological data via GEO meteorological satellites.

As shows Fig. 7.26, modern satellite meteorological observation and transmission system can implement fixed DVB-RCS/DVB-S/S2/S2X antennas with Receivers (Rx) for receiving direct readout data and even for retransmitting this and AWS data by Transmitters (Tx). On the other hand, Mobile Interactive Terminals (MIT), such as ships, land vehicles and aircraft, with VSAT satellite antennas can provide the same services. Besides, special portable VSAT devices can do all mentioned services. At the end it will be important to conclude that all data for weather broadcasting can be sent via DVB-RCS GEO satellites to fixed and mobile users via Internet (VDVoIP, IPTV, Fax) and Terrestrial Telecommunication Network (TTN).

The Eumetcast broadcast system of Environmental Data is a multi-service dissemination system based on standard DVB TV technology GES with special antenna systems. It uses commercial communication GEO satellites to multicast files (data and products) to a wide user community, so Eumetcast is Eumetsat's contribution to Geonetcast. This concept is fully supporting the family of DVB based satellite receiving GES with ground antenna systems for the meteorological user community using DVB commercial telecommunication satellites like Eutelsat, SES-6 and AsiaSat-4. Moreover, the Meteorological Satellite Centre (MSC) of

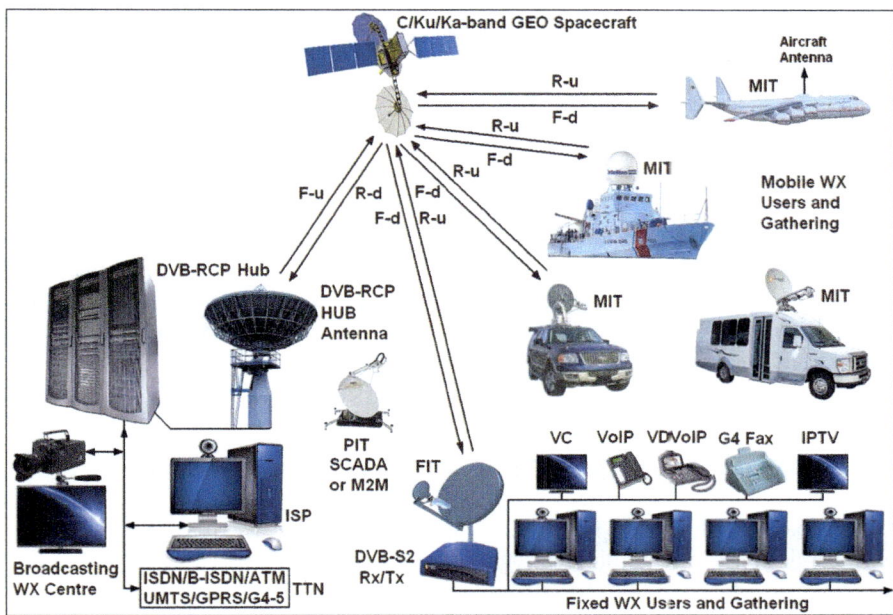

Fig. 7.26 DVB-RCS Weather (WX) Satellite Network. *Acronyms*: *F-d* forward downlink, *F-u* forward uplink, *R-d* return downlink, *R-u* return uplink, *ISP* Internet Service Provider, *FIT* Fixed Interactive Terminal, *PIT* Portable Interactive Terminal, *MIT* Mobile Interactive Terminal, *VC* videoconference, *VoIP* voice over IP, *VDVoIP* voice, data and video over IP (Courtesy of Manual: by Ilcev)

Japanese agency JMA is using own commercial communication satellite JCSAT-2A/2B for dissemination meteorological data and images at C-band, which weather service is known as HimawariCast. Thus, this agency uses own multifunctional satellite MTSAT-2 for the same purpose, because this satellite has onboard communication, meteorological and GNSS/CNS transponders and is using the L, Ku and Ka-bands.

Within the current Eumetcast system configuration, the server side is implemented at the Eumetcast uplink site (Usingen, Germany), and the client side installed on the individual Eumetcast reception stations. The telecommunication providers supply the DVB multicast distribution. Encoded LRIT/HRIT data and product files are transferred via a dedicated communication line from Eumetsat to the uplink facility where they are transmitted to a GEO communication satellite for broadcast to user receiving stations. Each receiving station decodes the signal and recreates the user data/products according to a defined directory and file name structure. In its current configuration, EUMETCast operates two turn-around systems. The turnaround service provider receives the DVB satellite signal from one satellite and retransmits it, without unpacking the DVB C-band turn-around service for EUMETCast Africa from its uplink site in Fucino (Italy) and Globecast provides the C-band turn-around service for Eumetcast Americas from its uplink facility in

Paris (France). Further information about Eumetcast and its services is provided on the Eumetsat Web page or in the Technical Document TD-15. However, as the missions are encrypted, every user needs to have a license agreement with Eumetsat or its national weather service.

The same scenario can be used for new proposed Global Meteorological Broadcasting (GMB) such as retransmission of meteorological LRIT/HRIT data, products and images via commercial GEO satellites transponder and SCP stations to customers received by DRS and transmissions of meteorological data from ground DCP via same GEO satellites to the DRS infrastructures. This proposal named as Meteocast system is developed by the Space Science Centre (SSC) for Research and Postgraduate Studies in Space Science at Durban University of Technology (DUT) and contains two scenarios shown in Fig. 7.26.

7.6.4.1   Meteocast Forward DVB-RCS Uplink Antenna Systems

The Meteocast Forward DVB-RCS Uplink Antenna Systems has to provide retransmission of meteorological data and images from Broadcasting WX Centre of Direct Readout Service (DRS) and send them to the DVB-RCS Hub at C, Ku or Ka-band via Internet and TTN sites. Then, the Hub server encapsulates and sends DRS data to the GEO communication satellite Rx by the forward uplink Hub reflector dish antenna. The communication GEO satellite broadcast data received from the uplink antenna to the receiving VSAT antenna as portable, fixed and mobile users (PIT, FIT and MIT) by the forward downlink. The VSAT receiving antenna and Low-noise Block Downconverter (LNB) unit receive the data from the GEO communication satellite and send the information to the DVB-S2/S2X Rx. The Meteocast professional client uncapsulates the data for storage in any directory on the PC memory, which complete solution of Forward DVB-RCS Uplink is shown in Fig. 7.26.

The VSAT Rx then converts the data into IP packets and sends them to the computers (PC) integrated in Local Area Network (LAN), which sample of GOES configuration is depicted in Fig. 7.27. In such a way, all wanted meteorological information by Meteocast users can be send in any real time and space via Hub with different ground antenna infrastructures to GEO communication satellites. In this sections will be introduced Advantech Hub (GES) and four ground antennas:

1. **Hub Stations** – The Hub station of Advantech producer is shown in Fig. 7.28 (Middle). The features of Advantech next generation DVB-RCS/RCS2 VSAT Hub Discovery 300 ground station are as follows: Multi-transponder and multi-satellite support; Up to 1 Gb/s of 5×200 Mb/s Forward Links capacity with 5:1 redundancy; Up to 3×240 Mb/s Return Links per rack capacity; Geographical redundancy and satellite roaming; Support for up to 45,000 VSAT terminals; Optimized for IP and multimedia content including MPEG4 video; Open standard design of DVB-RCS/DVB2 and DVB-S2/S2X for return links; DVB-S, DVB-S2 and new DVB-S2X for forward links; The VSAT Hub is frequency

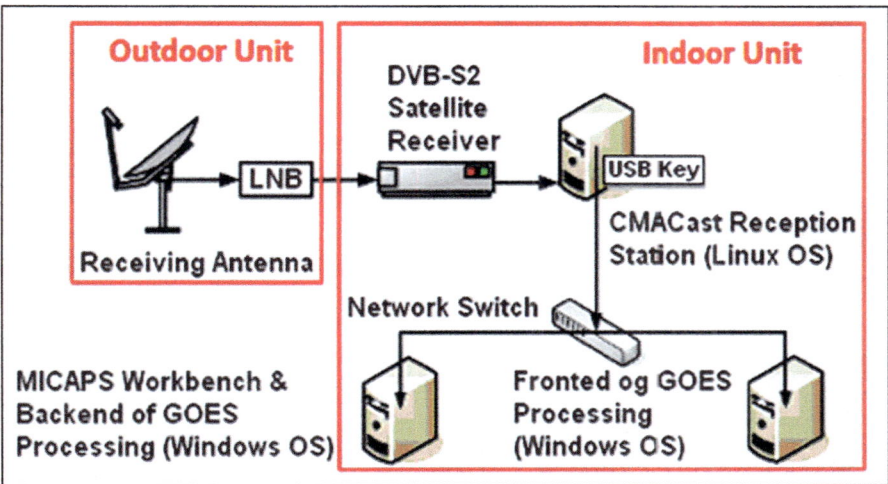

Fig. 7.27 GOES DVB-S2 Rx and antenna (Courtesy of Manual: by GOES)

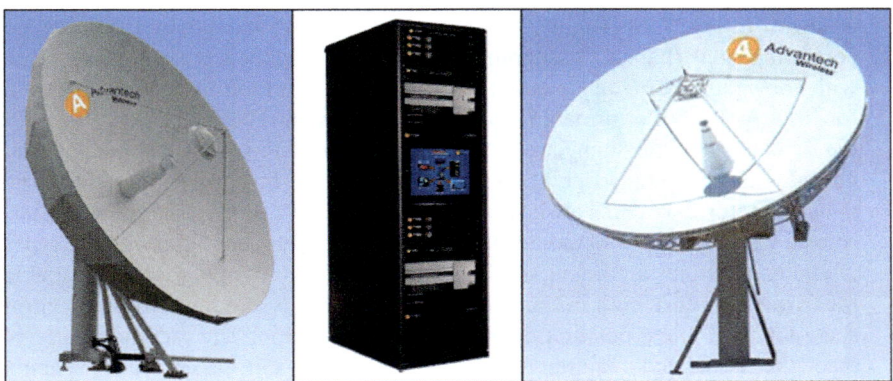

Fig. 7.28 DVB-RCS hub with GES multiband antennas (Courtesy of Manuals: by Advantech)

independent that its onboard processors can be operated in any frequency band (e.g., C, X, Ku and Ka-band or hybrids of these); The switching mechanism, on the return link, between the DVB-RCS TDMA system and the DVB-S/S2/S2X TCC SCPC modes is customer controlled and can be commanded by the Hub Operator; and so on.

2. **Ground C/X-band 12 m Hub Antenna** – The A-Line Series Advantech transceiving Hub 3912T antenna model shown in Fig. 7.28 (Left) is using to transmit all DVB-RCS forward uplinks information to the ground fixed and mobile users via GEO satellite Tx antenna and to receive all DVB-S2 return downlinks information from ground users VSAT via GEO satellite Rx antenna. This Hub A-Line Series model 3912T, 12 m antenna system, designed and

manufactured with CAD, can be applied via Intelsat and other satellite communication GEO systems. It consists of dual shaped Gregorian reflectors, a frequency reuse feed network with corrugated horn and an elevation-over-azimuth limit motion kingpost pedestal. The backup structure for the reflector, the Hub station connecting the main reflector with mount and the pedestal provides the guaranteed pointing accuracy required in normal operation. The main reflector diameter consists of 80 precision stretch formed aluminum panels riveted with the rings and radials in three rings. The main characteristic of this Hub antenna are high gain, low sidelobes, low cross polarization, capable for frequency reuse both in transmit and receive bands, high driving/control accuracy with angle position display in high resolution. This Hub antenna at C-band is using receiving frequencies of 3.625–4.200 GHz and optional of 3.400–4.200 GHz, while it also uses transmitting frequencies of 5.850–6.425 GHz and optional of 5.850–5.650 GHz. At X-band this antenna is using receiving frequencies of 7.25–7.75 GHz and transmitting frequencies of 7.9–8.4 GHz

3. **Ground C/X-band 16 m–20 m Hub Antenna** – The A-Line Series models 3920TCX and 3920TC of 20 metres and 3916TCX and 3916TC of 16 metres antenna systems are designed and manufactured with CAD and can be applied to the newly updated Intelsat standard-A and other systems of Hub stations, which is shown in Fig. 7.28 (Right). These types of Hub ground antenna systems are consisting dual shaped Cassegrain reflectors, a frequency reuse feed network with corrugated horn, an elevation-over-azimuth limit motion kingpost pedestal for limit motion or a turntable mount for full motion. The backup structure for the reflector, the hub connecting the main reflector with mount and the pedestal provides the guaranteed pointing accuracy required in normal operation. The main reflector diameter consists of precision stretch formed aluminium panels riveted with the rings and radials in three rings. The characteristics of these types of antenna systems are high gain, low sidelobes, low cross polarization, capable for frequency reuse both in transmit and receive bands and high driving/control accuracy with angle position display in high resolution. The radiation patterns meet the associated requirements of Intelsat, FCC and CCIR for 2° spacing location of GEO satellites. This Hub antenna at C and X-band is using the same receiving and transmitting as previous antenna.

4. **Ground C/Ku/DBS-band 9 m Hub Antenna** – This high performance Advantech A-Line Series 9 m antenna can operate at C, Ku or DBS-band for worldwide applications. The 9 m Hub antenna system designed and manufactured with CAD can be also applied to the newly updated Intelsat standard B or E and other type of GES or Hub terminals. This antenna system consists of dual shaped Cassegrain reflectors, a frequency reused feed network with corrugated horn, an elevation-over-azimuth limit motion kingpost pedestal. The backup structure for the reflector, the hub connecting the main reflector with mount and the pedestal provides the guaranteed pointing accuracy required in C band and Ku band operations. The main reflector consists of 48 precision stretch formed aluminum panels riveted with the rings and radials in three rings. Antenna system is characteristic of high gain, low sidelobes, low cross

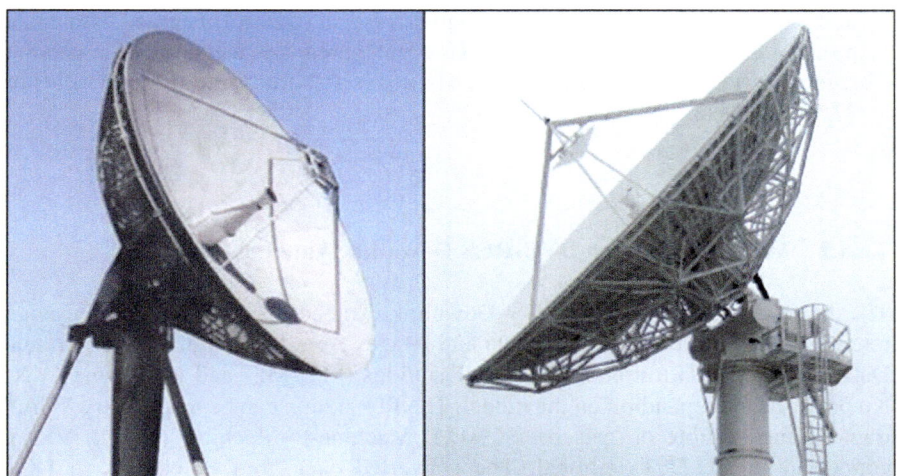

Fig. 7.29   GES C/Ku DBS and Ku-band antennas (Courtesy of Manual: by Advantech)

polarization, capable for frequency reuse both in transmit and receive bands, high driving/control accuracy with angle position display in high resolution. The radiation patterns meet the associated requirements of Intelsat, FCC and CCIR for 2° spacing location of geostationary satellites.

These Advantech A-Line Series 9 m antennas depicted in Fig. 7.29 (Left) can operate at the same C-band frequencies. This Hub antenna at Ku-band is using receiving frequencies of 10.70–12.75 GHz and it uses transmitting frequencies of 13.75–14.50 GHz. However for operations at DBS-band this antenna is using receiving frequencies of 10.70–12.75 GHz and transmitting frequencies of 17.3–17.8 GHz.

5. **Ground Ka-Band 13.5 m hub Antenna** – The Advantech A-Line Series 3913TKATHA model of 13.5 m antenna system designed and manufactured with CAD meets the most stringent requirements of high throughput and power Ka-band GEO satellite up and down links, which is depicted in Fig. 7.29 (Right). The antenna design meets the CCIR 580-5 mandatory requirements and consists of dual shaped Cassegrain reflectors, RF subsystem, with advanced satellite tracking system, antenna structure subsystem, safety devices and important motorization kits for providing control of AZ, E, and Pol angles. It also provides axis and foundation anchoring kits. For the backup structure for the reflector, the hub is connecting the main reflector with mount and the pedestal provides the guaranteed pointing accuracy required in normal operation. The main reflector antenna diameter includes back panels for enclosure with incorporated deicing system and temperature-controlled heaters. The main purpose of this unit is to ensure temperature conformity on reflector assembly and prevent deformations cause by temperature changes. This reflector antenna system is characteristic of high gain, low sidelobes, low cross polarization, capable for frequency reuse

both in transmit and receive bands, high driving and control accuracy with angle position display in high resolution. This Hub antenna is using only Ka-band at receiving frequencies of 18.1–19.7 GHz and it uses transmitting frequencies of 27.5–29.6 GHz.

7.6.4.2   Meteocast Return DVB-RCS Downlink Antenna Systems

The Meteocast Return DVB-RCS Downlink Antenna Systems has to provide retransmission of meteorological data and images from fixed, mobile and portable Data Collection Platforms (DCP) VSAT stations (FIT, MIT and PIT) using C, X, Ku or Ka-band, depending on the model. The PIT station can be an ordinary VSAT transceiving station or can be SCADA Machine-to-Machine (M2M) VSAT connected to the DCP terminal. The converted data from IP packets in DCP terminal VSAT Tx sends them to the GEO communication satellite Rx by the return uplink VSAT dish antenna. Then, the GEO communication satellite Tx sends received DCP data to the VSAT Hub antenna by the return downlink. Finally, the VSAT Hub (GES) is distributing this data directly to the BWC or to ground users via Internet and TTN, which Return DVB-RCS Uplink solution shown in Fig. 7.26.

The samples of VSAT transceiving stations that can be used for both GEO Meteocast forward and return uplink configurations are shown in Fig. 7.30: Advantech VSAT Ku/Ka 8200 Series M2 M/IoT/SCADA Ultra-Compact All-Outdoor transceiving antenna (Left); Satnet S4120 ACM DVB-RCS VSAT fixed and mobile transceiver terminal (Middle); and Ku /Ka 8000 Series Outdoor VSAT transceiving antenna (Right).

1. **VSAT Ku/Ka 8200 Series Antenna** – Advantech Ku/Ka 8200 Series of Ultra-Compact All-Outdoor VSAT transceiver-routers antenna was specifically designed for M2 M, SCADA, Telemetry and Internet of Things (IoT) over satellite, which is shown in Fig. 7.30 (Left). It can be combined with all other Advantech VSAT terminals within the same multiservice ASAT system according to the application requirements. The outbound channel rate is up to 80 Mb/s (typical 45 Mb/s) of this antenna is and inbound channel rate is up to 1 Mb/s (burstable) and 160 Kb/s (typical), and this VSAT antenna can be used for fixed and mobile VSAT transceivers. It is supporting transmitting frequencies at Ku-band of 13.75–14.5 GHz and at Ka-band of 29.5–30 GHz, while is supporting receiving frequencies at Ku-band of 10.75–12.75 GHz (dual band and dual LNB) and at Ka-band of 18.2–20.3 GHz.

2. **Satnet S4120 ACM DVB-RCS VSAT Transceiver Terminal** – This Advantech Satellite Networks VSAT terminals are DVB-RCS/RCS2 compliant, which is shown in Fig. 7.30 (Middle). The S4120 design is derived from the S4100 with the addition of DVB-S2 with ACM/VCM capability. It is optimized to achieve high-performance and quick response time for professional, enterprise and governmental applications in an economical fashion. The terminal has

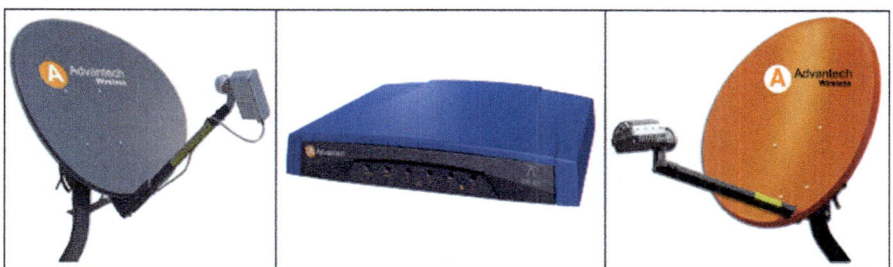

Fig. 7.30 VSAT transceivers with antenna (Courtesy of Manual: by Advantech)

been designed with all key IP features to fulfill all the needs of an enterprise for fixed and mobile applications. The attractive design and small form-factor make it ideal as a cost-effective, desktop high-speed solution. This VSAT terminal offers powerful connectivity directly to the LAN/WAN environment or directly to a host computer. A truly professional solution, it is an out-of-the-box, ready-to-go, cost-effective broadband solution for SOHO and enterprise use. As a replacement for a cable or DSL users connections, as a backup solution to provide continuity of operations or to provide broadband to remote regions, the S4120 performance provides bandwidth-on-demand. Designed to support unicast or broadcast traffic up to 36 Mb/s on the forward link (hub to remote terminal), with the choice of standardized DVB-S2 (CCM, VCM, ACM) or DVB-S transmissions and up to 6 Mb/s transmission on the return link (remote terminal to hub) the S4120 is ideally suited for all business needs. The attractive design and small form-factor make it ideal as a cost-effective, desktop high-speed solution. Thousands of the S4120 VSAT terminals can populate a DVB-RCS compliant any Advantech Hub ground terminal in a teleport. The DVB-RCS compliance allows for other vendor's terminals to interoperate with the Advantech SatNet DVB-RCS terminals in the same network or with other vendor's hubs. This VSAT terminal is supporting connections of antennas at C, Ku, Ka and X-Band.

3. **VSAT Ku /Ka-8000 Series Outdoor VSAT Terminal** – This Advantech family of 8000 Series VSAT antenna terminals are supporting the most powerful economical terminals available for High Throughput Satellite (HTS) GEO constellations as well as traditional satellite systems, which is illustrated in Fig. 7.30 (Right). This antenna is able to transmit in modes such as MF-TDMA, BM-FDMA or DVB-S2/S2X and receive DVB-RCS/RCS2 waveforms. It is supporting the same Ku and Ka-band transmitting and receiving frequencies as VSAT Ku/Ka 8200 Series Antenna.

4. **VSAT Mobile Antennas** – In the similar way, the same Advantech or Hughes VSAT stations can be installed onboard of Mobiles. These antennas are also able to provide GEO Meteocast forward and return uplinks via VSAT transceivers such as Satnet S4120 ACM DVB-RCS VSAT Transceiver Terminal and similar VSAT terminals of other producers. The mobile VSAT antennas of Orbit

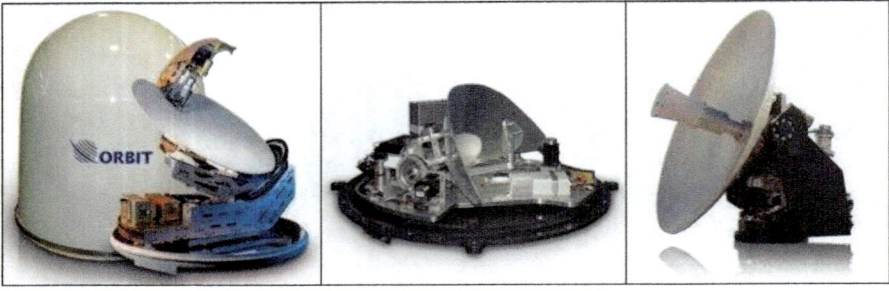

Fig. 7.31 Mobile DVB-S2 VSAT antennas (Courtesy of Manual: by Orbit)

producers are VSAT shipborne antenna illustrated in Fig. 7.31 (Left), in Fig. 7.31 (Middle) is depicted vehicleborne antenna and in Fig. 7.31 (Right) is shown airborne antenna.

All above stated VSAT antennas onboard mobiles can be used for receiving weather bulletin, weather forecast and tropical storms warnings, while they can be implemented as well as readout receiving antennas and for transmitting weather data from Data Collection Platform (DCP) via conventional GEO satellites. With some additional frequency bands DCP data can be also sent via metrological GEO satellites.

7.6.5 Meteocast Receiving Broadcasting GEO DVB-RCS Antennas

As stated before, customers and different users can now receive data from Eumetsat via a TV broadcasting satellite Eutelsat-10A as an alternative way to either direct reception from the Meteosat transponders or using the Internet to download the data. This WX service is called Eumetcast and it is provided by producer of a Telllicast server. Just like direct reception, the data is only broadcast once in the certain time and day by schedule via DVB-RCS transmissions.

In addition, the Eumetcast services provided by Eumetsat satellite constellation include data from Meteosat-10 (MSG-3), Meteosat-9 (MSG-2), Foreign Satellite Data (FSD), NOAA 19 MODIS (Terra/Aqua) and MetOp. The MSG-3 is GEO weather satellite for Europe, which is providing 12 spectral channels. Thus, FSD includes hourly images from GEO satellites around the world, such as the US GOES-W and GOES-E stationed over the Americas, Meteosat-7 provides the Indian Ocean Data Coverage (IODC), and Japanese MTSAT-1R covers Asia and Australia.

To receive this Eumetcast information will be necessary to provide DVB-S2 VSAT Rx stations with adequate antennas. The DVB-S2 C-band antenna is shown

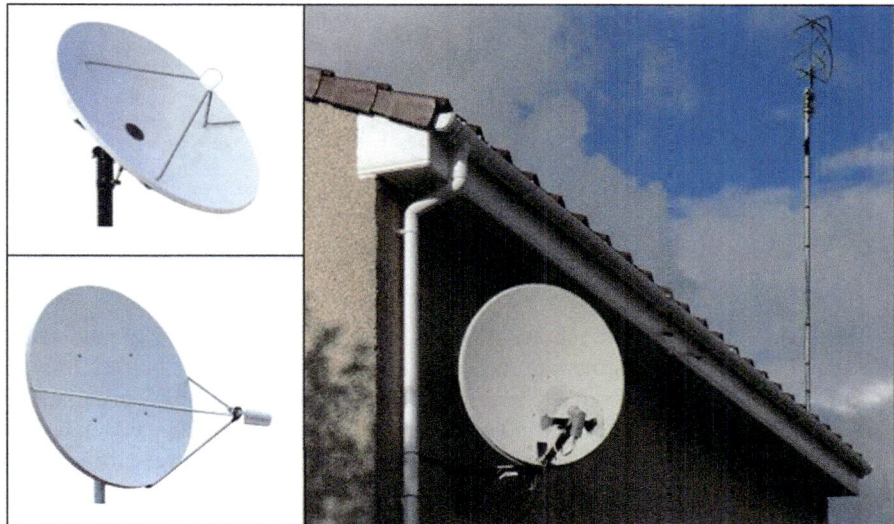

Fig. 7.32 DVB-S2 receiving VSAT antennas (Courtesy of Manual: by Dartcom)

in Fig. 7.32 (Left Above), Ku-band receiving antenna is shown in Fig. 7.32 (Right) and in Fig. 7.32 (Left Below) is shown Ka-band receiving satellite dish for GEO Eumetcast and QFH antenna for PEO satellite APT service over Europe. Thus, subsets of this data are also broadcast via C-band satellites providing coverage to Africa and some of the Americas.

For receiving only VSAT of LRIT or HRIT data on C-band can be used 2.4 m, 3.0 m or 3.7 m parabolic dishes and LNB, which sample of C-band LNB is shown in Fig. 7.33 (Left). The TechniSat SkyStar 2 internal PCI DVB-S2 receiver card for EUMETCast C-band service, which can be installed in PC is shown in Fig. 7.33 (Middle) and TBS 6983 internal PCIe DVB-S2 receiver card for C and Ku-band HimawariCast is shown in Fig. 7.33 (Right).

To provide an universal system for receiving service from the Broadcasting GEO DVB-RCS Stations will be necessary to establish a global casting Meteocast service proposed by the DUT Research Centre in Space Science. The Meteocast global casting of meteorological LRIT and HRIT data has to integrated existing Meteosat, HimawariCast and other GEO meteorological satellite casting in unique and interoperable Meteosat global system.

7.6.6   *User Earth Stations (UES) Antennas Onboard Mobiles*

The UES antenna terminals can be installed onboard mobiles, such as ships, land vehicles and aircraft. Aircrafts can use the same equipment as ships or vehicles use, but antenna has to be flat or installed in the tail of aircraft.

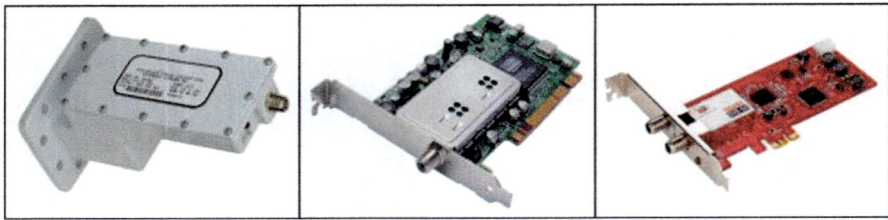

Fig. 7.33 DVB-S2 LNB and PC receiving cards (Courtesy of Manual: by TechniSat)

The Weather Decision Support (WDS) infrastructures are special ground satellite antenna and receivers that are receiving from meteorological satellites images directly broadcast to the fixed and mobile ground stations. These WDS installations can be some locations on the ground or onboard vessels and vehicles, which task is to process and produce received local and hemispheric weather images for government, corporate, personal and private customers.

7.6.6.1 Shipboard Satellite WDS Receiving 0.61 m L/S-Band Antenna

The WDS shipboard receiver is receiving data and producing local and hemispheric weather images from the meteorological satellites that are directly broadcast to the vessel and other customers. In addition to satellite data, the WDS system automatically receives, processes and displays meteorological and maritime forecast data tailored to the vessel's operating areas of interest. The TeraScan L/S-band shipboard system utilizes a 0.61 m flat plate phased array antenna mounted on a 3-axis positioner installed inside a 0.90 m radome. The radome size has been carefully matched to the antenna to minimize weight and overturning moment. In Fig. 7.34 (Left) is illustrated shipboard satellite weather receiving 0.61 m antennas, which is able to receive WDS shipboard Graphical User Interface (GUI), while in Fig. 7.34 (Right) is depicted image of WDS Shipboard GUI.

Ideal for shipboard receiving operations is the 0.61 m WDS antenna system automatically tracks satellites while in ship motion. It is designed to receive and process data from current PEO meteorological satellites. Thus, the WDS system comes with specialized software that automatically receives oceanographic products. The GUI display as a special human computer interface was developed with the mariner user in mind for more easy operations during any surrounded weather conditions.

The key features of both L and S-band shipboard satellite weather receiving antenna are as follows: (1) High reliability for mission critical operations with; (2) Theater-wide coverage; (3) Cost effective maintenance and low power consumption; (4) Unique feed and compact design; (5) Multi-mission with L and S-band capability; (6) Oceanographic Specific GUI; and (7) Automated data processing.

Fig. 7.34 Shipboard satellite weather receiving 0.61 m antenna and WDS shipboard GUI (Courtesy of Manual: SeaSpace)

This WDS antenna is able to receive cloud top temperature characteristics and Infrared (IR) cloud data with blue color enhancements. Here is necessary to underline that meteorologists use color-enhanced imagery as an aid in satellite interpretation. The colors enable them to easily and quickly see features that are of special interest. Usually they look for high clouds or areas with a large amount of water vapor. In an IR image cold clouds are high clouds, so the colors typically highlight the colder regions valued in pixels. The intensity of a pixel is recorded as a digital number (for example, in these images the numbers range from 0 to 255).

Both models WDS6100L (L-band) and WDS6100S (S-band) have common components, such as feed, antenna pedestal, reflector, RF electronics, radome (top and base), basic spares kit, and O&M manual, which samples of shipborne and ground antennas in radomes are shown in Fig. 7.35 (Left and Right), respectively. In addition, both type of WDS6100 antenna contains L-band or S-band feeds, antenna pedestals, reflectors, RF electronics, radomes (top and base), basic spares kits and Operation and Maintenance (O&M) Manual.

The L-band receiving antenna is providing on frequency band 1680–1710 MHz the following telemetries: NOAA does HRPT, FengYun does Chinese HRPT (CHRPT), EU does MetOp, PEO Eumetsat is doing AHRPT and it providing LRIT data service as well. Its antenna gain value is 17 dBic, aperture efficiency is 70%, noise temperature is 0.7 at 23 °C, total system noise temperature is 85.7 K and system G/T is 0.26 dBK.

The L-band receiving antenna is providing on frequency band 2200–2270 MHz telemetries, such as Real Time Data (RTD) for solutions of the US Defense Meteorological Satellite Program (DMSP). The gain of this antenna is 20 dBic, system G/T is 2.69 dBK and other values are the same to the L-band antenna.

7.6.6.2   Shipboard Satellite Weather Receiving 1.5 m L/S-Band Antenna

This type of receiver and antenna also produces WDS for local and hemispheric weather images from meteorological satellites that are directly broadcast to the

Fig. 7.35 Shipboard and ground-based WDS6100 L and S-band antennas (Courtesy of Manual: SeaSpace)

vessel. Same as previous solutions, this WDS system also automatically receives, processes and displays meteorological and maritime forecast data tailored to the certain vessel's operating areas of interest. The TeraScan L/S-band Shipboard System utilizes a 1.5 m antenna reflector mounted on a 3-axis positioner installed inside a 2.0 m radome. The radome size of the antenna has been designee in the way to reduce weight, overturning moment and site preparation requirements and costs. Low-loss, impedance-matched radomes enable operation in extreme environments without sacrificing RF performance. Each antenna system is perfectly balanced in all three axes and exhibits no inherent backlash.

The key features of L/S-band shipboard satellite antenna are: (1) High tracking accuracies, low velocity and acceleration requirements; (2) No keyhole losses at any elevations and completely balanced system; (3) 3-axis with Elevation over Cross (4) Elevation over Azimuth configuration; (5) Low-loss radome enclosure for all installations; and (6) Integrated tracking receiver with Pulse Width Modulation (PWM) servo control system.

The PWM mode or Pulse Duration Modulation (PDM) is a mode of modulation technique used to encode a message into a pulsing signal. Although this modulation technique can be used to encode information for transmission, its main use is to allow the control of the power supplied to electrical devices, especially to inertial loads such as motors.

There are no pointing and tracking errors associated with torsional stiffness since there is no strong wind loading on the tracking system and the velocities and accelerations are kept extremely low. The antenna automatically tracks satellites while vessel is in motion, which is shown in Fig. 7.36 (Left) and type of vessel with antenna onboard (Right). It is designed to receive and process data from current L and S-band PEO meteorological satellites. This antenna system includes

Fig. 7.36 Shipboard 2.0 m radome and type of WDS vessel (Courtesy of Manual: SeaSpace)

specialized software that automatically receives and processes data into oceano-graphic products, with a GUI display developed for use onboard vessels.

The 1.5 m shipboard L-band and L/S-band antennas contain L or L/S-band feed, antenna pedestal, reflector, RF electronics, antenna radome (top and base), basic spares kit and O&M manual. L-band input frequency is 1670–1720 MHz S-band input frequency 2166–2315 MHz. Thus, RF Performance G/T at S-band is 5.5 dBK minimum at 2240 MHz and G/T at L-band is 5.0 dBK minimum at 1700 MHz. However, input signal power is −90 to −50 dBm, input frequency is 126–154 MHz and demodulation is BPSK-PSK.

The system is Supporting the following telemetries: NOAA HRPT, METOP and DMSP RTD, and also is supporting data encoding such as Non-Return to Zero (NRZ) formats, NRZ-S, NRZ-M and Phase Encoded Format (Biphase-L). This solution uses Linux OS, such as Red Hat Enterprise Linux (RHEL) and Community Enterprise Operating System (CentOS).

In telecommunication, NRZ line is a binary code in which ones are represented usually by a positive voltage, while zeros are represented usually by a negative voltage, with no other neutral or rest condition.

7.6.6.3 Shipboard Satellite Weather Receiving 2.4 m L/S/X-Band Antenna

The 2.4 m multi telemetry WDS is configured with a tracking antenna and receiver for the reception of real-time weather images from meteorological satellites that are directly broadcast to the vessel. Real-time products include sea surface temperature for research, fishing, ice products for detecting ice edges and navigation support in the Polar Regions. The day-night bands from Suomi-NPP (National Polar-orbiting Partnership) allow low-light detection at night and are useful for identifying illegal fishing.

Fig. 7.37 Shipboard 2.4 m radome and iceberg mapping in Antarctic (Courtesy of Manual: SeaSpace)

Infrared/visible products can be used for fog detection, storm tracking, course plot decisions and help to save fuel and avoid rough seas, while ocean chlorophyll maps can be used for research, pollution tracking and fisheries. The system includes the necessary hardware and TeraScan software for the automated reception, processing, and visualization of products.

This solution includes WDS specialized software that automatically receives and processes data into oceanographic products, with a GUI display developed for users onboard special vessels. The TeraScan data system performs automated image navigation and geolocation for every major remote sensing satellite. A World Vector Coastline database and the TeraNav interactive navigation tool are included with each system. In Fig. 7.37 is depicted (central radome) shipboard 2.4 m antenna (Left) and mapping of icebergs image in Antarctic (Right).

The key features of L/S-band shipboard satellite weather receiving antenna are as follows:

1. High tracking accuracies and stabilized positioner with continuous azimuth rotation;
2. No keyhole losses at any elevation;
3. 3-axis, Elevation over Cross-Elevation over Azimuth configuration;
4. Completely balanced system and low maintenance cost and power consumption;
5. Operation in extreme environments, Program Track and Autotrack Modes;
6. High reliability for mission critical operations, easy to install, operate and maintain;
7. End-to-End system solutions – Turn Key; and
8. Automated capture and processing 24/7 and remote access control.

This solution is using Linux OS (RHEL and CentOS). Its input signal power is −90 to −50 L-band and −60 to −20 dBm at X-band. Input frequency is 126–154 MHz (L and S-Band), 720 MHz (X-Band) and demodulation types are: BPSK, SQPSK, USQPSK and QPSK.

7.6.6.4   Antennas for TeraScan Satellite Acquisition System (TacSAS)

This WDS data solution is ideal for commercial vehicle and military field operations, from mission planning to ordnance targeting service. The Tactical Satellite Acquisition System (TacSAS) model THM provides both automated reception and processing of direct-broadcast weather and environmental data from the NOAA and DMSP series of Low Earth Orbit (LEO) satellites. Where time, transportation and manpower are at a premium, the entire dual-telemetry system fits into a HMMWV-sized vehicle and is easily deployed into a fully functional ground station within 10 min. The TacSAS provides joint mission interoperability, such as for naval, ground and air forces, and can now utilize identical equipment to exploit real-time satellite meteorological data, from imagery to vertical soundings. Thus, the TacSAS model THM is designed to complement current deployments of TacSAS-FM terminals that receive and process data from GEO satellites.

However, in the same way this solution can be deployed for commercial land vehicles, which vehicleboard 0.61 m outdoor antenna is shown in Fig. 7.38 (Left) and ruggedized indoor laptop mounted in vehicle for WDS processing shown in Fig. 7.38 (Right).

The key features of L/S-band shipboard satellite weather receiving antenna are:

1. High Resolution Data: Uses data acquired from the NOAA and MetOp (Meteorological Operation) satellites. Ground pixel size is 1 km for visible and IR imagery, and 40 km for microwave soundings;
2. Wide Area Coverage: Data products include satellite accurate and precise imagery and vertical profiles, covering greater than a 1000 km radius, with profiles from ground level up to 20 km altitude;
3. LRIT Option: LRIT data is available when not acquiring NOAA/MetOp data;
4. Low Observability: Less than 1 m in height, fully camouflaged, and fully passive (no RF emissions);
5. Proven Automated Performance: Timely and automatic generation and dissemination of data products for real-time decision making;
6. Interactive Display. An intuitive graphical user interface includes the WDS interface and the TeraVision interface. Users can customize data viewing options; and
7. No Consumables: Delivers timely and accurate weather satellite data without the use of expendables such as balloons.

In Fig. 7.39 (Left) is illustrated vehicleboard system in the field or may be any private or commercial vehicle and in Fig. 7.39 (Right) is illustrated NOAA-16 satellite Advanced Microwave Sounding Unit (AMSU) Wind Speed Image.

The size of L-band satellite antenna is 0.61 m for receiving NOAA HRPT and MetOp data and to process them in TeraScan Satellite Acquisition Module (TSAM) processor. The system also possesses GPS antenna and receiver, which tracking antenna is integrated with Selective Availability and Anti-Spoofing Module (SSASM).

Fig. 7.38 Vehicleboard 0.61 m L-band antenna and ruggedized laptop mounted in vehicle for WDS processing (Courtesy of Manual: SeaSpace)

Fig. 7.39 THM vehicleboard system in the field and NOAA-16 AMSU wind speed image (Courtesy of Manual: SeaSpace)

7.6.6.5 Land HRPT/AHRPT Antenna System

The Dartcom HRPT/AHRPT Receivers and Antenna System provides very reliable, high performance land-based and marine mobile solutions for receiving, archiving, processing and displaying data from NOAA and MetOp satellites, and optionally from FengYun-3 (L-Band) and DMSP (S-Band).

A number of land-based antennas are available (1.8 m, 1.2 m and radome-enclosed 1.5 m) or for marine use a radome-enclosed 1.3 m antenna with active stabilization, which is illustrated in Fig. 7.40 (Left). Ingested weather data can be viewed and processed using the Dartcom IPAD/MicroPro software.

Fig. 7.40   Dartcom 1.3 m active-stabilized marine antenna (Courtesy of Manual: SeaSpace)

Moreover, these outputs are also available for popular image processing software   packages such as PCI Geomatica, ERDAS IMAGINE and ENVI/IDL, as well as standard interchange formats such as NOAA level 1B, EPS level 0 and GeoTIFF.

7.6.6.6   Marine HRPT/AHRPT Antenna System

This antenna is designed to track PEO satellites on moving vessels using a state-of-the-art active-stabilized X-Y pedestal to compensate for pitch, roll and yaw, which is shown in Fig. 7.40 (Right). It has continuous axis movement to eliminate cable wrap problems without slip rings or rotary joints. Together with the pedestal's high speed and accuracy this ensures no "cone of silence" (data loss at high elevations).

The antenna has a 0.35 F/D ratio and 24.7dBi gain (S-band 27dBiC) to achieve a system G/T of better than 3.4dBK at 1.7GHz and 4.7 dB/K at 2.252GHz, both at 5° elevation, and a bit error rate of better than 1:106 from 3.5° elevation angle. The reflector is a 1.3 m diameter prime focus aluminium parabolic dish finished in light gray paint (RAL 7044).

An Integrated Feed Downconverter (IFD) unit is mounted at the focal point in a hermetically sealed unit, which more details can be found in the Land-based antennas section for specifications. The dish and pedestal assembly is mounted inside a weather-tight glass fiber radome and base with access hatch. The Antenna Control Unit (ACU) is located below decks in a 19″ equipment racks with the receiver rack, ingest PC, UPS and network switch. The ACU provides fully automatic control of the pedestal using an advanced stabilization algorithm. In fact, full diagnostics and maintenance facilities are available on a color TFT screen. This antenna is ISO9001/CE certified and has been designed to meet or exceed Military Standard (MIL-STD) specifications.

Fig. 7.41 TX320 GOES transmitter (Courtesy of Brochure: Campbell Scientific)

7.6.6.7   GEO Data Collection Platform (DCP) with Antenna

Climatologic and synoptic meteorological surface data collected via DCP is used to create weather prediction models that describe the atmosphere and its changes over time, can be sent by TX32 Satellite transmitter via GOES satellite, which is illustrated in Fig. 7.41.

Thus, before that meteorological data needs to be collected, build a functional and exhaustive meteorological database for accurate models requires high quality observations of a variety of relevant measurements, which process is shown in Fig. 7.42, AWS system with antenna (Left) and its components (Right). The Automatic Weather Station (AWS) for DCP can be installed in offshore or onshore locations and onboard ships as well, which main task is to collect local meteorological data and send via GEO satellite to some Direct Readout Station (DRS). Each DCP station is specialized for the following tasks:

1. Hydrological measurement data is essential in the prediction and solution of flood, water pollution, drought and erosion problems. Some areas have plenty of water while other areas, afflicted by drought, go wanting.
2. Coastal weather stations produce meteorological data for sea weather forecasts and to warn ships and offshore operators about severe weather conditions. The same data is also used in regional and global computer models to help predict atmospheric changes and monitor ocean climate and the state of the oceans.
3. Agrometeorological AWS systems measure in-site weather conditions in parallel with local forecasts. The objective is to produce weather data that enables optimal timing and control of all field operations. The data is gathered, stored, viewed and analyzed to enable more profitable decisions with less risk in farm management.
4. Weather can create significant disruptions in urban areas. Heavy rains can cause severe flooding, snow and freezing rain can disrupt transportation systems, and major storms with accompanying lightning, hail and high winds can cause power failures.

The main variables of hydrological measurements in AWS are: Precipitation; Water level (rivers, lakes, reservoirs, wells); Water temperature; Snow depth;

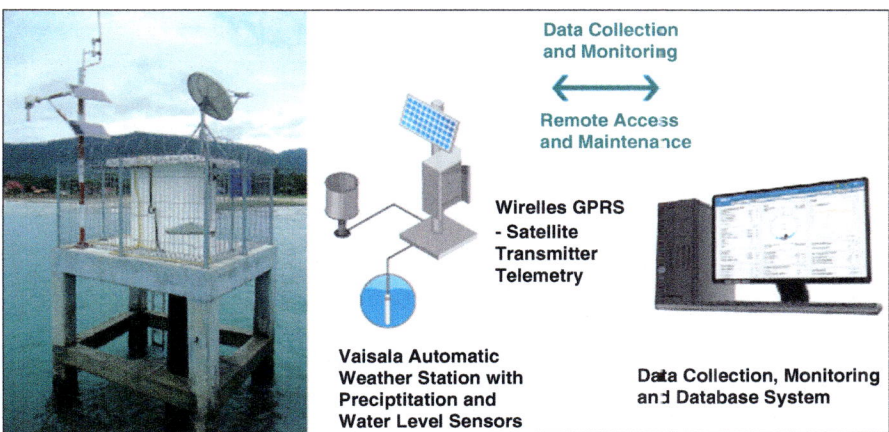

Fig. 7.42 Automatic weather station and components with antennas (Courtesy of Brochure: Vaisala)

Waterflow; Evaporation; Soil moisture; Ambient water quality, and so on. Otherwise, the sensors used in Vaisala AWS fulfill WMO recommendation for accuracy.

As stated earlier, at the end of this section is important to conclude that for sending DCP meteorological data via existing GEO (Inmarsat) or LEO (Iridium, Globalstar and Orbcomm) communication satellites can be implemented small satellite transceivers, while via SCADA or M2 M can be also sending and even controlling ground DCP meteorological stations.

7.7 Propagation and Interference Consideration

Propagation and interference characteristics are very important for providing quality and reliability of satellite propagation channels. The Quality of Service (QoS) can be expressed in terms of the Bit Error Rate (BER) performance, which depends on the Carrier-to-Noise Density Ratio (C/N_0), while the service reliability is manifested in the relation of service availability. Thus, the intervening medium between UES within the network and satellites is termed a transmission channel. The fixed satellite services have two constant channels between a minimum of two UES using the same spacecraft, which have many different characteristics, which need to be taken into account during the system design examination.

The common satellite channel environment affects radiowave propagation in changeless ways. The different parameters influenced are mainly path attenuation, polarization and noise. The factors to be considered are gaseous absorption in the atmosphere, absorption and scattering by clouds, fog, all precipitation, atmospheric turbulence and ionospheric effects. Thus, several measurement techniques serve to

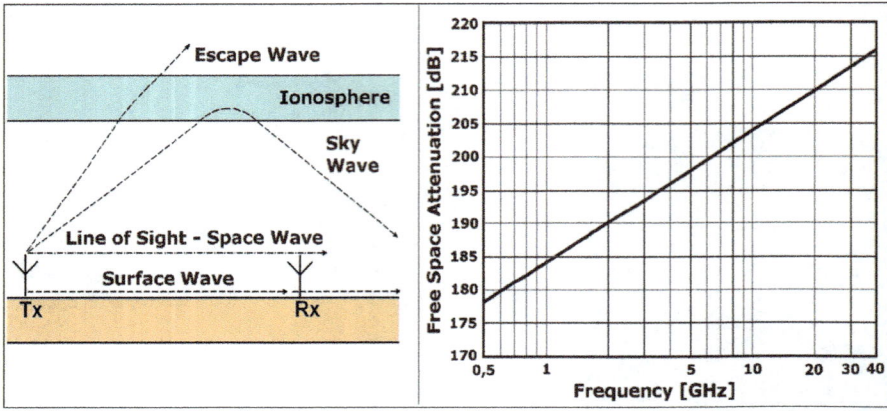

Fig. 7.43 Radiowave modes of propagation and free space loss (Courtesy of Books: by Tetley/ Ohmori)

quantify these effects in order to improve reliability in the system design. Because these factors are random events, satellite system designers usually use a statistical process in modeling their effects on radiowave propagation.

7.7.1 Radiowave Propagation

Radiowaves travel in space as electrons at the speed of light approximately 300,000,000 m/s. When radiowaves are propagated from a transmitter antenna, they form three modes of propagation path: surface, sky and space wave propagation, shown in Fig. 7.43 (Left). Radiowaves are generally transmitted from an antenna omnidirectionally but transmission can be also modified in a directional path by using a directional dish or antenna arrays.

The surface wave is a radiowave that is modified by the nature of the terrain over which it travels. The surface wave propagation predominates at all radio frequencies up to 3 MHz for VLF, LF and MF bands. Hence, sky waves are severely influenced by the action of free electrons, called ions, in the upper atmosphere, known as the ionosphere and are caused to be attenuated and reflected and possibly returned to Earth. In fact, sky wave propagation predominates in the HF band between 3 and 30 MHz.

Above 30 MHz, the predominant mode of propagation of radiowaves is by the space wave. This wave, when propagated between 30 and 300 MHz into the troposphere by an Earth VHF or UHF radio station, is subject to deflection by variations in the refractive index structure of the air through which it passes, which will cause the radiowave to follow the Earth's curvature for short distances up to about 100 km. However, space waves above 300 MHz propagated upwards, away from the troposphere, may be termed free space waves and are primarily used for satellite communications.

Whilst all transmitting antenna systems produce one or more of the three main modes of propagation, one of the modes will predominate. The predominant mode may be equated to the frequency used, as all other constraints remain constant. Therefore, for the purposes of this explanation of propagated radiowaves, it is assumed that the mode of propagation is dependent upon the frequency used in an adequate system, for that is the only parameter, which may be changed by an operator.

The simplest situation involving the transmission of information by EM waves is when the Tx antenna transmits directly to the Rx antenna, without any obstacles in the path through which the EM waves must travel. This situation occurs when both antennas are located at a relatively short distance on the Earth's surface with full mutual visibility. As the distance between the two antennas increases due to the Earth's curvature, the EM waves are no longer in a direct path (line-of-sight).

In order to increase the path length the antenna can be fitted on a higher mast but only to the extent that the costs do not become prohibitive. In this way, the problem can be solved by means of forthcoming stratospheric platforms or communications satellite payloads. Since these satellites are placed at very high altitudes, their reach is far superior to that of the higher towers that can be built on the Earth's surface, see altitudes and other parameters illustrated in Table 7.6.

When two antennas situated on the Earth are very far apart and there is no visibility between them, stratospheric platforms or satellites can be used to relay the signals. There is otherwise visibility between the Tx antenna and the spacecraft (platforms) and vice versa, between spacecraft (platforms) and Rx antenna. At this point, the ways these signals will travel on their line-of-sight depend on propagation characteristics and conditions of interference in the hypothetical environment.

Because communications satellites are located at higher altitudes relative to the Earth's surface than stratospheric platforms and masts, they cover a wider area, which is beneficial for ocean spaces, continents and for countries with large dimensions.

7.7.2   Propagation Loss in Free Space

It is necessary to define the loss between Tx and Rx antenna separated by a distance from the transmission medium, assumed to be a vacuum and the antenna is isotropic. It may serve either as Tx or Rx antenna in the radiation field where it is situated.

Considering that the isotropic radio antenna radiates signals in all directions of its spherical surface with total power flow P_T in Watts, including receiving antenna gain, at a sufficiently large distance (r) in meters away from the centre to the surface of the sphere, the power flux of hypothetical antenna per unit area through any portion of the spherical surface must be as:

Table 7.6 Altitude location of stratospheric platforms and spacecraft

System	Altitude (h) in km	Angle in degrees	Distance (d) in km
International flight	10	3.313	368.4
Stratospheric platforms	30	5.549	617.1
Amateur satellites	150	12.320	136.2
LEO satellites/low	780	27.008	3003.6
LEO satellites/high	2000	40.438	4497.1
MEO satellites	10,000	67.095	7461.7
GEO satellites	35,600	81.268	9037.8

$$P_f = P_T/4\pi r^2 \tag{7.27}$$

The amount of power that the receiving antenna absorbs in relation to the RF power density of the EM field is determined by its effective aperture, which is defined as the area of the incident EM wave front that has a power flux equal to the power dissipated in the load connected to the receive antenna output terminals. Following the previous (7.27) equation the receiving power at a receiving antenna can be expressed by:

$$P_r = P_T(\lambda/4\pi r)^2 = P_T/L_f \tag{7.28}$$

where λ = wavelength and L_f = free space transmission loss, which can be given in dB as is presented in the following relation:

$$L_f = 10 \log_{10}\left(P_T/P_r = 32,40 + 20 \log f_{MHz} + 20 \log r_{km}\right. \tag{7.29}$$

where f = frequency of the emitted radio field in dB and r = distance in km between Tx and Rx antenna. Thus, free space loss is related to operating frequency and transmission distance. In Fig. 7.43 (Right) is illustrated free space loss as a function of frequency in GHz for a distance of 36,000 km.

7.7.3   Atmospheric Effects on Propagation

This study deals with propagation effects in all regions of the atmosphere and free space, including the Earth's ionosphere. Namely, most of the Earth's weather precipitation and hydrometeors occur in the troposphere, which is the non-ionized region from the Earth's surface up to a height of about 15 km above the surface at the Equator. The thickness of the troposphere decreases towards the poles. Therefore, propagation effects in the troposphere tend to increase in importance as the RF increases above 1 GHz. For satellite systems, the effects of reflection from the Earth's surface are critically important at even lower RF.

At frequencies below about 1 GHz, the most important region of the Earth's atmosphere is the ionosphere, the ionized region above the stratosphere, within which low RF propagation effects are quite strong. Propagation effects within the ionosphere have an influence on the terrestrial and Earth space paths from VLF to SHF bands. However, it should be borne in mind that for an RF band range below a certain number of GHz, the ionospheric propagation effects can be quite important and for frequencies above a certain number of GHz, tropospheric effects may be negligible. At this point, radiowave signals passing through the atmosphere or over the surface of the Earth begin to lose strength. The decrease of signal due to the medium through which it passes is called attenuation.

7.7.4 Propagation Effects of the Troposphere

The troposphere extends upwards from the Earth's surface to a height of approximately 15 km, where it meets the stratosphere. More exactly, at the boundary between the two spaces there is a region called the tropopause, which possesses a different refractive index to each neighboring layer. The effect exhibited by the tropopause on a space wave is to produce a downward bending action, causing it to follow the Earth's curvature. Hence, the bending radius of the radiowave is not as severe as the curvature of the Earth, nevertheless, the space wave will propagate beyond the visual horizon. In practice, the radio horizon exceeds the visual horizon by approximately 15%.

7.7.4.1 Attenuation due to Atmospheric Gases

The different types of gases present in the atmosphere may attenuate the electromagnetic waves, which is caused by the molecular absorption of the atmospheric constituents and is strongly frequency dependent. The main contributors to this attenuation below 70 GHz are water vapor and oxygen. Absorption increases as elevation angle is reduced, which at any frequency is a function of temperature, pressure, humidity of the atmosphere and the elevation angle of the satellite. At sufficiently high frequencies, EM waves interact with the molecules of atmospheric gases to cause attenuation.

These interactions occur at resonance radio frequencies and are apparent in plots of zenith (90° elevation angle) attenuation versus frequency, as shown in Fig. 7.44 (Left). This is the theoretical total estimated one-way attenuation for vertical paths through the atmosphere, where solid curves are for moderately humid atmosphere, dashed curves represent the limits for 0% and 100% relative humidity, V is vertical polarization and A denotes limits of uncertainty, for a vertical Earth space path as a function between 1 and 200 GHz at 45° North latitude using US standard atmosphere. It may be noted that the specific frequency bands where the absorption is

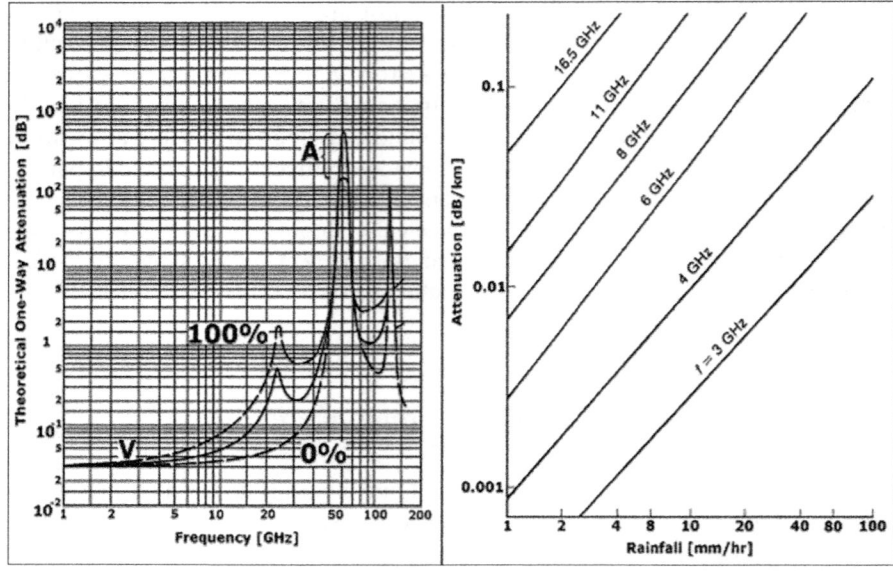

Fig. 7.44 Theoretical one-way and rainfall attenuations (Courtesy of Books: by Richharia/Gagliardi)

high are about 22.23 GHz caused by vapor (left peak in the curve) and centered between 53.50 and 65.20 GHz caused by oxygen (right peak in the curve).

Meanwhile, in the frequency range of current interest about 1–18 GHz the zenith one-way absorption is in the range of 0.03–0.20 dB and 0.35–2.30 at 5° elevation. The corresponding upper limit for 100% humidity is 0.7 dB at zenith and 8 dB at 5° elevation.

7.7.4.2   Attenuation by Precipitation and Hydrometeors

Hydrometeors are condensed vapors existing in the atmosphere, such as cloud, fog, rain, hail and snow. The last three forms atmospheric water as precipitation and produce transmission impairments and attenuation by absorbing and scattering the energy of radiowaves.

1. **Clouds and Fog** – Cloud and fog may cause attenuation but much less than that caused by light rain of about 10 mm/h. Clouds and fog are suspended water droplets, usually less than about 0.10 mm in diameter, whose effect is significant only for systems operating above 10 GHz. It is important to note that attenuation to radiowaves depends on the liquid water content of the atmosphere along the propagation path. Clouds have liquid contents from 0.05 to 5 g/m^3 and their formation shape can be due to a variety of atmospheric processes, which result in cloud layer types for each of three cloud heights: low, middle and high.

Thunderstorm cumulo-nimbus high clouds cause the maximum attenuation. The fog liquid content is in the order of about 0.40 g/m^3 and typically fog extends from 2 to 8 km. The attenuation from fog is negligible for satellite communications.

2. **Rain** – Rain precipitation is the most significant contribution to atmospheric attenuation, which is caused by radiowave absorption and scattering from raindrops. First of all, to evaluate this additional rain loss, it is necessary to obtain the expected rainfall rate in mm/hr. for the region of the telecommunication link. In such a way, in Fig. 7.44 (Right) are shown six curves, which can be used to read off dB losses per EM path length at the operating frequencies.

Namely, these curves are generated from a combination of empirical data and mathematical models that fit this data. Rainfall rate distribution is inhomogeneous in space and time and can impair a satellite communications link at frequencies above 10 GHz, as well as increase the noise temperature and impairment, or cross-polarization discrimination. Rain gauge records show short intervals of higher rain rates occurring in longer periods of lighter rain. Various precipitation methods have been proposed for predicting rain attenuation statistics calculations from rainfall rate measurements near the signal path. The slant path rain attenuation prediction is based on the estimation of the attenuation exceeded at 0.01% of the time $A_{0.01}$ from the rainfall rate $R_{0.01}$ (mm/h), exceeding at the same time percentage.

The effective path length is given by the formula:

$$A_{0.01} = \gamma_R \cdot L_e = \gamma_R \cdot L_s \cdot r_{0.01} \quad [dB] \tag{7.30}$$

Where γ_R (dB/km) = specific attenuation, L_e is the effective path length, L_s = slant path length and $r_{0.01}$ = path reduction factor. The spatial structure of rain can be modeled by an equivalent rain cell of uniform rainfall rate with a rectangular cross-section of equivalent length L_0 and effective height $h_R - h_S$ in the plane of the path, shown in Fig. 7.45 (Left).

The relation for slant path length for elevation angles above 5° is as follows:

$$L_s = h_R - h_S / \sin\theta \quad [km] \tag{7.31}$$

and for elevation angles less than 5° is:

$$L_s = 2(h_R - h_S) / \left[\sin^2\theta + 2(h_R - h_S)/R_e \right]^{\frac{1}{2}} + \sin\theta \quad [km] \tag{7.32}$$

where θ = path elevation angle and R_e = effective radius of the Earth (8500 km). Using simple trigonometry relations the horizontal projection is given by the formula:

$$L_G = L_s \cos\theta \quad [km] \tag{7.33}$$

The reduction factor has the following relation:

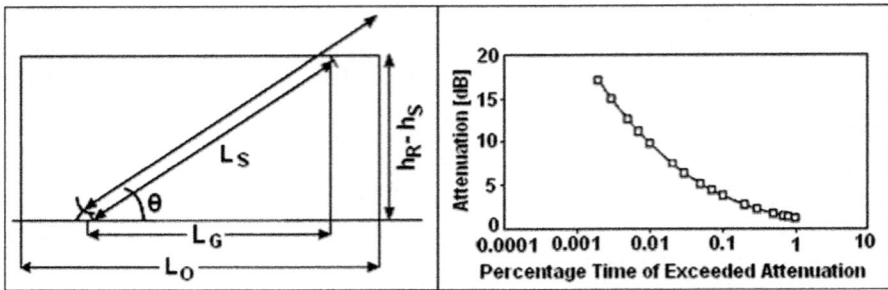

Fig. 7.45 Equivalent rain cell of rainfall rate and rain attenuation statistics (Courtesy of Handbook: by ITU)

$$r_{0.01} = 1/1 + L_G/L_0 \qquad (7.34)$$

For example, the specific attenuation using the frequency-dependent coefficients k and α gives the following expression:

$$\gamma_R = k(R_{0.01})^\alpha \ [\text{dB/km}] \qquad (7.35)$$

Values of the frequency-dependent coefficients are known, while values of some unknown frequencies can be obtained by interpolation using logarithmic scales for frequency and k and a linear scale for α. For linear and circular polarization, these parameters can also be obtained by the following equations:

$$k = \left[k_h + k_v + (k_h - k_v) \cos^2\theta \, \cos 2\tau \right]/2$$
$$\alpha = \left[k_h\alpha_h + k_v\alpha_v + (k_h\alpha_h + k_v\alpha_v) \cos^2\theta \, \cos 2\tau \right]/2k \qquad (7.36)$$

where θ = path elevation angle and τ = polarization tilt angle relative to the horizon ($\tau = 45°$ for circular polarization).

The attenuation value predicted for a percentage in the range of 0.001 to 1% can be obtained by the following equation:

$$A_p = A_{0.01}0.12 \cdot p^{-(0.546+0.043 \ \log \ P)} \qquad (7.37)$$

For percentages of the time outside the 0.001–0.1% is a relatively complicated procedure, while the predicted attenuation exceeded for 0.01% for an average year is $A_{0.01} = 9.8$ dB. Finally, the estimated attenuation exceeded for any other percentages of the time P may be calculated from $A_{0.01}$ by the expression:

$$A_p = A_{0.01}(P/0.01)^{-\alpha} \ [\text{dB}] \qquad (7.38)$$

The cumulative distribution of the rain attenuation statistics is shown in Fig. 7.45 (Right).

3. **Hail and Snow** – Hail depolarization often precedes or follows rain and the presence of hail above rain is one factor that causes wide variations in the instantaneous value of Cross Polarization Discrimination (XPD), which may accompany a given attenuation. Hail, with its ice meteors, causes a small loss due to scattering but not due to absorption.

Dry snow likewise does not absorb radiowave energy, however, the water content of wet snow dropping slowly through the ray path causes more absorption than would arise if the same amount of water were falling as rain.

7.7.4.3 Site Diversity Factors

Path diversity in satellite systems involves the provision of alternate propagation paths for signal transmission, with the capability to select the least-impaired path when conditions warrant. In this case, implementation of path diversity requires the deployment of two or more interconnected Earth terminals at spatially separated sites, hence the use of the term "site diversity".

This combination is considerable when an GES is being used for a commercially or militarily important purpose, where an outage is considered more serious, e.g., as the head end of a cable system, the probability of an outage can be reduced by having a second GES a few kilometers away and using whichever has the better signal at moment. Therefore, this is because heavy rain, in particular, tends to be quite localized and would probably not impact both stations at the same time.

Without an ability to increase the margin (e.g., through the application of power control via additional transmit gain and power, or increased resource allocation in the TDMA frame), there are basically three types of diversity schemes that can be used by satellite systems to overcome impairments at a given GES:

1. **Orbital Diversity** – This diversity is different from site diversity in that only one GES site is used. To achieve a measure of diversity, the GES uses two antennas that can access different satellites simultaneously and is not necessarily the diversity-interconnecting link between sites that is required for site diversity. However, to obtain significant decorrelation of concurrent attenuations along the two paths, the angle between the two paths at the GES must be large. Inasmuch as this angle is large, at least one or even both of the links will be at a relatively low elevation angle and therefore encounter a greater degree of impairment than at higher elevation angles. In general, the achievable diversity gain is very small, with values of about 2–3 dB in the 14/11 GHz bands.
2. **Frequency Diversity** – Path losses caused by particulates on the path increase as the frequency increases, particularly for rain. In effect, at 6/4 GHz, attenuation due to rain is negligible; at 14/11 GHz it can be significant in high rainfall regions of the world; at 30/20 GHz it is the dominant link impairment nearly everywhere. Thus, if it is possible to switch transmissions to a lower-frequency, significant increases in availability might be achieved. This capability requires that both frequency bands (the higher, impaired one and the lower one to which

the communication channels are to be switched) be simultaneously available at the GES in question. There must be spare capacity available in the lower frequency spectrum whenever needed, implying that significant spare capacity must be provided if the link is a high-capacity channel and that the complete network need to be under dynamic control. Namely, both elements require significant investment. Should such dynamic network control features be in place and the additional capacity in the lower frequency band be available on-call, frequency diversity can therefore undoubtedly provide large increases in availability.

3. **Time Diversity** – Severe rain events do not usually last long at a given location. These features can be used in any communication RF link that does not require interaction between the caller and the receiver. The service can be acceptable if let say Fax is successfully sent without any error within a 2-h period. The delay in sending the Fax can be considered a form of time diversity. This feature could also be used with advantage to determine the capacity requirement of a given RF link for optimal economic performance. If a link is sized for a maximum anticipated capacity it will have excess capacity for most of the time. If some transmissions can be delayed and sent, let say, at off-peak times, the capacity requirements can be reduced. The time delay could therefore be used either at times of peak capacity (i.e., the equivalent of call-blocking) or when the GES is undergoing a severe rain event.

7.7.5 Clear-Sky Effects on Atmospheric Propagation

Radio signals traveling through the atmosphere layers suffer attenuation even during fine weather. In any event, the clear-sky attenuation is mainly the result of absorption of energy from the transmission by water vapor and oxygen molecules; although there are other modes of clear-sky effect that have influences on propagation.

1. **Defocusing and Wave-Front Incoherence Contribution** – Several expressions have already been evaluated and are provided in No 2.3.2 of ITU-R P.618 recommendation to estimate the defocusing (beam-spreading) losses on paths at very low elevation angles. Otherwise, the loss is implicitly accounted for in the prediction methods for low-angle fading found in articles No 2.4.2 and 2.4.3 of the Recommendation.

   Hence, small-scale irregularities of the refractive index structure of the atmosphere cause incoherence in the wave front at the receiving antenna. This results in both rapid radio signal fluctuations and an antenna-to-medium coupling loss that can be described as a decrease of the antenna gain. In practice, signal loss due to wave-front incoherence is therefore probably only significant for large-aperture antennas, high frequencies and elevation angles below 5°. Measurements made in Japan with 22 m antenna suggest that at 5° elevation angle the loss is about 0.2 to 0.4 dB at 6/4 GHz, while measurements with a 7 m antenna at 15.5 and 31.6 GHz gave losses of 0.3 and 0.6 dB, respectively, at a 5° elevation angle.

2. **Scintillation and Multipath Influence** – Small-scale irregularities in the atmospheric refractive index cause rapid amplitude variations. Thus, tropospheric effects in the absence of precipitation are unlikely to produce serious fading in space telecommunication systems operating at a frequency below about 10 GHz and at elevation angles above 10°. At low elevation angles and at a frequency spectrum above about 10 GHz, tropospheric scintillations can, on occasion, cause serious degradations in performance. The atmosphere scintillation measure model includes frequency, elevation angle and antenna diameter, but also including meteorological parameters, can be used to account for regional and seasonal dependencies.

3. **Propagation Delays** – Additional propagation delays superimposed on the delay due to free space propagation are produced by refraction through the troposphere precipitation and the ionosphere. Thus, the jonospheric delay time above 10 GHz is less than this of troposfere.

4. **Angle of Arrival Values** – The gradient of the refractive index of the atmosphere causes a bending of the radio ray and the angle of arrival varies from that calculated based on the geometry of the path. Since the relative index varies largely with altitude, the angle-of-arrival variation is much greater in the elevation than in the azimuth angle. In addition, turbulent irregularities of the refractive index can give rise to angle-of-arrival scintillations. Both of these effects decrease markedly with elevation angle and are generally insignificant for elevation angles above 10°. The effects are independent of frequency.

7.7.6 Transionospheric Propagation

Radiowaves at frequencies of VHF and above are capable of penetrating the ionosphere and therefore, they provide transionospheric telecommunications. The ionosphere consists in a layer somewhere between 80 and 150 km altitude, where the density of the atmosphere is very low. Radiation from the Sun ionizes some molecules and it takes a long time for them to be neutralized by other ions. The concentration of ions varies with height, time of day, the season and in what part of its 11-year sunspot cycle the Sun happens to be.

7.7.6.1 Faraday Rotation and Group Delay

Ionospheric effects are significant for frequencies up to about 10 GHz and are particularly important for GEO and Non-GEO satellite constellations operating below 3 GHz. At the frequencies used for satellite transmission, signals pass right through and are subject to negligible refraction, less than 0.01° at 30° elevation.

The Total Electron Content (TEC) accumulated through the transionospheric transmission path results in the rotation of the linear polarization of the signal carrier and a time delay in addition to the anticipated propagation path delay. Given knowledge of the TEC, Faraday rotation and group delay can be estimated for communication applications. Therefore, this delay is known as the group delay, while the rotation of the linear polarization of the carrier is known as Faraday rotation. The TEC, denoted by N_T, can be evaluated by the formula:

$$N_T = \int_s N_e(s)\, ds \quad [electrons/m^2] \tag{7.39}$$

where N_e = electron density [electrons/m^2] and s = propagation path length through the ionosphere [m]. Typically, N_T varies from 1 to 200 TEC units (1 TEC unit = 10^{16} el/m^2). N_T has typical values in the range of 10^{16} and 10^{18} el/m^2. Even when the precise propagation path is known, the elevation of N_T is difficult to determine because N_e is highly variable in space and time. When propagating through the ionosphere, a linearly polarized wave will suffer a gradual rotation of its plane of polarization due to the presence of the geomagnetic field and the anisotropy of the plasma medium. Namely, this trend slows down the signal because the Earth's magnetic field penetrates the ionosphere when ions (charged particles), subject to the alternating electric field of a signal, tend to gyrate around the local line of force. The magnitude of Faraday rotation will depend on the frequency of the radiowave, the geomagnetic field strength and the electronic density (concentration) of the plasma as:

$$\Phi = N_T\left(KM/f^2\right) = 2.36 \times 10^2 B_E N_T f^2 \quad [radians] \tag{7.40}$$

where $K = 2.36 \times 10^4$ [MKS units]; M = value of (B_E secϕ) at 420 km of height; B_E = longitudinal component of the Earth's magnetic induction along the ray path [Tesla]; ϕ = zenith angle of the ray and f = frequency [Hz]. Typical values of Φ as a function of frequency for representative TEC values are shown in Fig. 7.46. Hence, the occurrence of Faraday is well understood and can be predicted with a high degree of accuracy and compensated for by adjusting the polarization tilt angle at the GES. The GPS and similar satellite navigation systems, which use the 1–2 GHz frequency spectrum and depend on measuring the travel time of EM signals, has to correct for this effect.

The presence of charged particles in the ionosphere slows down the propagation of radio signals along the path and produces a phase advance. Thus, the time delay in excess of the propagation time in free space is called the group delay and is given by:

$$T_g = 1.34 \times 10^{-7} N_T/f^2 \quad [sec] \tag{7.41}$$

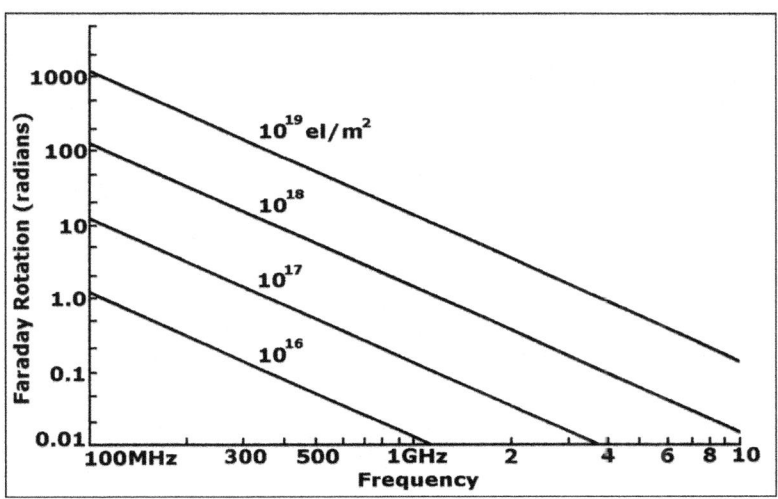

Fig. 7.46 Faraday rotation as a function of TEC and RF (Courtesy of Handbook: by ITU)

Accordingly, the time delay with reference to propagation in vacuum is an important factor to be considered for digital communication and navigation positioning systems.

7.7.6.2 Ionospheric Scintillation

Ionospheric effects are important at frequencies below 1 GHz, although they may even be important at frequencies above 1 GHz and are dependent on location, season, solar activity (sunspots) and local time. Thus, ionospheric scintillation occurs as short-term, rapid signal fluctuations and is mainly caused by irregularities in the ionosphere ranging from altitudes of 200–600 km. In fact, the frequency-dependence depends on the ionospheric conditions but the attenuation varies approximately at the same rate as the square of the wavelengths. The effect is greater for lower frequencies and at lower latitudes; while high latitude areas near the Arctic polar region bounded between ±20° are susceptible to intense scintillation activity. In the L and S-band, this effect can be ignored at medium latitudes except during periods of solar activity. When the Sun is very active, L-band enhancement and fading of 6 dB and −36 dB, respectively were observed even at 37° latitude. Scintillation activity is at a maximum during the night, lasting from 30 min to a number of hours.

7.7.6.3 Other Ionospheric Effects

1. **Dispersion** – When transionospheric radio signals occupy large bandwidth the propagation delay, being a function of frequency, introduces dispersion. The differential delay across the bandwidth is proportional to the integrated electron density along the ray path. Hence, for an integrated electron content of 5×10^{17} el/m^2, a signal with a pulse length of 1 μs will sustain a differential delay of 0.02 μs at 200 MHz, while at 600 MHz delay would be only 0.00074 μs.
2. **Refraction** – When radiowaves propagate obliquely through the ionosphere layer they undergo refraction, which produces a change in the direction of arrival of the ray.
3. **Absorption** – For equatorial and mid-latitude regions, radiowaves of frequencies above 70 MHz will assure penetration of the ionosphere without significant absorption, while for frequencies below 70 MHz the ionospheric absorption loss is significant.
4. **Doppler Frequency Shift** – The special effects of frequency change due to the temporal variability of the ionosphere layer upon the apparent frequency of the carrier (the Doppler shifted carrier). For example, at f = 1.6 GHz (GPS system), the observed frequency change Δf at high latitude is: $\Delta f/f < 10^{-9}$.

7.7.6.4 Sky Noise Temperature Contributions

The mechanism that causes absorption of energy from a wave passing between space and the Earth also causes the emission of thermal noise at RF. In fact, some radio noise is added to the emission reaching the receiver, whereas the Earth itself radiates noise, which can enter the transmission path via satellite or the GES receiving antennas. Therefore, the radio noise emitted by all matter, while used as a source of information in radio astronomy and remote sensing, may be a limiting factor in communication services. Otherwise, sources of radio noise of interest on Earth-to-space paths are the atmosphere, clouds, rain, extraterrestrial sources and noise from the surface of the Earth. Prediction methods are given in the ITU-R P.372 Recommendation. The thermal noise power N, available from a blackbody having a noise temperature of the source T [K], measured in bandwidth B [Hz], is given by the form:

$$N = kTB \ [W] \tag{7.42}$$

where k = Boltzmann's Constant. The special power density N_0 of noise from source is:

$$N_0 = N/B = kT \ [WHz^{-1}] \tag{7.43}$$

In considering the level of noise received at an GES or satellite from sources external to the environment, it is convenient to identify a brightness temperature T_B

for each separate source and a coefficient (η), which represents the efficiency that receiving antenna captures noise from that source. The noise temperature component (t) due to the identified source is given:

$$t = \eta T_B \ [K] \tag{7.44}$$

Therefore, the total noise entering the system from all of these sources, expressed as a noise temperature, can be obtained by summing all the component noise temperatures.

7.7.6.5 Environmental Noise Temperature Sources

The GES antennas in GEO satellite infrastructures are typically designed and sited so that the main lobe does not intersect the local terrain or obstructions, such as mountains or large buildings. Side lobes are also minimized to reduce the effect of the Earth's temperatures on the system performance. However, in satellite systems, the antenna beam may pass through vegetation and be obstructed by buildings or mountain terrain. Measurements suggest that the impact of the additional terrestrial noise is greater when the antenna has a low internal noise temperature, that is, for less directive antennas. Although these obstructions will raise the noise temperature seen by the antenna, they will also cause shadowing or multipath effects, which are likely to be more significant in the total link performance.

Industrial man-made sources of noise affect VHF and UHF frequencies for all but the quietest rural areas. Unlike other noise effects, there is a polarization dependence in that the vertical component is higher that the horizontal. In general, the median level of noise will decrease linearly with log (f). There are significant variations with location and time and little data are available to develop models to predict levels. Thus, considering the noise received at an GES from the ground, the sea etc. and buildings nearby, T_B typically lies between 100 and 250 K, approaching the lower limit at sea at low angles of elevation angles and the upper limit on land. The relation for this noise temperature component t_{gr} is same as for t.

If the angle of elevation of the main beam and the gain of the antenna are both high and the antenna design is good, with well-suppressed side lobes and little sub reflector spillover, the corresponding value of η will be small and t_{gr} may be no more than 20 K.

If, however, the gain of the antenna is very low, as is typical of GES, η may reach 0.5 and t_{gr} may exceed 100 K. On the other hand, satellite antennas directed towards the Earth, having sufficient gain for the main lobe to be filled by the Earth, will also receive terrestrial noise with $\eta \approx 1$, while T_B will be about 210 K if land occupies a large fraction of the beam footprint. However, except in the atmospheric absorption bands, T will be somewhat less, perhaps as little as 160 K, if the sea occupies a large part of the footprint.

7.7.6.6 Atmospheric Noise Temperature Elements

The noise temperature of a satellite-based antenna is dominated by the high temperature emitted by the Earth, which fills, or mostly fills, the main beam of the antenna. Additional noise from precipitation or other variables is insignificant in this case. For a global beam, the noise temperatures are dependent both on frequency and on the position of the satellite with relation to the major landmasses of the Earth.

The ground antenna observes the relatively cool sky and therefore, the presence of clouds and rain can significantly raise the noise temperature of the antenna. In general, the brightness temperature of the atmosphere due to the permanent gases and rain if it is present and seen by an antenna, is given as:

$$T_B = T_m \left(1 - 10^{-A/10} \right) \ [K] \tag{7.45}$$

where T_m = effective temperature of the attenuating medium (atmosphere, clouds, rain), typically about 270 K and A = total attenuation due to the medium. The effect of rain on a satellite downlink is not just the attenuation but also decrease in C/N due to the higher noise temperature seen in rainy conditions, compared to clear sky conditions. In some cases, the noise temperature increase can have more effect on the link than the attenuation itself.

7.7.6.7 Galactic and Other Interplanetary Noise Effects

Noise from interplanetary sources, particularly the Sun, the Moon and from the galactic background, is well understood and the effect on the total extraterrestrial noise temperature of a system can be calculated with the following relation:

$$T_B = T_g \times 10^{-A/10} \ [K] \tag{7.46}$$

where T_g = temperature of any interplanetary radio sources, including background Galactic noise (about 3 K above 3 GHz). Thus, the brightness temperature of the Sun decreases with increasing frequency, from about 10^6 at 30 MHz to 10^4 at 10 GHz under quiet conditions. At 20 GHz, an antenna of 2 m diameter and a beam width of about 0.5° would have an increase in noise temperature of about 8.100 K with a quiet Sun. The Sun and Moon each sub tend an angle of about 0.5°, so that if the antenna beam is significantly larger than that, the effect t of the Sun or Moon is averaged with a larger portion of relatively cool sky.

7.7.7 Path Depolarization Causes

The atmosphere behaves as an anisotropic medium for any radio propagation. Consequently power from one polarization is coupled to its orthogonal component, causing interference between the channels of a dual polarized system.

In this sense, depolarization or cross-polarization may occur when EM waves propagate through media that are anisotropic, namely asymmetrical with respect to the incident of polarization. Meanwhile, depolarization in the form of Faraday rotation of the plane of linear polarization occurs in the ionosphere because of presence of the Earth's magnetic fields. At this point, the resulting impairments are typically circumvented by using circular polarization at frequencies below 10 GHz, for which the effect can be significant. Depolarization is often the most significant path impairment for 6/4 GHz satellite systems and can be the limiting performance factor for some 14/11 GHz satellite paths, especially at lower path elevation angles in moderate rain climates.

On the other hand, depolarization in precipitation is caused by differential attenuation and phase shifts that are induced between orthogonal components of an incident wave by anisotropic hydrometeors. Orthogonally polarized radiowaves propagating in a medium that causes only differential phase shift are depolarized but maintain orthogonality. If the medium induces differential attenuation, the waves are also deorthogonalized.

7.7.7.1 Depolarization and Polarization Components

The importance of depolarization for satellite communications systems depends on a few components: f = frequency signal, geometry of path (θ = elevation angle and τ = tilt angle of the received polarization), local climatic factors (severity of the rain) and sensitivity to cross-polar interference (whether the system employs frequency reuse).

The EM waves comprise both the electric and magnetic field vectors. Therefore, these two components travel in the direction of the transmission path and are orthogonal, while the orientation of the electric field vector defines the polarization of the transmitted waves. In general, as the wave progresses in time the tip of the electric vector traces an ellipse in a plane perpendicular to the propagation direction. A representative polarization ellipse of representative elliptically polarized radiowave is displayed in Fig. 7.47.

Two important parameters in this figure are the axial ratio and τ = the inclination tilted angle with respect to the reference axis. The polarization ellipse may be tilted at an angle (τ) with respect to the particular coordinate frame. Thus, the general form of a polarized wave, when viewed perpendicular to the direction of travel, is elliptical in shape.

The polarization state of a wave is completely specified by its polarization ellipse, i.e., the amplitudes of the major axis (E_{max}), the minor axis (E_{min}) and the

Fig. 7.47 Generalized
elliptical waveform
(Courtesy of Book: by
Sheriff)

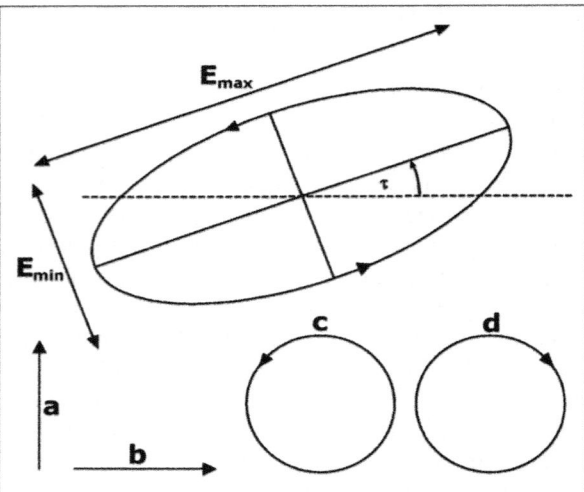

sense of rotation of the vector also defines the axial ratio by using the following expression:

$$A_R = 20\log (E_{max}/E_{min}) \ [dB] \tag{7.47}$$

In satellite communications four types of polarization are employed, such as: (**a**) Vertical Linear Polarization (VLP); (**b**) Horizontal Linear Polarization (HLP); (**c**) Left-Hand Circular Polarization (LHCP) and (**d**) Right-Hand Circular Polarization (RHCP). The direction of the travel is symbolical "into the paper".

Horizontal and vertical polarizations are defined with respect to the horizon, LHCP has an anticlockwise and RHCP has a clockwise rotation when viewed from the antenna in the direction of travel.

Therefore, if E_{max} and E_{min} are equal in magnitude, the polarization state can be RHCP or LHCP, depending on the sense of rotation and in the case where E_{max} is nonzero and E_{min} is zero, the value of electric vector maintains a constant orientation defined by E_{max} and the polarization state is said to be linear.

The polarization quantity of interest for frequency re-use satellite communication systems is the Cross-Polar Isolation (XPI), defined as the decibel ratio of the (desired) co-polar power received in a channel to the (undesired) cross-polar power received in that same channel.

In practice, however, XPI is difficult to measure because the cross-polarized components cannot be distinguished from noise in the co-polar channel. The quantity usually measured is the Cross-Polar Discrimination (XPD), defined as the ratio of the co-polarized power received in one channel to the cross-polarized power detected in the orthogonal channel, both arising from the same transmitted signal.

Moreover, theory predicts that XPD and XPI components are equivalent for most practical situations. In fact, a polarized wave will comprise the wanted polarization together with some energy transmitted on the orthogonal polarization. The degree of the coupling of energy between polarizations is given by:

$$XPD = 20\log \left| E_{cpr}/E_{xpr} \right| \ [dB] \tag{7.48}$$

where parameters E_{cpr} = received co-polarized electric field strength and E_{xpr} = received cross-polarized electric field strength. For the case of coexisting circular polarized waves the XPD can be determined from the axial ratio by the expression:

$$XPD = 20 \ \log \ (A_R + 1/A_R - 1) \ [dB] \tag{7.49}$$

7.7.7.2   Relation between Depolarization and Attenuation

The ITU provides a step-by-step method for the calculation of hydrometer-induced cross-polarization, which is valid for frequencies within the range 8 GHz < f < 35 GHz and for elevation angles less than 60°. In addition to operating frequency and elevation angle, the attenuation due to rain exceeded for the required percentage of time p and the polarization tilt angle τ, with respect to the horizontal, also needs to be known. The XPD due to rain is given by the following equation:

$$XPD_{rain} = C_f \ - C_A + C_\tau + C_\theta + C_\sigma \ [dB] \tag{7.50}$$

where C_f = frequency-dependent term or 30 logf [GHz]; C_A = rain-dependent term or V(f) $\log A_p$; C_τ = polarization improvement factor or $-10\log [1-0.484 (1 + \cos 4\tau)]$; C_θ = elevation angle-dependent term or $-10\log (\cos \theta)$ and C_σ = canting angle term or 0.0052σ. In the above, the canting angle refers to the angle at which a falling raindrop arrives at the Earth with respect to the local horizon.

Terms τ, θ and σ are expressed in degrees, while σ has value of 0°, 5°, 10° and 15° for 1, 0.1, 0.01 and 0.001% of the time, respectively. Taking ice into account, the XPD not exceeded for p% of time is given by:

$$XPD_p = XPD_{rain} - C_{ice} \ [dB] \tag{7.51}$$

where C_{ice} = ice depolarization or XPD_{rain} x (0.3 + 0.1 logp)/2.

7.7.8 Surface Reflection and Local Environmental Effects

Surface reflections and local environmental effects are important for satellite communications because such factors generally tend to impair the performance of satellite communications links, although signal enhancements are also occasionally observed. Local environmental effects include shadowing and blockage from objects and vegetation near the UES.

At this point, surface reflections are generated either in the immediate vicinity of the UES terminals or from distant reflectors, such as mountains and large industrial infrastructures. In fact, the reflected transmission signal can interfere with the direct signal from the satellite to produce unacceptable levels of signal degradation.

The impact of the impairments depends on the specific application, namely in the case of typical UES or GES links, all measurements and theoretical analysis indicate that the specular reflection component is usually negligible for path elevation angles above 20°. Moreover, for handheld terminals, specular reflections may be important as the low antenna directivity increases the potential for significant specular reflection effects. For satellite system links, design reflection multipath fading, in combination with possible shadowing and blockage of the direct signal from the satellite, is generally the dominant system impairment.

7.7.9 Reflection from the Earth's Surface

Prediction of the propagation impairments caused by reflections from the Earth's surface and from different objects (buildings, hills, mountains vegetation) on the surface is difficult because the possible impairment scenarios are quite numerous, complex and often cannot be easily quantified. Thus, the basic features of surface reflections and the resultant effects on propagating signals can be understood in terms of the general theory of surface reflections, as summarized in the following classification:

1. **Specular Reflection from a Plane Earth** – The specular reflection coefficient for vertical polarization is less than or equal to the coefficient for horizontal polarization. Polarization of the reflected waves is different of that from the incident wave if the incident polarization is not purely horizontal or purely vertical.
2. **Specular Reflection from a Smooth Spherical Earth** – Here, the incident grazing angle is equal to the angle of reflection. The amplitude of the reflected signal is equal to the amplitude of the incident signal multiplied by the modules of the reflection coefficient.
3. **Divergence Factor** – When rays are specularly reflected from a spherical surface, there is an effective reduction in the reflection coefficient, which is actually a geometrical effect arising from the divergence of the rays.

4. **Reflection from Rough Surface** – In many practical cases, the surface of the Earth is not smooth. Namely, when the surface is rough, the reflected signal has two components: one is a specular component, which is coherent with the incident signal, while the other is a diffuse component, which fluctuates in amplitude and phase with a Rayleigh distribution.
5. **Total Reflected Field** – The total field above a reflecting surface is a result of the direct field, the coherent specular component and the random diffuse component.
6. **Reflection Multipath** – Owing to the existence of surface reflection phenomena signals may arrive at a receiver from multiple apparent sources. Thus, the combination of the direct signal (line-of-sight) with specular and diffusely reflected waves causes signal fading at the receiver. The resultant multipath fading, in combination with varying levels of shadowing and blockage of the line-of-sight components, can cause the received signal power to fade severely and rapidly for UES and is really the dominant impairment in the satellite service.

Bibliography

1. Books

1. Ackerman, S.A., Knox, J.A.: Meteorology – Understanding the Atmosphere. Jones & Bartlett Learning, Burlington (2015)
2. Aragon-Zavala, A., et al.: High-Altitude Platforms for Wireless Communications. Wiley, Chichester (2008)
3. Asrar, G.R., Hurrell, J.W.: Climate Science for Serving Society. Springer, Dordrecht (2013)
4. Bajorski, P.: Statistics for Imaging, Optics and Photonics. Springer Boston (2011)
5. Berlin, P.: Geostationary Applications Satellite. Cambridge University Press, Cambridge, UK (1988)
6. Blonstein, L.: Communications Satellites, The Technology of Space Communications. Heinemann, London (1987)
7. Bowman, K., Yang, P.: Satellite Meteorology and Atmospheric Remote Sensing – An Introduction. Wiley, Chichester (2009)
8. Buglia, J.: Introduction to the Theory of Atmospheric Radiative Transfe. Langley Research Center, Hampton (1986)
9. Calcutt, D., Tetley, L.: Satellite Communications, Principles and Applications. Elsevier, Oxford (2004)
11. Chen, H.S.: Space Remote Sensing Systems. Academic, Washington, DC (2014)
12. Chuvieco, E., Huete, A.: Fundamentals of Satellite Remote Sensing. CRC Press – Taylor & Francis Publishers, Boca Raton (2009)
13. Conway, E.D.: An Introduction to Satellite Image Interpretation. Johns Hopkins University Press, Baltimore (1997)
14. Dalgleish, D.I.: An Introduction to Satellite Communications. IEE, Peter Peregrinus, London (1989)
15. Denegre, J.: Thematic Mapping from Satellite Imagery – An International Report. Pergamon Press, Oxford, UK (2013)
16. Dobrzykowski, S.: Advanced Topics in Satellite Meteorology and Remote Sensing. CreateSpace Independent Publishing Platform, North Charleston (2015)
17. Elbert, B.R.: Ground Segment and Earth Station Handbook. Artech House, Boston/London (2001)
18. Evans, B.G.: Satellite Communication Systems. IEE, Peter Peregrinus, London (1991)
19. Everett, J.: VSAT- Very Small Aperture Terminals. IEE, Peter Peregrinus, London (1992)

© Springer International Publishing AG 2018
S.D. Ilčev, *Global Satellite Meteorological Observation (GSMO) Theory*,
DOI 10.1007/978-3-319-67119-2

20. Feher, K.: Digital Communications, Satellite Earth Station Engineering. Prentice-Hall, Englewood Cliffs (1983)
21. Ferretti, A.: Satellite InSAR Data – Reservoir Monitoring from Space. EAGE Publishers, Houten (2014)
22. Fujimoto, K.: Mobile Antenna Systems Handbook. Artech House, London (2008)
23. Gagliardi, R.M.: Satellite Communications. Van Nostrand Reinhold, New York (1984)
24. Galic, R.: Telekomunikacije satelitima. Skolska knjiga, Zagreb (1983)
25. Georgiev, C., Santurette, P.: Weather Analysis and Forecasting – Applying Satellite Water Vapor Imagery and Potential Vorticity Analysis. Academic, Washington, DC (2016)
26. Ghasemi, A., et al.: Propagation Engineering in Wireless Communications. Springer, Boston (2016)
27. Gordon, G.D., et al.: Principles of Communications Satellites. Wiley, Chichester (1993)
28. Grace, D., Mohorcic, M.: Broadband Communications via High-Altitude Platforms. Wiley, Chichester (2011)
29. Group of Authors: Earth Observations from Space: The First 50 Years of Scientific Achievements. National Academic Press, Washington, DC (2008)
30. Group of Authors: Handbook on Satellite Communications. ITU, Geneva (2002)
31. Group of Authors: Radiowave Propagation Information for Predictions for Earth-to-space Path Communications – Handbook. Radiocommunication Bureau, ITU, Geneva (1996)
32. Group of Authors: Use of Radio Spectrum for Weather, Water and Climate Monitoring and Prediction. ITU, Geneva (2008)
33. Haan, S.: Meteorological Applications of a Surface Network of GPS Receivers. Universiteit Utrecht, Utrecht (2008)
34. Handlemann, E., Whitcomb, J.: NPOESS Activity Book. NASA, Washington, DC (2007)
35. Heng, O.: An Introduction to Contemporary Remote Sensing. McGraw-Hill, New York (2012)
36. Hou, A.Y., et al.: The Global Precipitation Measurement Mission. American Meteorological Society, Boston (2014)
37. Hutchison, K.D., Cracknell, A.P.: Visible Infrared Imager Radiometer Suite – A New Operational Cloud Imager. CRC Press, Boca Raton (2005)
38. Huurdeman, A.A.: Guide to Telecommunications Transmission Systems. Artech House, New York/Boston/London (1997)
39. Ilcev, D.S.: Global Aeronautical Communications, Navigation and Surveillance (CNS). Theory, vol. 1. AIAA, Reston (2013)
40. Ilcev, D.S.: Global Aeronautical Communications, Navigation and Surveillance (CNS). Applications, vol. 2. AIAA, Reston (2013)
41. Ilcev, D.S.: Global Mobile Satellite Communications for Maritime, Land and Aeronautical Applications. Springer, Boston (2005)
42. Ilcev, D.S.: Global Mobile Satellite Communications for Maritime, Land and Aeronautical Applications. Theory, vol. 1. Springer, Boston (2016)
43. Ilcev, D.S.: Global Mobile Satellite Communications for Maritime, Land and Aeronautical Applications, vol. 2. Springer, Boston (2017)
44. Jorge, F., Carrasco, J.F., et al.: Satellite and Automatic Weather Station Analyses of Katabatic Surges Across the Ross Ice Shelf. Wiley, Chichester (2013)
45. Kadish, J.E., et al.: Satellite Communications Fundamentals. Artech House, Boston/London (2000)
46. Kantor, L.Y., et al.: Sputnikovaya svyaz i problema geostacionarnoy orbiti. Radio i svyaz, Moskva, USSR, Moscow (1988)
47. Karthik, N.T., Minden, G.: Earth Observation System Satellite Communication Characteristics. NASA, Lawrence (2002)
48. Kelkar, R.R.: Satellite Meteorology. BS Publications, Hyderabad (2007)
49. Kidder, S.Q., Ham, T.C.: Satellite Meteorology – An Introduction. Academic, Sand Diego (1995)

50. Kramer, H.J.: Observation of the Earth and Its Environment: Survey of Missions and Sensors. Springer, Boston (2002)
51. Lasaponara, R., Masini, N.: Remote Sensing and Digital Image Processing. Springer, Boston/New York (2012)
52. Liu, G.: Satellite Meteorology. Department of Meteorology, Florida State University, Tallahassee (2005)
53. Liu, J.G., Mason, P.J.: Image Processing and GIS for Remote Sensing – Techniques and Applications. Wiley, Chichester (2016)
54. Levis, C., et al.: Radiowave Propagation – Physics and Applications. Wiley, Chichester (2010)
55. Lubin, D., Massom, R.: Polar Remote Sensing – Atmosphere and Oceans, vol. I. Springer, Boston (2006)
56. Lubin, D., Massom, R.: Polar Remote Sensing – Ice Sheets, vol. II. Springer, Boston (2006)
57. Maini, A.K., Agrawal, V.: Satellite Technology – Principles and Applications. Wiley, Chichester (2007)
58. Maral, G., et al.: Satellite Communications Systems. Wiley, Chichester (2009)
25. Menzel, P.: Remote Sensing with Meteorological Satellites. University of Wisconsin Academic Press, Medison (2012)
59. Milligan, T.: Modern Antenna Design. Wiley, Chichester (2005)
60. Modest, M.F.: Radiative Heat Transfer. Academic, Washington, DC (2003)
61. Njoku, E.G.: Encyclopedia of Remote Sensing Earth Sciences Series. Springer, New York (2000)
62. Ohmori, S., et al.: Mobile Satellite Communications. Artech House, Boston/London (1998)
63. Ohring, G.: Achieving Satellite Instrument Calibration for Climate Change ($ASIC^3$). Workshop Report, NOAA, Virginia (2007)
64. Paine, D.P., Kiser, J.D.: Aerial Photography and Image Interpretation. Wiley, Chichester (2012)
65. Pascall, S.P., et al.: Commercial Satellite Communications. Focal Press, Oxford (1997)
66. Passow, M.J., Kiehl: Heating the Atmosphere. Presentation, Dwight Morrow High School, Lamont-Doherty Earth Observatory, Englewood (2010)
67. Pelton, E., et al.: Handbook of Satellite Applications. Springer, Boston (2017)
68. Pidwirny, M., Jones, S.: The Fundamentals of Physical Geography. University of British Columbia Okanagan, Kelowna (2006)
69. Pidwirny, M.: Understanding Physical Geography. Our Planet Earth Publishing, Kelowna (2015)
70. Pratt, T., et al.: Satellite Communications. Wiley, Hoboken (2002)
71. Qu, J.J., et al.: Satellite-Based Applications on Climate Change. Springer, Boston (2015)
72. Raizer, V.: Advances in Passive Microwave Remote Sensing of Oceans. CRC Press – Taylor & Francis Publishers, Boca Raton (2017)
73. Rao, P.K., et al.: Weather satellites: Systems, Data and Environmental Applications. American Meteorological Society, Boston (1990)
74. Raghavan, S.: Radar Meteorology. Springer, Boston (2003)
75. Rees, G.: The Remote Sensing Data Book. Cambridge University Press, Cambridge, UK (1990)
76. Richards, J.A.: Remote Sensing with Imaging Radar. Springer, Boston (2009)
77. Richharia, M.: Satellite Communications System – Design principles. Macmillan, Basingstoke (1995)
78. Robinson, P.J., Henderson-Sellers, A.: Contemporary Climatology. Eoutledge, London (1999)
79. Roddy, D.: Satellite Communications. McGraw Hill, New York (2006)
80. Schott, J.R.: Remote Sensing. Oxford University Press, Oxford (1997)
81. Shang, L.D., Gong, J.: Geospatial Technology for Earth Observation. Springer, Boston (2009)

82. Sheriff, R.E., et al.: Mobile Satellite Communication Networks. Wiley, Chichester (2001)
83. Solimini, D.: Understanding Earth Observation: The Electromagnetic Foundation of Remote Sensing. Springer, Boston (2016)
84. Solovev, V.I., et al.: Svyaz na more. Sudostroenie, Leningrad (1978)
85. Sportisse, B.: Fundamentals in Air Pollution – From Processes to Modelling. Springer, Dordrecht (2008)
86. Stacey, D.: Aeronautical Radio Communication Systems and Networks. Wiley, Chichester (2008)
87. Tan, S.Y.: Meteorological Satellite Systems. Springer, New York (2014)
88. Tetley, L., et al.: Electronic aids to navigation – Position fixing. Edward Arnold, London (1991)
89. Ulaby, E.T., et al.: Microwave Radar and Radiometric Remote Sensing. NASA, California Institute of Technology (CIT), Pasadena (2014)
90. Vaughan, R.A.: Remote Sensing Applications in Meteorology and Climatology. Springer, Boston (2012)
91. Wallace, J., Hobbs, P.: Atmospheric Science – An Introductory Survey. Academic, Washington, DC (2006)
92. Wang, W.-Q.: Near-Space Remote Sensing. Springer, New York (2011)
93. Wendisch, M., Yang, P.: Theory of Atmospheric Radiative Transfer. Wiley, Chichester (2012)
94. Wertz, J.R.: Spacecraft Attitude Determination and Control. Springer, Boston (2002)
95. Zhilin, V.A.: Mezhdunarodnaya sputnikova sistema morskoy svyazi – Inmarsat. Sudostroenie, Leningrad, USSR (1988)

2. Papers

1. Antonini, M., et al.: Stratospheric Relay: Potentialities of New Satellite-high Altitude Platforms Integrated Scenarios. IEEEAC, Montana, USA (2003)
2. Benesch, W.: 60 Years Operational Satellites up to 2030. Deutscher Wetterdienst, Offenbach am Main (2010)
3. Blyenburgh Van, P.: UAV Systems: Global Review. Avionics 2006 Conference, 1–52, Amsterdam (2006)
4. Branski, F.: Pioneering the Collection and Exchange of Meteorological Data. WMO Bulletin, Geneva (2010)
5. Davis, G.: History of the NOAA Satellite Program. NOAA Satellite and Information, Silver Spring (2000)
6. Dong, C., et al.: An Overview of a New Chinese Weather Satellite FY-3A. American Meteorological Society, Boston (2009)
7. Faller, K.: MTSAT-1R: A Multifunctional Satellite for Japan and the Asia-Pacific Region. Proceedings of the 56th IAC 2005, Fukuoda (2005)
8. Gatenby, P.V., et al.: Optical intersatellite links. ECE J. 3(6). IEE, Stevenage (1991)
9. Group of Authors: Advanced VHF Ground Station for NOAA Weather Satellite APT Image Reception. Acta Technica Napocencis, Cluj-Napoca, Romania (2012)
10. Group of Authors: Airships. Keck Institute for Space Studies, Pasadena (2013)
11. Ilcev, D.S.: Satellite DVB-RCS standards for fixed and mobile commercial and military applications. Microwave J. Norwood (2009)
12. Ilcev, D.S.: Stratospheric Communication Platforms (SCP) as an alternative for space program. Aircraft Eng. Aerospace Technol. AEAT J. 83(2), 105–111 (2011)
13. Ilcev, D.S., Sibiya, S.S.: Weather observation via stratospheric platform stations. IST-Africa Conference, Lilongwe, Malawi, 2015

14. Ilchenko, M.E.: Application of high-altitude platform systems in regions of disaster and emergency. 14th International Crimean Conference on Microwave and Telecommunications Technology, Sevastopol, 2004.

15. Kamel, A.: Japanese Advanced Meteorological Imager (JAMI). Space Systems, Loral, Palo Alto (2003)

16. Kiehl, J., et al.: Earth's Global Energy. National Center for Atmospheric Research, Boulder (2009)

17. Knapek, M., et al.: Optical High-Capacity Satellite Downlinks via High-Altitude Platform Relays. German Aerospace Center DLR, Wessling (2008)

18. Liew, S.C.: Radiative Transfer Equation. National University of Singapore (NUS), Centre for Remote Imaging, Sensing and Processing (CRISP), Singapore (2003)

19. Pace, P., et al.: An integrated satellite HAP-terrestrial system architecture: Resource allocation and traffic management issues. IEEE 59th Vehicular Technology Conference, 2004.

20. Pranajaya, F.M., et al.: Nanosatellite Tracking Ships. Space Flight Laboratory (SFL), Toronto (2011)

21. Uesawa, D.: Status of Japanese meteorological satellites and recent activities of MSC. Proceedings of the 2006 Eumetsat Meteorological Satellite Conference, Helsinki, 2006

22. Zaharov, V., et al.: Smart antenna application for satellite communications with SDMA. J. Radio Electron. 2. Iztapalapa, Mexico (2001)

23. Widiawan A.K., et al.: Coexistence of high altitude platform station, satellite and terrestrial systems for fixed and mobile services. International Workshop on HAP Systems, Athens, 2005

3. Manuals

1. Asmus, V.V.: Status of current and planned Russian Meteorological Satellite Systems. SRC Planeta

2. Ayecka: Advanced DVB-S2 Receiver. Hod HaSharon (2015)

3. Evans, C.: Tiros-1/2 Ground Station. Camp Evans Project Diana Site Wall (2010)

4. CHEOS: The Space Earth Observation of China. NSMC, CMA, Beijing (2012)

5. Colozza, A., Dolce, J.L.: High-Altitude, Long-Endurance Airships for Coastal Surveillance. NASA, Cleveland (2005)

6. Davis, G.: History of the NOAA Satellite Program, Silver Spring (2011)

7. Wind, D.: The History and Science of Hurricanes. Oxford University Press, Oxford, UK (2005)

8. Eumetsat: TD 18 Metop Direct Readout AHRPT Technical Description. Darmstadt (2013)

8. Eumetsat: An Introduction to EUMETSAT Polar System. Darmstadt (2010)

9. Eumetsat: A New Geostationary Meteorological Satellite System for the 21s t Century. Darmstadt (2009)

10. Eumetsat: International Data Collection System User's Guide. Darmstadt (2009)

11. Eumetsat: Meteosat Receiving Multi Satellite Data from the EUMETCast DVB-S2 Service. Darmstadt (2009)

12. Eumetsat: TD 16 – Meteosat Data Collection and Distribution Service. Darmstadt (2009)

13. Eumetsat: The Meteosat System – Satellite Ground Segment Mission – Global Coordination. Darmstadt (2000)

14. Festo, D.: Satellite Orbits, Coverage and Antenna Alignment. Quebec (2015)

15. GAO: Geostationary Weather Satellites. Government Accountability Office, Washington, DC (2013)

16. GSFC: GOES I-M Data Book. National Aeronautics and Space Administration, Greenbelt (1996)

17. GSFC: GOES-N Series Data Book. National Aeronautics and Space Administration, Greenbelt (2009)

18. Holm, T.: Landsat: Building a Future on 40 Years of Success. EROS Centre, Alameda (2013)
19. Ilcev, D.S.: Waether Observation via Stratospheric Platform Systems (SPS). Manual, SSC, DUT (2016)
20. Ilcev, D.S.: Global Mobile CNS. Manual, DUT (2011)
21. Ilcev, D.S.: Satellite Meteorological Observation System. Manual, DUT (2015)
22. Ilcev, D.S.: Space Meteorology. Manual, SSC, DUT (2017)
23. ISRO. INSAT 3-D Observing Weather from Space. ISRO, Bangalore (2014)
24. Jayaraman, V.: Oceansat-2. ISRO, Bangalore (2014)
25. JMA: MTSAT Multifunction Transport Satellite. Japan Meteorological Agency, Tokyo (2015)
26. ISRO: INSAT 3-D and Cartosat-1. NRSC, Bangalore (2015)
27. Jones, J.: GNSS and Weather Forecasting. Met Office, Potsdam (2013)
28. Kiesiläinen, S.: Improving Data Collection and Management Process of Weather Observation. Master thesis, Helsinki Metropolia University of Applied Sciences, Helsinki (2011)
29. Kongsberg: MEOS 3.0 m to 5.0 m L/S/X-band Antenna. Tromso (2014)
30. Kongsberg: MEOS GEO DVB-A2 HimawariCast. Tromso (2015)
31. Kongsberg: MEOS PEO and GEO Geostationary Ground Station. Tromso (2015)
32. Krovotyntsev, V.: Perspectives of Development of the Russian Meteorological Satellite Constellation. Roshydromet, SRC Planeta, Moscow (2009)
33. Laing, A.: Introduction to Tropical Meteorology – Chapter 2: Remote Sensing. Pennsylvania State University, University Park (2011)
34. Malgorzata, V.W.: Data Acquisition and Integration in Remote Sensing. Oktatasi Hivatal, Budapest (2010)
35. McGinnis, F.D., Dries, M.: Meteorological Satellite Communications. NASA/Eumetsat, Washington, DC (2009)
36. Minihold, R., Dieter, B.D.: Introduction to Radar System and Component Tests. Rohde & Schwarz, München (2012)
37. NASA: Earth Observations from Space – The First 50 Years of Scientific Achievements. National Academies Press, Washington, DC (1991)
38. NOAA: Future of NOAA's Direct Readout and Direct Broadcast Services. Washington, DC (2012)
39. NOAA: GOES-N/O/P The Next Generation. NESDIS, Suitland (2008)
40. NOAA: NOAA-N Satellite. Suitland (2004)
41. NOAA/NESDIS: Antenna and RF Systems Capabilities Handbook. Suitland (2008)
42. NOAA/NESDIS: User's Guide for Building and Operating Environmental Satellite Receiving Stations. Suitland (2009)
43. Peet, M.M.: Spacecraft Dynamics and Control. Arizona State University, Tempe (2010)
44. Peitso, P.: Space Weather Instruments and Measurement Platforms. Master thesis, School of Electrical Engineering, University Finland, Espoo (2013)
45. Reckeweg, M., Rohner, C.: Antenna Basic. Rohde & Schwarz, München (2015)
46. Roscosmos: Main Components of Russian Electro GEO Satellite. Moscow (2015)
47. Roscosmos: Status of Current and Future Russian Satellite Systems. Roshydromet, Moscow (2010)
48. Roshydromet: High-elliptical Orbits Satellite System Arctica. SRC Planeta, Moscow (2014)
49. Roshydromet: Status of Current and Future Roshydromet Satellite Programmes. State Research Centre on Space Hydrometeorology "Planeta", Moscow (2009)
50. Rublev, A.: Development and Application Perspectives of Space-Based Components in Roshydromet Observation Network. Roshydromet (State Research Centre PLANETA), Moscow (2014)
51. Schmetz, J.: An Introduction to Meteosat Second Generation (MSG). American Meteorological Society, Washington, DC (2002)
52. Schmid, J.: The SEVIRI Instrument. ESA/ESTEC, Noordwijk (2011)
53. Sea Launch: User's Guide. Long Beach (1996)

54. Sihle, S., Sibiya, S.S.: Integration of Satellite System and Stratospheric Communication Platforms (SCP) for Weather Observation. Doctoral thesis, Durban University of Technology (DUT), Durban (2016)
55. Salas, F.R., Maidment, D.R.: Worldwide Exchange of Water Data. Center for Research in Water Recourses, University of Texas, Austin (2014)
56. Singh, D.: Status Report on Current and Future Geostationary Indian Satellites. Satellite Meteorology, New Delhi (2014)
57. Tang, Y., et al.: FY-3A Meteorological Satellite and the Applications. CMA, Beijing (2014)
58. NOAA: User's Guide for Building and Operating Environmental Satellite Receiving Stations. NESDIS, Suitland (2009)
59. Valero, D.S.: Rule-Based System Architecting of Earth Observation Satellite Systems. Doctoral thesis, Massachusetts Institute of Technology (MIT), Cambridge (2012)
60. West, R.: Soil Moisture Active and Passive Mission (SMAP). NOAA, Washington, DC (2014)
61. WMO: Guide to the Observing System. World Meteorological Organization, Geneva (2013)
62. Xuebao, W.: Fengyun Bulletin. NSMC, CMA, Beijing (2010)

4. Brochures

1. Advantech: Discovery 300 Next Generation DVB-RCS VSAT Hub. Montreal (2015)
2. Advantech: Fixed, Mobile and Portable VSAT Units. Montreal (2012)
3. Advantech: Ground Satellite Antennas. Montreal (2006)
4. Advantech: Millennium Gen II Series High Capacity VSAT Hub. Montreal (2015)
5. Advantech: Theoretical Performance of DVB-S2 and DVB-S. Montreal (2009)
5. Advantech: Discovery 300 – Next Generation DVB-RCS VSAT Hub DVB-S2/S2X and DVB-RCS/2. Montreal (2008)
6. Advantech: ASAT II System Multiservice and Multi-Waveform VSAT Platform. Montreal (2010)
7. Advantech: VSAT DVB-RCS, DVBSCPC/RCS, DVB-S/DVB-S2 Terminals. Montreal (2010)
8. Adler, B.R., Huffman, G.: Global Precipitation Climatology Project. GSFC/NASA, Climate Guide, UCAR/NCAR, Boulder (2015)
9. Carana, S.: Arctic warming due to snow and ice demise – sea ice. Albedo Change http://albedochange.blogspot.co.za/ (2014)
10. COMET: Satellite Monitoring. UCAR, Boulder. http://www.goes-r.gov/users/comet/volcanic_ash/tools/navmenu.php_tab_1_page_2.0.0.htm (2011)
11. Dartcom: GVAR System. Yelverton, Devon (2015)
12. Dartcom: HRPT/AHRPT System. Yelverton, Devon (2015)
13. Dartcom: HRPT, CHRPT and SeaWiFS System. Yelverton, Devon (2015)
14. Dartcom: LRIT/HRIT Multi Satellite System. Yelverton, Devon (2015)
15. Dartcom: LRIT/HRIT System C/Ku-Band Hardware. Yelverton, Devon (2015)
16. Dartcom: Polar Orbiter Ingester. Yelverton, Devon (2015)
17. Dartcom: X-Band EOS System. Yelverton, Devon (2015)
18. George, M., et al.: IASI – A Space Sounder for Atmospheric Composition Monitoring. LATMOS, Earthzine (2012)
19. Grayzeck, E.: ATS Spacecraft. NASA, Washington, DC. http://nssdc.gsfc.nasa.gov/nmc/spacecraftSearch.do (2015)
20. Group of Authors: Eumetsat Satellites. Eumetsat, Darmstadt. http://www.eumetsat.int/website/home/Satellites/index.html (2015)
21. GT&T: VSAT Equipment. Louvain-La-Neuve (2003)
22. Gunter's Space Page: Intelsat-1, Internet. http://space.skyrocket.de/doc_sdat/intelsat-1.htm (2015)

23. Harris: GOES-R Rebroadcast (GRB) Access. Melbourne (2015)
24. Harris: GOES-R Rebroadcast (GRB) and GOES VARiable (GVAR) Service. Melbourne (2015)
25. Harris: IntelliEarth Global Weather Solutions. Melbourne (2015)
26. Harris: Weather Ground Processing. Melbourne (2015)
27. Harris: Web-based Access and Retrieval Portal (WARP). Melbourne (2015)
28. Heidorn, C.K.: TIROS-1: First Eye in the Sky. The Weather Doctor, San Clemente. http://www.islandnet.com/~see/weather/almanac/arc2009/alm09sep.htm (2010)
29. Hughes: Hughes HX260 Mesh/Star Broadband Router. Germantown (2014)
30. Hughes: DVB-RCS C, Ku and Ka-band System. Germantown (2008)
31. Hughes: HUB HN System. Germantown (2012)
32. Jones, J.: GNSS and Weather Forecasting. Met Office, Potsdam (2013)
33. Launius, R.: The First Commercial Space Activity: Communication Satellites. Telstar-1, Internet. https://launiusr.wordpress.com/2014/06/13/the-first-commercial-space-activity-communications-satellites/ (2014)
34. NASA: Helios Prototype Flying Wing. Washington DC. http://www.nasa.gov/centers/dryden/multimedia/imagegallery/Helios/ED03-0152-60.html (2009)
35. NASA: History of the Blue Marble. Earth Observatory, Washington, DC. http://landsat.gsfc.nasa.gov/ (2014)
36. NASA: Landsat 4-8. NASA, Washington, DC. https://directory.eoportal.org/web/eoportal/satellite-missions/l/landsat-4-5 (2014)
37. NASA: Nimbus-1 and Tiros-9 Images. NOAA, Washington, DC. http://history.nasa.gov/SP-168/contents.htm (2014)
38. NSSDC: Spacecraft Catalog. NASA, Washington, DC. http://nssdc.gsfc.nasa.gov/nmc/masterCatalog.do?sc (2015)
39. Orbit: Solutions of Meteorological Antenna Systems. Netanya (2014)
40. Orbit: System Layout of Orbit Technology Group Antenna. Netanya (2015)
41. Rohner, C.: Antenna Basic. Rohde & Schwarz, München
42. Rossow, B.W.: International Satellite Cloud Climatology Project. NASA, Washington, DC. http://isccp.giss.nasa.gov/about/ (2015)
43. Schmidt, K.: World's First Weather Satellite Launched 50 Years Ago Today. Rockville (2010)
44. SCISYS: Adaptable Satellite Telemetry Receiver. Bochum (2014)
45. SCISYS: Data Acquisition Software for PEO/GEO Observation Missions. Bochum (2014)
46. SCISYS: DCP Direct Readout Ground Station. Bochum (2014)
47. SCISYS: DSR II Multi-mission Satellite Data Receiver. Bochum (2014)
48. SCISYS: EUMETCast/GEONETCast DVB User Station. Bochum (2014)
49. SCISYS: GMC User Station Monitoring and Control. Bochum (2014)
50. SCISYS: GOES and MTSAT LRIT/HRIT User Station. Bochum (2014)
51. SCISYS: High Performance L and X-Band Down-converter. Bochum (2014)
52. SCISYS: L and X-Band Ground Station PEO Direct Reception System. Bochum (2014)
53. SCISYS: PEO 2.4/2.8/3.0 X/Y Tracking Satellite Antennae. Bochum (2014)
54. SeaSpace: Ground GEO 3.7m Antenna. Poway
55. SeaSpace: Ground GOES-R Station. Poway
56. SeaSpace: Mobile LEO Satellite Direct Readout. Poway
57. SeaSpace: PEO 1.5m L-band Ground Station. Poway
58. SeaSpace: Shipboard Satellite Weather Receiving Antenna – 0.61m (L-band). Poway
59. SeaSpace: Shipboard Satellite Weather Receiving Antenna – 1.5m L/S-band. Poway
60. SeaSpace: Shipboard Satellite Weather Receiving Antenna – 2.4m X/L/S-Band. Poway
61. Solar Ship: Planeship UAV. SS, Toronto. http://www.wired.com/2011/10/its-a-bird-its-a-plane-its-a-solar-ship/ (2011)
62. SSEC: 40 Years of Geostationary Satellite Research and Observations', Space Science and Engineering Center. Madison. http://cimss.ssec.wisc.edu/geo-40th/timeline.htm (2006)
63. TAO: TAO Stratospheric Platforms Project. Tokyo (2006)

64. US: SumbandilaSat. University of Stellenbosch, Cape Town (2007)
65. Yulsman, T.: Historic First Weather Satellite Image. NASA, Washington, DC (2013)
66. Zak, A.: Kliper Spaceplane. Roscosmos. http://www.russianspaceweb.com/kliper_fuselage. html (2010)

5. Web Sites

1. Advantech Wireless: http://www.advantechwireless.com/
2. ASTRIUM.: http://www.intelligence-airbusds.com/en/4446-oil-spill-detection and http:// apogeospatial.com/astrium-services/
3. Campbell Scientific: https://s.campbellsci.com/documents/ca/product-brochures/tx320_br.pdf and https://www.campbellsci.com/tx320
4. China Meteorological Administration (CMA): http://www.cma.gov.cn/en2014/
5. CNES.: https://cnes.fr/en
6. Comet: https://www.meted.ucar.edu/hydro/basic_int/flash_flood/navmenu.php?tab=1& page=4.2.0 and https://www.meted.ucar.edu/tropical/textbook/ch3/images/globalsys.jpg
7. Dartcom: www.dartcom.co.uk
8. DLR: http://www.dlr.de/caf/en/desktopdefault.aspx/tabid-6232/10272_read-23601/
9. ESA: www.esa.int/
10. ESA/KNMI: http://images.slideplayer.com/18/6130410/slides/slide_31.jpg
11. Eumetsat: https://www.eumetsat.int/
12. Gloria: http://gloria.helmholtz.de/
13. GOES: www.goes.noaa.gov
14. GT&T (GlobalTT): www.globaltt.com
15. Harris: www.harris.com
16. IEEE Official Event of IEEE VIS Data 2014: http://www.viscontest.rwth-aachen.de/data.html
17. INSAT.: www.isro.gov.in/spacecraft/communication-satellites
18. ISRO.: www.isro.gov.in/
19. Japan Meteorological Agency (JMA): http://www.jma.go.jp/en/gms/
20. Kliper Spaceplane: www.russianspaceweb.com/kliper.html
21. Kongsberg: www.kongsberg.com
22. LATMOS – IASI: A Space Sounder for Atmospheric Composition Monitoring: https:// earthzine.org/2012/10/24/iasi-a-space-sounder-for-atmospheric-composition-monitoring/
23. MTSAT – Himawari-8/ Himawari-9: www.jma.go.jp/jma/jma-eng/satellite/
24. NASA.: https://www.nasa.gov/
25. NCAR.: https://www.rap.ucar.edu/~djohnson/satellite/geo_coverage2.html
26. NOAA.: www.noaa.gov
27. Orbit: http://orbit-cs.com/
28. POES.: https://poes.gsfc.nasa.gov/
30. Rohde & Schwarz: www.rohde-schwarz.com
31. Roshydromet: http://www.meteorf.ru
32. Roscosmos: http://en.roscosmos.ru/
33. SeaSpace: www.seaspace.com
34. SCISYS.: www.scisys.de
35. Vaisala: http://www.vaisala.com/en/products/automaticweatherstations/Pages/default.aspx; www.vaisala.com and www.vaisala.com/requestinfo

Printed by Printforce, the Netherlands